Structures Strengthened With Bonded Composites

Woodhead Publishing Series in Civil and Structural Engineering

Structures Strengthened With Bonded Composites

Zhishen Wu

Yufei Wu

Mohamed F. M. Fahmy

ELSEVIER

WP
WOODHEAD
PUBLISHING
An imprint of Elsevier

Woodhead Publishing is an imprint of Elsevier
The Officers' Mess Business Centre, Royston Road, Duxford, CB22 4QH, United Kingdom
50 Hampshire Street, 5th Floor, Cambridge, MA 02139, United States
The Boulevard, Langford Lane, Kidlington, OX5 1GB, United Kingdom

Notices
Knowledge and best practice in this field are constantly changing. As new research and experience broaden our understanding, changes in research methods, professional practices, or medical treatment may become necessary.

Practitioners and researchers must always rely on their own experience and knowledge in evaluating and using any information, methods, compounds, or experiments described herein. In using such information or methods they should be mindful of their own safety and the safety of others, including parties for whom they have a professional responsibility.

To the fullest extent of the law, neither the Publisher nor the authors, contributors, or editors, assume any liability for any injury and/or damage to persons or property as a matter of products liability, negligence or otherwise, or from any use or operation of any methods, products, instructions, or ideas contained in the material herein.

British Library Cataloguing-in-Publication Data
A catalogue record for this book is available from the British Library

Library of Congress Cataloging-in-Publication Data
A catalog record for this book is available from the Library of Congress

ISBN: 978-0-12-821088-8 (print)

ISBN: 978-0-12-821089-5 (online)

For information on all Woodhead Publishing publications
visit our website at https://www.elsevier.com/books-and-journals

Publisher: Matthew Deans
Acquisitions Editor: Gwen Jones
Editorial Project Manager: Isabella C. Silva
Production Project Manager: Vijayaraj Purushothaman
Cover Designer: Vicky Pearson Esser

Typeset by MPS Limited, Chennai, India

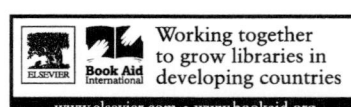

Working together
to grow libraries in
developing countries

www.elsevier.com • www.bookaid.org

Contents

Preface

Strengthening or stiffening of civil infrastructure using externally bonded fiber-reinforced polymer (FRP) composites such as FRP plates, sheets, and grids has become a promising structural strengthening technique. It has also gained much popularity recently, particularly in rehabilitation of structures/infrastructures due to the worldwide deterioration problems and earthquake hazards. Their high strength-to-weight ratio, durability in adverse environments, and high fatigue performance make them a good choice for civil engineering applications. Moreover, FRP sheets can easily be externally bonded to the surface of structural members. However, the brittle behavior of FRP materials, low FRP reinforcing ratio, debonding behavior, and other undesirable failure modes often limit the expected gains in structural performance and leave a large portion of potentials of FRP materials unused. Besides, the practice of the FRP bonding technique was being widely adopted without a thorough examination of the many parameters involved. Facing these problems, considerable studies have been conducted recently and good achievements have been achieved through these active efforts.

This book aims to present a comprehensive account of strengthening of concrete, reinforced and prestressed concrete, masonry, steel, and other composite structures using externally bonded FRP composites with emphasis on giving a systematic and fundamental investigation on bonding and debonding behavior of FRP—concrete interface and structural performances of FRP-strengthened structures with a combination of experimental, theoretical, and numerical studies, developing the analytical models and design codes, and some emerging and innovative strengthening techniques. FRP strengthening of different kinds of structures including bridge, tunnel, building and historic structures, and underwater constructions is discussed in the book. The main topics covered include the following:

- Fundamental behavior of FRP and bonding technique
- Bond characteristics and debonding mechanism of FRP-to-concrete interface
- FRP-strengthened tensile members
- Flexural strengthening of structures
- Shear and torsional strengthening of structures
- FRP strengthening of concrete columns
- Reinforcing spalling resistance of concrete structures with bonded FRP composites

Clear classification and discussions are made on various interfacial failure modes from both microscopic and macroscopic viewpoints with a combination of experimental, theoretical, and numerical studies. After clarifying the interfacial failure mechanisms, a comprehensive design guideline including the optimum design of strengthening composites and FRP-strengthened structures is described. Efforts are also made to develop effective FRP strengthening with prestressed and new composite materials (newly developed PBO and Dyneema fiber sheets).

Symbols

A	aramid
ε_f^{eff}	the effective tensile strain.
V_w^f	the jacket contribution
P_u	the ultimate load
μ_R	response ductility factor of pier.
a	beam shear span
A_g	the gross cross section of the element.
ag	peak ground acceleration.
A_s	cross-sectional area of reinforcing bars
A_{sl}	area of longitudinal steel reinforcement
A_{st}	area of stirrups bars reinforcement
A_w	total cross-sectional area of shear reinforcement in space
B	member width (unit: mm)
b	beam width
b^*	t-section
b_c	width of concrete prism
b_f	width of FRP composites
BF	debonding failure
BFRP	basalt fiber reinforced polymer
b_w	web width
C	carbon
CA	interface between concrete and adhesive
cc	concrete cover to the main column reinforcement.
C_C	concrete crushing
CFRP	carbon fiber reinforced polymer
COV	coefficients of variation
C_R	factor depending on the bilinear factor r.
cu	neutral axis depth
D	the column dimension in the loading direction
d	effective depth of element
d_b	bar diameter of the main column reinforcement
D_e	an equivalent column diameter
d_f	the effective depth of FRP sheet
DFRP	dyneema fiber reinforced polymer
D_n	the dead (permanent) load
d_s	debonding length of reinforcing bar
d_{st}	diameter of stirrups steel
E_c	the modulus of elasticity of the concrete
E_f	the elastic modulus of FRP.
E_j	the elastic jacket modulus

E_r	the double modulus of steel at the stress level.
ESDOF	equivalent single degree of freedom system
E_t	FRP sheet stiffness
F	flexural
f	stress at any strain ε
f_{cc}	ultimate strength of confined concrete under a maximum lateral pressure.
f_{co}	unconfined concrete compressive strength.
f'_{cd}	design compressive strength of concrete (unit: N/mm^2)
f_{cu}	compressive strength after 28-day
f_f	the ultimate tensile strength of FRP.
$f_{f,max}$	the maximum stress
$f_{f,min}$	the minimum stress
f_{fud}	design tensile strength of continuous fiber sheet
f_{fuk}	characteristic value of tensile strength of continuous fiber sheet (unit: N/mm^2)
f_h	represents the horizontal stress level
f_{ju}	the ultimate unidirectional tensile strength
f_l	the confinement strength of FRP.
FL	flexural failure
f_r	the applied stress range
FRP	fiber reinforced polymer
$f_{s,crit}$	the critical buckling stress
f_{sy}	yield strength of the main reinforcement
f_t	the tensile strength of concrete
f_{wyd}	design tension yield strength of shear reinforcement
f_{yst}	yield strength of steel stirrups reinforcement
G	glass
G_e	energy release rate
G_f	interfacial fracture energy
G_f^{II}	the total fracture energy
GFRP	glass fiber reinforced polymer
G_{fs}	the fracture energy
G_{IIp}	the irreversible work
h	total height of element
h_f	height of the FRP sheet
I_{pier}	the pier upgrading index
K	the unloading stiffness
K_1	column first stiffness
K_b	the initial bond interface stiffness
k_e	tie-by-tie arching-effect coefficient
$k_{f,y}$	the effectiveness coefficients for the FRP transverse confining system
$k_{st,y}$	the effectiveness coefficients for the steel transverse confining system
$K_\mu(t0)$	elastic aging coefficients
L	bonded length
l_{av}	average crack spacing
L_c	the effect of crack spacing
L_e	effective bonding length
l_{max}	the maximum crack spacing
l_{min}	minimum crack spacing
L_n	the live load.

L_s	the shear span.
LVDT	linear variable displacement transducer
M	the maximum moment
M_0	decompression moment
M_c	elastic compliance of concrete
M_d	design bending moment
M_n	ultimate bending moment
n	number of spliced bars along p.
N	the number of load cycles
N_a	the normal pressure
NA	not reported
N_b	the number of bars (or pairs of spliced bars).
N'_d	design axial compressive force
n_f	Number of plies of continuous fiber sheets
p	the perimeter line in the column cross section along the lap-spliced bar location.
P	the load carried on pin through loading
PBO	polyparaphenylenl benzobisoxazole
P_d	the bond strength contribution
P_f	the frictional bond strength
ϕ_f	flexural capacity reduction factor
PGA	peak ground acceleration
P_n	ultimate load
P_r	the applied load range P_{max}-P_{min}.
P_u	the maximum bonding strength
ϕ_v	shear capacity reduction factor
r	(K_2/K_1) bilinear factor defined.
R	stress ratio
r	coefficients of correlation
R^2	coefficient of determination
R_{Icf}	fracture resistance of plain concrete member
S	the spacing between stirrups bars.
S_f	spacing of continuous fiber sheets (unit: mm)
S'_f	Width of continuous fiber sheet (unit: mm)
S_h	height of the steel plate
S_{max}	the maximum stress
S_{min}	the minimum stress
s_p	is the softening slip
s_r	crack spacing
S_r	stress range, $S_{max} - S_{min}$.
s_s	spacing of shear reinforcement
STR	steel tensile rupture failure
STR-D	steel tensile rupture failure followed by FRP
S_{w1}, S_{w2}	the width of the steel plates in the short and long directions of rectangular column.
SY	steel yield failure
T	the peel force per unit width of FRP sheet
τ_{b0}	the average bond stress along the steel bar
Tf	the torsion capacity of FRP

τ_f	bond stress
t_f	the total thickness of FRP
TR	tensile rupture failure
T_s	the torsion capacity of RC element
U	stain energy density
u	deflection
u_m	the maximum deflection
V_{anch}	the shear force when the anchorage lap-splice reach their development capacity.
V_{buckl}	the shear force when compression bars reach instantaneous buckling conditions at the critical section.
V_c	shear contribution due to concrete
V_f	shear contribution of the FRP
V_{iflex}	M_u/L_s is the seismic shear force
V_{mu}	maximum shear force when a member reaches the existing flexural load-carrying capacity M_u
V_n	the design shear capacity
V_o	column shear demand
V_p	the shear capacity contribution of axial load
V_s	shear contribution due to shear reinforcing bar members.
V_{shear}	the nominal shear resistance.
W^*	orientation angle of FRP sheet = 45
W_f	width of FRP sheet
w_o	the crack opening width
Y	the ratio of sustained load to the tensile capacity
yc	the depth of the concrete compression zone
YPS	Yield Point Spectra
z	Lever arm length (generally set to $d/1.15$)
α	is the crack inclination angle
α_0	Coefficient used to calculate member ductility ratio
α_L, α_f and α_m	coefficients of thermal expansion of FRP sheets
α_{max}	the peel angle
α_s	angle formed by shear reinforcement about the member axis
β	is the fiber direction angle.
β_1	parameter defining a rectangular stress block in the concrete
β_c	the shear transfer coefficient of closed crack
βf	coefficient depending on the fracture energy of the FRP
β_t	the shear transfer coefficient of open crack
γ_b	member factor
γ_{bf}	member factor used for calculation of μ_{fd}
γ_{cf}	the debonding fracture energy per unit area of FRP sheet-concrete crack surface
γ_{mf}	material factor of continuous fiber sheet
δ	the slip deformation
δ_d^{ava}	the available ductility
δ_d^{tar}	target displacement ductility
ΔL	debonding length increment
δ_{res}	residual displacement of a pier after earthquake.

Δs	the total slip increment at the FRP-to-concrete interface element
Δt	the time increment
Δu	deflection increment
Δxi	the distance between gauge i and $i + 1$.
δ_y	yield displacement.
$\Delta \tau$	the shear stress increment
ε_{cc}	confined concrete failure strain
ε_{cu}	ultimate concrete strain
ε_d	dilation strains
$\varepsilon_{f,e}$	the effective strain of FRP sheet
ε_{fd}	debonding strain
ε_f	the tensile strain in FRP sheet
ε_{fu}	ultimate strain of continuous fiber sheet
ε_{ju}	the ultimate unidirectional tension failure strain
ε_o	ultimate strain compressive strength of concrete
ε_{su}	ultimate strain of steel hoops
ζ	damage accumulation factor
$\eta_\mu(t0)$	the viscosities of the dampers
λ	softening coefficient
μ	the coefficient of friction.
μ_{fd}	ductility ratio of members upgraded with continuous fiber sheets
μ_s	mechanical ratios of longitudinal reinforcements.
μ_{st}	mechanical ratios of transverse reinforcements.
ν	poisson's ratio
ρ	radius of the pin
ρ_f	the volumetric ratio of FRP to concrete.
ρ_{fv}	the volumetric ratios of FRP reinforcement.
ρ_j	the volumetric jacket reinforcement ratio
ρ_s	the ratio of longitudinal reinforcement.
ρ_s	volumetric stirrup ratio
ρ_{st}	volumetric ratio of steel hoops
ρ_{sv}	the volumetric ratios of stirrup reinforcement.
σ_0	the initial strength of FRP
σ_a	stress amplitude
σ_c	the average stress in concrete
σ_f	the FRP tensile stress
σ_{lat}	the pressure exerted upon the lateral surface of the bar by the cover.
σ_m	the mean stress
σ_r	the residual strength of FRP
τ_f	bond strength
Φ	the diameter of tensile reinforcement
φ_c	strain energy of concrete containing
φ_{dcf}	total debonding energies on all debonded interfaces between CFS and concrete
φ_{ds}	total debonding energies on all debonded interfaces between reinforcing bars and concrete
φ_s	total sliding energies on all debonded interfaces between reinforcing bars and concrete

$\boldsymbol{\Psi}_f$	the reduction factor for the FRP debonding strain
$\psi_{\mathbf{min}}$	the minimum ratio of the bonded length
$\boldsymbol{\Phi}_u$	the ultimate section curvature
$\boldsymbol{\Phi}_y$	section yield curvature
$\sigma^f_{lat,y}$	the transverse pressure in concrete owing to the jacket in the direction of lateral sway.

Fundamental behavior of fiber-reinforced polymers and their bonding technique

1

1.1 Fiber-reinforced polymer constituents

1.1.1 General

The fiber-reinforced polymer (FRP) composites primarily consist of a resin (polymer matrix material), reinforced with cloth, mat, strands, or any other form of fibers (reinforcing material). The matrix can comprise vinyl-ester, epoxy, or unsaturated resins. The fibers in the FRP composites can consist of carbon, aramid, glass, basalt, poly-p-phenylene-benzobisoxazole (PBO), or other types of polyethylene fibers. The commercially available externally bonded FRP systems can be categorized into "wet layup," "prepreg," and "precured" systems. The wet layup FRP systems consist of dry unidirectional or multidirectional sheets or fabrics impregnated with a saturated resin on-site, while the prepreg FRP systems consist of dry unidirectional or multidirectional sheets or fabrics preimpregnated with a saturated resin at the manufacturer's facility. Lastly, the precured FRP systems consist of a wide variety of composite shapes manufactured off-site. The factory-made laminates, such as thin unidirectional strips or plates (with thicknesses of the order of 1 mm) are one type of precured FRP systems. Other examples of precured systems include precured multidirectional grids and precured shell segments. The selection of FRP composites for different strengthening systems is a critical process.

The reinforcing fibers for FRP bonding systems are available in many forms, such as rovings (almost parallel bundles of continuous untwisted filaments), yarn (bundles of twisted continuous filaments), and short fibers. To make sheet-like elements with multiple reinforcing directions, the fibers are further fabricated to textile production specifications. As reinforcing elements, continuous fiber sheets with woven and nonwoven fabrics, grid or mesh, as well as reinforcing mats and fleeces have been developed. However, presently, continuous fiber sheets with woven and nonwoven fabrics are mainly used as reinforcements for FRP bonding systems. Figs. 1.1—1.3 show nonwoven fabrics of continuous fiber sheets, while Fig. 1.4 shows a woven fabric of continuous fiber sheets. In addition, Fig. 1.5 shows an FRP thin plate laminate, while Fig. 1.6 shows FRP grids. In the following sections, both short- and long-term behaviors of different FRP constituent materials are described in detail.

Compared with the traditional structural materials (steel, concrete, wood, aluminum alloy, etc.), FRPs possess superior mechanical (high specific strength),

Structures Strengthened with Bonded Composites. DOI: https://doi.org/10.1016/B978-0-12-821088-8.00001-1

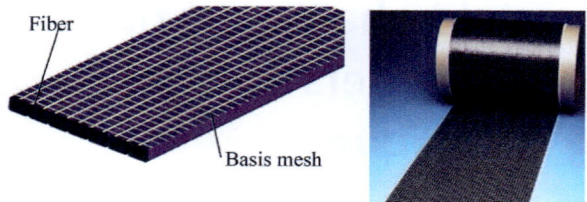

Figure 1.1 Nonwoven fabrics of continuous fiber sheets.

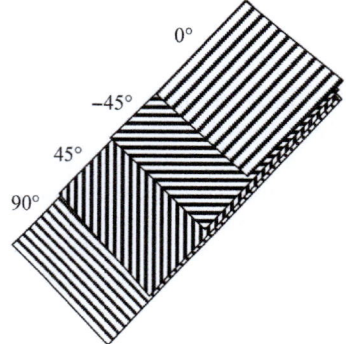

Figure 1.2 Layered nonwoven fiber sheet.

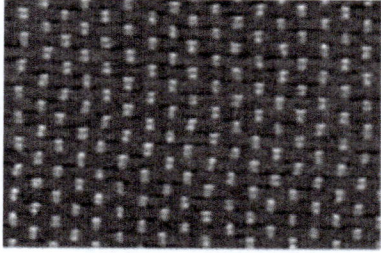

Figure 1.3 Unidirectional nonwoven fiber sheet.

physical (lightweight), and chemical (corrosion resistance, etc.) properties. In addition, they also possess other functional characteristics, such as insulation (except for carbon fiber composites, which have weak conductivity), and resistance to both high- and low-temperature environments. The major characteristics of FRPs are as follows (Wu et al., 2007a,b; Bakis et al., 2002; Keller, 2003; Hollaway, 2010; Ke, 2003).

- *Lightweight*: FRPs possess 1/4−1/5 the density of steel, and 70%−80% of the density of conventional plain concrete.

Figure 1.4 Bidirectional woven fiber sheet.

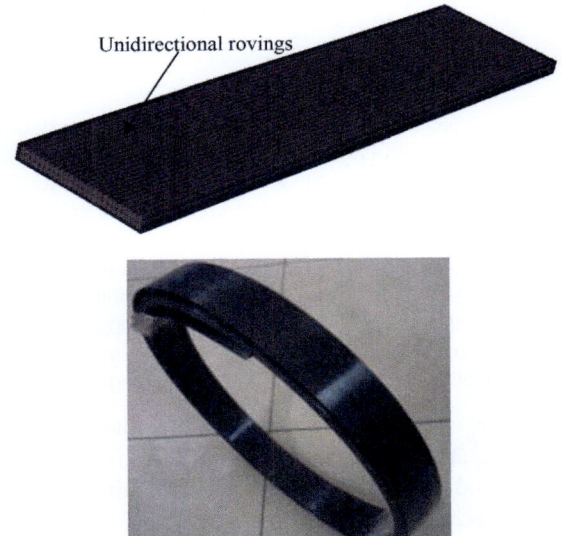

Unidirectional rovings

Figure 1.5 FRP thin plate laminate.

- *High strength*: The tensile strength of an FRP is 4—10 times higher than that of ordinary low carbon steel, and is equivalent to or even higher than that of a prestressed steel wire/ cable. It can be used as a lightweight and high-strength structural component to achieve relatively large spans or lightweights.
- *Excellence in corrosion resistance*: In general, the FRP composites can resist corrosion in acidic, alkaline, salty, and other kinds of environments. Among all the FRPs, the carbon FRP (CFRP) shows the strongest resistance, followed by the basalt FRP (BFRP) and the glass FRP (GFRP).
- *Superior fatigue resistance*: The maximum fatigue stress and the amplitude of the fatigue load are both higher in FRPs than in steel reinforcements. Thus the application of FRPs as dynamic load-bearing structural components could effectively enhance the fatigue life of structures.

Figure 1.6 Fiber-reinforced polymer grids.

- *Good designability*: Reinforcements with FRPs can come in a variety of types and characteristics and hence, hybridization could be realized by selecting different types of fibers and traditional materials (including steel, wood, bamboo, and other glulams). Thus the structural requirement for different engineering constructions could be achieved, while simultaneously reducing their comprehensive cost.
- *Versatility*: Various properties of FRPs, including their nonmagnetic property, ability to microwave, low thermal conductivity, and high energy absorption coefficient, could play specific roles in structures and facilities. Low magnetic steel, copper bars, and other materials could be replaced by FRP materials in the construction of infrastructures, such as the radar stations, geomagnetic observation stations, degaussing facilities, and medical nuclear equipment.

The wide application of CFRP in structural strengthening/repairing started in 1995 after the big earthquakes in Osaka and Kobe cities in Japan. A fast and effective reinforcement method was urgently needed for repairing the damaged bridges and buildings, and the technology involving carbon fiber sheet confinement played a great role in their seismic strengthening. Thus the CFRP seismic strengthening has become one of the mainstream technologies in structural seismic retrofitting. In addition, aramid fiber and other high ductility fibers, such as ultrahigh molecular-weight polyethylene (UHMWPE), polyethylene naphthalate two formic acid glycol, and polyethylene terephthalate (PET) are also used for structural rehabilitation. By the beginning of the 21st century, in addition to the traditional carbon fiber, aramid fiber, glass fiber, and green and high-performance basalt fiber have been developed. The basalt fiber as a structural reinforcing material could replace carbon, glass, and other fiber materials. BFRP has been widely used for its high cost-effectiveness, as well as excellent short- and long-term performances.

In recent years, more and more bridges have achieved significant improvements in their performance after being reinforced with FRPs. For example, the West Gate Bridge in Australia was strengthened by an externally bonded FRP, using more than 40-km length of carbon fiber sheet; the railway viaduct in Japan was reinforced after the 2011 earthquake of the Pacific coast of Tōhoku. The FRPs applied in these projects exhibited superior properties to steel. According to statistics, the CFRP, BFRP, and other high-performance fiber materials have been used in the reinforcement of nearly half of the critical components in the projects in China, such as the reinforcement of Jiangyin Yangtze River Bridge and the Beijing Great Hall of the People. Therefore the practical applications of FRPs have greatly promoted the development of the FRP industry.

Numerous research studies on the material properties, testing methods, and structural design methods of FRPs have been carried out, and the results have been reflected in the technical standards and guidelines for FRP applications in structures in various countries. For example, scientists in Japan developed the performance test and application methods for FRP materials in the mid-90s (JSCE, 1997), and put forth a method for the performance evaluation of fiber cloth and the sticking reinforcement method (JSCE, 2001). The design code for the application of FRP bars in concrete structures was established in Europe in 2001. Since then, Canada (ISIS Canada. ISIS-M04-01, 2001; Canadian Standard Association CSA. 2002), the United States (ACI Committee 440, 2007), and other developed countries have promulgated guidelines for the design and construction of structural concrete reinforced with FRPs. Being at the forefront in the fields of energy saving, environmental protection, and disaster prevention, the Japanese Society of Civil Engineering organized a domestic industry—university research task force in 2004 to set up a committee for the application of FRPs in civil engineering. In 2005, the committee prepared a clear outlook for the application prospects of FRPs in the fields of marine structures, ports, bridges, underground structures, and seismic and disaster prevention. The committee also pointed out that the people's understanding of the long-term performance of FRP-reinforced structures, the design of life cycles, and the internationalization of the industry were not adequate, and limited the extensive applications of FRPs in civil engineering. The international organization for standardization (ISO) TC71 working group recently developed the draft design guidelines for FRP-reinforced structures (covering both new and existing structures) on the basis of collection of opinions of various countries. With the continuous in-depth research and application, China has promulgated a series of products and application standards, including the "technical code for infrastructure application of FRP composites" (GB50608-2010), "basalt fiber composites for strengthening and restoring structures" (GB/T 26745-2011), "composite bars for civil engineering" (GB/T 26743), "pultruded fiber-reinforced polymer composite structural profiles" (GB/T 31539-2015), and "fiber-reinforced polymer composite grids for civil engineering" (GB/T 36262-2018). Moreover, the "fiber-reinforced composite cables and anchors for structures" (20150490-T-609) is under preparation. The continuous improvement of these standards and codes is expected to promote the applications of FRPs in the reinforcement of civil engineering structures.

1.1.2   Fibers

The typical fibers that are used in FRP bonding systems as reinforcements include carbon, glass, aramid, and basalt fibers. Some new types of fibers such as polyethylene fibers and PBO fibers are also exploited in FRP bonding systems. UHMWPE, PET, and some new types of plant fibers, such as flax fiber and bamboo fiber, have also been used. Different raw materials, chemical compositions, and production processes lead to differences in their performance. Some fibers possess high strength and high stiffness, but low elongation. Some have low strength and low stiffness, but high elongation. Identifying the right fibers according to the requirements is the key to FRP applications. Typical ranges of mechanical properties and comparison of other characteristics of fibers are given in Fig. 1.7 and Table 1.1. A more detailed description is given below.

1. *Carbon fibers*

Carbon fiber is a fibrous carbon material having a graphite crystal structure obtained by fibrillation of acrylic resin. The fibers are classified into polyacrylonitrile (PAN)-based, pitch-based, and rayon-based. The PAN fibers are made by carbonization of polyacrylonitrile through burning. Owing to their superb mechanical properties and chemical stability, carbon fibers have been the most widely used reinforcing fibers in FRPs. For example, the strength of T300 is 3530 MPa and its tensile modulus is 230 GPa; while its density is only 1.8 g/cm^3. Moreover, carbon fibers have good resistance to acidic, alkaline, salty, and ultraviolet environments. In addition, they can work at a high temperature of 600°C. Pitch fibers are fabricated by using refined petroleum or coal pitch that is passed through a thin nozzle and stabilized by heating. Rayon fibers are prepared by preoxidation and carbonization of natural fibers, such as wood, hemp, and cotton, which are treated by pulping, sulfonation, ripening, and spinning.

Based on their performance, carbon fibers can be divided into high-strength, middle-modulus, and high-strength, high-modulus, and super-high-modulus fibers. Among them, the PAN-based carbon fibers are the largest produced and most widely used. They are extensively applied in a variety of applications in view of their mechanical characteristics, such as high tensile strength and stiffness, and other characteristics such as low density, low coefficient of thermal expansion, high temperature resistance, and chemical stability.

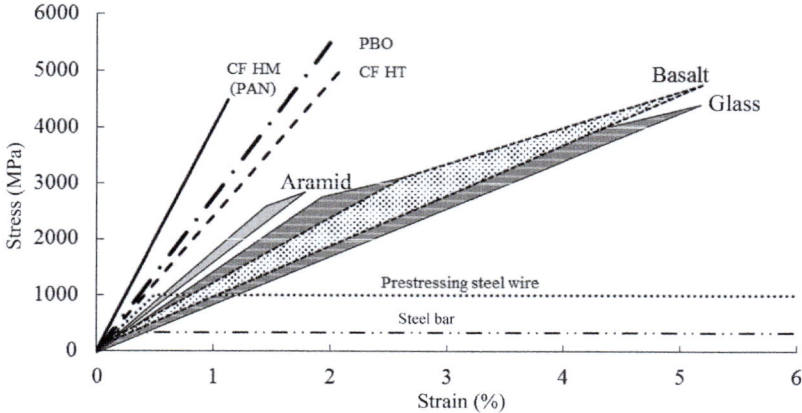

Figure 1.7 Tensile stress—strain relationship of different kinds of fibers.

Table 1.1 Summary of main parameters of different fibers.[a]

	Carbon			Aramid		Glass			Basalt	PBO	Polyethylene
	High-strength type	High-modulus type	High-modulus type	Aramid	Aramid	E-	S-	Alkali-resistant			
	PAN	PAN	Pitch	I	II	Glass	Glass	Glass			
Diameter (μm)	5–8	5–8	8–14	12	12	3.3–4.8	4.2–5.4	2–4.3	7–16	2.1–3.2	12
Density (kg/cm³)	1.5–1.8	1.8–2.0	1.8–2.2	1.45	1.39	2.5–2.6	2.5–2.6	2.7	2.65	1.54–1.56	0.97
Tensile strength (MPa)	3000–6000	2500–4500	2450–4000	2750	3040	3400–3700	4300–4900	3000–3500	3600–4800	5800	2600–4000
Tensile stiffness (GPa)	230–300	350–500	200–900	113	70	72–77	75–88	21–74	85–91	180–270	88–123
Elongation at failure (%)	2–2.5	0.4–1.5	1.4–2.1	2.3	4.4	4.8	5.2–7.0	2–3.0	2.65	2.5–3.5	3–5.0
Coefficient of thermal expansion (ppm/K)	−0.4	−0.7	−0.7	−6	−6	4.9	5.6	7.2	6–8	−6	−16
Hydrochloric acid	○	◎	◎	×	○	×	−	−		○	◎
Sulfuric acid	◎	◎	◎	×	○	×	−	−		○	○
Nitric acid	○	◎	◎	×	○	×	−	−		○	○
Alkali resistance	−	−	−	−	−	△	×	○		−	−
Sodium hydroxide	◎	◎	◎	○	○	△	−	○		○	○

(*Continued*)

Table 1.1 (Continued)

	Carbon			Aramid		Glass				PBO	Polyethylene
	High-strength type	High-modulus type	High-modulus type	Aramid	Aramid	E-	S-	Alkali-resistant	Basalt		
	PAN	PAN	Pitch	I	II	Glass	Glass	Glass			
Seawater	◎	◎	◎	○	○	△				○	○
Organic solvent resistance	—	—	—	—	—	—	—	—		—	—
Acetone	◎	◎	◎	◎	○	◎	—	—		◎	◎
Benzene	◎	◎	◎	◎	○	◎	—	—		◎	◎
Gasoline	◎	◎	◎	◎		◎	—	—		◎	◎

[a]The tensile performances mean the guaranteed values.

Companies, such as Toray and Teijin from Japan, Hexcel, and Zoltek from the United States, and SGL Group from Germany are reputed for carbon fiber production. At present, the global annual output of carbon fiber is about 8−10 t/y.

2. *Glass fibers*

Single filaments are produced by mechanically drawing molten glass streams. Then the filaments are usually gathered into bundles called strands or rovings. The strands may be used in a continuous form for filament winding. At present, three major types of glass are used to make fibers: (1) A-glass is a high alkali glass containing 25% soda and lime, which offers very good resistance to chemicals, but with relatively low tensile strength, stiffness, and electrical properties, when compared to other types of glass fibers. (2) E-glass is an electrical grade glass with low alkali content. More than 50% of the glass fibers used for reinforcements are of this type. (3) S-glass is a high-strength glass with about 30% higher tensile strength than the E-glass, but not resistant to alkaline environments. Generally, glass fibers have good insulation and high-temperature resistance. Another significant aspect of glass fibers is their low cost.

3. *Aramid fibers*

The structure of aramid fiber is anisotropic and gives relatively high strength and modulus in the longitudinal direction of the fibers. Aramid fibers can be classified into aromatic polyamide (I) and polyetheramide (II). Most of the aramid fibers used for reinforcements are of aramid I. The aramid fibers exhibit good toughness, damage tolerance, and fatigue characteristics.

Aramid fiber is a new high-tech synthetic fiber. Some major brands of aramid fibers in the present day include Nomex and kevlar of American Dupont; Conex and Technora of Japanese Tejin; and Terlon and Fenelon of Russia. Aramid fibers have excellent mechanical properties. For example, the tensile strength of aramid fibers can be as high as 3000 MPa, and the tensile modulus can be 70−110 GPa. They can work at a temperature of 250°C, while their density is only 1.3 g/cm^3. Aramid fibers have good toughness and shock resistance. However, in engineering applications, it is usually found that they suffer from severe stress relaxation, and are sensitive to UV. Moreover, the cost of aramid fibers is almost the same as that of carbon fibers. Considering all these factors, aramid fibers have not been widely used in engineering.

4. *Basalt fibers*

Basalt continuous fibers are regarded as an inert and naturally occurring material that is produced from basalt rock (volcanic rock). Its components include SiO_2, Al_2O_3, MgO, CaO, Fe_2O_3, FeO, TiO_2, K_2O, and Na_2O. Basalt fibers can be produced by drawing and winding fibers from the molten rock at a temperature between 1400°C and 1500°C. Basalt-based materials are environmentally friendly, energy saving, and nonhazardous natural green materials. As there is almost no emission in the production process, basalt fiber is recognized as "green material." The whole production process is simple and consumes much less energy, which is about 1/16 of that required for producing carbon fibers. Moreover, little CO_2 and other gases are discharged to the environment, thus establishing its feature of low-carbon emitting and environment-friendliness. The processes are shown in Fig. 1.8 and Table 1.2.

The basalt fibers have a density of 2.6 g/cm^3, strength of 3000−4500 MPa, and tensile modulus of 90−110 GPa. Basalt fibers can work over a temperature range of −260°C−850°C. In addition, basalt fibers have good resistance to chemical corrosion, as well as adiabatic insulation. The carbon, aramid, high molecular polyethylene, and basalt fibers have become China's four top high-tech fibers. Basalt fibers have many unique and excellent attributes, such as good mechanical properties, excellent temperature

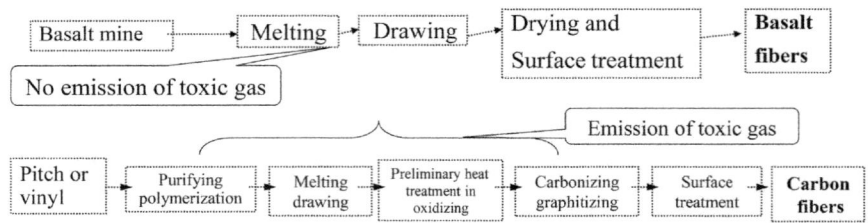

Figure 1.8 Emission comparison of Basalt fiber and carbon fiber during production.

Table 1.2 Energy consumption and emission of producing 1 kg fibers.[a]

Materials	Energy consumption (MJ kg^{-1})	CO$_2$ emission (kg kg^{-1})	NOX emission (kg kg^{-1})	SOX emission (kg kg^{-1})
Carbon fiber	478.5	29.713	2.010	0.068
Basalt fiber	30	0	0	0

[a]MJ kg^{-1} means the amount of energy consumption during per kg of basalt fiber production. kg kg^{-1} means the amount of gas emission during per kg of basalt fiber production.

performance, acid and alkali resistance, anti-UV, low moisture absorption, good insulation, good filtration under high temperature, antiradiation, and sound wave-transparency properties.

Based on their mechanical properties, basalt fibers can be divided into general, high-strength, and high-modulus types. According to their special properties, basalt fibers can be divided into alkaline-resistance and high-temperature-resistance fiber types.

China has a unique advantage in the research and development of basalt fibers. In the recent 10 years, based on the technical breakthroughs in the production processes (shown in Fig. 1.9), basalt fiber has made great progress in stable large-scale, high-technology production and production equipment. High-performance basalt fiber production technology has made great strides, making China a world leader. In China, the National and Local Unified Engineering Research Center for Basalt Fiber Production and Application Technology in Southeast University is leading the basalt fiber industry.

5. *PBO fibers*

Poly-*p*-phenylene benzobisoxazole or PBO, was first developed in the 1960s. In 1991, the American company Dows and the Japanese company Toyobo jointly researched PBO and achieved significant development in PBO's mechanical properties. The newly developed PBO fiber consisted of rigid-rod chain molecules of PBO. The strength and modulus of the organic fiber were superior to those of high-strength carbon fiber and almost double those of aramid fiber (Fig. 1.1). The density of a typical PBO fiber is about 1.56 g/cm^3 and its tensile strength is 5.8 GPa; while its tensile modulus is 280−380 GPa. Furthermore, it can work over a temperature range of 300°C−350°C, and its thermal decomposition temperature is up to 670°C. PBO fibers can absorb large amounts of energy during an impact, thus making them a promising alternative for prestressed tension engineering applications. However, their poor compression performance and aging resistance, together with a poor bonding tendency with the matrix resin, limit their use in advanced structural composites.

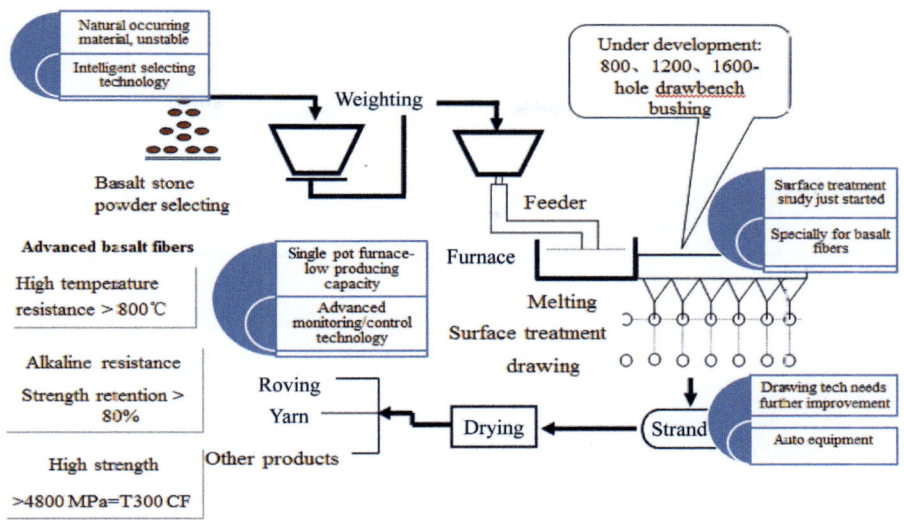

Figure 1.9 Basalt fiber production.

In addition, the PBO fibers demonstrate high resistance to fire and flame with a limited oxygen index (LOI) value of 68 in comparison to a value of 58 for carbon fibers and 30 for aramid fibers. Because of the high LOI number, the fiber does not burn. When the service temperature exceeds the fiber's decomposition temperature of 650°C, the toxic gases generated are 10 times lower than those from aramid fibers. The PBO fibers also exhibit high creep resistance. The retained stress in a relaxation test using sheets of 2-m long and 8-cm wide proves that the PBO fiber sheets similarly sustain loads as carbon fiber sheets do.

6. *Ultrahigh molecular-weight polyethylene fiber*

UHMWPE is a subset of the thermoplastic polyethylene. Also known as high-modulus polyethylene, it has extremely long chains, with a molecular mass usually between 3.5 and 7.5 million amu (Kurtz, 2004). The longer chains serve to transfer the load more effectively to the polymer backbone by strengthening the intermolecular interactions. This makes it a very tough material, with the highest impact strength of any thermoplastic presently made (Stein, 1988). UHMWPE is odorless, tasteless, and nontoxic. It embodies all the characteristics of high-density polyethylene, with the added traits of resistance to concentrated acids and alkalis, and numerous organic solvents (PE Material, 2017). It is highly resistant to corrosive chemicals except for oxidizing acids. Furthermore, it has extremely low moisture absorption and a very low coefficient of friction. It is self-lubricating (see boundary lubrication) and is highly resistant to abrasion, with some of its forms being 15 times more resistant to abrasion than carbon steel. Its coefficient of friction is significantly lower than that of nylon or acetal and is comparable to that of polytetrafluoroethylene (PTFE, Teflon); however, UHMWPE has better abrasion resistance than PTFE (Tong et al., 2006; Budinski, 1997). There are different types of polyethylene fibers, such as Dyneema fibers, Spectra900, and Tekmilon. The Dyneema

Table 1.3 Summary of matrix materials.

Thermosetting	Thermoplastics
Epoxy resin	Polypropylene
Nonsaturation polyester resin	Nylon
Phenolic resin	Poly carbonate
Vinyl ester resin	Epoxy
MMA resin	

fibers are one type of high-strength polyethylene fibers made by Gel Spinning Technology. These fibers have excellent alkali resistance, weather resistance, flexibility, lightweight, and negative expansion.

1.1.3 Matrix materials

Fibers, by themselves, have limited use in engineering applications as they cannot transmit loads from one to another. However, when they are embedded in a matrix material to form a composite, the matrix binds the fibers together, enabling the transfer of loads to the fibers, and thus, protecting them from environmental attack and damage due to handing. Moreover, the matrix has a strong influence on some mechanical properties of the composites, such as the transverse modulus and strength, and shear and compressive properties. Physical and chemical characteristics of the matrix, such as melting or curing temperature, viscosity, and reactivity with fibers, influence the choice of the fabrication process (Agarwal and Broutman, 1980). Both polymer and cement matrices can be used for structural composite materials. However, the matrix for an FRP composite is of a specific type of polymer material. The polymer matrix for FRP composites can either be of the thermosetting type or thermoplastic type as given in Table 1.3. Some commonly used polymeric matrix materials (resins) include epoxies, vinyl esters, and polyesters. Epoxies have, in general, better mechanical properties, higher temperature resistance, and better durability than vinyl esters and polyesters. However, vinyl esters and polyesters are cheaper. The polymeric matrix materials are viscoelastic and are sensitive to ultraviolet radiation. Some types of thermoplastic resins have unique advantages compared to thermosetting resins, such as reforming ability. However, their high cost and manufacturing conditions currently limit their applications to civil engineering applications. They have the potential to be used as matrix materials if the limitations can be overcome in the future. Table 1.4 gives some typical properties of commonly used matrix materials.

1.2 Characteristics of fiber-reinforced polymer composites

As defined earlier, a composite is a material composition consisting of fibers, matrix, and additives. Depending on the type of fibers used, the FRP composites

Table 1.4 Properties for matrix materials.

Material	Elastic modulus (GPa)	Tensile strength (MPa)	Ultimate tensile strain (%)
Polyester	2.1−4.1	20−100	1.0−5.5
Vinylester	3.2	80−90	4.0−5.0
Epoxy	2.5−4.1	55−130	1.5−9.0

Table 1.5 Summary of tensile properties of typical fiber-reinforced polymer sheets.

	Carbon		Aramid	Glass	PBO	Basalt	High-strength polyethylene
	HS type PAN	HM type PAN	II	E-Glass			
Tensile strain (%)	1.5	0.35	1.7	2.1	1.5	2.6	3.1
Tensile strength (MPa)	3500	1900	2060	1500	3500	2300	1840
Tensile stiffness (GPa)	230	540	118	73	240	90	60
Impact absorb energy (J)	4.7	−	7.67	−	19.4	−	−

are referred to as CFRP (carbon fiber-reinforced polymer), GFRP (glass fiber-reinforced polymer), AFRP (aramid fiber-reinforced polymer), BFRP (basalt fiber-reinforced polymer), and PFRP (PBO fiber-reinforced polymer). The typical volume fraction of fibers in different FRP composites is in the range of about 50%−65% for precured FRP plates or bars and 25%−35% for wet layup FRP sheets. A summary of the tensile properties of typical FRP sheets is shown in Table 1.5. The test specimens and layouts of tensile tests with the centric placement of unidirectional FRP sheets are in accordance with the size (a total length of 200 mm, width of 12.5 mm, and tab length of 50 mm) recommended by JSCE guidelines, as well as ACI-440 (ACI Committee 440, 2007) for upgrading concrete structures with the use of continuous fiber sheets.

The previously described properties of different FRP composites are obtained from a coupon test. However, with an increase in the specimen size, the mechanical properties can be lowered, which is referred to as the size effect of material as shown in Fig. 1.10.

Fig. 1.10 shows the size effect on the tensile strength of the CFRP sheets. It is seen that the tensile strength decreases with an increase in the length of the specimen. However, the dispersion of tensile strength becomes smaller for larger sized sheets. Fig. 1.11 compares the tensile strength of dry fiber sheets of PBO in lengths of 2 and 10 m to simulate the prestressing process. Because of the relatively low

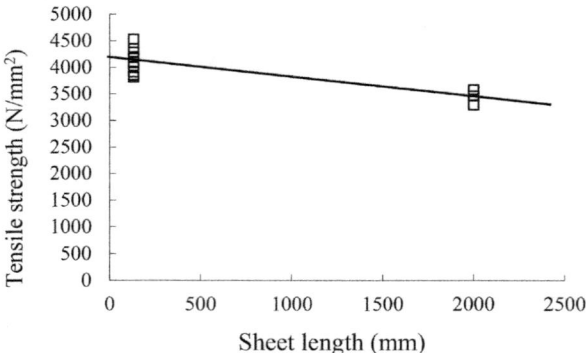

Figure 1.10 Size effect on tensile strength of carbon fiber-reinforced polymer sheets.

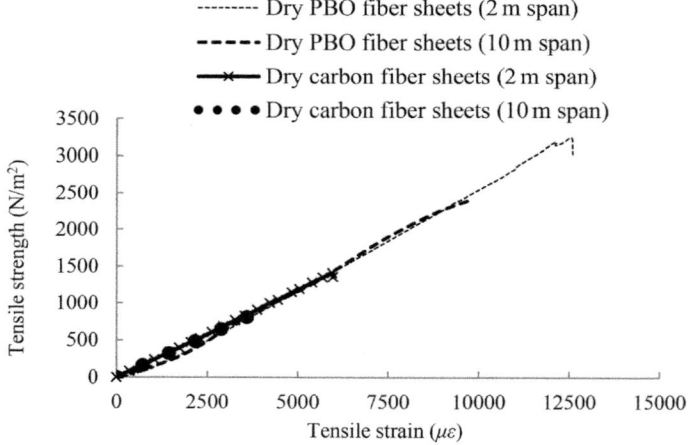

Figure 1.11 Tensile stress—strain curves of dry PBO and carbon fiber sheets (Wu et al., 2007a, b).

energy absorption behavior, the carbon fiber sheets without resin impregnation break at 40% of their tensile strength when stretched over a length of 2 m, and can sustain only 27% when stretched over a length of 10 m. On the contrary, the 2-m long dry PBO fiber can be tensioned to 90% of its capacity and the 10 m long fiber holds the load up to 60%. Although the FRP sheets have such size effect on their tensile strength, the reduction of strength should not be allowed in real applications of the sheets in strengthening the structures, as the sheets are bonded with structures to resist the external loads. The load transferred length between the sheets and structures is very limited even when the structures are cracked. Thus the size effect in externally strengthened structures is not significant.

1.2.1 Basic mechanical properties and behavior

The tensile strength, stiffness, and strain capacity of different kinds of continuous FRP sheets are shown in Table 1.5, based on the data of FORCA tow fiber sheets supplied by Nittetsu Composite Co. Ltd. and the basalt fiber sheet supplied by Jiangsu GMV Co., LTD. The proportion of fiber/matrix is an important factor for the strength of the composite. The volume ratio (V) and weight ratio (W) can, respectively, be written as follows:

$$v_c = v_f + v_m \tag{1.1}$$

$$V_f = \frac{v_f}{v_c}, V_m = \frac{v_m}{v_c} \tag{1.2}$$

$$w_c = w_f + w_m \tag{1.3}$$

$$W_f = \frac{w_f}{w_c}, W_m = \frac{w_m}{w_c} \tag{1.4}$$

where v and w represent the volume and weight, respectively; the subscripts c, f, and m denote composite, fiber, and matrix, respectively. Furthermore, the following relationship between the volume and weight apply:

$$V_f = \frac{\rho_c}{\rho_f} W_f \tag{1.5}$$

$$V_m = \frac{\rho_c}{\rho_m} W_m \tag{1.6}$$

where ρ_c, ρ_f, and ρ_m are the densities of the composite, fiber, and matrix respectively. Therefore it is known that the coefficient of thermal expansion for unidirectional FRP composite can be estimated from the thermal expansion coefficients and elastic modulus of each constituent material, using Eq. (1.7).

$$\alpha_L = \frac{E_f \alpha_f V_f + E_m \alpha_m (1 - V_f)}{E_f V_f + E_m (1 - V_f)} \tag{1.7}$$

where α_L, α_f, and α_m are coefficients of thermal expansion of the FRP composite, continuous fibers, and impregnated resin, respectively; E_f, E_m, and E_c are the elastic moduli of the continuous fibers, impregnated resin, and the composite, respectively. Moreover, the tensile stress and modulus of elasticity are given by

$$\sigma_c = \sigma_f V_f + \sigma_m V_m \tag{1.8}$$

$$E_c = E_f V_f + E_m V_m \tag{1.9}$$

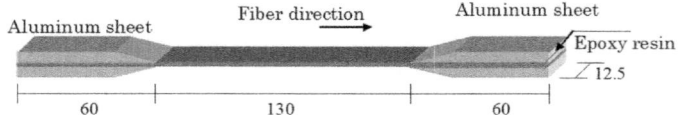

Figure 1.12 Fiber-reinforced polymer (FRP) sheet (or impregnated roving or FRP profiles).

Eqs. (1.8) and (1.9) are usually referred to as the "rule of mixtures" and are fundamental to composite theory.

1.2.1.1   Tensile test method

The tensile properties of traditional fiber cloth for strengthening are measured through a test on the FRP sheet made by fiber cloth impregnated in resin, according to existing standards such as GB50608 and JSCE-41. The dimension and anchorage of the specimen are shown in Fig. 1.12. Generally, only the cross-section of the fibers is considered for the strength measurement of the FRP sheet, without considering the contribution of the resin. Thus the test method can be used to measure the tensile properties of different types of fibers.

1.2.1.2   Tensile properties of fiber-reinforced polymer composites

Taking the tensile properties of the FRP sheet (e.g., a fiber cloth), the tensile properties of different types of FRP are introduced as follows.

Fig. 1.13 shows a comparison among different types of FRP and steel reinforcement in terms of their tensile properties. The tensile properties of FRP bars and profiles will be introduced in the following chapter on structural applications.

The figure illustrates different elastic stress–strain relationships of the FRPs and traditional steel.

1. Strength: in general, the strength of the FRPs ranges from 1500 to 4900 MPa without considering the resin strength, indicating a significantly higher strength than that of ordinary steel, which only ranges from 300 to 500 MPa.
2. Elastic modulus: a large difference exists among different FRPs; for example, the elastic modulus of high modulus fibers, such as carbon fibers can reach up to 230 GPa. The corresponding values of other fibers range from 50 to 110 GPa, which are lower than those of ordinary steel.
3. Ductility: This property also varies largely among different FRPs. High-modulus fibers, such as carbon fibers and PBO only have a strain capacity of 1.4% approximately, while that of Dyneema fiber is higher than 3%. Furthermore, some other fibers have a satisfactory balance in terms of strength, modulus, and strain capacity, for example, basalt fiber and aramid fiber.

Thus a reasonable use of various FRPs is the key to enhancing the structural behavior of existing structures.

The tensile strength of FRPs is generally dominated by fiber rupture. The failure initiates in the portion of fibers, which is subjected to maximum stress in the FRP

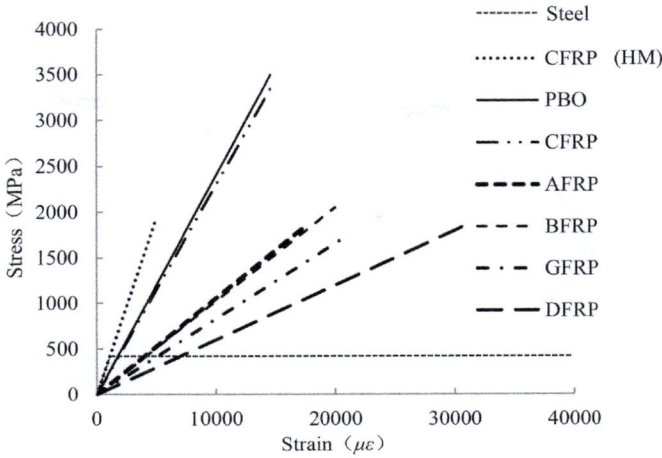

Figure 1.13 Tensile properties of different types of fiber-reinforced polymer composites.

material. The released stress from the ruptured fibers is transferred to nearby fibers. If these fibers can bear the relieved stresses caused by the ruptured fibers, the FRP can take further loading. Otherwise, the successive rupture will take place in the fibers, causing the ultimate failure of FRP.

1.2.2 Property enhancement by hybridization

1.2.2.1 Concept

In some cases, the application of FRP in structures cannot satisfy the overall structural integrity requirements. For instance, in the case of FRP with high strength and high elastic modulus, such as carbon FRP, it is difficult to achieve the desired ductility of the structure. This is owing to the elastic nature of the FRP materials and the inherent small failure strain of the carbon FRP. Meanwhile, carbon FRP is also more expensive than other types of FRP. On the contrary, the relatively inexpensive FRPs, such as basalt and glass FRPs with larger failure strain, can make the structures more ductile. However, their relatively low modulus usually restricts their applications owing to structural deformation requirements. To overcome the aforementioned limitations and enhance the utilization of various FRP composites, the hybridization of different types of fibers is introduced to overcome their shortcomings integrate their advantages, and consequently, achieve the best performance to cost ratio. The idealized the stress–strain relationships of hybrid FRP composites are shown in Fig. 1.14.

Fig. 1.14 shows an idealized stress–strain relationship for hybrid FRPs consisting of three types of FRP, namely high-modulus, high-strength, and high-strain capacity fibers. High-modulus fibers are designed to provide a large initial stiffness. The mixture of high-modulus and high-strength fibers presents a certain strain

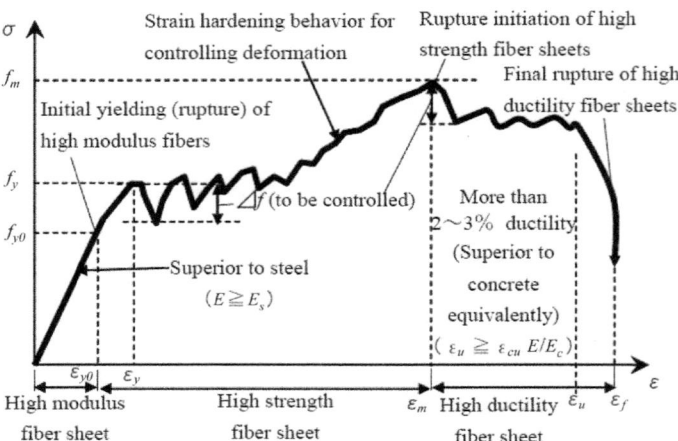

Figure 1.14 Idealized stress−strain behavior of hybrid fiber-reinforced polymer composites (Wu et al., (2006).
Wu, Z., Sakamoto, K., Iwashita, K., & Yue, Q. (2006). Hybridization of continuous fiber sheets as structural composites. *Journal of the Japan Society for Composite Materials*, *32*(1), 12−21.

hardening behavior until the rupture of the high-strength fibers, which may be used to control the deformation of structures with a good recoverability. In addition, high ductility can be achieved by mixing with high-strain capacity fibers in certain proportions.

Some fundamental and critical issues of the hybrid FRPs have been studied, including the stress drop at the rupture of high-modulus FRPs, creep behavior, and fatigue resistance. The observed stress drop owing to the gradual rupture of high-modulus fibers, shown as Δf in Fig. 1.14, should be monitored and controlled. Otherwise, the capacity of the FRPs would be greatly decreased and the strain hardening behavior would be difficult to realize. Focusing on this issue, different kinds of high-strength and high-modulus fiber sheets have been hybridized in different proportions to study the effect of the stress drop, as shown in Fig. 1.15. The control index is defined as the theoretical value of the stress drop divided by the stress, at the rupture of the lowest elongation fiber.

1.2.2.2 Effect

To get a clear insight into the mechanical behavior of hybrid fiber sheets, a series of uniaxial tension specimens was tested by composing different types of fiber sheets that included high-modulus, high-strength, and high-strain capacity fibers. The actual stress−strain relationship of hybrid FRP sheets is shown in Fig. 1.16, which conforms to the idealized behavior.

The previously described properties of hybrid composites can be also obtained for hybrid FRP tendons produced by pultrusion, as shown in Fig. 1.17. As apparent

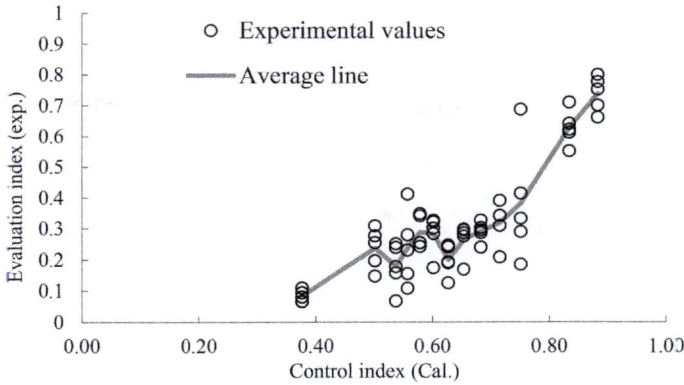

Figure 1.15 Comparison between control and evaluation indices of MC-SC hybrid specimens.

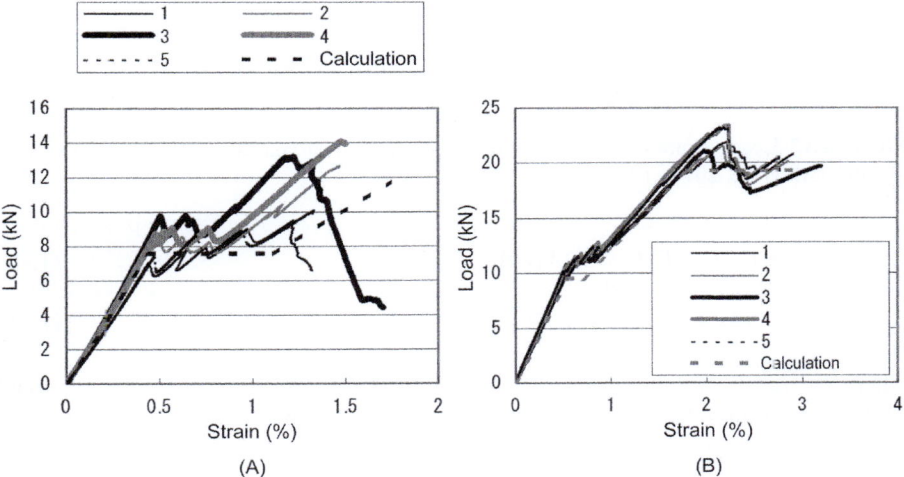

Figure 1.16 Relationship between load and strain of hybrid fiber-reinforced polymer sheets. (A) HM:HSC = 1:2. (B) HM:HSC:HDD = 1:1.5:4.

from the figure, with an increase in the proportion of ductile basalt fibers, the overall ductility of the hybrid FRP tendons is improved. Furthermore, a similar load--strain relationship as that of the steel reinforcement with a yielding platform is achieved, which can satisfy specific structural requirements during strengthening.

Following the rule of mixtures, the stress σ_{ave}—strain ε relationship, and the elastic modulus E_{ave} of hybrid composites can be represented by the following equations.

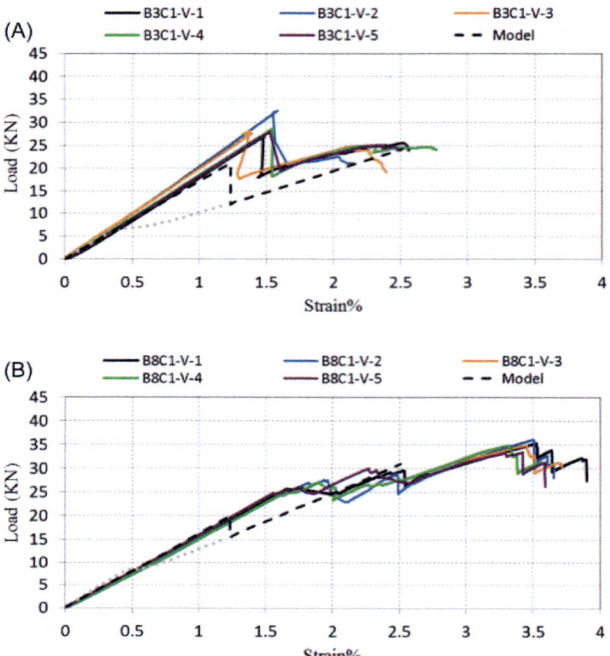

Figure 1.17 Load–strain relationship of hybrid basalt and carbon fiber-reinforced polymer tendons (Ali et al., 2014). (A) B3C1-V. (B) B8C1-V.

$$\sigma_{ave} = \varepsilon(E_1 A_1 + E_2 A_2 + E_3 A_3)/A \qquad (1.10)$$

$$E_{ave} = (E_1 A_1 + E_2 A_2 + E_3 A_3)/A \qquad (1.11)$$

where E_1, E_2, E_3 are the elastic moduli and, and A_1, A_2, A_3 are the cross-sectional areas of high-modulus, high-strength, and high-strain capacity fiber sheets, respectively. A is the entire cross-sectional area of the fiber sheets (i.e., $A = A_1 + A_2 + A_3$).

In fact, from hybridization with three different types of fibers, three distinct zones in the stress–strain curves are realized. The first zone, just before the rupture of high-modulus (HM) fiber sheets, represents the stiffness of the hybridization. Finally, the capability of high-strain capacity fiber sheets determines the ultimate rupture and the ductility of the hybridized FRP.

A refined model to predict the overall mechanical behavior of the hybrid FRP composites was developed, which was based on the study of interfacial stress transfer based on fracture mechanics.

To find the contribution of low elongation (LE) fibers to hybrid stiffness after its total rupture, it was necessary to study the interfacial stress/load transfer from the high-elongation (HE) fibers to the LE fibers. The specimen length (l_t) was divided

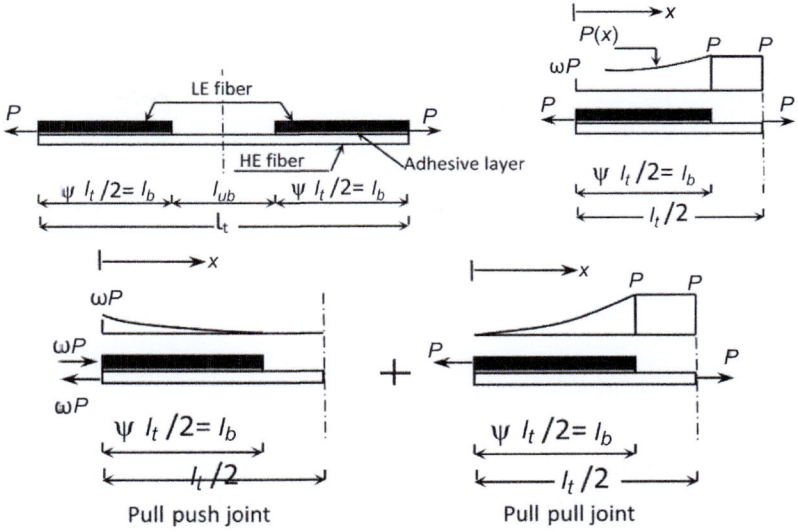

Figure 1.18 Mechanism of fiber fracture in the fiber-reinforced polymer (Ali et al., 2014).

into two parts: bonded part (l_b), where the LE fibers were still bonded to the HE fibers; and the unbonded part (l_{ub}), where the bonds between the LE and the HE fibers were totally delaminated. In the bonded part, some loads would be transferred through the adhesive layer (resin matrix) from the HE fibers to the LE fibers; then, the load carried out by the LE fibers would decrease from the left end of the specimen-half to its right end. By increasing the load, debonding at the interfacial face would propagate at the right end of the specimen-half, and the bonded length ratio ($\psi = l_b/l_t$) would decrease until the total debonding was achieved. It was assumed that the shear stress—slip relation was linear before the occurrence of the interfacial fracture, and the value of the shear stress would suddenly drop to zero when the value of slip exceeded its maximum value without taking into account the softening behavior. Therefore it was assumed that the specimen would behave linearly for the applied loads until its total debonding. Consequently, by using the superposition method, the load distribution between the LE and the HE fibers could be simply divided into two common adhesive joints: pull—pull joint, and pull—push joint, as illustrated in Fig. 1.18. Here, ω is the percentage of the load that would be transferred to the LE fibers at the left end of the specimen.

1.2.2.3 Design method

The prediction of the load—strain relation of hybrid FRP composites can be described through the following steps;

1. *Before the LE fiber rupture:*

Following the rule of mixtures, the modulus of elasticity (E) and the pseudoyielding load (P_y) can be predicted from the following equations:

$$E = E_1 v_1 + E_2 v_2 \tag{1.12}$$

$$p_y = f_1 A_1 + \varepsilon_1 E_2 A_2 \tag{1.13}$$

where v_1 and v_2 are the volume ratios of HE and LE fibers, respectively; and f_1 is the tensile strength of the LE fibers.

2. *After LE fiber rupture:*

Find the minimum ratio of the bonded length ratio ψ_{min} using Eq. (1.14).

$$\psi_{min} = \frac{2}{\lambda l_t} \operatorname{arctanh}\left(\frac{\delta_f \lambda}{0.9 \times \varepsilon_1}\right) \tag{1.14}$$

where δ_f is the maximum relative displacement between adherents; ε_1 is the failure strain of the HE fiber; and l_t is the total length of specimen.

Assuming the value of the bonded length ratio in this range, $\psi_i = (1.0 : \psi_{min})$

From Eq. (1.15), find the maximum load required to make further debonding at each value of the selected (ψ_i).

$$p_{dep} = \frac{2G_f E_1 A_1 \lambda}{\tau_f \tanh(\lambda \psi l_t / 2)} \tag{1.15}$$

where G_f is the interfacial fracture energy; E_1 is the modulus of elasticity of the HE fiber; A_1 is the cross-sectional area of the HE fiber; τ_f is the shear strength of adhesive layer; and ψ is the ratio of the bonded length to the total length of the specimen.

Next, utilize Eq. (1.16) to find the strain corresponding to the applied load;

$$\varepsilon = \frac{p(1 - \psi)}{E_1 A_1} + \frac{\tau_f^2 p.S}{G_f \lambda^2 E_1^2 A_1 l_t}\left[\frac{\tanh(\lambda \psi l_t / 2)}{A_1 \lambda} + \frac{\psi l_t}{2A_1 \beta}\right] \tag{1.16}$$

where S is the width of the interfacial surface; β is the ratio of the LE fiber to the HE fiber stiffness.

Repeat steps as long as ψ_i is less than ψ_{min}.

Draw the relationship between the load and the strain.

1.2.3 Temperature-dependent behavior

1.2.3.1 High-temperature performance

As an FRP consists of fibers and resin, the mechanical behavior of the FRP under elevated temperatures is highly dependent on the thermal properties of the fibers and the resin. Many fibers are usually resistant to high temperatures up to 400°C−800°C. However, for most of the commonly used matrices, the glass transition temperature (T_g) is below 60°C for hand layup applications, and below 120°C for pultrusion. Thus the temperature-dependent behavior of FRPs is important for their application in building construction, especially under high temperature conditions.

1. *Thermal properties of matrix*

The matrix in an FRP (epoxy or vinyl ester resin) is vulnerable to high temperature because of the poor behavior of the resin at high temperatures compared with that of the fibers. The typical viscoelastic behavior of resin at a high temperature can be described in terms of three key parameters, namely storage modulus, loss modulus, and damping coefficient (TAN) as shown in Fig. 1.19. The storage modulus denotes the elastic part and the loss modulus represents the viscoelastic part. On the other hand, TAN stands for the ratio of loss modulus to storage modulus. When the temperature reaches or exceeds the T_g of the resin, the viscosity increases and elasticity decreases rapidly. Thus T_g is a threshold value determined by these three parameters.

The thermosetting polymer is the most commonly used matrix among the impregnating resins for structural strengthening. The most common thermosetting polymer is the epoxy resin, with T_g ranging from 40°C to 60°C. The typical thermodynamic properties of epoxy resin are shown in Fig. 1.20. As the temperature increases from 30°C to 120°C, the

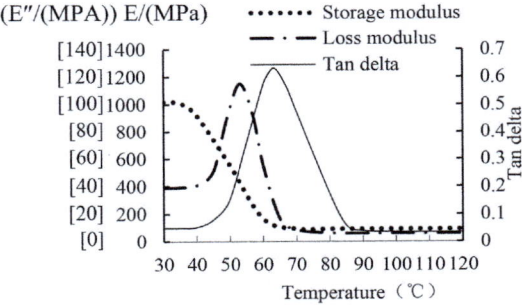

Figure 1.19 DMA properties of epoxy resin (Cao et al., 2009).

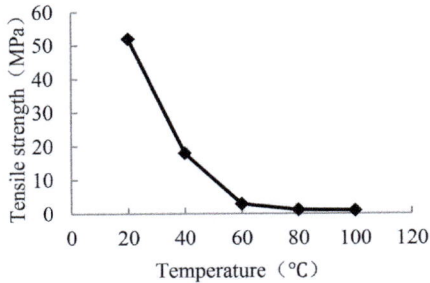

Figure 1.20 Tensile strength of epoxy resin at high temperature.

storage modulus decreases rapidly from a maximum at 38°C to 10% of the initial value at 60°C, after which, the value remains constant. On the other hand, the loss modulus increases with temperature. It decreases rapidly after reaching its peak value at 52°C and then, tends to be stable. Epoxy resin is commonly used in FRP sheets, while for FRP bars or profiles, the most commonly used matrix is high temperature curing epoxy resin or vinyl ester resin, with T_g reaching up to approximately 130°C. Thus the FRP products manufactured through pultrusion at a high temperature have better high-temperature performance than the FRP sheets. However, the temperatures under conflagration are far higher than the aforementioned T_g of the matrix resin. Thus the evaluation of the strength retention of the FRPs at high temperatures is necessary.

The thermal tolerance capacity of epoxy resin is characterized by T_g, which is obtained by dynamic thermomechanical analysis (DMA) tests. The value of T_g represents a critical temperature when epoxy resin softens and its tensile stiffness decreases under the influence of the temperature. For details of the DMA tests, ISO 6721 and JIS K 7244 can be referred to. The relationship between the storage modulus and the service temperature and the representative behavior whose shape is similar to a staircase are shown in Fig. 1.21. The upper and lower baselines and the points on the straight line where the slope of the curve becomes maximum are defined as T_g. In addition, the maximum temperatures of the loss modulus and the Tan Delta are also defined as T_g, as shown in Figs. 1.22 and 1.23. In

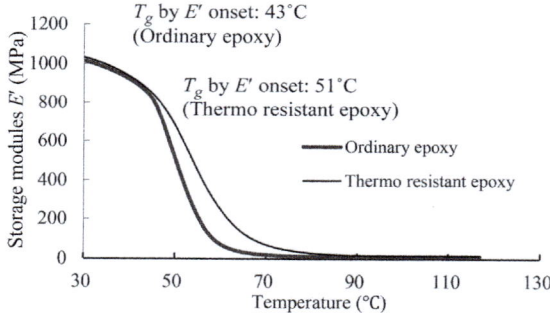

Figure 1.21 Relationship between storage modulus and temperature.

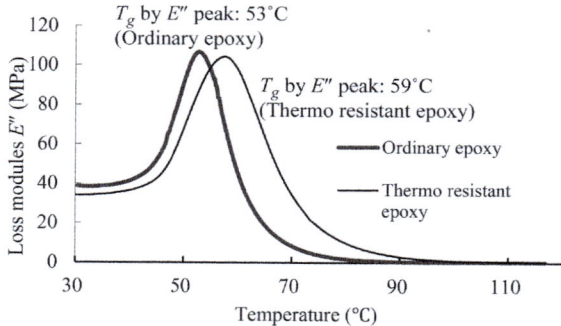

Figure 1.22 Relationship between loss modulus and temperature.

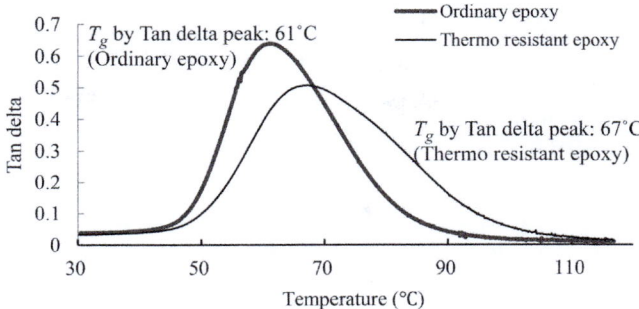

Figure 1.23 Relationship between Tan Delta and temperature.

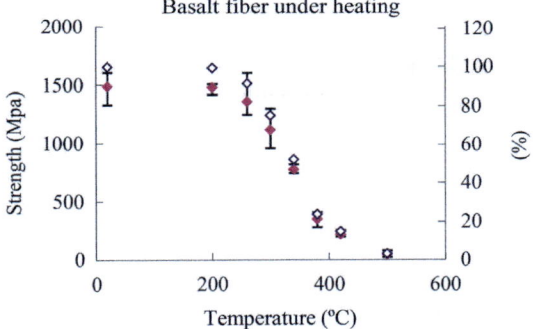

Figure 1.24 Tensile strength of basalt fibers under heating.

this result, it is observed that the T_g value obtained from the storage modulus of the newly developed thermal resistant epoxy resin is about 10°C higher than that of the ordinary epoxy.

2. *Tensile behavior of fibers under elevated temperatures*

The degradation of the tensile strength of the carbon fiber bundles under elevated temperatures of up to 500°C, which depicts the degradation trend of fibers under high temperatures. It is observed that when the temperature is over 100°C, the degradation of the tensile strength of the carbon fibers is almost linear with respect to an increase in the temperature. Under a temperature of 500°C, the tensile strength remains over 50% of its original strength under ambient temperatures.

The degradation of the tensile strength of the basalt fiber bundle under elevated temperatures of up to 500°C are shown in Fig. 1.24, which is similar to the behavior of carbon fibers up to 300°C. Subsequently, the tensile strength drops significantly down to almost zero under a temperature of 500°C. The temperature of 350°C corresponds to a strength retention of 50%, which could be useful for design considerations. The strength degradation after heating is shown in Fig. 1.25, which shows a relatively slower degradation compared to that under heating.

The aforementioned temperature-dependent behavior of basalt fibers is consistent with the results by Militký et al. (2002), as shown in Fig. 1.26. For most of the conditions, the

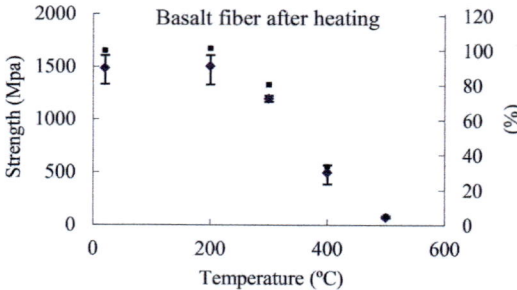

Figure 1.25 Tensile strength of basalt fibers after heating.

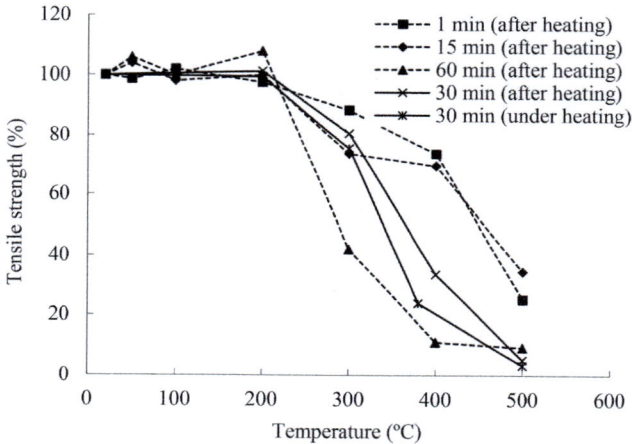

Figure 1.26 Strength comparison between the test results and from Militký et al. (2002).

basalt fiber shows obvious strength degradation after 300°C and loss of most of the strength at 500°C.

However, owing to different sources of basalt fibers with different chemical compositions, some types of basalt fibers can maintain much higher strength under elevated temperatures. Sim and Park (2005) reported their results on basalt fibers and carbon fibers under elevated temperatures of up to 600°C. The results in Fig. 1.27 show that the basalt fibers could retain more than 90% of their tensile strength at a temperature of 600°C, whereas carbon and glass fibers retained only 60% of their tensile strength.

Thus a significant difference in the temperature-dependent behavior of basalt fibers indicates that the properties should be identified based on the source of the fibers, for example, from China or Russia.

The high-temperature-resistant basalt fibers have high strength retention up to 70% at 400°C for 2 h, and no bonding due to the melting of the fibers takes place in the fibers at 900°C for 2 h; whereas, under identical circumstances, bonding can be observed in glass fibers, as shown in Fig. 1.28.

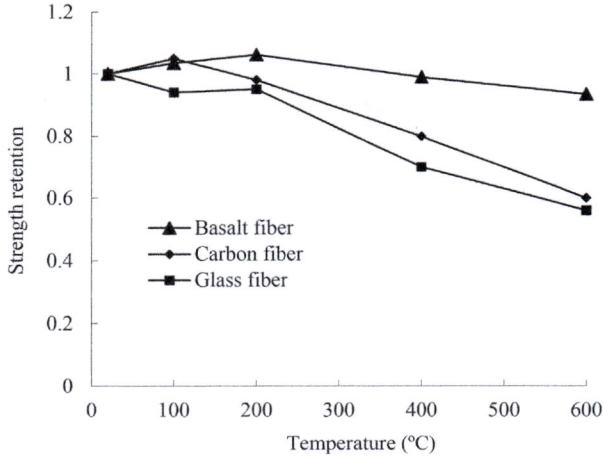

Figure 1.27 Strength variation with respect to heat exposure (Sim and Park, 2005).

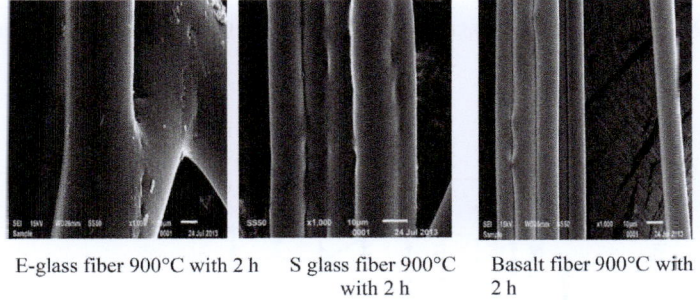

E-glass fiber 900°C with 2 h   S glass fiber 900°C with 2 h   Basalt fiber 900°C with 2 h

Figure 1.28 Surface morphology of fibers after high temperature treatment. E-glass fiber 900°C with 2 h. S glass fiber 900°C with 2 h. Basalt fiber 900°C with 2 h.

3. *Tensile property of FRP under elevated temperature*
a. *Single type of FRP sheets*

It is seen in Fig. 1.29 that different FRP sheets exhibit similar degradation trends under elevated temperatures. CFRP and BFRP do not have any strength degradation below a temperature of 200°C. GFRP shows a slight degradation even at a temperature of 200°C. After 300°C, all the FRP sheets exhibit obvious strength degradation and lose almost all of the strength, except CFRP.

When the fiber sheets are impregnated with high temperature resistance resin (TG-1009), the tensile property is different, as shown in Fig. 1.30. After being impregnated with TG-1009, the tensile behavior remains at a high level up to 300°C, which is 25% higher than that of the specimen with ordinary epoxy resin. At 400°C, the degradation of tensile strength is obvious owing to the decomposition of carbon fibers. At a temperature

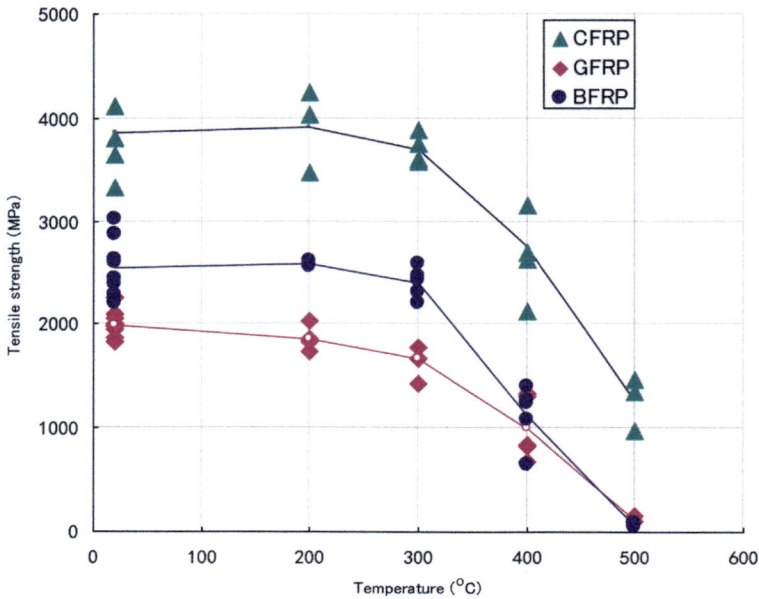

Figure 1.29 Tensile property comparison among different fiber-reinforced polymer sheets.

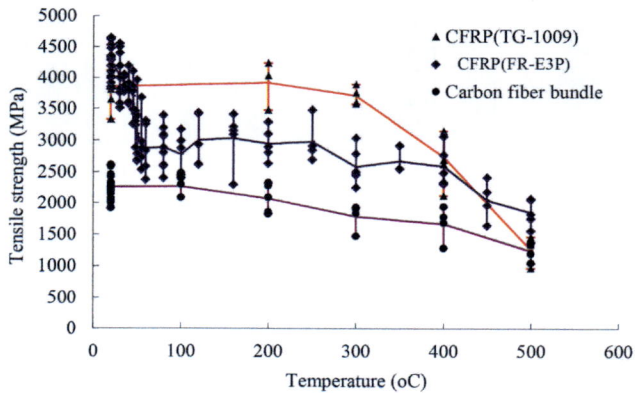

Figure 1.30 Comparison of tensile strength degradation among fiber bundle and fiber-reinforced polymer with different resin.

of 500°C, the tensile strength of CFRP with TG-1009 degrades to that of the fiber bundles, which indicates a total loss of function of the resin in FRP. In contrast, although CFRP with ordinary epoxy resin does not show high strength retention compared to that with TG-1009, it does show a higher strength than the fiber bundle at all temperatures, which demonstrates the residual function of resin after its T_g.

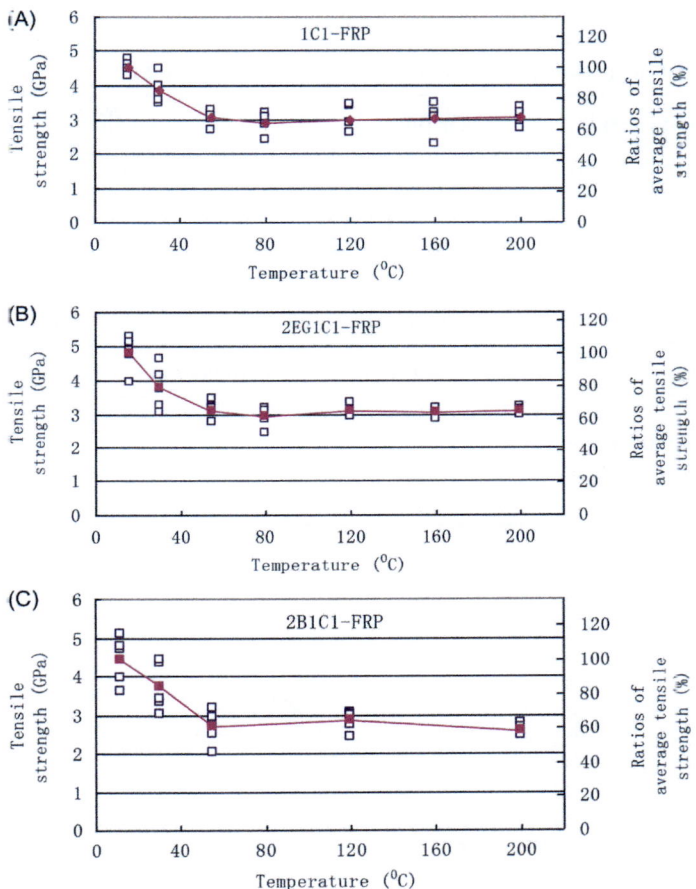

Figure 1.31 Hybrid fiber-reinforced polymer (FRP) sheet under elevated temperatures (Cao et al., 2009). (A) Carbon FRP. (B) Hybrid C/Glass FRP. (C) Hybrid C/Basalt FRP.

b. *Hybrid FRP sheets under elevated temperatures*

The hybridization effect on the tensile behavior of the CFRP sheets under elevated temperatures was investigated by focusing on the hybridization of carbon and glass fiber sheets, and carbon and basalt fiber sheets. The degradation of their tensile strength is shown in Fig. 1.31. From a comparison of CFRP and two kinds of hybrid FRPs, the degradation trend is similar, which exhibits a major degradation at the T_g and remains constant up to 200°C. The strength retention of hybrid FRP is similar to that of the CFRP sheets, whereas the variation of strength retention is much smaller than that of CFRP. This fact indicates that although hybridization cannot improve the strength retention under high temperatures, it does contribute to the stability of strength, which can enhance the design strength of FRP under high temperatures.

Table 1.6 Summary of the results of tensile tests.

Span of the test specimen	130 mm		2 m	
Type of epoxy resin	Ordinary	Thermal	Ordinary	Thermal
Tensile strength (MPa)	4127	4318	3444	3391
Elastic modulus (GPa)	254.4	289.7	231.4	227.9
Fracture strain (%)	1.622	1.491	1.488	1.488

Table 1.7 Comparison of temperature coefficient of expansion between resin and fiber.

Material type	Temperature coefficient of expansion ($\times 10^{-6}/°$C)
Epoxy	45−65
Carbon fiber (fiber direction)	−0.6 to −0.2
Basalt fiber (fiber direction)	6.5−8
Glass fiber (fiber direction)	6−8

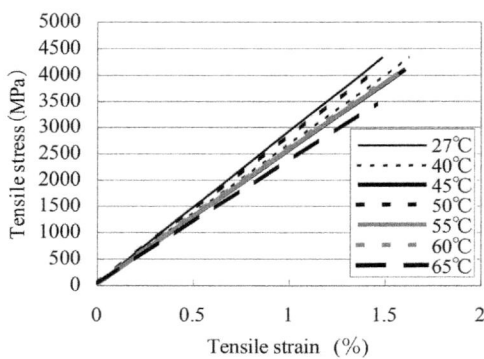

Figure 1.32 Tensile stress−strain curve for 130 mm specimens (ordinary epoxy).

c. *Size effect of FRP at elevated environmental temperatures*

Apart from the coupon test of the 130 mm specimens, specimens with 2 m length were also investigated under high temperatures by a rubber heater directly attached on the sheet surface. The results are compared in Table 1.6.

Two types of epoxy resins were used in the tests. The average strength, stiffness, and fracture strain are shown in Table 1.7. The relationships between the tensile stress and strain for the 130 mm specimens with different types of epoxy resins under different temperatures are shown in Figs. 1.32 and 1.33. From the results, it is evident that the environmental temperature does not affect the tensile stiffness. On the other hand, the relationships between the tensile strength and environmental temperature,

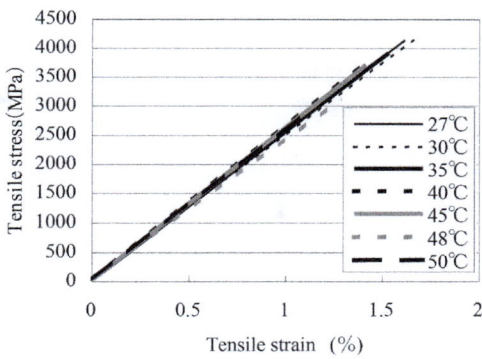

Figure 1.33 Tensile stress—strain curve for 130 mm specimens (thermoresistant epoxy).

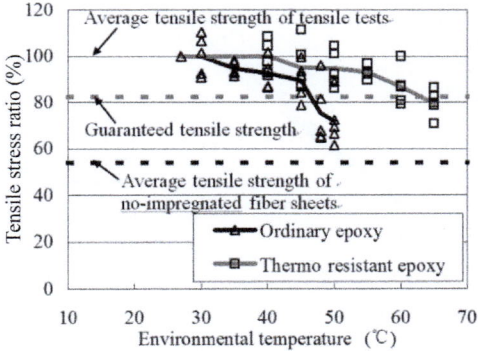

Figure 1.34 Tensile stress ratio—temperature curve for 130 mm specimens.

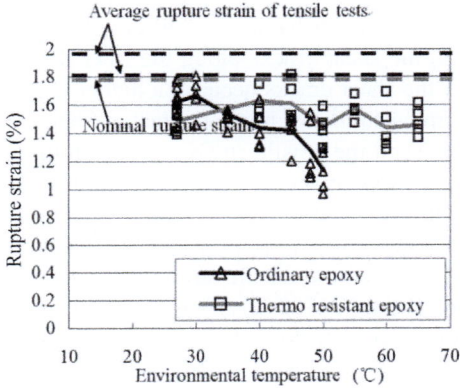

Figure 1.35 Rupture strain—temperature curve for 130 mm specimens.

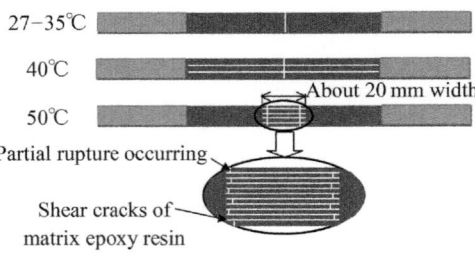

Figure 1.36 Rupture modes under different levels of temperature.

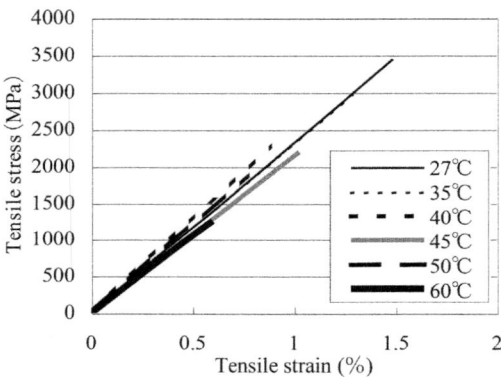

Figure 1.37 Tensile stress—strain curve for 2 m specimens (ordinary epoxy).

and strain capacity and environmental temperature are shown in Figs. 1.34 and 1.35, respectively. In these figures, the tensile strength and strain capacity values decrease suddenly at temperatures higher than T_g. Therefore these values of the FRP sheets are the same as those of nonimpregnated (dry) fiber sheets at temperatures higher than T_g plus 10°C. Moreover, rupture behavior at different temperatures is shown in Fig. 1.36. In this figure, the shear cracks along the fiber direction and the local ruptures around the end of the shear cracks occur at a higher temperature higher than T_g. This rupture mode is similar to that of nonimpregnated fiber sheets.

The relationships between the tensile stress and the strain of 2 m specimens with different types of epoxy resins under different temperatures are shown in Figs. 1.37 and 1.38. Compared with the experimental results of 130 mm specimens, it appears that the environmental temperature did not affect the tensile stiffness. On the other hand, the relationships between the tensile strength/strain capacity and the environmental temperature are shown in Figs. 1.39 and 1.40. In these figures, the tensile strength and the strain capacity values decreased linearly with the environmental temperature.

Thus the tensile strength and strain capacity values of small-dimension FRP specimens remained approximately unchanged at temperatures lower than T_g and

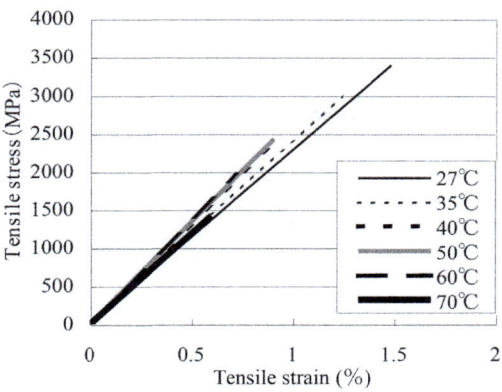

Figure 1.38 Tensile stress—strain curve for 2 m specimens (thermoresistant epoxy).

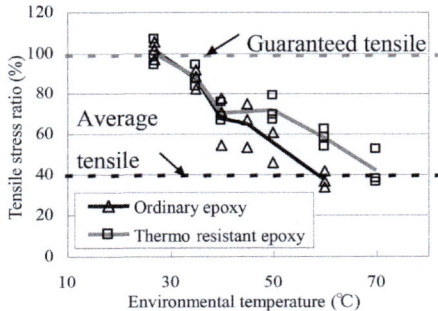

Figure 1.39 Tensile stress ratio—temperature curve for 2 m specimens (ordinary epoxy).

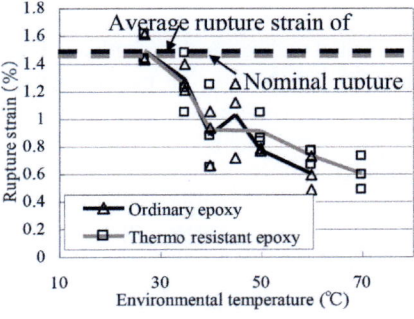

Figure 1.40 Rupture strain—temperature curve for 2 m specimens.

decreased drastically at temperatures higher than T_g. In contrast, the corresponding values of large-dimension FRP specimens decreased linearly with the environmental temperature. The reason for the previously mentioned phenomenon was

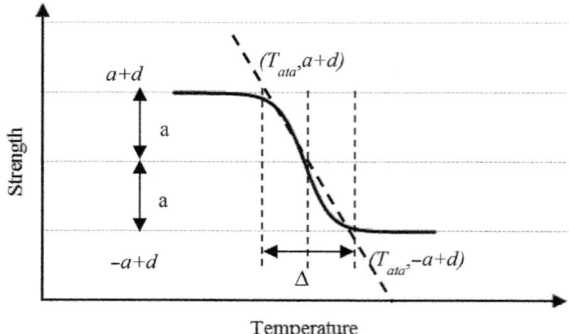

Figure 1.41 Mechanical property model of fiber-reinforced polymer at high temperature (Cao et al., 2011).

attributed to the increased probability of a larger specimen containing a flaw large enough to lead to failure.

4. *Degradation mechanism of FRP at a high temperature and the design method*

Based on the previous analysis, within a certain temperature range, the mechanical properties of FRP at high temperatures are closely related to the high-temperature performance of the matrix resin. Thus the prediction of residual strength at high temperature should consider the T_g of the resin. Fig. 1.41 shows the concept of strength degradation of FRP at high temperatures. The reduction within the T_g range and the stable residual strength in the later phase are the keys for fitting. The fitting formula is established according to that concept

$$\sigma = \frac{1}{2}(\sigma_0 - \sigma_r)\tanh\left[-\frac{1}{\Delta T/2}\left(T - (T_g + \Delta T/2)\right)\right] + \frac{1}{2}(\sigma_0 + \sigma_r) \qquad (1.17)$$

in which, σ_0 and σ_r are the initial and residual strengths of FRP, respectively, which are obtained from experiments. T_g and ΔT can be measured through the DMA test. The residual strength of FRP at any temperature can be predicted using the previous equation.

The previously described model at high temperatures is suitable for the region near the T_g of the resin in FRP (300°C in general). When the temperature is further increased, the thermal decomposition temperature will be reached, and a further decrease in the mechanical properties of FRP will take place. The degradation laws are shown in Figs. 1.42 and 1.43.

A difference in degradation law at higher temperatures exists among FRPs with different fibers. For a low-modulus fiber, for example, the basalt fiber, when the temperature reaches the thermal decomposition temperature (T_d) range, the strength significantly decreases until failure. For a high-modulus fiber, for example, the carbon fiber, when the temperature reaches the T_d range, a satisfactory strength can be retained even though a decrease in the strength may take place. Thus an allowable

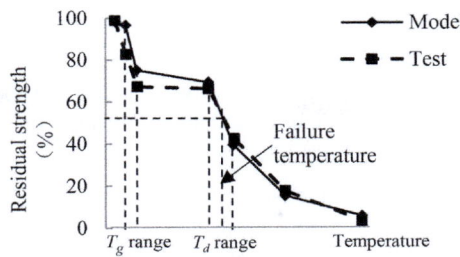

Figure 1.42 Theoretical and experimental results of basalt fiber-reinforced polymer sheet with epoxy resin.

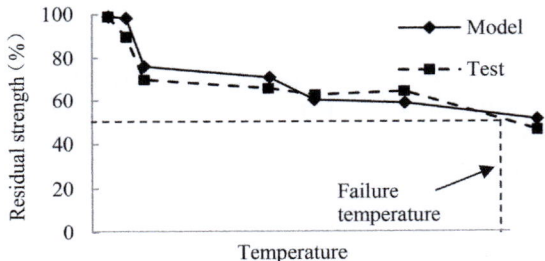

Figure 1.43 Theoretical and experimental results of carbon fiber-reinforced polymer sheet with epoxy resin.

residual strength is a key for the performance design of FRP at a high temperature. If residual strength is defined to be $50\% \times f_u$ (where f_u is the tensile strength), an allowable failure temperature can be explicitly proposed for BFRP and CFRP. During the design, structural safety can be guaranteed by controlling the temperature of FRP in structures below that allowable failure temperature.

In summary, the degradation law of FRP at high temperature is divided into four stages:

1. When the temperature is lower than the T_g, the resin exists as a glassy state. In this state, the properties of resin can be considered identical to those in the state at a normal temperature.
2. When the temperature is higher than the T_g, the elastic modulus and shear property decrease rapidly and the viscidity increases. This state is maintained within a certain temperature range. According to the existing research, a large loss of modulus in this stage causes a significant strength decrease of the FRP.
3. When the temperature further increases, the modulus of the resin tends to be stable and the viscidity keeps increasing. In this stage, the FRP also maintains a stable strength.
4. When the temperature is higher than the thermal decomposition temperature, thermal decomposition and oxygenolysis occur. A large amount of toxic gas is released. Combustion and carbonization may take place if the temperature is further increased. In this stage, the strength of the FRP further decreases. The FRP can be treated as fiber bundles after the complete decomposition of the resin.

5. *Enhancement of high-temperature performance of FRPs*

In the practical applications of FRPs, the working temperature can be guaranteed through construction measures or protective layers. Additionally, the high-temperature performance of FRPs can be enhanced through the following method.

a. *Hybridization*

Hybridization enhances the high-temperature performance of FRPs in terms of the stability of strength at high temperatures, rather than the value of the strength. Through the hybridization of low-modulus and high-modulus fibers, the stresses at rupture in a part of high-modulus fibers can be alleviated effectively; thus, the stability is enhanced. The hybridization of carbon fibers with basalt or glass fibers can effectively reduce the scattering of strength at high temperatures, as shown in Fig. 1.44. With a decrease in the strength scattering, the design stress of the FRP at high temperatures can be enhanced.

b. *Modification of the matrix*

The adoption of high-temperature-resistant resins (HTRR) can enhance the performance of FRPs within a certain temperature range. Additionally, owing to the requirement of cost in civil engineering, it is unrealistic to adopt the resin used in aerospace engineering. Thus based on the commonly used epoxy resin, a modified

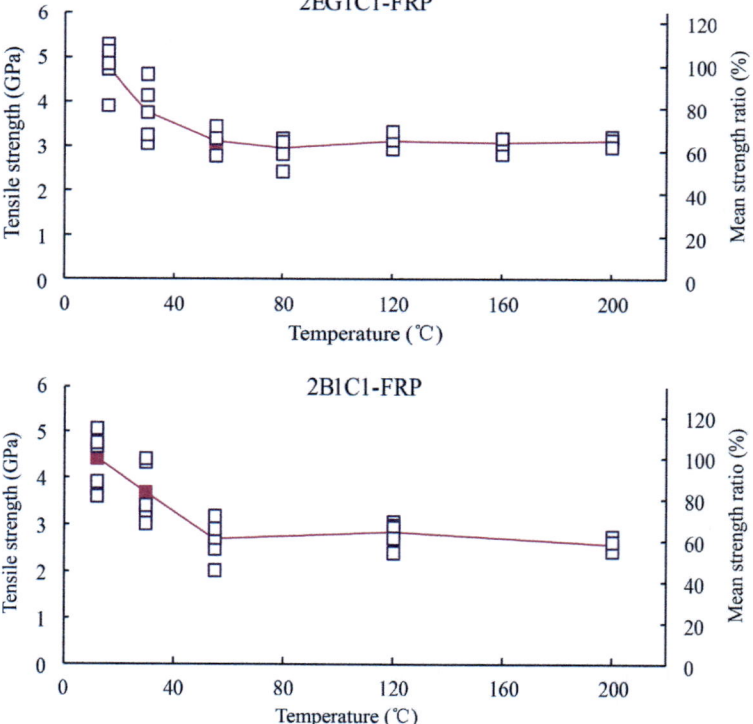

Figure 1.44 Effect of hybridization on high-temperature performance of fiber-reinforced polymer (Cao et al., 2009).

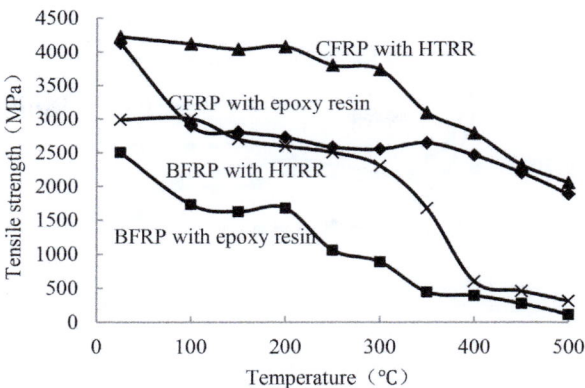

Figure 1.45 High-temperature performance of fiber-reinforced polymer with high-temperature-resistant resin (HTRR).

resin with a high T_g of up to 200°C can be obtained, which effectively enhances the high-temperature performance of the FRP and controls the cost. Fig. 1.45 shows the properties of the FRP with a modified resin. Within 300°C, the strength shows a little decrease. However, at a higher temperature, the strength decreases significantly owing to the carbonization of fibers and thermal decomposition of the resin.

Another modification of the matrix is by adding montmorillonite to enhance the bond behavior of the fiber and the matrix. The methods of developing the properties of montmorillonite include four aspects: (1) the mixing method is very important because montmorillonite is a type of nanoscale material. It should be sufficiently dispersed in the resin. (2) A continuous mixing at a constant temperature of 70°C is necessary. (3) The addition of water or an organic solvent, such as methyl alcohol is necessary. (4) The existing research studies have reported that the optimal amount of montmorillonite is 10%. The addition of montmorillonite has a significant effect on the high-temperature performance of BFRP. However, when the temperature exceeds 300°C, the degradation is controlled by the carbonization of the fiber and the resin, and the function of montmorillonite cannot be developed, as shown in Fig. 1.46. Furthermore, montmorillonite has no obvious effect on CFRP because the carbon fiber matches the resin so perfectly that the bond behavior cannot be further enhanced by montmorillonite. Thus the selection of enhancement methods should consider the matching of the fibers and the matrix, which has limitations.

The combination of the above two methods is also an effective enhancement method. The high-temperature performance of CFRP sheets can be enhanced through the addition of montmorillonite to HTRR, as shown in Fig. 1.47. Within 300°C, high strength retention of CFRP is achieved. Thus the functions of HTRR and montmorillonite are both developed. However, when the temperature is higher than 300°C, no enhancement effect can be observed. The strength retention decreases to 50% of that at a temperature of 500°C.

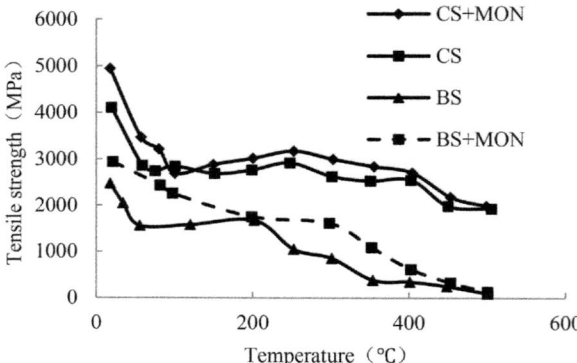

Figure 1.46 High-temperature performance of fiber-reinforced polymer with modified matrix by montmorillonite (CS: carbon fiber sheet; BS: basalt fiber sheet; MON: montmorillonite) III HTRR + modification.

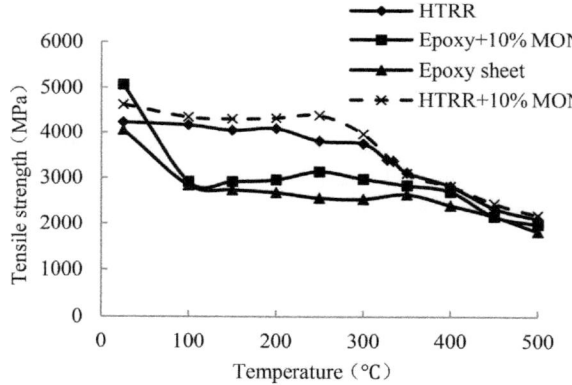

Figure 1.47 High-temperature performance of carbon fiber-reinforced polymer sheet.

1.2.3.2 Low-temperature performance

A large number of civil engineering construction projects must be carried out in low-temperature environments. Research studies on FRP-reinforced concrete materials and structures show that the strength and ultimate bearing capacity of concrete, steel bar, and their interfaces are increased at low temperatures; however, the toughness is decreased, and the difference is significant at room temperatures.

Mechanical properties of FRP at low temperatures. The performance degradation at low temperatures of the fibers is very limited. As shown in Fig. 1.48, the basalt fiber is frozen at −50°C for 1−5 h, and its strength is almost unchanged. When the FRP material is used for strengthening a structure in the cold region, its mechanical properties will degrade with a decrease in the temperature, which is

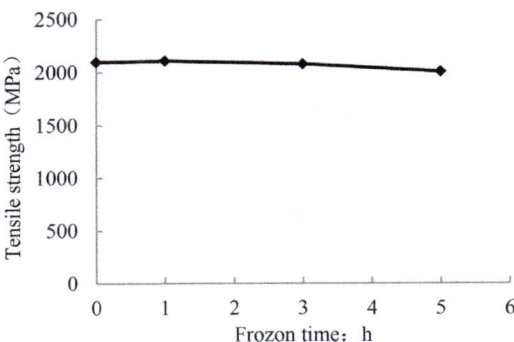

Figure 1.48 Strength of basalt fiber after frozen at −50°C.

mainly caused by the temperature incompatibility between the resin and the fibers in FRP. In general, the temperature expansion coefficient of the FRP resin matrix is about one order of magnitude higher than that of the fiber, as shown in Table 1.7. Owing to the different shrinkages of the resin and the fibers, the fibers prevent the resin from shrinking further, thus resulting in a certain residual stress in the interface between the two. Owing to the negative thermal expansion coefficient of the carbon fibers and the aramid fibers, the residual stress between the carbon fibers and the resin at the same conditions can be much larger than the residual stress at the interface between the glass fibers or the basalt fibers and the resin (Rivera and Karbhari, 2002). Another unfavorable factor for the application of FRP in cold regions is the embrittlement of the resin matrix, which increases the modulus of elasticity of the resin at low temperatures, resulting in an increased stiffness, but reduced ultimate deformation (Karbhari, 2007). At a low temperature of −10°C to −40°C, the tensile strength in the fiber direction of the FRP sheet will be degraded, while the tensile strength in the vertical direction of fiber will be increased because of the low-temperature resin hardening.

In addition, the temperature cycling at low temperatures increases the amount and density of resin microcracks owing to the residual stress. The tensile strength of a glass fiber-reinforced continuous winding pipe is improved at low temperatures; however, the tensile strength of the corresponding GFRP specimen is significantly reduced (Kuz'min et al., 1989). It is also found that FRP in concrete at low temperatures has better performance than that of FRP exposed directly to a low temperature. At the same time, thick FRP bridge panels are less susceptible to low temperatures than thin FRP materials, such as an FRP sheet (Dutta et al., 2006).

At present, the study on the performance of FRP at low temperatures is still relatively scarce, especially in this study, we only consider the role of low temperature in isolation (Kuz'min et al., 1989). The actual structures are mostly under the coupled effect of environmental humidity, load, and low temperature, which may further reduce the performance of the FRP material at low temperatures. It is necessary to carry out multifactor coupling research of FRP materials under low temperatures. The research only studies the durability of the existing materials, and how to improve the performance of the FRP materials at low temperature is not considered. The prediction model and the design method of the FRP material under low temperatures has not yet been established.

1.2.3.3 Performance under freezing and thawing cycles

The freezing and thawing cycle is one of the most important factors affecting the service life of concrete structures in cold regions. The long-term freezing and thawing cycles can lead to spalling of surface concrete and corrosion of the reinforcing steel bars. An FRP material used in a cold region to strengthen a concrete structure also faces this challenge. On the one hand, the differences in the temperature coefficients of expansion of the fibers and the resin may cause interface stress and debonding in the freezing and thawing cycle. On the other hand, the resin matrix can easily absorb the water vapor in the environment, and the resulting water swelling and shrinkage during the freeze—thaw cycle reduces the interface property between the resin matrix and the fiber—resin matrix. In the long-term, in the water environment, the properties of the resin matrix, interface, and glass fiber degrade, the brittleness of the resin matrix increases, and the ductility decreases at low temperatures. In addition, the presence of the structural service load will produce or expand microcracks on the resin matrix and interface, thus making it easier for the water vapor in the external environment to invade and accelerate the freeze—thaw cycle of the FRP material degradation. In cold areas, especially in areas affected by the freeze—thaw cycle, the FRP reinforcement system will be affected by this. The embrittlement of the resin matrix will lead to the brittle failure mode of the FRP and FRP—concrete interfaces. Furthermore, an increase in the rigidity will reduce the effectiveness of the resin transfer stress, including that between the fibers and the FRP, and the concrete. In addition, the use of deicing salt in cold regions can also accelerate the degradation of the resin matrix by the freeze—thaw cycle.

Freezing and thawing cycle test method with a center temperature control within $8 \pm 2°C$ to $-17 \pm 2°C$, as shown in Fig. 1.49 was designed to test the performance of the material and the structures in freezing and thawing cycles. After the

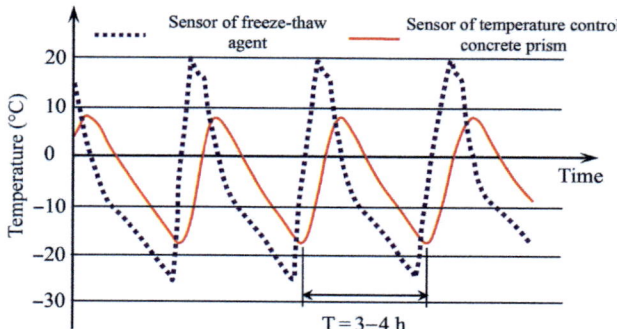

Figure 1.49 Temperature change during freeze—thaw cycle test [GB-T50082-2009].

freeze—thaw cycle, the surface of the FRP sheet did not change compared to that before the freeze—thaw cycle, and the failure modes of the specimens after a monotonous tensile loading did not change significantly. The tensile strength and elongation of the FRP sheets decreased with an increase in the number of freeze—thaw cycles, and the elastic modulus of FRP sheets did not decrease. The tensile strength of the BFRP sheets did not significantly decrease after 200 freeze—thaw cycles, which was higher than that of the CFRP and GFRP sheets. The elastic modulus of the FRP sheets increased first, and then decreased slightly after the freeze—thaw cycles, and the growth rate of the BFRP sheet was larger than that of the CFRP and GFRP sheets. After the freezing and thawing cycles, the relative strain of the BFRP sheet decreased slightly, which was superior to CFRP and GFRP, as shown in Fig. 1.50.

As the basalt fiber FRP has good antifreeze—thaw performance, the use of BFRP and CFRP hybrid is conducive to further enhancing the freeze—thaw resistance of the CFRP performance. The FRP sheets exhibited better freeze—thaw resistance and mechanical stability than the single FRP sheets, as shown in Fig. 1.51.

1.2.4   Time-dependent behavior

The importance of fatigue and creep behavior of the FRP composites becomes significant when structures strengthened with FRP are under dominant cyclic loads or a high level of sustained load. The evaluation of FRP under different levels of cyclic and sustained stresses and the prediction of corresponding fatigue strength and allowable stress are the key for FRP applications.

1.2.4.1   Fatigue behavior

1. *Mode of FRP failure under fatigue load*
   Failure phenomenon under different levels of stresses

The fatigue failure is the failure of a structure, component, or material under cyclic stress or strain, which is usually less than the ultimate load strength or strain. It is important to understand the different mechanisms of fatigue failure of FRP.

Similar to the failure mode of the FRP composites under static loads, the typical damage pattern in fatigue tests under high levels of stress is shown in Fig. 1.52, which displays fibers breaking (Zhao et al., 2016). This phenomenon indicates that the fatigue limit of FRP under a high level of stress is controlled by the most critical fibers, which can induce a continuous fracture along the cross-section of the FRP after their initial fracture occurs.

In most of the applications, the cyclic stresses on the FRP are not high enough to cause a fracture of the fibers directly in the composites. The FRP composites exhibit matrix-cracking accumulation under cyclic loading as shown in Fig. 1.53. The microcracks accumulate rapidly in the early stages of the fatigue tests. It can be seen in Fig. 1.53A and B that the number and width of the matrix cracks become larger with an increase in the number of cycles. Then the accumulation rate slows

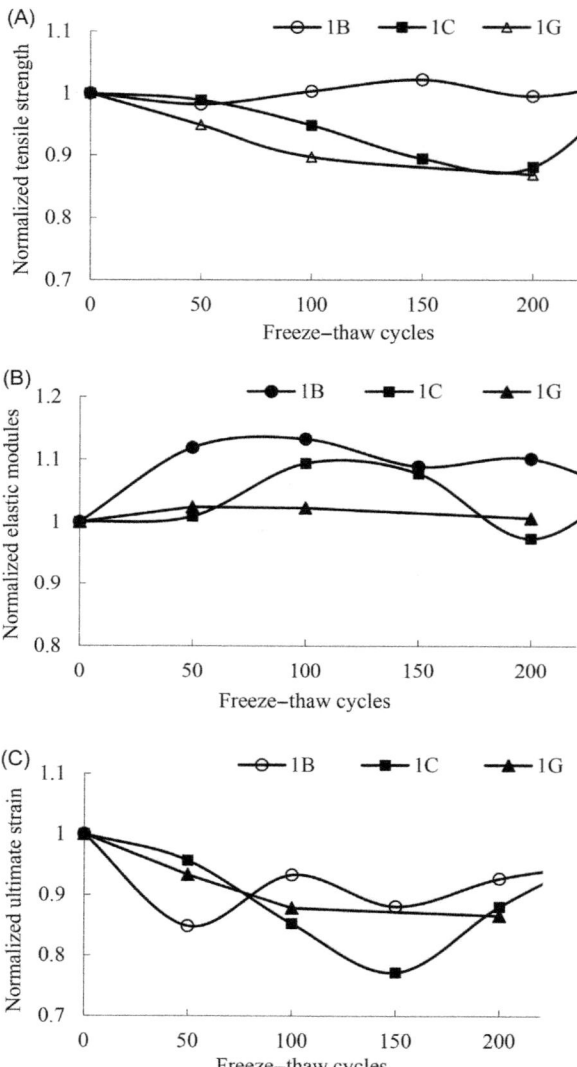

Figure 1.50 Comparison of mechanical properties of basalt fiber-reinforced polymer, carbon fiber-reinforced polymer, and glass fiber-reinforced polymer sheets after freezing—thawing cycles (Shi et al., 2014). (A) Normalized tensile strength. (B) Normalized elastic modules. (C) Normalized ultimate strain.

down with an increasing number of cycles. The specimen failed at 380,699 cycles and induced an interfacial debonding, as shown in Fig. 1.53D.

Fig. 1.54 presents a typical SEM image of damage observed for a different number of fatigue cycles at a low-stress level. No matrix cracks and interface debonding

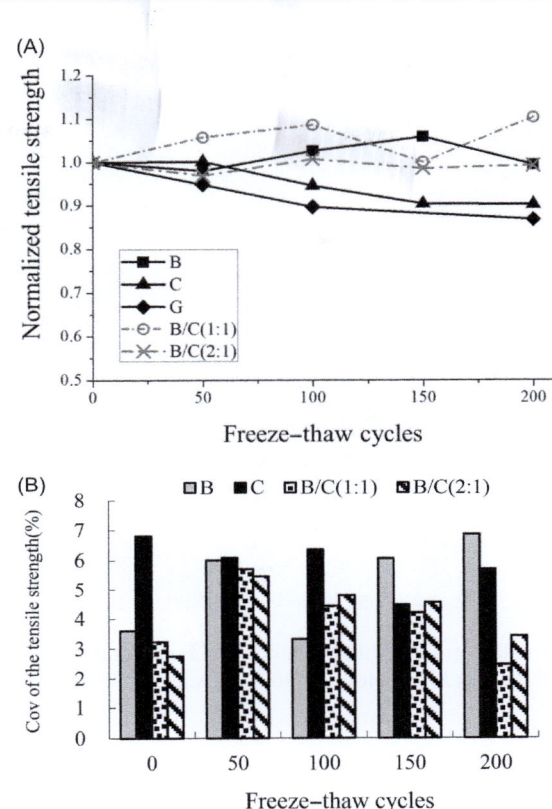

Figure 1.51 Tensile properties of fiber-reinforced polymer sheets under freeze—thaw cycles (Shi et al., 2014). (A) Tensile strength. (B) Cov of the tensile strength.

can be observed from the beginning of the fatigue loading shown in Fig. 1.54A. However, the accumulation of the microcracks developed slowly and finally formed discontinuous interfacial debonding when they reached the interface of the fibers and the matrix, as shown in Fig. 1.54B and C.

The previous description of the failure modes of the FRP under fatigue load is the general characteristic of FRP. For specific types of fibers and resin, the fatigue behavior and failure modes may change depending on the bonding performance, mechanical properties of the resin, and the modulus of the fibers. Detailed identification of the mechanism for each type of FRP composite under fatigue loading is required for specific applications (Wu et al., 2010). Fig. 1.55 shows that fatigue failure modes are different for different FRP specimens. The CFRP and PBO composites demonstrated several longitudinal cracks when the fatigue cycles were increased (Fig. 1.55A), whereas the GFRP and BFRP composites exhibited transverse cracks (Fig. 1.55B). This observation implies that the tensile moduli of the fibers may have influenced the fatigue failure mode. The high moduli of the CFRP

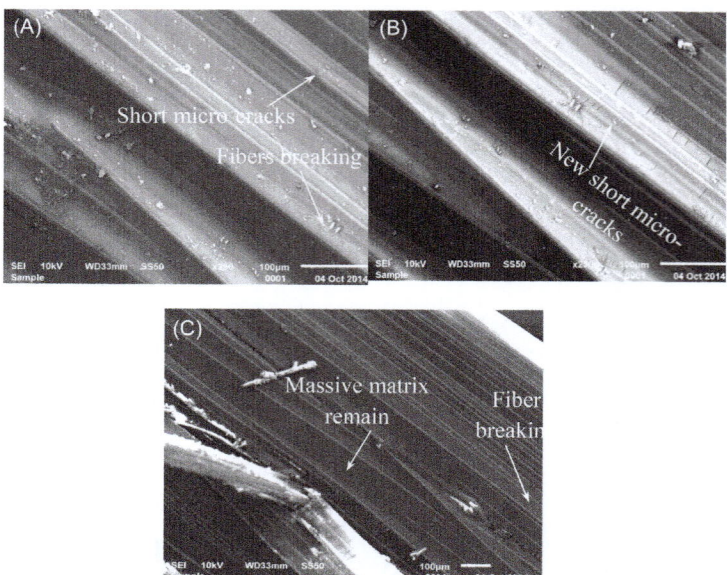

Figure 1.52 Typical failure mode of fiber-reinforced polymer under high level of cyclic stress (Zhao et al., 2016). (A) 519 cycles. (B) 50,007 cycles. (C) Failure at 102,642 cycles.

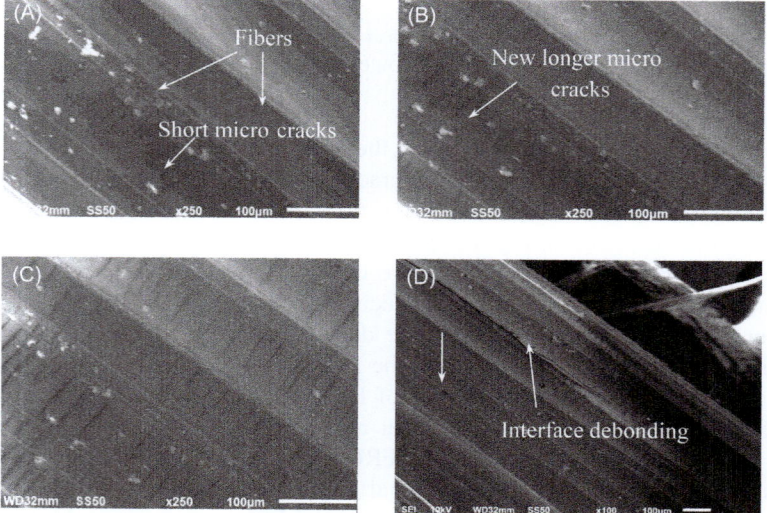

Figure 1.53 The matrix cracking in basalt fiber-reinforced polymer (Zhao et al., 2016). (A) 549 cycles. (B) 10,000 cycles. (C) 300,000 cycles. (D) Failure at 380,699 cycles.

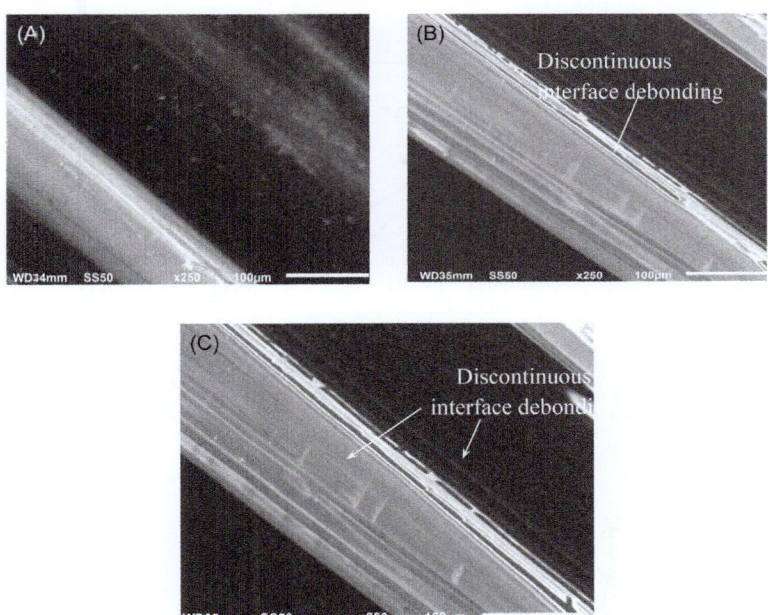

Figure 1.54 The interface debonding in basalt fiber-reinforced polymer (Zhao et al., 2016). (A) 508 cycles. (B) 5,000,000 cycles. (C) 10,000,000 cycles.

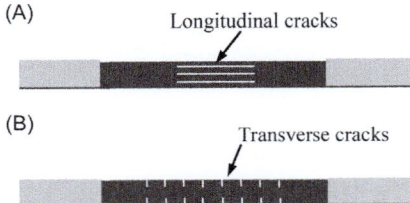

Figure 1.55 Failure modes of tested fiber-reinforced polymer (FRP) coupons (Wu et al., 2010). (A) carbon FRP and poly-*p*-phenylene-benzobisoxazole composites. (B) Glass FRP and basalt FRP composites.

and PBO composites (Table 1.5) effectively transferred the fatigue stresses along the fibers; however, the relatively low moduli of the GFRP and BFRP composites spread the stresses to the epoxy adhesive so that the transverse cracks propagated over the fibers embedded in the adhesive.

2. *Fatigue limit*

As mentioned earlier, the fatigue damage pattern would be different with an increase in the number of fatigue cycles. A bilinear phenomenological fatigue model had a better accuracy of the prediction than a linear S−N model (Zhao et al., 2016), as shown in Fig. 1.56A. This phenomenon is similar to the different degradation rates of low-cycle

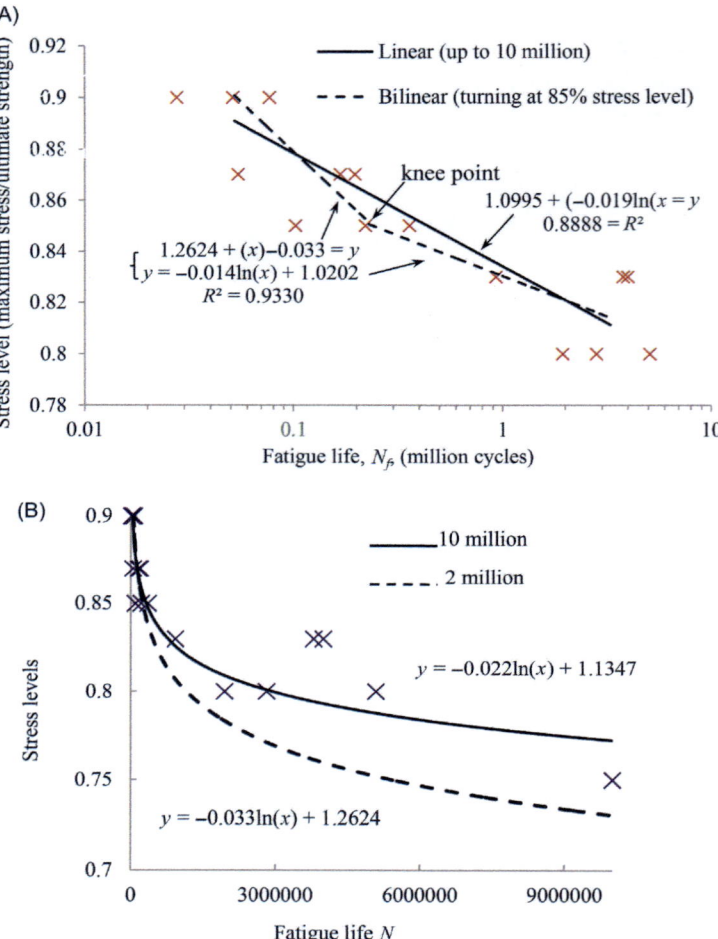

Figure 1.56 S−N curves and life prediction (Zhao et al., 2016). (A) linear and bilinear S−N curve. (B) Fatigue life prediction using data form different cycles.

fatigue loads and high-cycle fatigue loads observed in unidirectional GFRP composites. Moreover, the prediction with data up to 2×10^6 cycles is lower than that with all the data, as shown in Fig. 1.56B, which indicates that the degradation rate of the BFRP specimens under fatigue load slows down after 2×10^6 cycles. Using 10 million fatigue data evaluations, the fatigue strength of BFRP can be more accurate and can increase from the original 74% of f_u to 80% of f_u.

3. *Residual modulus*

The FRP fatigue residual modulus is the modulus of the FRP after a certain number of fatigue cycles. The fatigue failure of the tested specimens occurred when the total accumulated damage reached a critical limit. All the stiffness degradation curves could be characterized by a drop in the initial part of the life, followed by a plateau region, and

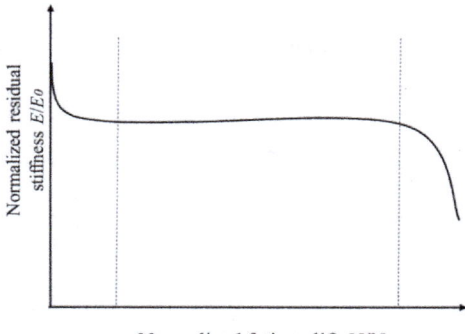

Figure 1.57 A typical damage-related degradation pattern.

then succeeded by a much faster drop during the last stage of the life, as shown in Fig. 1.57. This is a typical damage-related degradation pattern of the fiber-reinforced composites. However, the magnitude and rate of stiffness degradation are affected by different FRP materials, different fatigue stress amplitudes, etc.

4. *Fatigue damage mechanism*

The transverse matrix cracking occurs during fatigue loading. Wu et al. also reported that the BFRP sheet composites exhibited macrotransverse cracks under cyclical loading (Wu et al., 2010). This damage pattern differed from its counterpart of unidirectional CFRP composites. In unidirectional CFRP composite fatigue tests, the longitudinal matrix cracks are usually dominant under fatigue cycles (Wu et al., 2010). This can be the result of the different tensile moduli of the fibers. For BFRP composites, the relatively low modulus of the basalt fibers makes the fatigue stresses spread to the epoxy adhesive, whereas the high modulus of the CFRP composites (usually around 230 GPa) effectively transferred the fatigue stresses along the fibers.

The SEM images (Figs. 1.52−1.54) also indicate that three types of damages with different damage propagation patterns under different stress levels occurred during fatigue loading. This speculation allows the creation of a fracture diagram under fatigue loading as shown in Fig. 1.58. In region I of Fig. 1.58, that is, high strain region, such as the one above 85% in this test, the fracture of the samples took place in a manner of progressive fiber breaking similar to the static tensile fracture. Small matrix cracks occurred and the weakest fibers broke within the first few cycles. Because of the small crack stress concentration and fiber strength distribution, more fibers around the small cracks and the breaking fibers degraded and failed with increasing numbers of fatigue cycles, eventually causing massive and progressive composite failure. In region II, for example, 80%−85% in this test, the stress intensity was not big enough to cause progressive fiber breaking because the stress level was relatively low. The short matrix cracks propagated transversely over a few fibers or along the interface of the fiber and the matrix resulting in the initiation of progressive interface debonding. The fiber damage was induced and fiber strength degraded owing to the wear caused by the initiation of long matrix cracks and the fiber/matrix interfacial debonding. Eventually, the weakened fibers broke and the damage grew, leading to a total fracture with fatigue loading. On the other hand, the residual strength increased or recovered to its original value in region III, that is, below the fatigue limit. The discontinuous fiber/matrix interface debonding and the fiber

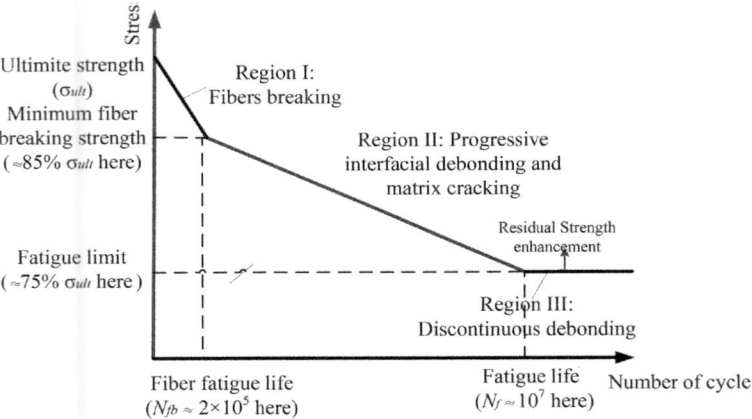

Figure 1.58 Schematic drawing of the S—N curve of the fiber-reinforced polymers with damage patterns (Zhao et al., 2016).

straightening during the fatigue took place in region III, which allowed fibers to exhibit their intrinsic strength.

5. *Fatigue behavior of FRP sheets*

   To evaluate the fatigue behavior of FRP sheets under cyclic loading, the specimens were prepared according to the tensile strength test. The fatigue load was applied usually in a sinusoidal waveform with a frequency of 5 Hz to avoid any temperature increase of the specimen induced by high frequencies. To verify the fatigue limit for FRP composites, usually, a stress ratio of 0.1 (S_{min}/S_{max}) is adopted to maximize the effect of cyclic loading instead of according to the actual stress ratio in the structures.

   Fig. 1.59 shows the fatigue response of the tested specimens up to 2×10^6 cycles (i.e., $\log(2 \times 10^6) = 6.3$). To better evaluate the fatigue characteristics of each FRP composite, a regression method (Eq. 1.1), proposed by Lorenzo and Hahn (1986) was adopted.

   The R-squared values of each regression line are listed in Table 1.8, which represents the precision of the regression prediction. The predicted values of the BFRP were the most reliable whereas the CFRP displayed relatively worse predictions. The regression lines of other FRP sheets maintained moderately high reliability.

   The CFRP and PBO composites exhibited the best performance under the cyclic loads with a loss rate of 2.6% and 3.6%, respectively, whereas the GFRP and BFRP composites showed relatively low fatigue resistances, represented by a loss rate of 6.2% and 7.1%, respectively. It should be noted that the coupons, which survived at 2×10^6 cycles, indicated by the horizontal arrows in Fig. 1.59, were not included in the curve fitting. The maximum applied load ratio (S at 2×10^6 in Table 1.9) indicates the fatigue limit of the FRP composites to achieve at least 2×10^6 cycles of the fatigue life. For the CFRP and PBO composites, the fatigue limits were 83.7% and 76.7% of the monotonic tensile capacity, respectively. The limits for the GFRP and BFRP composites were, however, 61.3% and 55.0%, respectively.

   Fig. 1.60 compares the load-carrying capacity of each FRP composite before and after the fatigue load. An average of the residual capacity of the coupons that survived at 2×10^6 cycles was given to represent the capacity of the FRP coupons that were

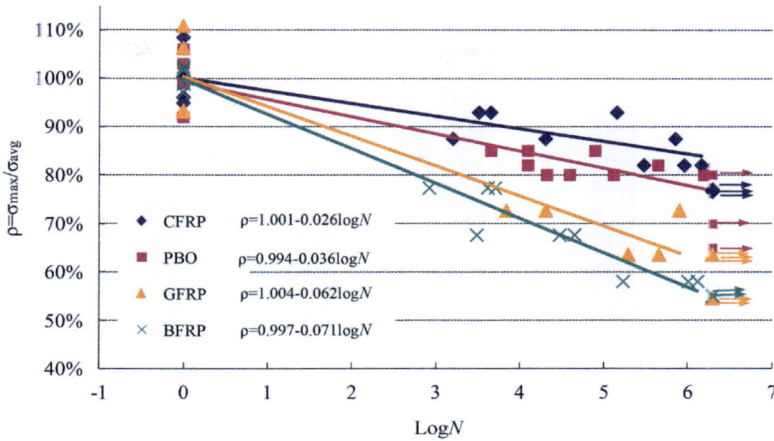

Figure 1.59 Fatigue behavior of fiber-reinforced polymer composites (Wu et al., 2010).

Table 1.8 Fatigue constants of regression lines (Wu et al., 2010).

FRP		CFRP	PBO	GFRP	BFRP
Constants	S'	1.001	0.994	1.004	0.997
	S''	2.6%	3.6%	6.2%	7.1%
R-squared value of regression		0.70	0.84	0.86	0.96
S at 2×10^6		83.7%	76.7%	61.3%	55.0%

Table 1.9 Reduction coefficient for fatigue of concrete members strengthened by fiber-reinforced polymer according to several standards.

Type of FRP	CFRP	AFRP	GFRP	BFRP
JSCE2001	$\mu = 0.7$	–	–	–
ACI440.2R-08	$0.55f_u$	$0.3f_u$	$0.2f_u$	–
ISIS2008	$0.65f_u/0.6f_u$	–	–	–
NRC-DT200	$\eta_1 = 0.5$	–	–	–
Experiment and reliability analysis	–	–	–	$0.53f_u$

submitted to the fatigue. The capacity drop due to the fatigue load was 16% and 39% for the CFRP and GFRP composites, respectively, whereas the capacity drops of the PBO and BFRP composites were 23% and 45%, respectively.

Fig. 1.61 shows a change of the tensile moduli of FRP composites, depending upon the normalized fatigue cycles (fatigue cycle/fatigue life). The damage represented by the reduced modulus was permanent. The fatigue failure of the tested coupons occurred when the total accumulated damage reached a critical limit. Although there were notable scatters in the

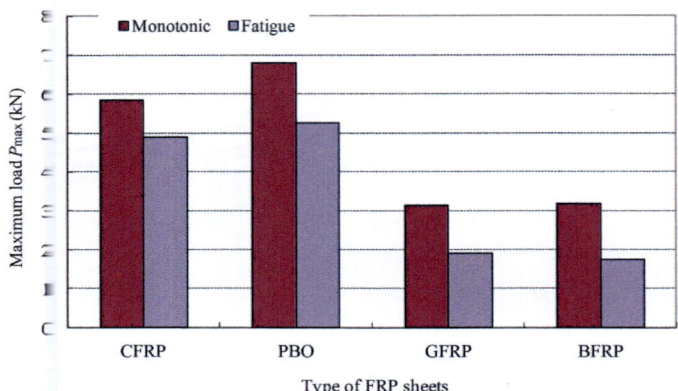

Figure 1.60 Comparison of monotonic and fatigue capacity of fiber-reinforced polymer composites (Wu et al., 2010).

modulus reduction, the critical limit was approximately 60%−80% of the initial modulus for all types of fibers. Contrary to other composites that showed a gradual decrease in the tensile modulus when the fatigue cycles increased (Fig. 1.61), the PBO composite showed an abrupt drop in the tensile modulus (i.e., about 30% decrease) within the first 1000 cycles, as shown in Fig. 1.61B. This phenomenon can be explained by the fact that the polymer structure of the PBO fibers was different from that of other fibers.

6. *Fatigue life prediction model of FRP*

The adoption of methods for the interpretation of static and fatigue data to interpolate between the experimental data (modeling) and extrapolate beyond that for the prediction of the expected material behavior is a demanding task that depends on the examined material and the thermos-mechanical loading conditions. Deterministic or stochastic theoretical models can be employed for this purpose. The use of the selected models (S−N curves, constant life diagrams, residual strength models, residual stiffness models, etc.) permits interpretation of the fatigue data and estimation of the fatigue life of the materials, theoretically under any applied loading pattern.

- *Prediction model based on the S−N curve*
   August Wöhler, as far back as the 1850s, conceived the idea of representing the cyclic stress against the number of cycles to failure to quantify the results of his experimental program. The only input required with regard to the experimental data consists of pairs of numbers of cycles up to failure and the corresponding alternating stress or strain parameter. The S−N or e−N curve of the material is then determined under the applied loading condition.

$$N\sigma^m = \text{Const}$$ (1.18)

or

$$\sigma = \sigma_0 N^{-1/r}$$ (1.19)

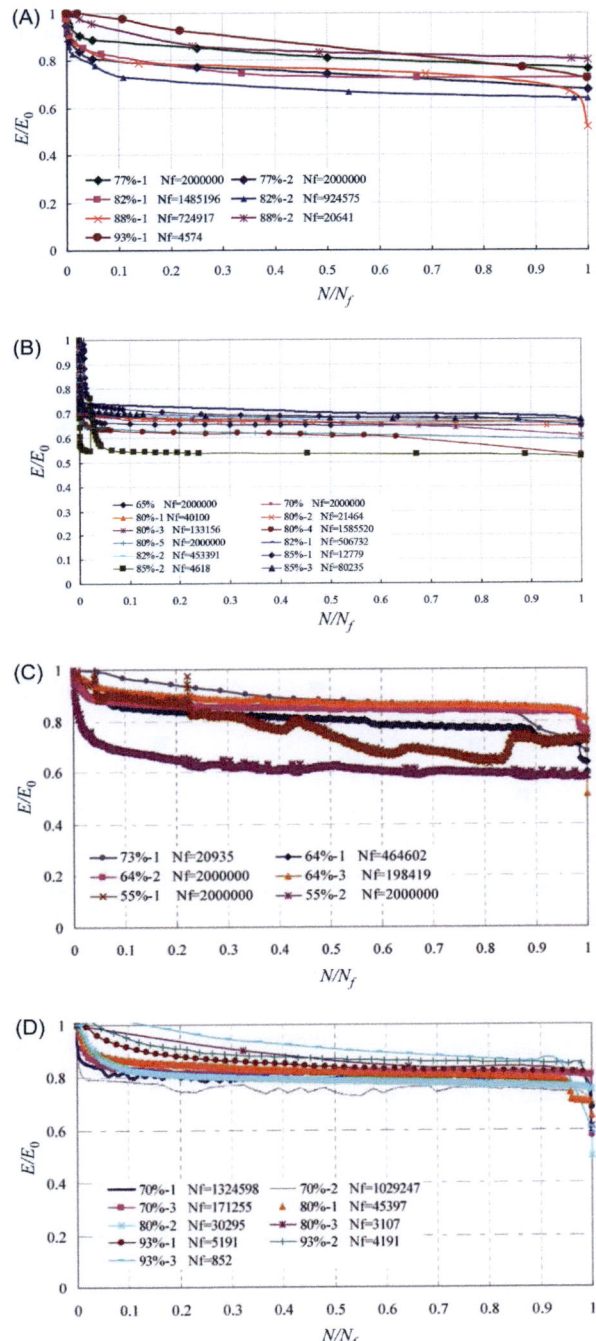

Figure 1.61 Modulus degradation of different fiber-reinforced polymer (FRP) sheets (Wu et al., 2010). (A) Carbon FRP. (B) PBO FRP. (C) Glass FRP. (D) Basalt FRP.

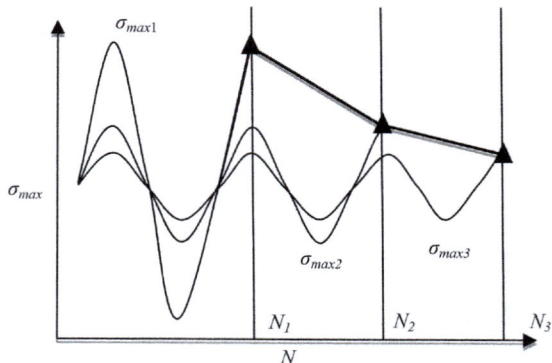

Figure 1.62 Schematic diagram of S−N curve.

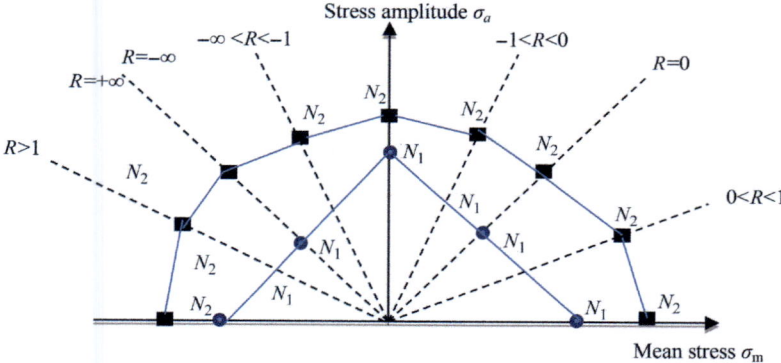

Figure 1.63 Constant life diagram curves.

The method of estimating the S−N curves based on constant amplitude fatigue data is schematically shown in Fig. 1.62. Three different tests are presented, each corresponding to one stress level and resulting in different numbers of cycles to failure. As expected, the lower the stress level, the longer the fatigue life of the examined material. The interpolation between the collected fatigue data results in the S−N curve of the material under the selected fatigue conditions namely R-ratio, frequency, environment, etc.

- *Constant life diagram*

The constant life diagrams (CLDs), as shown in Fig. 1.63, are linear or nonlinear interpolation schemes between experimentally obtained S−N curves in the mean stress (σ_m)− stress amplitude (σ_a) coordinate system. These diagrams are widely used for demonstrating the effect of the stress ratio on fatigue life and generalizing the predictions for an arbitrary stress ratio. Lines through the origin represent data for a specific stress ratio.

- *Residual strength model and residual stiffness model*

The degree of damage in a polymer matrix composite material can be followed by measuring the decrease in a relevant damage metric, usually the residual strength

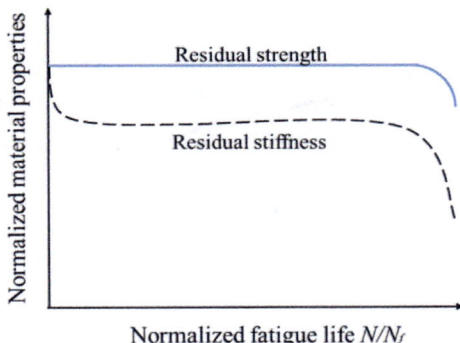

Figure 1.64 Fiber-reinforced polymer strength and stiffness degradation model under constant amplitude fatigue load.

or residual stiffness. A theory based on the residual strength degradation assumes that the damage is accumulated in the composite and failure occurs when the residual strength decreases to the maximum applied cyclic stress level, as shown in Fig. 1.64.

7. *FRP fatigue performance improvement method*
a. *Hybridization effect on fatigue behavior*

As discussed in Table 1.4, hybrid FRP sheets exhibited lower CVs of fatigue life under fixed applied load, which is mainly attributed to the low stress of the glass/basalt fibers in the hybrid fibers. Because of the difference in the modulus between glass/basalt and carbon fibers, glass/basalt fibers would maintain lower stress. During the fatigue loading, the initiated and progressive cracks in the resin would affect the simultaneous stressing of the carbon fibers. Once the fractures of some carbon fibers occurred, the surrounding fibers would fracture in succession, which could lead to the premature failure of the entire FRP sheets. In contrast to this type of failure, in hybrid FRP sheets, the low-stress fibers can prevent the continuous fracture of high-stress fibers after the partial damage of the resin, which results in a steadier fatigue life under the same cyclic load. A similar phenomenon of the hybridization effect on a variation was also observed in the tensile test on hybrid carbon and glass/basalt FRP sheets at high temperatures (Cao et al., 2009).

In addition to the variation of fatigue life, fatigue life itself is enhanced by the hybrid effect, as can be observed from Fig. 1.65 and Table 1.4, which compare hybrid C1G1 and C1B1 FRP sheets with CFRP, GFRP, and BFRP sheets. For hybrid C1B1 FRP sheets, a pronounced enhancement of fatigue resistance was realized through hybridizing carbon and basalt fibers; in contrast, a slightly lower fatigue resistance was seen for C1G1 FRP sheets. From both theoretical and experimental points of view, the low-modulus glass/basalt fibers have low stress during fatigue loading, which can contribute to the steady fatigue life. As a result, fatigue life should be enhanced. However, the experimental results reveal an inverse effect between these two hybrid FRP sheets. This difference between C/G and C/B FRPs may be explained by the experimental observation. The C/G and C/B FRPs exhibited different failure modes. For the C/G FRP, obvious delamination between carbon and the glass fiber sheets occurred during the experiment, which led to the final failure of the specimens. The initiation of this delamination between carbon and

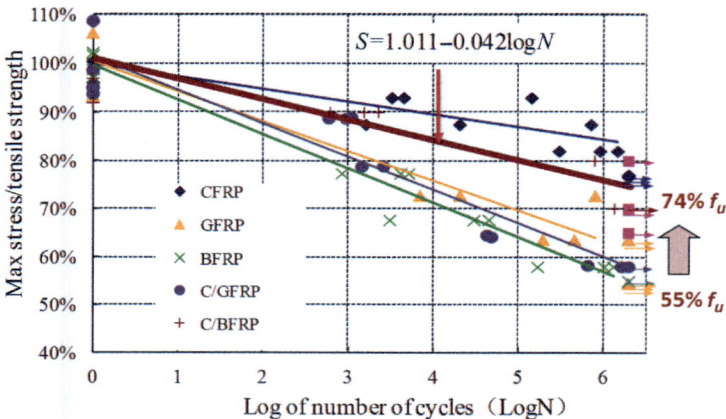

Figure 1.65 Fatigue behavior enhancement by hybridization (Wu et al., 2010).

the glass fibers could be explained by two aspects: one is the existing high difference in the stress owing to their different moduli, which may cause debonding between the fibers and the matrix at the interface of the two fiber sheets; the other is the roughness of fiber surface, which determines whether fibers could work simultaneously under high interfacial stress. The smooth surface of the E-glass fibers was reported by a former study (Militký et al., 2002), and it greatly affects the bonding between the two types of fiber sheets. As a result, the high shear stress induced by different moduli of the two fibers ultimately causes final delamination between the two fiber sheets. In contrast, no delamination occurred in the hybrid C/B FRP, which affirmed sufficient bonding between the basalt fiber surface and the matrix. This may be attributed to the rough surface of the basalt fibers (Militký et al., 2002), which allow them to bond more reliably. Therefore a pronounced enhancement of fatigue resistance was achieved through hybridizing carbon and basalt fiber sheets instead of hybridizing carbon and E-glass fibers.

b. *Matrix toughening/high ductility resin*

According to the FRP fatigue damage mechanism, resin cracking and interfacial debonding affect the fatigue properties of the FRP; therefore, it can improve the fatigue properties of the FRP by toughening or interfacial modification. The use of toughened resin can effectively increase the fatigue life of the FRP fatigue stress threshold; in other words, the resin crack may be less likely to develop under the same fatigue stress when toughened with the same damage size, or longer fatigue life. For example, when 10 wt.% of silica particles are toughened, the fatigue life of GFRP toughened at the same stress level increases by about 13%. This is owing to the dispersion of silicone elastomeric rubber particles limiting the development of resin damage (as shown in Fig. 1.66), thus increasing the fatigue properties of the resin, and thereby increasing the overall fatigue properties of the GFRP. In the same way, an interfacial modification can improve the properties of fiber and the interface, and make the damage mode change from a distributed and gradual interface debonding damage to a small-scale crack, control the damage size and expansion rate, and improve the fatigue life. For example, with the use of GFRP

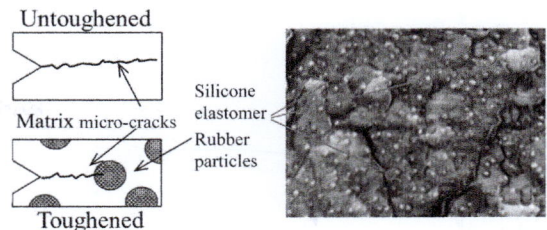

Figure 1.66 Matrix toughening (Shi et al., 2015a,b).

impregnated with maleic anhydride interface modified polypropylene resin, the fatigue life was 2−10 times longer than that of the control specimen.

8. *Requirements for the fatigue behavior of FRP in design standards*

In the structural design, for the concrete members strengthened by FRP, ACI440.2R-08 recommends allowable stress levels of $0.2 \times f_u, 0.3 \times f_u,$ and $0.55 \times f_u$ for glass, aramid, and carbon FRPs, respectively to eliminate the effects of fatigue. JSCE (JSCE2001) recommends a reduction coefficient of $\mu = 0.7$ to consider the effect of fatigue loading on the fracture energy of the interface between FRP and concrete. Canadian ISIS (ISIS 2008) restricts the maximum stress level under fatigue load to avoid the rupture of FRP composites. For example, the stress levels of carbon FRP composites are limited within $0.65 \times f_u$ and $0.6 \times f_u$, for bridges and buildings, respectively. However, there is no restriction for fatigue stress. *fib* group (*fib* 2001) made a design philosophy for nonquantitative fatigue and creep. The design guidance of FIB also proposed some key regulations for the long-term behavior of concrete members bonded with FRP. NRC-DT200 (NRC, 2004) in Italy proposed a reduction coefficient (η_1) considering fatigue and creep for design. For CFRP, η_1 is 0.8 for creep and 0.5 for fatigue.

For a BFRP bar, considering the common stress range in civil engineering, the maximum fatigue stress and stress range are predicted to be $0.53 \times f_u$, and $0.04 \times f_u$, respectively, based on reliability analysis. These stresses are close to those of CFRP and higher than those of AFRP and GFRP. Thus BFRP has good potential as a highly efficient prestressing component. The reduction coefficients for the fatigue of concrete members strengthened by FRP are shown in Table 1.9.

1.2.4.2   Creep behavior

The creep behavior of FRP introduced in this section is mainly based on the research studies on FRP bars, which have an identical mechanism as those of FRP sheets.

1. *Creep characteristics of FRP*
   • *Creep rupture stress*

   Creep rupture stress denotes the stress-causing failure after a specified period from the initiation of a sustained load. The creep rupture stress is a key index for FRP

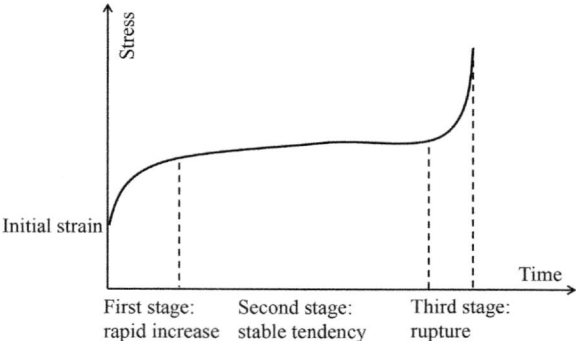

Figure 1.67 Creep development curve.

Table 1.10 Creep rupture stresses.

Types	CFRP	AFRP	GFRP	BFRP
Creep rupture stress	$0.70f_u$	$0.55f_u$	$0.29f_u$	$0.54f_u$

materials subjected to a sustained load. The ratio of sustained load to the tensile capacity has a semilogarithmic relationship with time, as shown in the following equation:

$$Y = a - b\log T \tag{1.20}$$

where Y is the ratio of sustained load to the tensile capacity; a and b are empirical constants; T is time, h. The previously mentioned formula can be used to predict the creep life or creep rupture stress of an FRP under a certain sustained load. The typical creep development curve is shown in Fig. 1.67.

The creep rupture stresses of several types of FRP are listed in Table 1.10, in which, the values of CFRP, AFRP, and GFRP refer to the clauses in ACI-440.4R and the research on GFRP by Yamaguchi; the value of BFRP is based on the study of the creep behavior of BFRP. In terms of creep rupture stress, GFRP has the lowest value, while CFRP possesses the highest value among several types of FRP. Since the creep rupture stress determines the upper limit of the design stress, a high creep rupture stress is necessary for the prestressing component. Thus GFRP is not suitable for prestressing applications because the creep rupture may take place at a relatively low sustained load.

° *Creep rate*

The creep rate denotes the percentage increment rate of the creep strain of an FRP compared with its initial strain. The creep strain directly determines the relaxation rate of the FRP because creep and relaxation have a similar mechanism that is, a viscoelastic deformation of the material. The relaxation rate is a key factor controlling the prestressing effect of the FRP. Thus the research results of creep rate are expected to reflect the relaxation rate.

As shown in Fig. 1.68, the creep rate of the FRP increased with an increase in the temperature because of the softening of the matrix resin under high temperatures, which

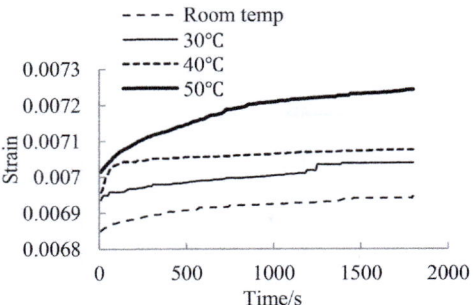

Figure 1.68 Creep curves at different temperatures.

lowers its capability to transfer stresses to internal fibers. Especially, when the temperature reached 50°C, the increase in the creep strain became significant.

In addition, the creep rate of the FRP is also affected by a corrosive medium because the ions in acid, alkali, or salt solution can infiltrate into the FRP material through microcracks and cause corrosion on the fiber—matrix interface. Furthermore, a sustained load can accelerate the infiltration of ions, making the corrosive effect more serious. The sensitivity of FRP to corrosion differs among different types of FRP. Generally, the corrosive medium has little effect on the creep behavior of CFRP but may cause some degradation in some other FRPs. For instance, an AFRP bar has a creep rate of 3% under a stress level of $0.4f_u$ for 3000 h in the ambient, while the corresponding values in alkali and acid solutions are 9% and 60% higher than that in the ambient, respectively. These results show that the acid solution causes significant damage to the bonding between the aramid fibers and the matrix resin. The results of creep tests on BFRP in salt solution indicate that its creep rate remains approximately 3%, regardless of the existence of the salt solution. The salt solution does not affect the creep behavior of the BFRP, except for some slight increases in the creep strain.

2. *Conventional FRP composites*

In prestressed structural members typically, with prestressing levels of around 30%—60% of their tensile strength, the creep behavior usually controls their applicability. High sustained loads may cause large relaxation of prestressing force or even creep rupture of the FRP materials. This phenomenon is different from the creep of steel reinforcement in reinforced concrete, which can only become dominant in the condition of extremely high temperatures. Thus the creep behavior of an FRP should be identified and creep rupture should be avoided when applying the FRP for structural strengthening. However, for typical types of FRP composites, namely glass, aramid, and carbon, the creep rupture stress is limited to 0.2, 0.3, and 0.55 times their tensile strength, respectively according to ACI-440.1R-06 (ACI Committee 440, 2007). This limitation leads to the fact that only carbon FRP can be used for prestressing, whereas glass and aramid FRP cannot be sufficiently used in prestressed applications.

The typical strain—time relationship of CFRP, GFRP, and Hybrid C/G FRP sheets under a sustained load is shown in Fig. 1.69 (Sasaki, 2008). It is shown that the first creep will initiate and develop rapidly after maintaining the load for about 40—50 h, and then a stable and slow increase of creep could be observed, namely, the secondary creep. If the tensile strains of the FRP sheet exceed these values (for CFRP, 14,000με; for C1G1, 15,000με; and for C1G2, 18,000με), the creep strains will develop sharply after a short

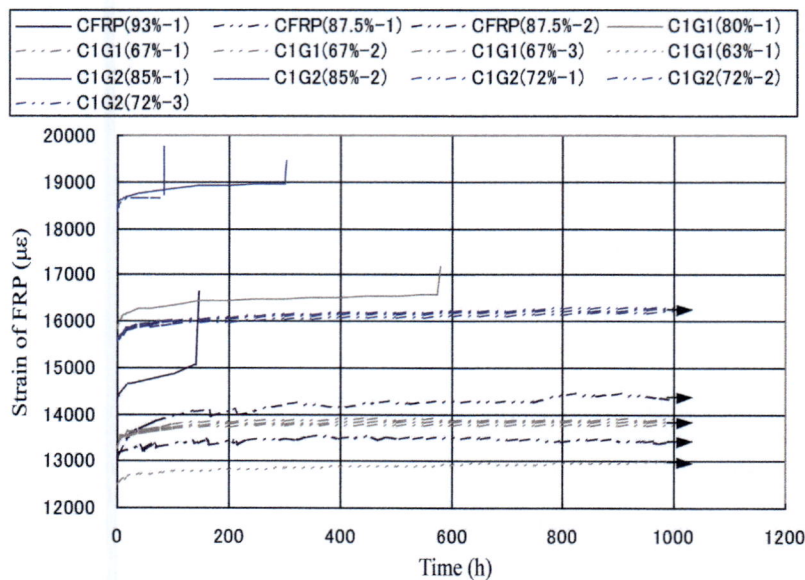

Figure 1.69 Creep behaviors of different fiber-reinforced polymer (Sasaki, 2008).

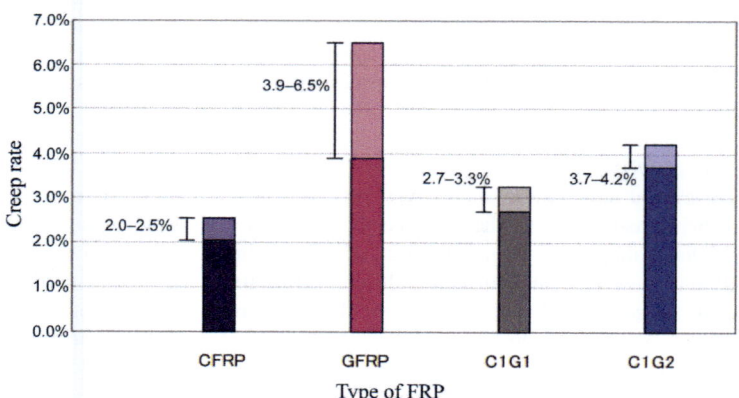

Figure 1.70 Creep rate of different fiber-reinforced polymer composites (Sasaki, 2008).

time (80−200 h) until the fracture of the FRP sheets. Considering these creep behaviors of the FRP sheets, the pretension of the FRP sheet was adopted to release the first creep, through which the stable secondary creep could be observed and evaluated, while the tertiary creep is controlled by limiting the tensile strain below the above values. In this way, the maximum and minimum values of the secondary creep of different FRP sheets are compared in Fig. 1.70, where the first creep of the GFRP sheets is used as a reference. From this figure, the development of the secondary creep became slow for C1G1 and

C1G2 compared with GFRP, which indicates that the effect of hybridization can efficiently control the creep development.

3. *Basalt FRP composites*

Owing to their potential advantages in prestressing applications, the creep behavior of BFRP composites is specially investigated.

a. *Strain development*

The typical strain curves with respect to time ($\varepsilon-T$) of the BFRP composite are plotted in Fig. 1.71. Under stress levels of 70%, 68%, and 65%, all of the specimens fractured with an increase in the strain. An initial rapid increase of the creep strain could be observed during the first stage. Then the rate of strain increase gradually decreased with respect to time, and finally transitioned into the second stage, characterized by a stable and slow rate of creep increment. The second stage tended to be a long-term stage and specimens at some stress levels (e.g., 50% of f_u) were capable of maintaining the second stage without a further rapid strain increase stage, that is, the so-called creep rupture stage. For the specimens under higher stress levels (e.g., 68% of f_u), the third stage took place with a sudden increase in the strain up to failure. It should be mentioned that the

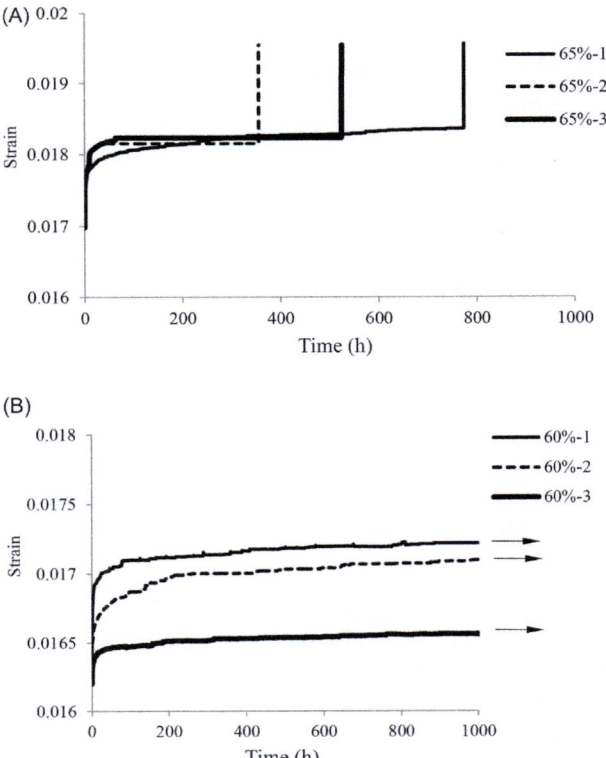

Figure 1.71 Typical $\varepsilon-T$ curves of basalt fiber-reinforced polymer (Wang et al., 2014). (c) $\varepsilon-T$ curves at stress level of 65%f_u. (d) $\varepsilon-T$ curves at stress level of 60%f_u.

Table 1.11 Creep rate of BFRP specimens (Wang et al., 2014).

Stress level	50%	60%	65%	68%	70%
Mean value	3.58%	3.70%	6.51%	14.19%	15.17%
SD	1.33%	1.45%	0.96%	0.90%	0.54%
CV	37%	39%	15%	6%	4%

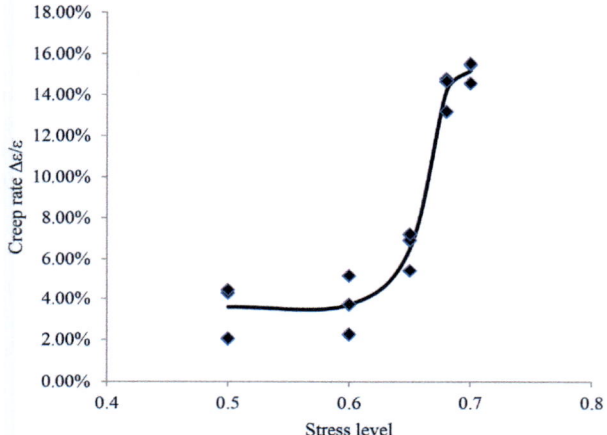

Figure 1.72 Relationship of creep rate and stress level (Wang et al., 2014).

$\varepsilon-T$ curves of the specimens at a stress level of 70% of f_u shows three different stages featured by a shorter second stage, where the creep rupture occurred after a short time of sustained constant load. The typical three stages of creep development are evident for the specimens at the stress levels of 68%f_u and 65%f_u. For the specimens under even lower stresses such as 60% of f_u and 50% of f_u, their $\varepsilon-T$ curves show only the first and second stages of strain increase, in which the symbol "→" denotes the specimens without creep rupture after 1000 h. Overall, it can be seen that with an increase in the stress level, the scatter of the $\varepsilon-T$ relationship became more significant.

b. *Creep rate*

The creep rate, which is defined by the ratio of the increased strain up to the end of the second stage to the initial strain of specimens at different stress levels, is shown in Table 1.11 and Fig. 1.72. It is shown that under low-stress levels, the creep rates maintain a relatively low level; however, the variation is large. While the creep rates increase greatly, the relevant variation decreases with respect to the increasing stress level. This observation demonstrates the mechanism of defect expansion inside the FRP as described earlier. Under a low-stress level, the expansion of defects in the FRP is still in progress, and therefore, large differences are observed. However, under high levels of stress, all of the specimens ruptured within 1000 h, indicating that the defect expansion is sufficient for each. Thus relatively low variations are found.

Table 1.12 Residual tensile strength of specimens that did not creep rupture (Wang et al., 2014).

Number	60% f_u	50% f_u
Mean value of strength retention	93.8%	95.0%
CV	0.61%	0.32%

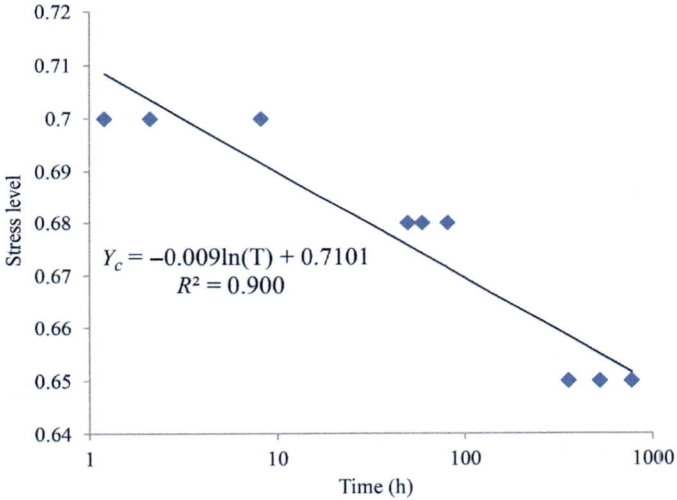

Figure 1.73 The approximation line for $Y_c - \log T$ (Wang et al., 2014).

c. *Residual tensile strength*

For specimens that did not fail after 1000 h of sustained loading, further loading up to failure was performed. The results of all the six specimens under stress levels of 50% and 60% of f_u are shown in Table 1.12, which shows that the residual strength of these specimens remained at a high level close to the static tensile strength. Furthermore, the variation of the residual strength was limited to a much lower level compared to the static tensile strength results. This further proves the creep deterioration mechanism described previously. It shows that insufficient expansion of defects in the FRP will not substantially affect the tensile behavior of the FRP. On the other hand, the stress redistribution of fibers in the FRP in the first stage of creep development benefitted the simultaneous load-carrying capacity of the fibers in the FRP, resulting in a lower variation induced by the nonuniformity of the fibers.

d. *Prediction of creep rupture stress*

To determine the creep rupture stress, the experimental data of sustained stress and rupture times were analyzed using a semilogarithmic graph. A linear fitting was calculated from the data using the least-squares method. The results are shown in Fig. 1.73.

By substituting $T = 10^6$ h into the equation in Fig. 1.73, $Y_c = 0.59$ was obtained, indicating that the BFRP tendons could sustain a stress of 59% of their tensile strength without creep rupture during the service lifetime of approximately 114 years. The accuracy of the prediction represented by $R^2 = 0.9$ is also acceptable.

Considering the reliability analysis, the lower limit value for the confidence band according to the prescribed lifetime (10^6 h) was calculated as 0.52 (the value could be enhanced to 0.54 through pretension treatment as mentioned later), which means that with a reliability of 95%, the BFRP tendons could carry a sustained stress of 52% of their tensile strength without creep failure in 114 years. This result was only 12% less compared to the prediction by the mean values indicating a relatively small scatter of the rupture time of the BFRP tendons under different levels of stresses. It should be mentioned that in ACI-440.1R the creep rupture stresses of GFRP, AFRP, and CFRP were 0.2, 0.3, and 0.5, where a safety factor of 1/0.6 was considered. With the same safety factor, the creep rupture stress of BFRP tendons could be only 0.35, based on a mean value prediction of 0.59, which was much less than the predicted value of 0.52, based on a statistical analysis. Thus it indicates that the limitation of creep rupture stress was relatively conservative in ACI-440.1R. In the future, with the accumulation of creep data of the BFRP tendons, the creep rupture stress of BFRP tendons will become more accurate and evident. Based on the analysis in this study, the creep rupture stress of BFRP tendons is recommended to be 0.52 for practical applications. It should be noted that a large amount of data is still necessary for a more reliable and representative estimation of the creep rupture behavior of BFRP because of the limited existing data.

e. *Creep strain rate control*

Several methods were proposed to control the creep behavior of FRP composites, which mainly focused on the hybridization of different fibers. For instance, Maksimov and Plume (2001) explored the creep behavior enhancement by FRP hybridization and showed that the creep strain of AFRP could be well restricted by hybridization with GFRP. Sasaki (2008) investigated the hybridization of GFRP and CFRP and revealed that the creep strain of GFRP can be lowered from the original 5% to 3.5%. A similar mechanism of hybridization with higher modulus fibers played an important role. Moreover, Acha et al. (2007) considered that the creep deformation was directly affected by the interfacial properties, which could be improved through the addition of coupling agents or chemical modification of the fibers. In addition, the modification of the matrix in FRP was also investigated to control creep strain. Li et al. (2012) revealed that adding a carbon nanotube in the resin could reduce the creep strain rate of CFRP by an appropriate amount. Although the earlier studies could control the creep behavior of FRP composites, they were still restricted by some practical reasons, such as low performance to cost ratio, stability, and quality control of overall mechanical properties.

In addition to the control methods from the materials themselves, it may be a practical and effective method through external pretension referred to the treatment of low relaxation of high-strength steel wires. The pretension process with tempering was widely adopted to lower the relaxation of high-strength steel wires, in which the mechanism optimizes the microstructure of steel (Libby, 1990). For FRP composites, as no creep is observed for the tensioned fibers themselves, while significant creep occurred in the resin (Ascione et al., 2012), it indicates if the fibers in FRP are sufficiently straight, the creep strain of the entire FRP material will be very limited (Zhou and Yang, 1985). However, owing to the limitations of production technology, the unevenness of the fiber rovings, such as local bending and skewness, cannot be avoided during the process of pultrusion. This is the main

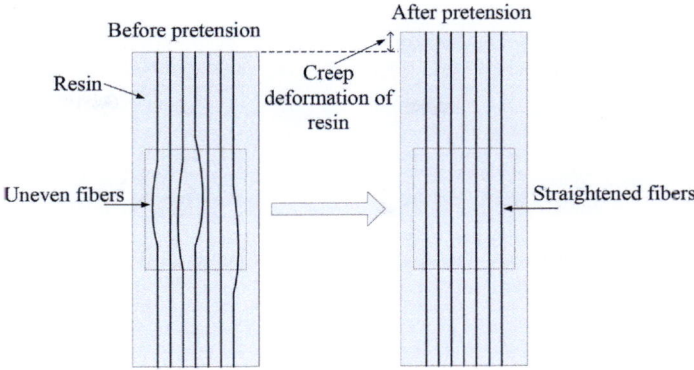

Figure 1.74 Mechanism of creep strain control by pretension (Wang et al., 2016).

cause of the rapid stress increment in the first stage of the FRP tendon creep (Everett, 1996; Paluch, 1996). As revealed in Table 1.12, the residual strength of BFRP specimens experienced a sustained load for 1000 h can achieve more stable results compared to the original strength before the creep test. This phenomenon indicates that through sustained loading the fibers in the FRP can be straightened which results in a more uniform load carrying of each fiber and enhancement of overall stability of strength. Furthermore, the research by Spathis and Kontou (2012) also showed that the rate of creep strain is related to the internal stress in a polymer material. With creep deformation of resin, the internal stress can be relieved by resin relaxation and transferred to fibers (Soudki, 1998). Thus it can be estimated that the unevenness of fibers can be adjusted by means of applying pretension along the direction of the longitudinal fibers. The resin in the FRP continues to deform by creep when the FRP is subjected to a sustained load, which provides a possibility for the fibers to interact in the resin. During this process, the fibers tend to be straightened owing to the creep of the resin, and the previous unevenness of the fibers can be adjusted. The mechanism on the improvement of internal fiber distribution in the FRP through the pretension process is shown in Figs. 1.74 and 1.75. Owing to the improved collaboration of the fibers after pretension, the creep rate of BFRP can be significantly controlled (Fig. 1.76A), and the creep rupture stress of BFRP increases from $0.52 \times f_u$ to $0.54 \times f_u$ (Fig. 1.76B).

1.2.5 Durability behavior

Numerous studies on the durability of carbon and glass fiber composites, mainly emphasizing the behavior under alkali, acid, and salt environments, have been reported in the literature. Because the initial application of FRPs in civil infrastructure was to replace the steel reinforcement in concrete structures, the durability of FRPs under an alkali environment is of significant interest. In general, CFRP exhibits superior performance in resisting different corrosive effects, whereas GFRP exhibits relatively weak resistance to alkaline and acidic environments. For the

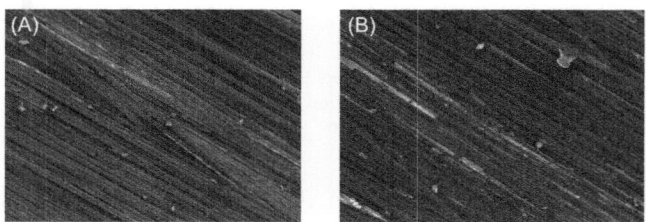

Figure 1.75 SEM images before and after pretension (Shi et al., 2017). (A) Before pretension. (B) After pretension.

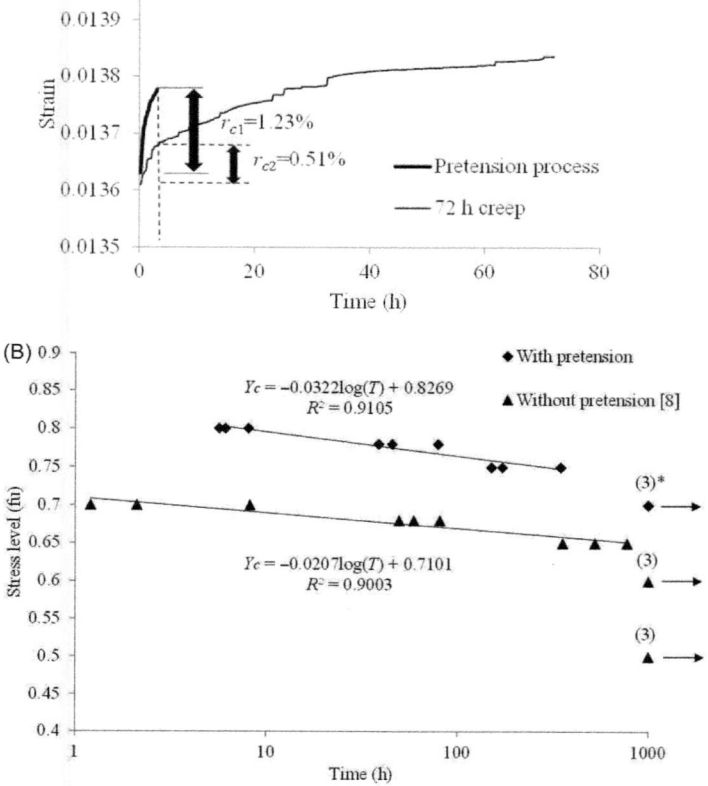

Figure 1.76 72 hCreep strain rates and creep rupture stress before and after pretension. (A) Creep strain−time curves (Wang et al., 2016). (B) Creep rupture stress (Shi et al., 2015a,b).

newly developed BFRP, fewer durability studies have been performed compared to those on other FRP composites. Within the limited literature on the durability of BFRP, most of them emphasized on the alkali resistance and moisture absorption. Few studies investigated the degradation of BFRP sheets under seawater and analyzed the corrosion mechanism. No particular study on salt resistance behavior of BFRP tendons/bars can be found in the existing literature. Although BFRP sheets were reported to exhibit high resistance to salt corrosion among various corrosive environments (Wu et al., 2014), the degradation of BFRP tendons should still be further investigated. Because the differences between the FRP sheets and the tendons not only lie in the production technology (hand layer-up to pultrusion), but also in the matrix types (epoxy to vinyl ester), relevant forming temperatures (ambient temperatures to 150°C), and fiber volume fraction, those differences can affect the degradation behavior of BFRP under corrosive environments. Thus in this section, the degradation of fibers, resins, and composites will be discussed separately.

1.2.5.1 Fiber degradation

Under different solutions, the basalt, carbon, and glass fibers exhibited different types of changes in their surface conditions. The basalt fibers showed no obvious changes in their surface, except for a lower level of glassiness that developed in the salt and water solutions. No obvious changes were observed in the surfaces of the carbon and glass fibers exposed to the same solutions. However, in samples stored in the alkaline solution, the fracture of a few basalt fibers was observed after 1 month of immersion, and the fracture of more fibers was observed after 2 months of immersion. The surfaces of the basalt fibers also changed from their original gold to a white color, and the fibers became hardened. The surfaces of the glass fibers became coarse, and partial fracture of the fibers occurred. Unlike the basalt and glass fibers, the carbon fibers exhibited no fracture after 2 months of immersion. Partial fracture of the basalt fibers exposed to the acid solution was also observed, whereas no fracture was observed in the carbon and glass fibers exposed to the acid solution.

Fig. 1.77 shows that the surface condition of the carbon fibers after 66 days of exposure to the four types of solutions at 55°C. In addition to the precipitates attached to the surface of the fibers, no etching or pitting corrosion was observed, indicating a strong resistance of the carbon fibers to different corrosive elements to which they were exposed in this study. This phenomenon was also consistent with the tensile strength retention of the carbon fibers.

Fig. 1.78 shows the corrosion of the glass fibers after 66 days of exposure to four types of solutions at 55°C. Compared with the surface condition of the basalt fibers, the etching corrosion of the glass fibers by water was more severe, but the corrosion resulting from salt and acid exposure was relatively greater. Severe pitting corrosion of the fiber itself, as a result of alkaline exposure, very similar to that of the basalt fibers, was observed. This corrosion of the fiber itself caused the loss of strength and modulus demonstrated in the tensile property degradation tests. The

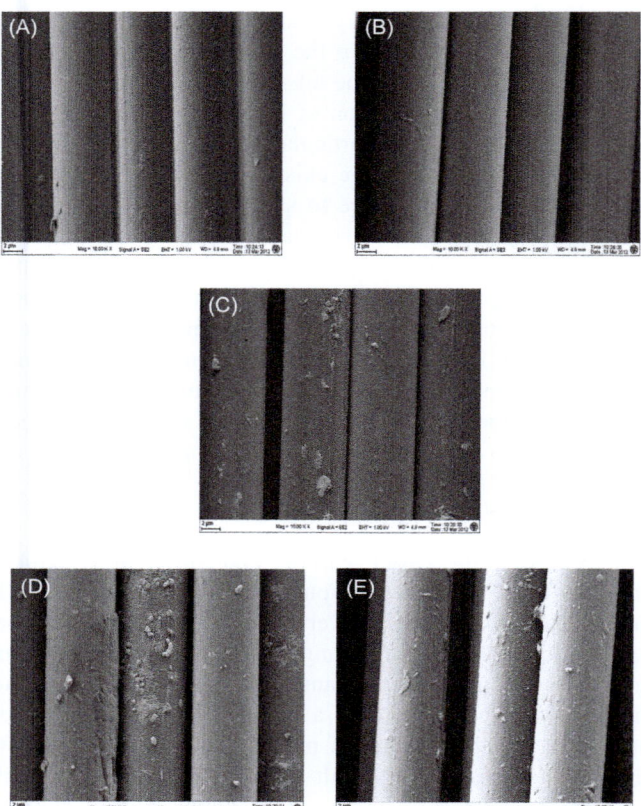

Figure 1.77 SEM images of carbon fibers after different corrosion (Wu et al., 2014). (A) Original fibers. (B) Water corrosion. (C) Salt corrosion. (D) Alkaline corrosion. (E) Acid corrosion.

basalt fibers showed similar microphenomenon as E-glass fibers under various corrosive solutions.

From the previously mentioned SEM images of different fibers under corrosive solutions, it can be seen that the carbon fibers could resist all kinds of corrosions without obvious changes in the fibers. In contrast, the glass and basalt fibers showed relatively good resistance to salt, acid, and water corrosions, but weaker resistance to alkaline corrosion.

1.2.5.2   Matrix degradation

An observation of the surface showed that the color of all of the resin castings turned from their original dark green color to a light green color after immersion in the alkaline, acid, salt, and water solutions for 66 days. These observations indicate that the corrosive effects could penetrate the internal resin casting. The degradation

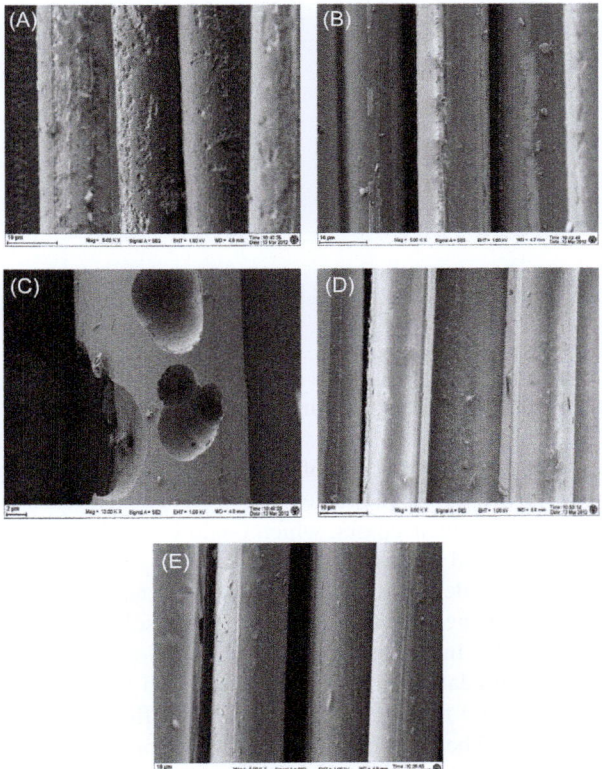

Figure 1.78 SEM images of corrosion of glass fibers after different types of exposure (Wu et al., 2014). (A) Original fibers. (B) Water corrosion. (C) Salt corrosion. (D) Alkaline corrosion. (E) Acid corrosion.

of the strength and modulus of the resin casting is illustrated in Fig. 1.79. The tensile strength decreased as a result of the exposure to water and acid solutions, whereas exposure to the alkaline and salt solutions resulted in little or no degradation of the tensile strength. This result indicates that the water and acid solutions penetrated more into the resin, which resulted in more internal effects in the resin after the drying of the penetrated solution. Those effects accelerated the development of the damage and the strength reduction in the resin. However, those effects did not influence the modulus of the resin casting because the damage was localized.

1.2.5.3 Fiber-reinforced polymer Composites degradation

Surface condition. The original BFRP laminate impregnated with epoxy resin was black in color. After corrosion, the BFRP laminate changed its color from the original black to gold, which was similar to the color of the basalt fibers themselves.

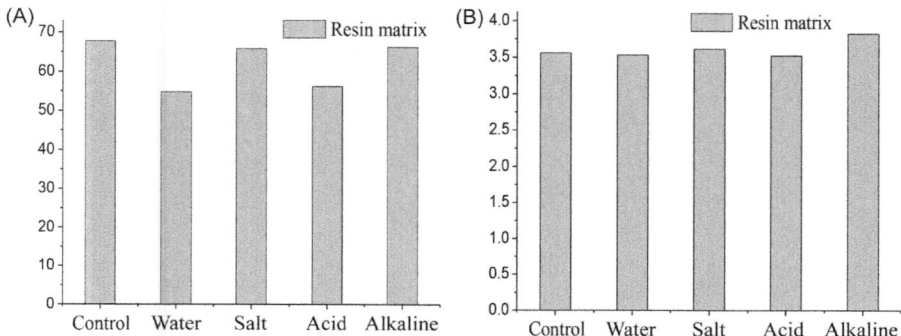

Figure 1.79 Degradation of resin casting as a result of exposure to different corrosive solutions (Wu et al., 2014). (A) Strength retention. (B) Modulus retention.

This change indicated that the bonding of the resin and fibers was damaged during the corrosion. Unlike the BFRP laminate, the CFRP and GFRP laminates did not exhibit obvious changes in their colors, which could be explained by relatively better bonding of the fibers and the resin.

Tensile properties. The degradation of the tensile strength and elastic modulus of the different FRP laminates in the four types of solutions are illustrated in Fig. 1.80. On the whole, in all of the types of solutions, the BFRP composites exhibited much less degradation in their tensile strength than the basalt fibers, and the variation in their elastic modulus also exhibited a different trend compared with that of the basalt fibers. The tensile strength retention of the BFRP exposed to the alkaline solution was much higher than that of the basalt fibers, and the higher temperature accelerated the degradation rate. Exposure to the acid solution was observed to result in significant degradation of strength at the higher temperature, which was similar to the trend observed for the fibers. However, at a lower temperature, the strength of the BFRP laminate increased as the length of exposure increased. A similar phenomenon was observed for the BFRP laminates exposed to water and salt solutions at a high temperature. The elastic modulus of BFRP exposed to the different solutions varied between 97% and 105% compared with its original modulus, indicating that the degradation of the composites was localized and did not induce a change in the stress—strain relationship before the final failure.

The differences in the degradation among the BFRP, CFRP, and GFRP composites after 66 days of exposure are illustrated in Fig. 1.81. The BFRP and GFRP laminates exhibited greater improvements in their tensile strength retention after exposure to acid and alkaline solutions than the respective fibers, primarily as a result of the delayed penetration of the corrosive solutions into the resin. The enhancement of the tensile strength of the BFRP and GFRP laminates exposed to salt and water solutions was similar to that observed for their fibers; however, the amplitude of the enhancement was relatively small. The fibers and composites of the CFRP exposed to different types of solutions retained their tensile strength.

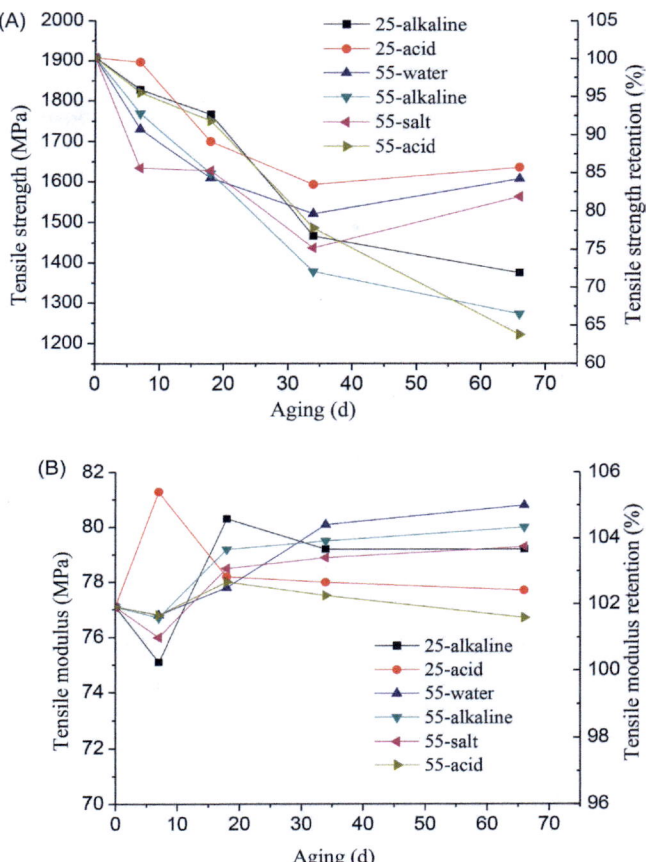

Figure 1.80 Degradation of basalt fiber-reinforced polymer composites (Wu et al., 2014). (A) Tensile strength. (B) Elastic modulus.

Unlike the reductions observed in the tensile strengths, the elastic moduli of all of the FRP composites remained constant, regardless of the influence of different corrosive solutions. These results also indicate that the corrosion of the FRP composites was localized and could only affect their ultimate strength rather than their modulus.

SEM image analysis. Because the surfaces of different FRP composites after different types of corrosive exposure exhibited similar conditions, owing to their impregnation with the same type of epoxy resin, the BFRP composites were selected as representatives of the surface change, as shown in Fig. 1.82. Although different types of corrosive elements resulted in different types of corrosion in the epoxy resin, they did not substantially affect the mechanical behavior of the resin, as indicated in the last section, and as shown in Fig. 1.83. The SEM images of the fractured portions of the specimens shown in Fig. 1.83B−D indirectly reveal that

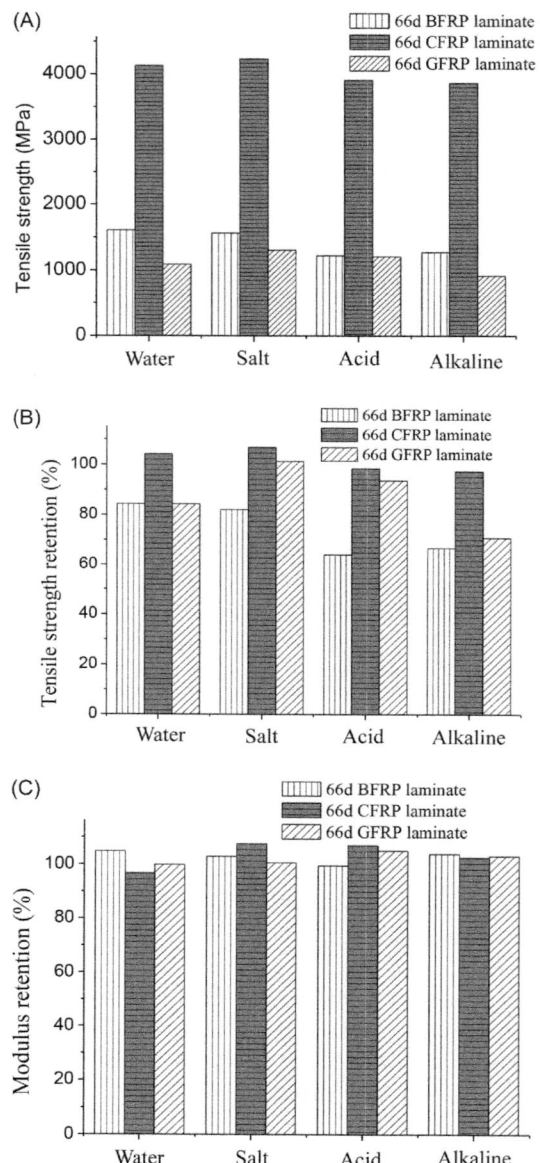

Figure 1.81 Comparison of mechanical degradation among different fiber-reinforced polymer composites (Wu et al., 2014). (A) Tensile strength. (B) Tensile strength retention. (C) Modulus retention.

the solutions damaged the interface between the resin and the fibers, and even damaged the fibers themselves. These figures show that the pulled-out fibers had smooth surfaces, with little attached resin, and that some fibers were partially

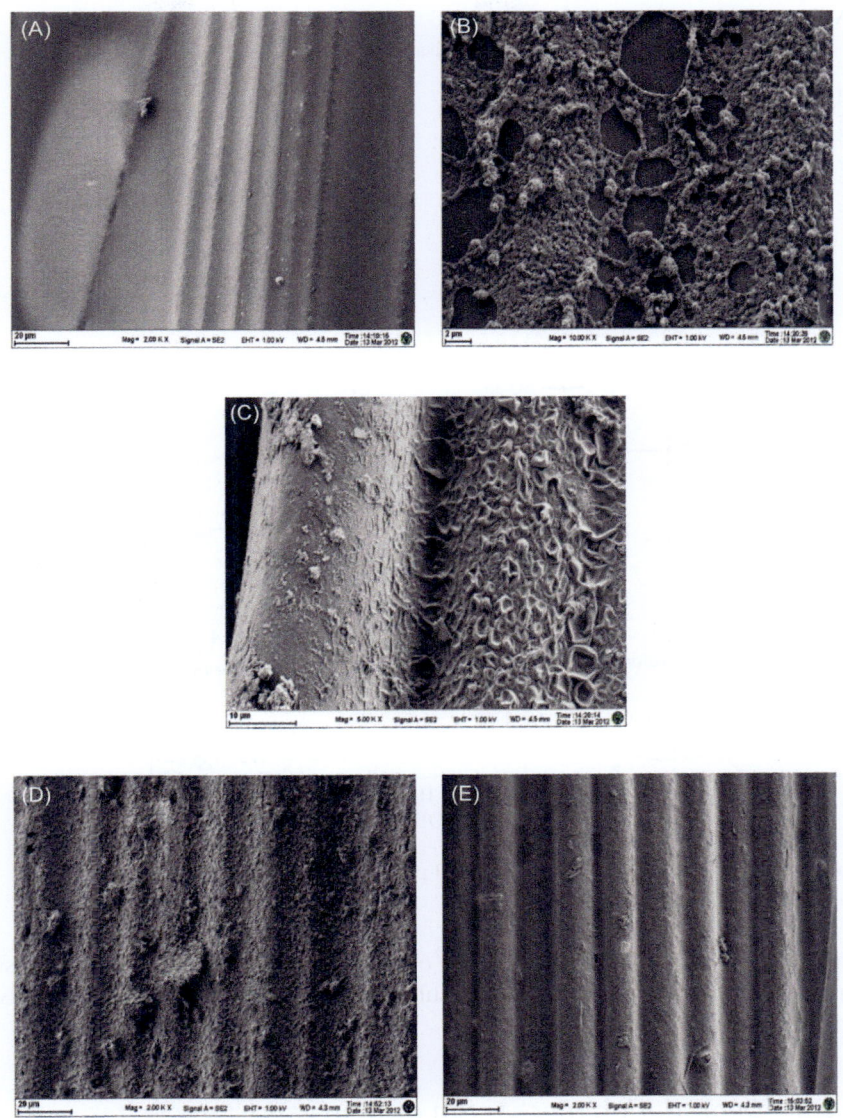

Figure 1.82 SEM images of fiber-reinforced polymer composites after corrosion (Wu et al., 2014). (A) The original. (B) Water corrosion. (C) Salt corrosion. (D) Alkaline corrosion. (E) Acid corrosion.

corroded by the alkaline solution. This phenomenon indicates the difference in the surface condition of the original fibers (bonded with resin) compared to those pulled out of the composites. Although different degrees of corrosion of the FRP could be observed, similar severe damage to the fibers was not observed, which

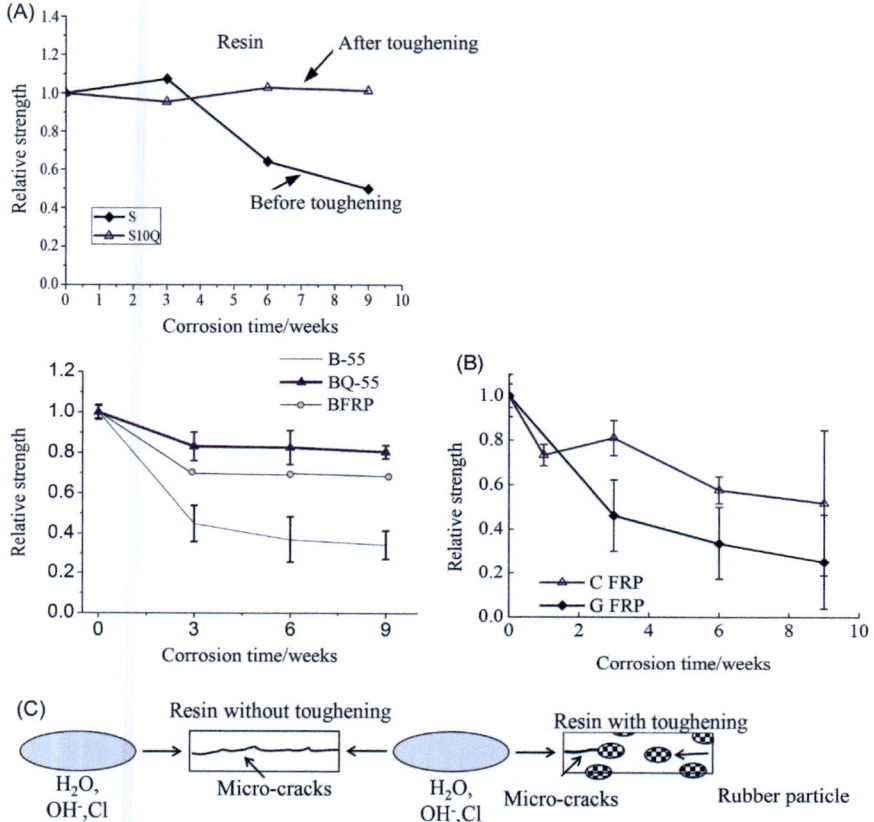

Figure 1.83 Modification for fiber-reinforced polymer (FRP) sheets. (A) Basalt FRP. (B) Glass FRP. (C) Modification mechanism.

indirectly indicates that the resin resisted or delayed the penetration of the corrosive solutions into the composites, thus resulting in the enhancement of the tensile properties.

1.2.5.4   Degradation mechanism of fibers and composites

Based on the results of the experiments on the tensile properties of the fibers, matrix, and composites described earlier, as also from the results of the mass loss weighing, SEM imaging, and energy spectrum analysis, the degradation trend and mechanism of basalt fibers and composites can be understood.

The degradation of basalt fibers by water and salt is essentially caused by the etching of the solution. This type of corrosion results in a certain loss of fiber mass. Localized etching damage accelerates the propagation of the damage in the fibers and consequently, reduces the tensile strength. The elastic modulus of the corroded fibers is not changed because the etching damage is localized, and the fibers can still perform as a whole to support an external load before the fracture of the fibers. According to the degradation mechanism described earlier, the degrees of degradation in the basalt fibers exposed to salt and water solutions are quite similar.

Unlike the etching corrosion described earlier, the basalt fibers were greatly damaged by the OH$^-$ ion in the alkaline solution, which significantly weakened the cross-sectional area of the fibers, and resulted in the complete loss of strength. For the same reason, the elastic modulus of the fibers was diminished by the weakening of the cross-section.

The basalt fibers exposed to the acid solution displayed a moderate degree of corrosion in comparison to those of the fibers exposed to the water, salt, and alkaline solutions. It was difficult to identify significant damage owing to the H + ions on the surface of the fibers, as they exhibited only numerous circular plot pitting. The damage or change in the chemical composition of the fibers could be concluded from the results of the mass loss weighting, which indicated a much larger loss of weight of fibers than that observed in the fibers exposed to the salt and water solutions. Furthermore, the decrease in the content of metal elements in the fiber detected in the energy spectrum analysis also indicated changes in the chemical composition of the fibers, which was likely to be the main reason for the substantial decrease in the tensile strength and elastic modulus.

In comparison to the basalt fibers, the basalt FRP composites exhibited less degradation as a result of the exposure to the different corrosive solutions. As for the fibers, the strength retention was approximately 50% (in water) and 60% (in salt), while for the FRP composites, the strength retention was 80% of the original tensile strength. This improvement was attributable to the protection of the matrix and indicated strong resistance to corrosion. This strong corrosion resistance delayed the penetration of the corrosive solution and corrosion of the internal fibers and improved the tensile properties of the composites. However, this improvement was limited by the relative poor bonding strength of the interface between the fibers and the matrix. The damage to the interface between the fibers and the matrix resulted from the reaction between the corrosive solutions and the coupling agent of the fibers. For instance, the surface of the basalt fibers was treated with the coupling agent γ-aminopropyl triethoxy (KH550), which could form a Si−O−Si chemical bond with the fibers. However, the Si−O−Si chemical bond was unstable; it could be hydrolyzed in corrosive solutions and form silanol. This chemical reaction would damage the interface between the fibers and the matrix, and result in the fracture of the composites by interfacial debonding.

1.2.5.5 Enhancement of corrosion resistance

1. *Matrix toughening*

The toughening of epoxy can increase the corrosion resistance of the FRP. In this process, the toughening particles will be uniformly dispersed in the epoxy matrix, producing a strong cross-linking with the epoxy. On the one hand, it improves the toughness of the FRP. On the other hand, it can prevent the formation of microcracks and thus, decrease the diffusion rate of H_2O and block the invasion of the corrosive media. Fig. 1.83 shows the strength of the FRP with matrix toughening. After corrosion in an alkaline solution at 55°C, the strength increases significantly. In addition, the bonding between the fibers and the matrix is strong. The matrix destruction causes a decrease in tensile strength. As for the pull-out fibers, significant amounts of resin can be found on the surfaces. In addition, the basalt fibers show good alkaline resistance, as shown in Fig. 1.83. The alkaline resistance is between those of the general BFRP (B-55) and the modified BFRP (BQ-55). The toughness effect can also be found in GFRP. The bonding strength between the matrix resin and the glass fibers is not as good as in BFRP, that is, the increase is not as obvious as in BFRP.

2. *Matrix modification*

Organic montmorillonite (OMMT) modifies the epoxy by inhibiting water penetration. From Fig. 1.84, it is seen that in pure water, the strength of resin without OMMT corrosion after 10 days was only 80% of the initial strength, and the fracture strain decreased sharply to 50%. After 90 days, the strength was about 60% of the initial strength and the fracture strain was only about 40%. While with 1 and 5 wt.% OMMT, the decreases were largely narrowed. Although the initial strength with OMMT was lower than that of the resin without OMMT, the properties after corrosion show a little decrease. In NaOH solution, the epoxy without OMMT decreased to 85% of the initial strength, while the epoxy with OMMT showed no degradation even after 90 days. In conclusion, the OMMT could increase the corrosion resistance of FRP by 1% was proven as the best.

In addition, montmorillonite has also proved to be effective in UV resistance. Under the UV light, the strength of the epoxy showed an obvious degradation and brittle destruction. Adding hindered amine light stabilizer (HALS) reduced the degradation, while the amount of HALS itself showed little effect. Fig. 1.85 shows that the strength degradation of the epoxy with and without HALS under UV as corrosion time increased.

3. *Pretreatment of fiber surface*

In addition, pretreatment of the fiber surface could increase the alkaline resistance to a certain extent. According to Fig. 1.86 in SEM after 1 and 3 days corrosion at 80°C, with pretreatment, the surface of the fiber after corrosion was smooth, and no obvious defects were found, indicating the fibers had good alkaline resistance. From Fig. 1.87, it can be seen that the strength retention rate for the fibers with pretreatment had an obvious increase than that of the control fibers. However,

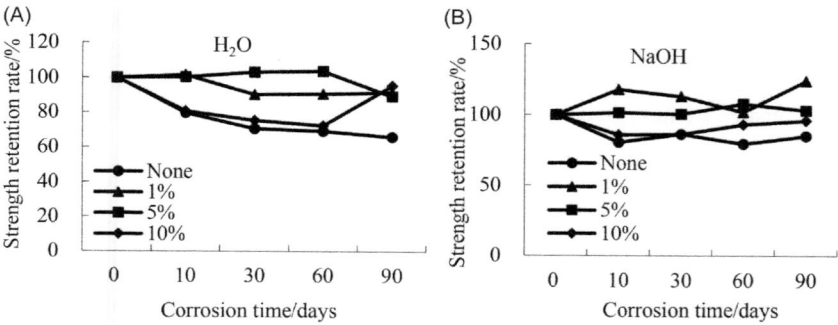

Figure 1.84 Modification for epoxy resin. (A) Water and (B) Alkaline solution.

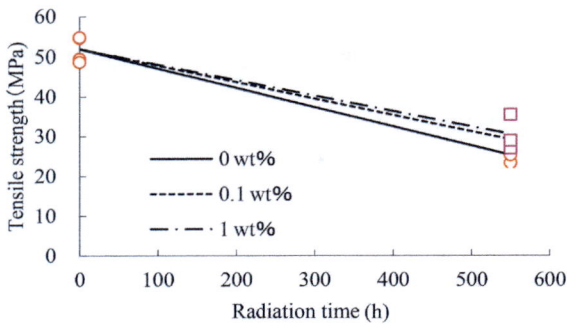

Figure 1.85 Relationship of radiation time and tensile strength for fiber-reinforced polymer sheet.

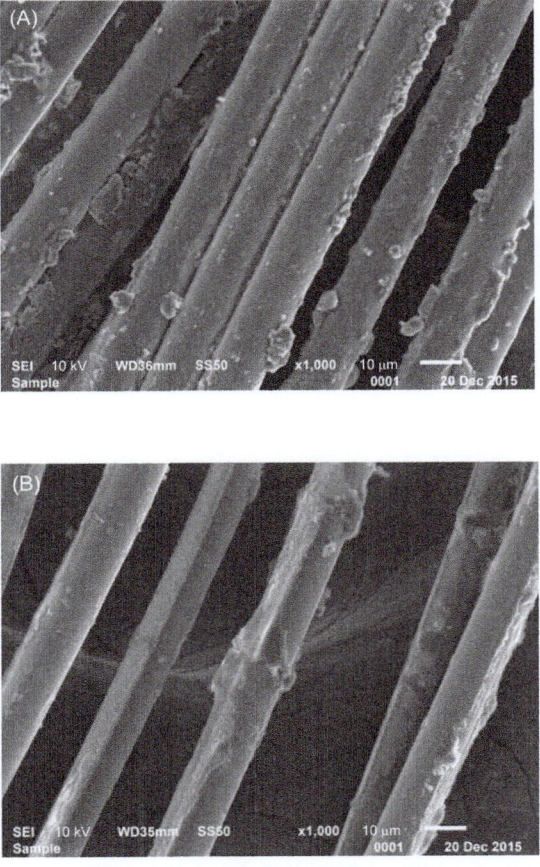

Figure 1.86 SEM for surface condition of fibers for different corrosion time. (A) 1 day/ 80°C. (B) 3 days/80°C.

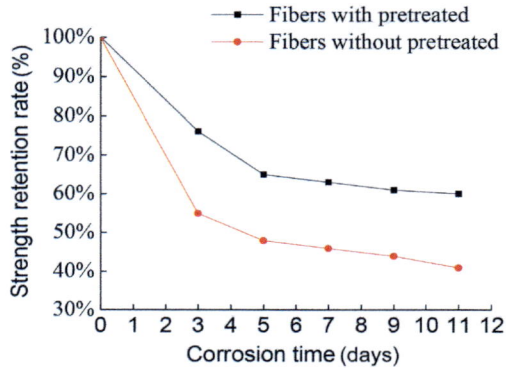

Figure 1.87 Strength retention rate for fibers with/without pretreated.

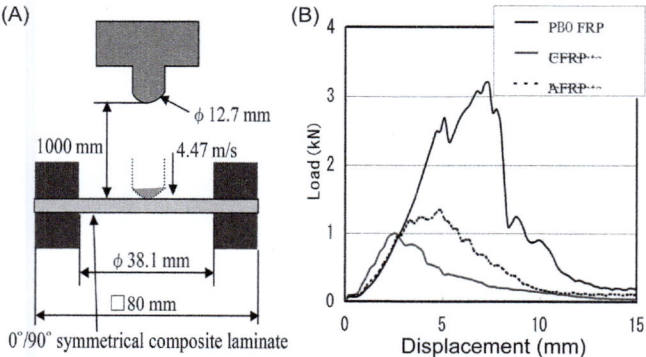

Figure 1.88 Impact tolerance test. (A) Experimental setup of impact test. (B) Results of impact tolerance test.

the bonding strength of the fibers and the matrix must be taken into account, and the produce processing must be optimally designed.

1.2.6 Impact behavior

In composites engineering, different impact velocity points result in different impact behaviors. As impact velocity less than 25 m/s, is called a low-velocity event. When the velocity is more than 10 m/s, the impact is called a ballistic event. When the velocity is more than 100 m/s, the impact is called a high-velocity event.

Impact testing is carried out according to GB/T 32377-2015 and ASTM D7136-12. Normally, it is performed with dropping weight. The objective of this test is to measure the amount of energy transferred to or lost from the striking body. The common weight impact test method is shown in Fig. 1.88A and the results of the test are shown in Fig. 1.88B referring to the datasheet (2005) supplied by Toyobo

Co. Ltd. Fig. 1.88 demonstrates that the PBO fibers possess much better energy absorption capacity than both carbon and aramid fibers, thus making it possible to obtain prestressed PBO sheets without resin impregnation.

1.2.6.1   Damage appearance and mechanism

There is no obvious boundary between the low-velocity and the high-velocity impact events. Under the impact, the FRP often exhibits many special behaviors that are different from those observed in studies under static conditions. According to the energy of impact, the appearance of damage can be divided into the following forms, as shown in Fig. 1.89.

Fig. 1.89A and B depicts the damage appearance under the low-energy impact, while Fig. 1.89C shows the same under a high-energy impact. When the energy is small enough, the impact does not damage the FRP. As the energy increases, the FRP firstly shows defects inside, as shown in Fig. 1.89A. With further increase in the energy, a number of cracks will appear on the surface of the FRP, as shown in Fig. 1.89B. When the energy is large enough, the FRP will be run through, in which case, the damaged area will be smaller than that in the aforementioned cases, as shown in Fig. 1.89C.

1.2.6.2   Comparison of impact resistance for FRP

Compression after impact (CAI) is an indicator for evaluating the toughness of FRP. Generally, when CAI < 138 MPa, the FRP is considered brittle. When CAI varies between 193 and 255 MPa, the FRP is considered to be a ductile composite. When CAI < 256 MPa, the FRP is considered to be super-high ductile. The CAI of some typical FRPs (T300/BMI, T700/BMI, IM7/Epoxy, IM7/BMI, T800/Epoxy, and AS4/PF) are 230, 260, 193−310, 214, 368, and 227 MPa, respectively (Zhang and Chen, 1999).

When a PBO fiber is impacted, it can absorb large amounts of energy. It shows a much higher impact resistance than carbon and high-modulus aramid fibers, as shown in Fig. 1.90. For the same impact condition, the composites reinforced with PBO fibers can bear the maximum impact load of 3.5 kN, absorbing about 20 J energy. While the maximum load that a T300 reinforced composite can bear is 1 kN, the energy it can absorb is 5 J. For high-modulus aramid fiber-reinforced composites, the maximum load is 1.3 kN, and the energy absorbed is a bit larger

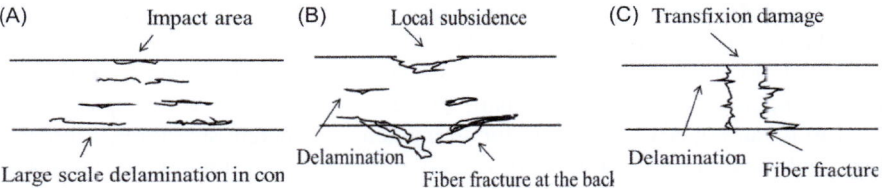

Figure 1.89 Impact appearance and mechanism (Weiwei, 2007).

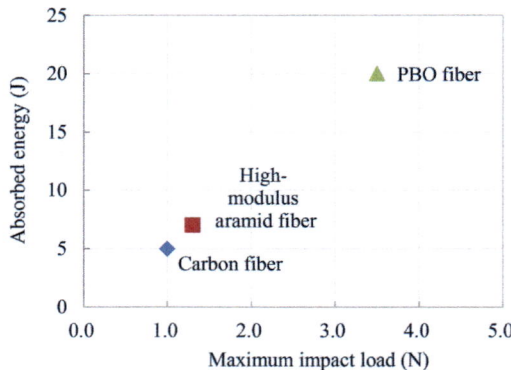

Figure 1.90 Impact load and energy absorb for carbon/aramid/poly-p-phenylene-benzobisoxazole fibers.

Table 1.13 Advantages and disadvantages of different kinds of FRP bonding technique.

	FRP sheets bonding technique	**FRP plates bonding technique**
Advantages	Shape of the FRP sheets can be easily designed according to the structures surface configuration	In case that excessive layers of FRP sheet, cost of this technique is cheaper than that of the FRP sheets bonding technique
Disadvantages	In case that excessive layers of FRP sheet, the cost of this technique is much higher than that of the FRP plates bonding technique	Shape of the FRP sheets cannot be changed

than that of the carbon fibers. As a result, the PBO is more widely used in structural construction and ballistic composites.

1.3 Fiber-reinforced polymer bonding technique for the concrete and steel structures

At present, FRP sheets and plates are widely used in externally bonded strengthening techniques for concrete structures. The details are shown in Table 1.13.

The deterioration of many RC or PC structures leads to the need for rehabilitation and upgradation. Using externally bonded prestressed FRP composites was found to be an effective technique for strengthening concrete structures. Moreover, other configurations of FRP composites are also used as the reinforcement in near-surface-mounted FRP reinforcement technique, in which the FRP tendons/bars are used besides the plates. In addition, as an alternative to the sheets/plates for structural strengthening, the FRP grids are gradually used for bond strengthening owing to their reliable bonding, convenient construction, and material savings compared to the FRP sheets/plates.

1.3.1 Installation procedure

A view of the execution and flow of the FRP bonding technique are shown in Figs. 1.91 and 1.92.

1. Preliminary preparation
   a. Removal and grinding of the weathered concrete layer
   b. Clean-up and drying
2. Restoration of concrete cross-section
   a. Chipping and removal of deteriorated concrete layer
   b. Mortar repair of concrete surface
   c. Resin injection into cracks
3. Primer coating
   a. Preparation of primer mix
   b. Brushing and curing
4. Surface preparation

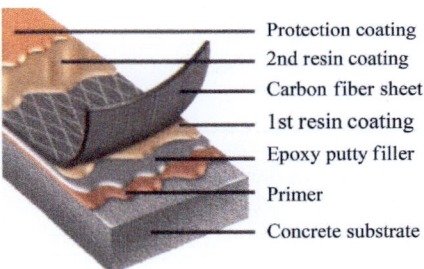

Protection coating
2nd resin coating
Carbon fiber sheet
1st resin coating
Epoxy putty filler
Primer
Concrete substrate

Figure 1.91 Schematic representation of cross-section.

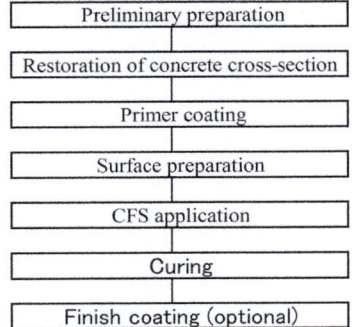

Preliminary preparation

Restoration of concrete cross-section

Primer coating

Surface preparation

CFS application

Curing

Finish coating (optional)

Figure 1.92 An execution flow of the fiber-reinforced polymer bond technique.

5. Carbon fiber sheet application
   a. Preparation of resin mix
   b. Roller brushing (under coat)
   c. Carbon fiber sheet adhesion
   d. Resin impregnation and defoaming
   e. Roller brushing (final coat)
   f. Replacement of impregnating resin
6. Curing
7. Finish coating (optional)
   a. Filling of concrete surface depressions with epoxy putty

1.3.2 Strengthening strategy

A strengthening strategy for the FRP bonding technique is shown in Fig. 1.93.

The following aspects of performance are increased by reinforcing with continuous fiber sheets:

1. Flexural load-carrying capacity
2. Shear load-carrying capacity

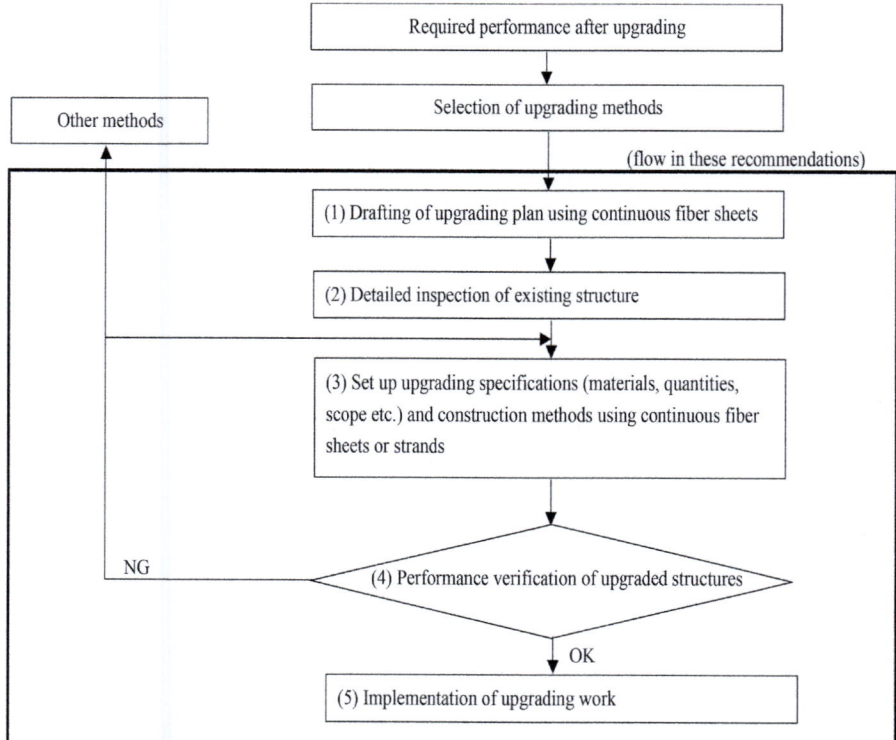

Figure 1.93 Upgrading using continuous fiber sheets.

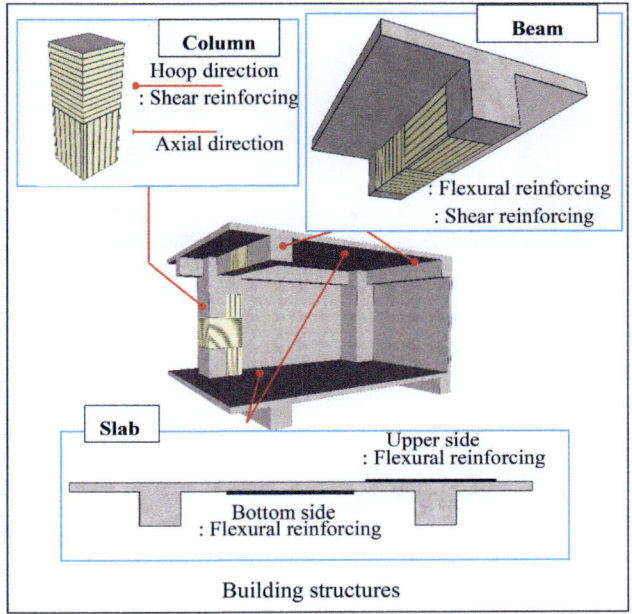

Figure 1.94 Example of applications to the reinforcing (building structures).

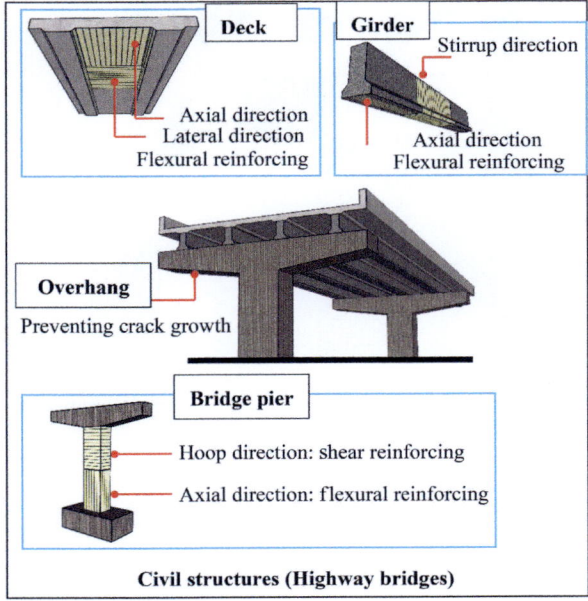

Figure 1.95 Example of applications to the reinforcing (highway bridges).

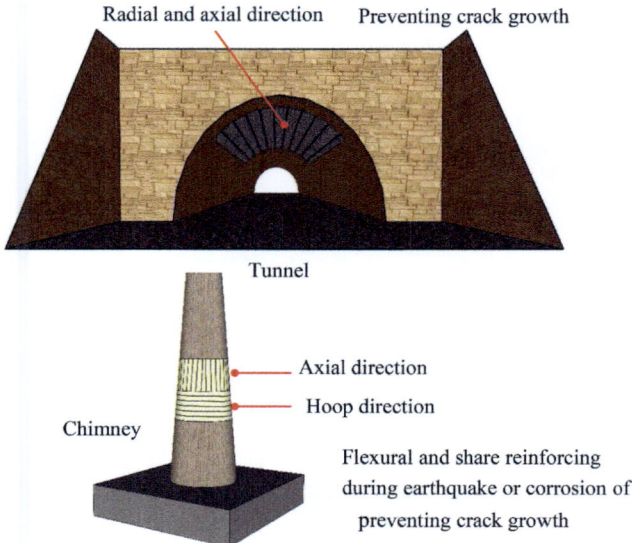

Figure 1.96 Example of applications to the reinforcing (tunnels and chimneys).

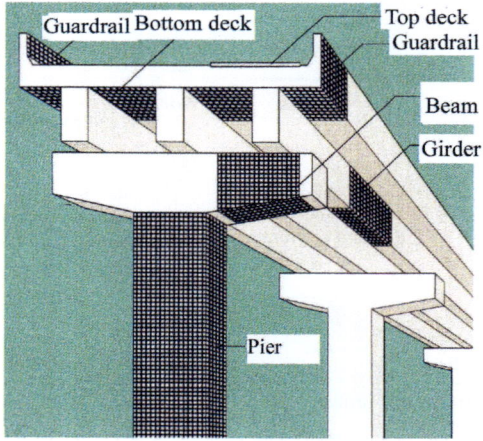

Figure 1.97 Bridge external strengthening with fiber-reinforced polymer grids.

3. Ductility
4. Fatigue capacity
5. Crack resistance
6. Spalling resistance

Some examples of the application of reinforcing are shown in Figs. 1.94−1.97.

References

20150490-T-609, 2018. Fiber Reinforced Composite Cables and Anchors for Structures.: Standards Press of China, Beijing.

Acha, B.A., Reboredo, M.M., Marcovich, N.E., 2007. Creep and dynamic mechanical behavior of PP—jute composites: effect of the interfacial adhesion. Compos. A: Appl. Sci. Manuf. 38 (6), 1507—1516.

ACI Committee 440. ACI 440.1R-06. Guide for the Design and Construction of Structural Concrete Reinforced with FRP Bars, American Concrete Institute, Farmington Hills, MI, 2007.

Agarwal, B., Broutman, L., 1980. Analysis and Performance of Fiber Composites. Wiley, New York, NY, p. 355.

Ali, N., Wang, X., Wu, Z., 2014. Integrated performance of FRP tendons with fiber hybridization. J. Compos. Constr. 18 (3), A4013007.

Ascione, L., Berardi, V.P., D'Aponte, A., 2012. Creep phenomena in FRP materials. Mech. Res. Commun. 43, 15—21.

Bakis, C.E., Bank, L.C., Brown, V.L., et al., 2002. Fiber-reinforced polymer composites for construction -state-of-the-art review. J. Compos. Constr. 6 (2), 73—87.

Budinski, K.G., 1997. Resistance to particle abrasion of selected plastics. Wear 203—204, 302.

Canadian Standard Association (CSA), 2002. CAN/CSA S806-02. Design and Construction of Building Components With Fiber Reinforced Polymers. Canadian Standards Association, Rexdale, Ontario, Canada.

Cao, S., Wang, X., Wu, Z., 2011. Evaluation and prediction of temperature-dependent tensile strength of unidirectional carbon fiber-reinforced polymer composites. J. Reinf. Plast. Compos. 30 (9), 799—807.

Cao, S., Wu, Z., Wang, X., 2009. Tensile properties of CFRP and hybrid FRP composites at elevated temperatures. J. Compos. Mater. 43 (4), 315—330.

Dutta, P.K., Lopez-Anido, R., Kwon, S.C., 2006. Fatigue durability of FRP composite bridge decks at extreme temperatures. Int. J. Mater. Prod. Technol. 28 (1—2), 198—216.

Everett, R.K., 1996. Quantification of random fiber arrangements using a radial distribution function approach. J. Compos. Mater. 30 (6), 748—758.

GB 50608-2010, 5060. Technical Code for Infrastructure Application of FRP Composites. Standards Press of China, Beijing.

GB/T 26743, 2012. Composite Bars for Civil Engineering. Standards Press of China: Beijing.

GB/T 26745-2011, 2674. Basalt Fiber Composites for Strengthening and Restoring Structures. Standards Press of China, Beijing.

GB/T 31539-2015, 3153. Pultruded Fiber Reinforced Polymer Composites Structural Profiles. Standards Press of China, Beijing.

GB/T 36262-2018, 3626. Fiber Reinforced Polymer Composite Grids for Civil Engineering. Standards Press of China, Beijing.

Hollaway, L.C., 2010. A review of the present and future utilisation of FRP composites in the civil infrastructure with reference to their important in-service properties. Constr. Build. Mater. 24 (12), 2419—2445.

ISIS Canada. ISIS-M04-01, 2001. Strengthening Reinforcing Concrete Structures With Externally-Bonded Fiber Reinforced Polymers. University of Manitoba, Winnipeg, Manitoba.

JSCE, 1997. Recommendation for design and construction of concrete structures using continuous fiber reinforced materials, research committee on continuous fiber reinforcing materials. Japan Society of Civil Engineers.

JSCE, 2001. Recommendation for Upgrading Concrete Structures With Use of Continuous Fiber Sheets. Japan Society of Civil Engineers.

Karbhari, V.M., 2007. Durability of Composites for Civil Structural Applications. Woodhead Publishing in Materials and CRC Press, England,Cambridge.

Ke, W., 2003. Survey of Corrosion in China. Chemical Industry Press, Beijing.

Keller, T., 2003. Use of Fiber Reinforced Polymers in Bridge Construction. IASBSE-AIPC-IVBH, Zurich.

Kurtz, S.M., 2004. The UHMWPE Handbook: Ultra-High Molecular Weight Polyethylene in Total Joint. Elsevier.

Kuz'min, S.A., Bulmanis, V.N., Struchkov, A.S., 1989. Experimental investigation of the strength and deformability of wound fiberglasses and organoplastics under low climatic temperatures. Mech. Compos. Mater. 25 (1), 49–53.

Li, Y., Shen, M., Chen, W., et al., 2012. Tensile creep study and mechanical properties of carbon fiber nano-composites. J. Polym. Res. 19 (7), 1–8.

Libby, J.R., 1990. Steel for Prestressing, Modern Prestressed Concrete. Springer, US, pp. 11–44.

Lorenzo, L., Hahn, H. T., 1986. Fatigue failure mechanisms in unidirectional composites. In Composite Materials: Fatigue and Fracture. ASTM International., Philadelphia.

Maksimov, R.D., Plume, E., 2001. Long-term creep of hybrid aramid/glass-fiber-reinforced plastics. Mech. Compos. Mater. 37 (4), 271–280.

Militký, Jiří, Kovačič, Vladimír, Rubnerova, Jitka, 2002. Influence of thermal treatment on tensile failure of basalt fibers. Eng. Fract. Mech. 69, 1025–1033.

Paluch, B., 1996. Analysis of geometric imperfections affecting the fibers in unidirectional composites. J. Compos. Mater. 30 (4), 454–485.

PE Material: Porex Porous Polyethylene for Plastic Filter Media. <www.porex.com/> (retrieved 14.02.17).

Rivera, J., Karbhari, V.M., 2002. Cold-temperature and simultaneous aqueous environment related degradation of carbon/vinylester composites. Compos. Part. B: Eng. 33 (1), 17–24.

Sasaki, T., 2008. Experimental Study on Long-term Performance of FRP and FRP Bonding (Technique Master Thesis), Ibaraki University, Japan.

Shi, J.W., Zhu, H., Wu, G., et al., 2014. Tensile behavior of FRP and hybrid FRP sheets in freeze–thaw cycling environments. Compos. Part. B: Eng. 60, 239–247.

Shi, J., Wang, X., Wu, Z., Zhu, Z., 2015a. Creep behavior enhancement of a basalt fiber-reinforced polymer tendon. Constr. Build. Mater. 94, 750–757.

Shi, J., Zhu, H., Dai, J., Wang, X., Wu, Z., 2015b. Effect of rubber toughening modification on the tensile behavior of FRP composites in concrete based alkaline environment. J. Mater. Civ. Eng. ASCE 27 (12), 04015054.

Shi, J., Wang, X., Huang, H., Wu, Z., 2017. Relaxation behavior of prestressing basalt fiber-reinforced polymer tendons considering anchorage slippage[J]. J. Compos. Mater. 51 (9), 1275–1284.

Sim, J., Park, C., 2005. Characteristics of basalt fiber as a strengthening material for concrete structures. Compos. Part B 36, 504–512.

Soudki, K.A., 1998. FRP reinforcement for prestressed concrete structures. Prog. Struct. Eng. Mater. 1 (2), 135–142.

Spathis, G., Kontou, E., 2012. Creep failure time prediction of polymers and polymer composites. Compos. Sci. Technol. 72 (9), 959—964.

Stein, H.L., 1988. Ultrahigh molecular weight polyethylenes (uhmwpe). Eng. Mater. Handb. 2, 167—171.

Tong, J., Ma, Y., Arnell, R.D., Ren, L., 2006. Free abrasive wear behaviour of UHMWPE composites filled with wollastonite fibres. Compos. Part. A: Appl. Sci. Manuf. 37, 38. Available from: https://doi.org/10.1016/j.compositesa.2005.05.023.

Wu, Z., Sakamoto, K., Iwashita, K., Yue, Q. 2006. Hybridization of continuous fiber sheets as structural composites. J. Japan. Soc. Compos. Mater. 32 (1), 12—21.

Wang, X., Shi, J., Liu, J., Yang, L., Wu, Z., 2014. Creep behavior of basalt fiber reinforced polymer tendons for prestressing application. Mater. Des. 59, 558—564.

Wang, X., Shi, J., Wu, Z., et al., 2016. Creep strain control by pretension for basalt fiber-reinforced polymer tendon in civil applications. Mater. Des. 89, 1270—1277.

Weiwei, D., 2007. Study on Impact Resistance of Glass Fiber/Unsaturated Polyester Resin Laminates. Tianjin Polytechnic University, in Chinese.

Wu, Z., Iwashita, K., Hayashi, K., et al., 2007a. Development of continuous fiber sheets as a prestressing material and externally bonded upgrading technique. J. Jpn. Soc. Compos. Mater. 33 (2), 72—75.

Wu, Z., Wang, X., Iwashita, K., 2007b. State-of-the-art of advanced FRP applications in civil infrastructure in Japan. Compos. Polycon 37, 1—17.

Wu, Z., Wang, X., Iwashita, K., et al., 2010. Tensile fatigue behaviour of FRP and hybrid FRP sheets. Compos. Part B: Eng. 41 (5), 396—402.

Wu, G., Wang, X., Wu, Z., Dong, Z., Zhang, G., 2014. Durability of basalt fibers and composites in corrosive environments. J. Compos. Mater. 49 (7), 873—887.

Zhang, B.Y., Chen, X.B., 1999. Improvement of impact resistance of carbon fiber reinforced resin composites. Fiber Compos. 12, 1—6. in Chinese.

Zhao, X., Wang, X., Wu, Z., et al., 2016. Fatigue behavior and failure mechanism of basalt FRP composites under long-term cyclic loads. Int. J. Fatigue 88, 58—67.

Zhou, Z., Yang, Y., 1985. Preliminary discussion on creep mechanism of fiber reinforced plastics. Glas. Fiber Reinf. Plast. Compos. 4, 012. in Chinese.

Bond characteristics and debonding mechanism of FRP−concrete interface

2

2.1 Interfacial fractures and debonding modes

There are many definitions to classify debonding modes. Based on available investigations, the failure modes related to interfacial fractures can be grouped into three categories (as shown in Fig. 2.1): intermediate crack-induced debonding (IC debonding), delamination at the cut-off point [fiber-reinforced polymer (FRP) delamination], and peeling-off owing to shear cracks (peeling-off). IC debonding and FRP delamination failures typically occur in beams with high shear resistance. In general, delamination of the concrete layer occurs along rebars and in beams of short laminate lengths, indicating that significant stress concentrations occur at the laminate anchorage zone. FRP end shear failure is an FRP delamination type of interfacial fracture. Shear crack-induced peeling-off of laminate generally occurs with longer laminate lengths, wherein significant shear cracks form. The peeling-off type of interfacial fracture is caused mainly by shear cracks of concrete and the unevenness or roughness of the concrete surface.

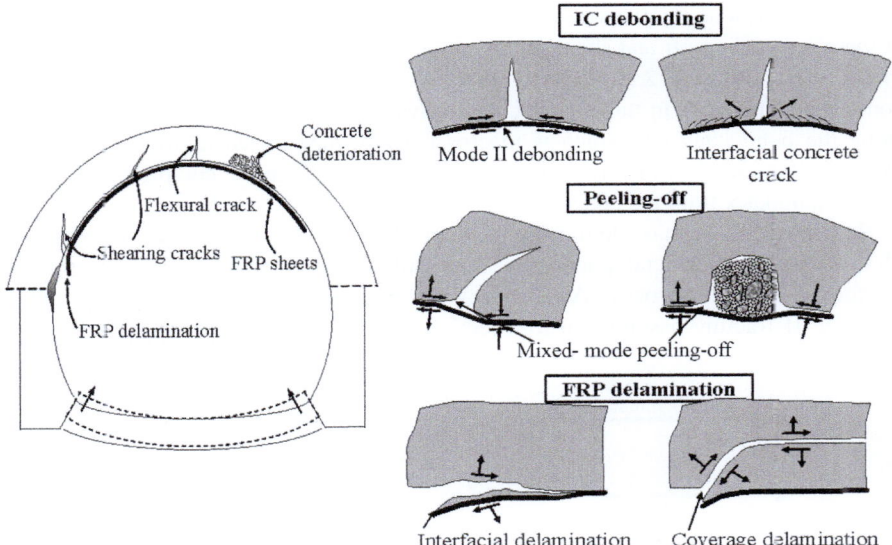

Figure 2.1 Different debonding modes in FRP-strengthened concrete structures.

Structures Strengthened with Bonded Composites. DOI: https://doi.org/10.1016/B978-0-12-821088-8.00002-3

In particular, IC debonding initiates at a major crack and propagates along the FRP—concrete interface. In reinforced concrete (RC) beams flexural-strengthened with FRP plates/sheets on the tension face, IC debonding may arise at a major flexural crack or flexural-shear crack (Wu et al., 1997; Wu and Niu, 2007). In RC beams shear-strengthened with FRP plates/sheets bonded to the beam sides, IC debonding can arise as a result of a shear crack (Chen and Teng, 2001). In IC debonding, the interface is dominated by shear stresses. Therefore the debonding failure is also referred to as Mode II fracture in the context of fracture mechanics. It should be noted that whereas the focus of this chapter is on FRP—concrete joints, the analytical solution is applicable to similar joints between thin plates of other materials (e.g., steel and aluminum) and concrete. Debonding failures of RC beams bonded with steel plates have also been studied extensively in the literature (Roberts, 1989; Oehlers, 1992). For a localized single crack pattern in IC debonding, the stress state of the FRP—concrete interface is similar to that of a shear test specimen in which a plate is bonded to a concrete prism and subjected to tension (Fig. 2.2). For a distributed crack pattern, the interfacial stress distribution is more complex and can be resolved with the smeared crack model based on the stress solution of a single-crack pattern (Niu and Wu, 2005). As a result, a large number of studies, both experimental and theoretical, have been carried out on simple shear tests on bonded joints, with the earlier works being concerned with steel plates bonded to concrete.

The enhanced performance of FRP-strengthened concrete elements is owing to the stress transfer from the concrete to the FRP reinforcement through the interfacial bond layer. Failures (in the form of interfacial debonding) in this transfer region may result in an abrupt fracture. This must be considered appropriately in design. There are different types of debonding phenomena in FRP-strengthened structures. Two primary ones of these are peeling-off and shear delamination of FRP sheets, classified as Mode I and Mode II fractures (Fig. 2.3), respectively. In most cases, the interfacial failure is shear-dominated. Therefore in this chapter, the analysis focuses primarily on the debonding along the bond interface, which is macroscopically represented as Mode II fracture. Debonding causes loss of force transfer and finally results in the complete detachment of FRP laminates from the concrete substrate.

Recent research has demonstrated that the understanding and modeling of FRP—concrete interfacial phenomena and failures may be improved by applying fracture mechanics theories. All the previous works were based on the assumption of Mode II fracture assumption irrespective of where the debonding occurs. These

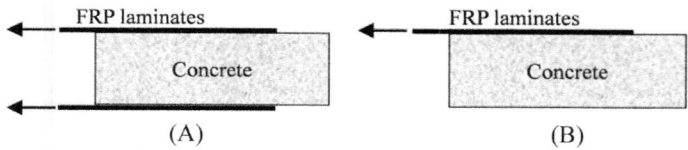

Figure 2.2 FRP-to-concrete joints. (A) Double-lap shear specimen. (B) Single-lap shear specimen.

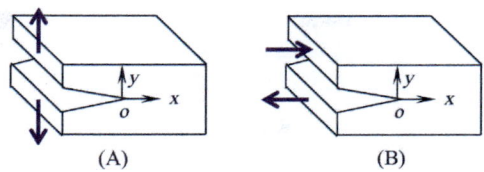

Figure 2.3 Fracture mode: (A) Mode I and (B) Mode II.

works provided a general understanding of interfacial fracture in FRP-strengthened concrete structures. Moreover, they can model the debonding behavior quantitatively, including the force transfer procedure and load-carrying capacity of FRP–concrete interface. However, if we focus on the microlevel fracture, particularly in the concrete adjacent to the interface, the simple Mode II fracture model is apparently not suitable. This is because a concrete crack generally initiates as Mode I fracture. Therefore the bonding/debonding behavior must be related to the cracking of concrete, which depends on the properties of the concrete, such as tensile strength and Mode I fracture energy.

Available studies indicate that the main failure mode of FRP–concrete joints in shear tests is concrete failure under shear. This generally occurs a few millimeters away from the concrete–adhesive interface (Chen and Teng, 2001). Therefore the ultimate load (i.e., the maximum transferable load) of the joint depends strongly on the concrete strength. The plate-to-concrete member width ratio also exerts a significant effect. Another highly important aspect of the behavior of these bonded joints is the effective bond length: an increase in the bond length beyond this does not result in a further increase in the ultimate bond load. This is a fundamental difference between an externally bonded plate and an internal reinforcing bar, for which a sufficiently long anchorage length can be determined so that the full tensile strength of the reinforcement could be achieved.

2.2 Stress transfer and fracture propagation of FRP–concrete joints

The adhesive-bonded joints, as shown in Fig. 2.4, may be considered as a simple and typical FRP-strengthened structure to understand the stress transfer and debonding behavior (Yuan et al., 2001; Wu et al., 2002). Fig. 2.4A depicts a bonded pull–push joint that is generally encountered in FRP/steel-strengthened beams. Fig. 2.4B depicts a double-lap shear test of FRP laminate-strengthened concrete structures. Fig. 2.4C depicts a single-lap shear test of FRP laminate-strengthened steel structures, and the two types of specimen are equivalent to the bonded pull–pull joint as shown in Fig. 2.4D. Omitting the bending effects, the adhesive layer can be regarded to be subjected mainly to shear (Mode II fracture). The thickness and width of the layers are t_1 and b_1 for Adherent 1, and t_2 and b_2 for Adherent 2,

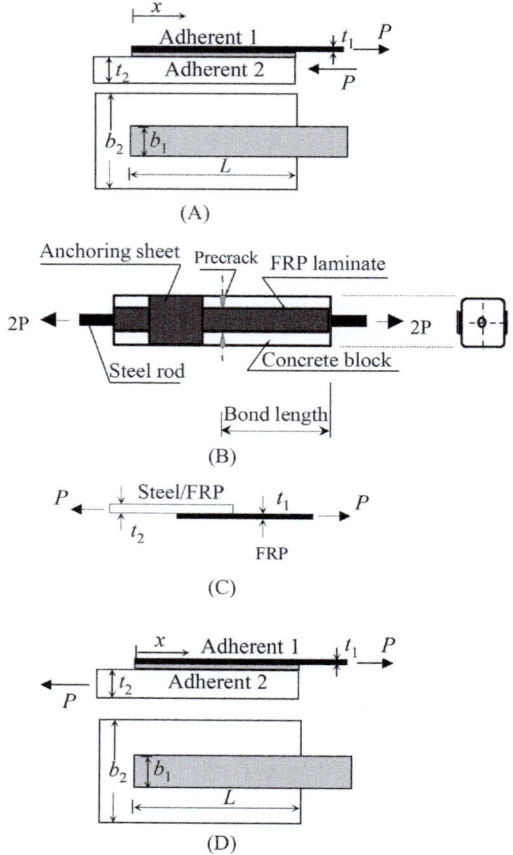

Figure 2.4 Different kinds of adhesive bonded joints (Wu et al., 2002). (A) Pull—push single-lap joint. (B) Pull—pull double-lap joint. (C) Pull—pull single-lap joint. (D) Equivalent bonded pull—pull joint.

respectively. L is the bond length. Young's moduli of Adherent 1 and Adherent 2 are E_1 and E_2, respectively.

2.2.1 Fundamental formulas

In what follows, two typical types of joint are studied, and four models of bond stress—slip law are introduced in detail to derive the shear stresses along the inter-face in the bond zone.

Before starting the derivations, we assume the following for simplicity of the problems:

- The adherents are homogeneous and linearly elastic.
- The adhesive is exposed only to shear forces.

- Bending effects are omitted.
- The normal stresses are uniformly distributed over the cross-section.
- The thickness and width of the adherents are constant throughout the bond line.

It should be noted that in such a model, the deformation of the adhesive layer represents the relative movement or slip of the adherents. Therefore the adhesive layer is also referred to as the interface. Based on these assumptions and on equilibrium considerations, the following fundamental equations can be obtained (Fig. 2.5):

Considering the element shown in Fig. 2.5, the equations of equilibrium for the adherents can be specified as

$$\frac{d\sigma_1}{dx} - \frac{\tau}{t_1} = 0 \tag{2.1}$$

For a pull–push joint

$$\sigma_1 t_1 b_1 + \sigma_2 t_2 b_2 = 0 \tag{2.2a}$$

For a pull–pull joint

$$\sigma_1 t_1 b_1 + \sigma_2 t_2 b_2 = P \tag{2.2b}$$

The slip deformation δ is defined as the relative displacement of adherents, which is a function of the location, x:

$$\delta = u_1(x) - u_2(x) \tag{2.3}$$

The constitutive equations for the adhesive layer and two adherents are expressed as

$$\tau = f(\delta) \tag{2.4}$$

$$\sigma_1 = E_1 \frac{du_1(x)}{dx} \tag{2.5}$$

$$\sigma_2 = E_2 \frac{du_2(x)}{dx} \tag{2.6}$$

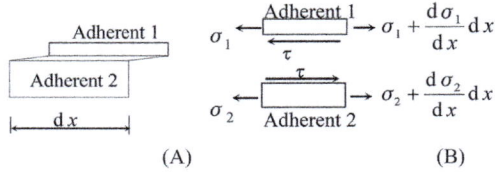

(A)   (B)

Figure 2.5 Deformation and stresses of bonded joint (Wu et al., 2002). (A) Deformation. (B) Shear stress.

Substituting Eqs. (2.2)–(2.6) into Eq. (2.1) and introducing the parameters local bond strength τ_f and interfacial fracture energy G_f yield the following:

$$\frac{d^2\delta}{dx^2} - \frac{2G_f}{\tau_f^2}\lambda^2 f(\delta) = 0 \tag{2.7}$$

For the pull−push joint

$$\sigma_1 = \frac{\tau_f^2}{2G_f t_1 \lambda^2}\frac{d\delta}{dx} \tag{2.8a}$$

For the pull−pull joint

$$\sigma_1 = \frac{\tau_f^2}{2G_f t_1 \lambda^2}\left(\frac{d\delta}{dx} + \frac{P}{b_2 E_2 t_2}\right) \tag{2.8b}$$

where

$$\lambda^2 = \frac{\tau_f^2}{2G_f}\left(\frac{1}{E_1 t_1} + \frac{b_1}{b_2 E_2 t_2}\right) \tag{2.9}$$

Eq. (2.7) is the governing differential equation of the bonded joint shown in Fig. 2.5 and can be solved if the local bond−slip model relating the local interfacial shear stress to the local shear slip (represented by $f(\delta)$) is defined. The interfacial fracture energy, which is the area under the local bond−slip curve, is introduced because once it is specified, it can be used regardless of the shape of the local bond−slip curve.

The four models of bond stress−slip relationship ($\tau-\delta$) (shown in Fig. 2.6), which are considered to be feasible representations of the nonlinear interfacial behavior, are introduced here (Yuan et al., 2001).

Model I:

$$f(\delta) = \begin{cases} \dfrac{\tau_f^2}{2G_f}\delta & \text{when } 0 \le \delta \le \delta_f \\ 0 & \text{when } \delta > \delta_f \end{cases} \tag{2.10}$$

As shown in Fig. 2.6A, the stress−slip relation is linearly ascending before the occurrence of interfacial fracture. The shear stress decreases abruptly to zero when the slip exceeds δ_f without considering softening behavior.

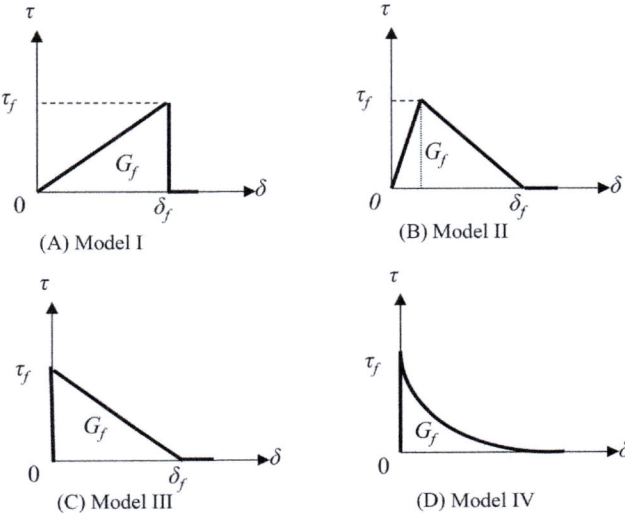

Figure 2.6 Different simplified shapes of $\tau-\delta$ curves (Yuan et al., 2001).

Model II:

$$f(\delta) = \begin{cases} \dfrac{\tau_f}{\delta_1}\delta & \text{when } 0 \le \delta \le \delta_1 \\[2mm] \dfrac{\tau_f}{\delta_f - \delta_1}\left(\delta_f - \delta\right) & \text{when } \delta_1 < \delta \le \delta_f \\[2mm] 0 & \text{when } \delta > \delta_f \end{cases} \tag{2.11}$$

As shown in Fig. 2.6B, the stress—slip relation is linearly ascending when the slip is smaller than δ_1. After the occurrence of interfacial fracture, the stress—slip relation is linearly descending within the range $\delta_1-\delta_f$. The shear stress reduces to zero when the slip exceeds δ_f.

Model III:

$$f(\delta) = \begin{cases} f(\delta) = \tau_f\left(1 - \dfrac{\tau_f}{2G_f}\delta\right) & \text{when } 0 \le \delta \le \delta_f \\[2mm] 0 & \text{when } \delta > \delta_f \end{cases} \tag{2.12}$$

Fig. 2.6C is a good approximation of Model α (the ascending linear part is omitted because δ_1 is generally significantly less than δ_f). It shows the linearly descending stress—slip relation when the slip is smaller than δ_f. Moreover, shear stress reduces to zero when the slip exceeds δ_f.

Model IV:

$$f(\delta) = \tau_f \exp\left(-\dfrac{\tau_f}{G_f}\delta\right) \tag{2.13}$$

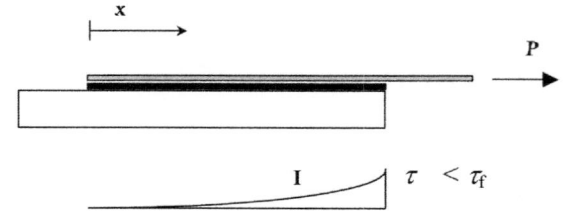

(A) Elastic stress state

(B) Initiation of softening at $x = L$ (point A in Fig. 2.8)

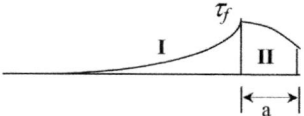

(C) Propagation of softening zone

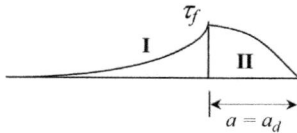

(D) Initiation of debonding at $x = L$ (point B in Fig. 2.8)

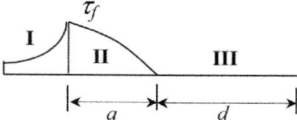

(E) Propagation of debonding

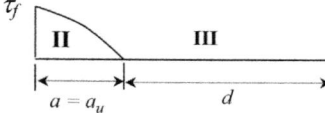

(F) Peak shear stress at $x = 0$ (point D in Fig. 2.8)

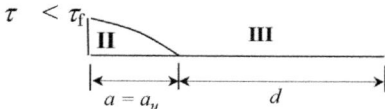

(G) Linear unloading

Figure 2.7 Interfacial shear stress distribution and propagation of debonding for a large bond length (Yuan et al., 2001).

As shown in Fig. 2.6D, the stress–slip relation exhibits an exponential softening behavior, and the shear stress gradually decreases to zero as the slip increases.

2.2.2 Pull–push joint

The theoretical solutions for the case of the pull–push joint (Fig. 2.4A) can be obtained as follows:

2.2.2.1 Model I: linear interface shear stress–slip relationship with an abrupt decrease in stress

Substituting the relationship as shown in Fig. 2.7A for the case of $\delta \leq \delta_f$ into Eq. (2.7) yields

$$\frac{d^2\delta}{dx^2} - \lambda^2\delta = 0 \tag{2.14}$$

with the boundary conditions

$$\sigma_1 = 0 \quad \text{at} \quad x = 0, \quad \text{and}$$
$$\sigma_1 = \frac{P}{b_1 t_1} \quad \text{at} \quad x = L \tag{2.15}$$

The solution of Eq. (2.14) for relative shear displacement, shear stress of adhesive layer, and normal stress of laminate can be expressed as

$$\delta = \frac{2G_f}{\tau_f^2} \frac{P\lambda\cosh(\lambda x)}{b_1\sinh(\lambda L)} \tag{2.16}$$

$$\tau = \frac{P\lambda}{b_1} \cdot \frac{\cosh(\lambda x)}{\sinh(\lambda L)} \tag{2.17}$$

$$\sigma_1 = \frac{P\sinh(\lambda x)}{b_1 t_1 \sinh(\lambda L)} \tag{2.18}$$

when $\tau = \tau_f$ at $x = L$, P attains its maximum value

$$P_{\max} = \frac{\tau_f b_1}{\lambda}\tanh(\lambda L) \tag{2.19}$$

For large values of L, Eq. (2.19) converges to

$$P_{\max} = \frac{\tau_f b_1}{\lambda} \tag{2.20}$$

Crack propagation occurs when $\delta > \delta_f$. It can be studied by decreasing the bonding length L.

2.2.2.2 Model II: $\tau-\delta$ relationship with linearly ascending and descending branches

Substituting the relationship as shown in Fig. 2.7B into Eq. (2.7), we obtain

$$\frac{d^2\delta}{dx^2} - \lambda_1^2\delta = 0 \quad \text{for} \quad 0 \leq \delta \leq \delta_1 \tag{2.21}$$

$$\frac{d^2\delta}{dx^2} + \lambda_2^2\delta = \lambda_2^2\delta_f \quad \text{for} \quad \delta_1 < \delta \leq \delta_f \tag{2.22}$$

where

$$\lambda_1^2 = \lambda^2\frac{2G_f}{\delta_1\tau_f} = \frac{\tau_f}{\delta_1}\left(\frac{1}{E_1t_1} + \frac{b_1}{b_2E_2t_2}\right)$$

$$\lambda_2^2 = \lambda^2\frac{2G_f}{(\delta_f - \delta_1)\tau_f} \tag{2.23}$$

$$= \frac{\tau_f}{\delta_f - \delta_1}\left(\frac{1}{E_1t_1} + \frac{b_1}{b_2E_2t_2}\right)$$

The solution is determined by substituting the boundary and continuous conditions:

$$\sigma_1 = 0 \quad \text{at} \quad x = 0 \tag{2.24}$$

$$\sigma_1 \text{ is continuous at } x = L - a \tag{2.25}$$

$$\delta = \delta_1 \text{ or } \tau = \tau_f \text{ at } x = L - a \tag{2.26}$$

$$\sigma_1 = \frac{P}{t_1b_1} \quad \text{at} \quad x = L \tag{2.27}$$

where a is the softening zone (microcrack) length. Finally, we obtain the following for $0 \leq \delta \leq \delta_1$, that is, $0 \leq x \leq L-a$:

$$\delta = \delta_1\frac{\cosh(\lambda_1 x)}{\cosh[\lambda_1(L - a)]} \tag{2.28}$$

$$\tau = \tau_f\frac{\cosh(\lambda_1 x)}{\cosh[\lambda_1(L - a)]} \tag{2.29}$$

$$\sigma_1 = \frac{\tau_f}{t_1 \lambda_1} \cdot \frac{\sinh(\lambda_1 x)}{\cosh[\lambda_1(L - a)]} \tag{2.30}$$

and the following for $\delta_1 \le \delta \le \delta_f$, that is, $L - a \le x \le L$;

$$\delta = (\delta_f - \delta_1)\left\{\frac{\lambda_2}{\lambda_1}\tanh[\lambda_1(L - a)] \times \sin[\lambda_2(x - L + a)] - \cos[\lambda_2(x - L + a)] + \frac{\delta_f}{\delta_f - \delta_1}\right\} \tag{2.31}$$

$$\tau = -\tau_f\left\{\frac{\lambda_2}{\lambda_1}\tanh[\lambda_1(L - a)]\sin[\lambda_2(x - L + a)] - \cos[\lambda_2(x - L + a)]\right\} \tag{2.32}$$

$$\sigma_1 = \frac{\tau_f}{\lambda_2 t_1}\left\{\frac{\lambda_2}{\lambda_1}\tanh[\lambda_1(L - a)]\cos[\lambda_2(x - L + a)] + \sin[\lambda_2(x - L + a)]\right\} \tag{2.33}$$

Substituting Eq. (2.27) into Eq. (2.33), we obtain

$$P = \frac{\tau_f b_1}{\lambda_2}\left\{\frac{\lambda_2}{\lambda_1}\tanh[\lambda_1(L - a)]\cos(\lambda_2 a) + \sin(\lambda_2 a)\right\} \tag{2.34}$$

It is apparent that P attains maximum value when $\frac{dP}{da} = 0$ for the general case or, for simplicity, when $\tau = 0$ at $x = L$ for large values of bonding length. Therefore a at the maximum load can be determined from the relation

$$\tanh[\lambda_1(L - a)] = \frac{\lambda_2}{\lambda_1}\tan(\lambda_2 a) \tag{2.35}$$

Substituting Eq. (2.35) into Eq. (2.34), we obtain

$$P_{\max} = \frac{\tau_f b_1}{\lambda_2} \cdot \frac{\delta_f}{\delta_f - \delta_1}\sin(\lambda_2 a) \tag{2.36}$$

The problem in this model is that a is defined as an implicit function and can be determined only by iteration. However, it can be demonstrated that for large values of L, Eq. (2.36) converges to Eq. (2.37). This is the case with Model I as well.

$$P_{max} = \frac{\tau_f b_1}{\lambda} \tag{2.37}$$

Crack propagation occurs when $\delta > \delta_f$. It can be studied by decreasing the bonding length L. For large values of L, by letting $\delta = \delta_f$ at the load end ($x = L$) in Eq. (2.31), we obtain the maximum value of a that is associated with the initiation of macrocrack (debonding):

$$a_{max} = \frac{1}{\lambda_2} \arctan\left(\frac{\lambda_1}{\lambda_2}\right) \tag{2.38}$$

2.2.2.3 Model III: $\tau-\delta$ relationship with only linearly descending branch

With regard to this model, there is no relative shear displacement until the maximum shear stress τ_f occurs at the load end ($x = L$). Moreover, the relative shear displacement increases gradually when the maximum shear stress shifts from the load end to the other end of a laminate. Substituting the relationship shown in Fig. 2.7C for the case of $\delta \leq \delta_f$ into Eq. (2.7), we obtain

$$\frac{d^2\delta}{dx^2} + \lambda^2\delta = \lambda^2 \frac{2G_f}{\tau_f} \tag{2.39}$$

At the cross-section without interface slip (no shear microcrack), the interfacial shear displacement and normal stresses in the steel/FRP and the concrete are equal to zero. At the microcrack tip, $\tau = \tau_f$, and $\delta = 0$. From the equilibrium of the steel/FRP element, we can also infer that $\sigma_1 = 0$ at the microcrack tip. Assuming the microcrack (softening zone) length to be a, we can specify the boundary conditions as follows:

$$\sigma_1 = 0, \quad \delta = 0 \quad \text{at} \quad x = L - a \tag{2.40}$$

The solution of Eq. (2.39) for the relative shear displacement, shear stress of adhesive layer, and normal stress of laminate can be expressed as

$$\delta = \frac{2G_f}{\tau_f}\{1 - \cos\lambda[x - (L - a)]\} \tag{2.41}$$

$$\tau = \tau_f \cos \lambda [x - (L - a)] \tag{2.42}$$

$$\sigma_1 = \frac{\tau_f}{\lambda t_1} \sin \lambda [x - (L - a)] \tag{2.43}$$

Therefore the interfacial shear displacement, shear stress, and normal stress in steel/FRP laminate are related to a. a can be determined from the value of P at the load end. Given that

$$\sigma_1 = \frac{P}{t_1 b_1} \quad \text{at} \quad x = L \tag{2.44}$$

we obtain

$$P = \frac{\tau_f b_1}{\lambda} \sin(\lambda a) \tag{2.45}$$

When $L \geq a_{\max} = \frac{\pi}{2\lambda}$, P attains maximum at $a = a_{\max}$, that is,

$$P_{\max} = \frac{\tau_f b_1}{\lambda} \tag{2.46}$$

Then, $\tau = 0$ at $x = L$.
When $L < a_{\max}$, P reaches maximum at $a = L$, that is,

$$P_{\max} = \frac{\tau_f b_1}{\lambda} \sin(\lambda L) \tag{2.47}$$

Then, $\tau = \tau_f \cos \lambda L \neq 0$ at $x = L$.
Crack propagation occurs when $\delta > \delta_f$. It can be studied by decreasing the bonding length L. a_{max} is associated with the initiation of macrocrack (debonding).

2.2.2.4 Model IV: $\tau - \delta$ relationship with only exponential softening branch

Substituting the relationship shown in Fig. 2.7D for the case of $\delta \leq \delta_f$ into Eq. (2.7), we obtain

$$\frac{d^2 \delta}{dx^2} - \lambda^2 \frac{2G_f}{\tau_f} \exp\left(-\frac{\tau_f}{G_f} \delta\right) = 0 \tag{2.48}$$

Similar to the case of Model III, we have the following boundary conditions:

$$\sigma_1 = 0 \quad \text{and} \quad \delta = 0 \quad \text{at} \quad x = L - a \tag{2.49}$$

Solving Eq. (2.48), we obtain

$$\delta = \frac{2G_f}{\tau_f} \ln\{\cosh[\lambda(x - L + a)]\} \tag{2.50}$$

$$\tau = \frac{\tau_f}{\{\cosh[\lambda(x - L + a)]\}^2} \tag{2.51}$$

$$\sigma_1 = \frac{\tau_f}{\lambda t_1} \tanh[\lambda(x - L + a)] \tag{2.52}$$

Again, the interfacial shear displacement, shear stress, and normal stress in steel/FRP laminate are related to a. a can be determined by the value of P at the load end. Given that

$$\sigma_1 = \frac{P}{t_1 b_1} \quad \text{at} \quad x = L \tag{2.53}$$

we obtain

$$P = \frac{\tau_f b_1}{\lambda} \tanh(\lambda a) \tag{2.54}$$

Therefore P attains maximum value at $a = L$, that is,

$$P_{max} = \frac{\tau_f b_1}{\lambda} \tanh(\lambda L) \tag{2.55}$$

which is identical to Eq. (2.19). Therefore for large values of L, Eq. (2.55) converges to Eq. (2.20). Rapid propagation of crack starts when P attains 97% of P_{max}. Considering that $\tanh(2) \approx 0.97$, the following equation can be derived from Eq. (2.55):

$$a_{max} = \frac{2}{\lambda} \tag{2.56}$$

2.2.2.5 Effective stress transfer (bond) length l_e

The effective stress transfer (bond) length l_e is defined as the length required to attain 97% of P_{max} for Models I, II, and IV, and as the length required to attain

P_{max} for Models III and IV. Considering that $\tanh(2) \approx 0.97$, the following equations can be derived:

$$l_e = \frac{2}{\lambda} \text{ for Models I and IV, and}$$

$$l_e = a + \frac{1}{2\lambda_1} \ln \frac{\lambda_1 + \lambda_2 \tan(\lambda_2 a)}{\lambda_1 - \lambda_2 \tan(\lambda_2 a)} \text{ for Model II}$$

$$\text{where } a = \frac{1}{\lambda_2} \arcsin \left[0.97 \sqrt{\frac{\delta_f - \delta_1}{\delta_f}} \right]$$

$$l_e = \frac{\pi}{2\lambda} \text{ for Model III}$$

(2.57)

2.2.2.6 Debonding propagation of FRP—concrete joints

A test reveals that the shear stress—slip relationship of Model II (Eq. 2.11) is the closest to reality. Therefore it is discussed in detail in the following and is shown in Fig. 2.7.

At marginal loads, there is no interfacial softening or debonding along the plate—concrete interface. Therefore the entire length of the interface is in an elastic stress state (State I) (Fig. 2.7A). This is true as long as the interfacial shear stress at $x = L$ is less than τ_f.

By solving Eq. (2.21) with these boundary conditions, the following expressions for the interfacial slip, interfacial shear stress, and axial stress in the plate are determined:

$$\delta = \frac{\delta_1}{\tau_f} \frac{P\lambda_1 \cosh(\lambda_1 x)}{b_p \sinh(\lambda_1 L)}$$

(2.58)

$$\tau = \frac{P\lambda_1}{b_p} \cdot \frac{\cosh(\lambda_1 x)}{\sinh(\lambda_1 L)}$$

(2.59)

$$\sigma_p = \frac{P\sinh(\lambda_1 x)}{b_p t_p \sinh(\lambda_1 L)}$$

(2.60)

During this stage of loading, the interfacial shear stress distribution is of the form shown in Fig. 2.7A, with the maximum bond shear stress being smaller than the local bond strength τ_f. The slip at the loaded end (i.e., the value of δ at $x = L$) is defined as the displacement of the bonded joint and is denoted by Δ. Applying this definition, the following load—displacement relationship can be obtained from Eq. (2.58):

$$P = \frac{\tau_f b_p}{\lambda_1} \cdot \frac{\Delta}{\delta_1} \tanh(\lambda_1 L)$$

(2.61)

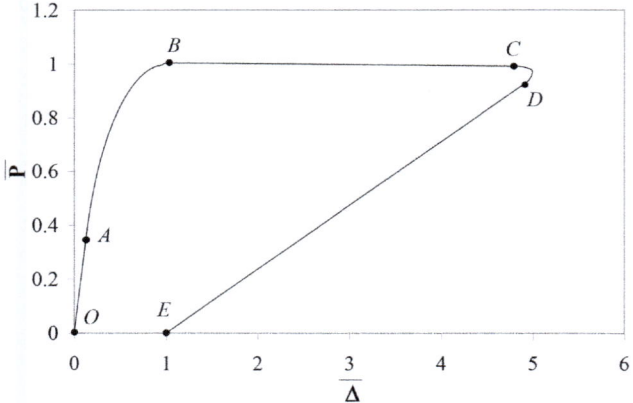

Figure 2.8 Typical full-range theoretical load—displacement curve.

By introducing the following two nondimensional parameters

$$\overline{P} = P / \left(\frac{\tau_f b_p}{\lambda} \right) \quad \text{and} \quad \overline{\Delta} = \frac{\Delta}{\delta_f} \tag{2.62}$$

Eq. (2.61) can be simplified to

$$\overline{P} = \overline{\Delta} \frac{\delta_f \lambda}{\delta_1 \lambda_1} \tanh(\lambda_1 L) \quad \text{for} \quad 0 \le \overline{\Delta} \le \frac{\delta_1}{\delta_f} \tag{2.63}$$

The typical full-range theoretical load—displacement curve for a bonded joint as shown in Fig. 2.5 is presented in Fig. 2.8. Eq. (2.63) depicts the linear load—displacement relationship in the elastic stage of loading, which is shown as segment OA in Fig. 2.8. The elastic stage of deformation of the bonded joint terminates when the shear stress attains the local shear strength τ_f at a slip of δ_1 at $x = L$ (Fig. 2.7B).

Once the shear stress attains τ_f at $x = L$ ($\Delta = \delta_1$), softening commences at the loaded end of the bonded plate. Therefore part of the plate—concrete interface enters the softening state (State II), whereas the rest remains in the elastic state (State I) as shown in Fig. 2.7C. P continues to increase as the length of the softening zone a increases. The ultimate load P_u is first attained at the end of this stage. The interfacial slip, interfacial shear stress, and axial stress in the plate are expressed as Eqs. (2.28)—(2.33). The following load—displacement relationship can be obtained:

$$\overline{P} = \frac{\lambda}{\lambda_2} \left\{ \frac{\lambda_2}{\lambda_1} \tanh[\lambda_1(L-a)]\cos(\lambda_2 a) + \sin(\lambda_2 a) \right\} \tag{2.64}$$

$$\overline{\Delta} = \frac{(\delta_f - \delta_1)}{\delta_f} \left\{ \frac{\lambda_2}{\lambda_1} \tanh[\lambda_1(L-a)]\sin(\lambda_2 a) - \cos(\lambda_2 a) + \frac{\delta_f}{\delta_f - \delta_1} \right\} \tag{2.65}$$

The distribution of the interfacial shear stress during the elastic-softening stage is illustrated in Fig. 2.7C. During this stage, the load—displacement curve plotted from Eqs. (2.64) and (2.65) is shown as segment AB in Fig. 2.8. Apparently, the joint softens during this stage and attains its ultimate load at the end of the stage.

During this stage of loading, debonding (or macrocracking or fracture) commences and propagates along the interface. At the initiation of debonding $\Delta = \delta_f$. By utilizing this condition, the corresponding value of a, denoted by a_d, can be obtained from Eq. (2.65) as

$$\frac{\lambda_2}{\lambda_1}\tanh[\lambda_1(L - a_d)]\sin(\lambda_2 a_d) - \cos(\lambda_2 a_d) = 0 \qquad (2.66)$$

For infinite bond lengths, Eq. (2.66) reduces to

$$a_d = \frac{1}{\lambda_2}\arctan\left(\frac{\lambda_1}{\lambda_2}\right) \qquad (2.67)$$

The interfacial shear stress distribution along the interface at the initiation of debonding is shown in Fig. 2.7D. As debonding propagates, the peak shear stress τ_f moves toward $x = 0$, that is, the unloaded end of the bonded plate. Depending on the location, the plate—concrete interface during debonding propagation is in one of three feasible stress states: elastic state (State I), softening state (State II), and stress-free debonded state (State III) (Fig. 2.7E). Assuming that the debonded length of the interface starting at the loaded end of the plate is d, Eqs. (2.28)—(2.33) continue to be valid when L is replaced by $(L - d)$. Therefore the load—displacement relationship can be expressed as

$$\overline{P} = \frac{\lambda}{\lambda_2}\left\{\frac{\lambda_2}{\lambda_1}\tanh[\lambda_1(L - d - a)]\cos(\lambda_2 a) + \sin(\lambda_2 a)\right\} \qquad (2.68)$$

$$\overline{\Delta} = 1 + \overline{P}\lambda d \qquad (2.69)$$

As the interfacial shear stress at $x = L - d$ is zero, the following equation relating a to d can be obtained

$$\frac{\lambda_2}{\lambda_1}\tanh[\lambda_1(L - d - a)]\sin(\lambda_2 a) - \cos(\lambda_2 a) = 0 \qquad (2.70)$$

Substituting Eq. (2.70) into Eq. (2.68) yields the following simplified form:

$$\overline{P} = \frac{\lambda}{\lambda_2\sin(\lambda_2 a)} \qquad (2.71)$$

This stage is represented by segment BCD of the load—displacement curve shown in Fig. 2.8. Point C corresponds to the deformation state at which the transferable load starts to reduce because the interfacial stress distribution is now truncated by the free end. At the end of this stage (point D), where $L - d = a_u$, the softening—debonding stage begins. The interface shear stress distribution at the end of this stage is shown in Fig. 2.7F.

The softening—debonding stage is governed by Eq. (2.22) with the following boundary conditions:

$$\sigma_p = 0 \quad \text{at} \quad x = 0 \tag{2.72}$$

$$\delta = \delta_f \quad \text{and} \quad \sigma_p = \frac{P}{t_p b_p} \quad \text{at } x = a \tag{2.73}$$

Thus the following solution can be obtained:

$$a = a_u = \frac{\pi}{2\lambda_2} \tag{2.74}$$

$$\delta = \delta_f - \frac{P\delta_f \lambda^2}{b_1 \tau_f \lambda_2} \cos(\lambda_2 x) \quad \text{at} \quad 0 \leq x \leq a_u \tag{2.75}$$

Eq. (2.74) illustrates that the softening zone length remains constant during the softening—debonding stage. During this stage, the maximum interfacial shear stress at $x = 0$ reduces with the load (Fig. 2.7G). The displacement at the loaded end can be obtained by solving Eq. (2.11) for the case of $\delta > \delta_f$, or directly and more simply by displacement superposition along the bonded joint. This yields the following load—displacement relationship:

$$\overline{\Delta} = 1 + \overline{P}\lambda(L - a_u) \tag{2.76}$$

Eq. (2.76) indicates that the displacement reduces linearly with the load as shown by segment DE of the load—displacement curve in Fig. 2.8.

As discussed earlier, the entire load—displacement curve (Fig. 2.8) consists of a number of distinct segments corresponding to distinct stages of loading. Of particular importance are point A at the initiation of interfacial softening and point B at the initiation of debonding (first attainment of the ultimate load). These two points can be obtained with reasonable convenience from an experimental load—displacement curve and can be used to identify the necessary parameters of the interface. Point C (the end of the plateau), point D (at which the peak shear stress reaches the unloaded end of the bonded plate), and point E (at complete debonding failure of the joint) are generally not obtainable from experiment. Therefore the present analysis provides an effective tool for determining the parameters of the bilinear bond—slip model from simple tests on bonded joints as illustrated later.

To determine the interfacial parameters of the bond—slip model for the pull—push joint, the points A (u_1, P_1) and B (u_2, P_2), as defined in Fig. 2.8, are identified on the experimental load—displacement curve. Based on the analytical solution presented earlier and by assuming that the axial stiffness of the plate is significantly smaller than that of the concrete (i.e., $(EA)_1/(EA)_2 \approx 0$), the following expressions can be obtained:

$$\delta_f = u_2, \quad \tau_f = \frac{P_{max}^2}{E_1 t_1 b_1^2 u_2}, \quad \delta_1 = u_1 \text{ or } \delta_1 = \frac{P_1^2}{E_1 t_1 b^2 \tau_f} \tag{2.77}$$

2.2.3 Pull—pull joint

For the case of the pull—pull joint (Fig. 2.4B—D), the theoretical solutions can be obtained as follows.

Without foregoing generality, we assume that $b_2 E_2 t_2 \geq b_1 E_1 t_1$. In the following, only the Model II interfacial bond stress—slip relationship is discussed. $\beta = (b_2 E_2 t_2)/(b_1 E_1 t_1)$ is a nondimensional parameter introduced for convenience. Two cases (case (1) with $b_2 E_2 t_2 = b_1 E_1 t_1$ and case (2) with $b_2 E_2 t_2 > b_1 E_1 t_1$) are distinguished for discussion.

Model α: $\tau - \delta$ relationship with linearly ascending and descending branches

In case (1), softening appears simultaneously at both ends as the load increases. The boundary and continuous conditions can be expressed as

$$\sigma_1 = 0 \quad \text{at} \quad x = 0 \tag{2.78}$$

$$\sigma_1 \text{ is continuous at } x = a \text{ and } x = L - a \tag{2.79}$$

$$\delta = \delta_1 \text{ or } \tau = \tau_f \text{ at } x = a \text{ and } x = L - a \tag{2.80}$$

$$\sigma_1 = \frac{P}{t_1 b_1} \quad \text{at} \quad x = L \tag{2.81}$$

Finally, we obtain the expressions of interfacial slip, shear stress in adhesive layer, and normal stress of FRP sheet as follows:

For $0 \leq x \leq a \ (0 \leq \delta \leq \delta_1)$

$$\delta = -\frac{\lambda_1}{\lambda_2} \delta_1 \tanh\left[\frac{1}{2}\lambda_1(L - 2a)\right] \sin[\lambda_2(x - a)] + (\delta_1 - \delta_f)\cos[\lambda_2(x - a)] + \delta_f \tag{2.82}$$

$$\tau = \tau_f \frac{\lambda_2}{\lambda_1} \tanh\left[\frac{1}{2}\lambda_1(L - 2a)\right] \sin[\lambda_2(x - a)] + \tau_f \cos[\lambda_2(x - a)] \tag{2.83}$$

$$\sigma_1 = -\frac{\tau_f}{\lambda_1 t_1} \tanh\left[\frac{1}{2}\lambda_1(L - 2a)\right] \cos[\lambda_2(x - a)] + \frac{\tau_f}{\lambda_2 t_1} \sin[\lambda_2(x - a)]$$
$$+ \frac{\tau_f}{\lambda_1 t_1} \cdot \frac{P}{\lambda_1 \delta_1 b_2 E_2 t_2} \tag{2.84}$$

For $a \leq x \leq L - a$ $(\delta_1 < \delta \leq \delta_f)$

$$\delta = \delta_1 \frac{\sinh[\lambda_1(L - a - x)] + \sinh[\lambda_1(x - a)]}{\sinh[\lambda_1(L - 2a)]} \tag{2.85}$$

$$\tau = \tau_f \frac{\sinh[\lambda_1(L - a - x)] + \sinh[\lambda_1(x - a)]}{\sinh[\lambda_1(L - 2a)]} \tag{2.86}$$

$$\sigma_1 = \frac{\tau_f}{t_1 \lambda_1} \frac{\cosh[\lambda_1(x - a)] - \cosh[\lambda_1(L - a - x)]}{\sinh[\lambda_1(L - 2a)]} + \frac{\tau_f}{t_1 \lambda_1} \cdot \frac{P}{\lambda_1 \delta_1 b_2 E_2 t_2} \tag{2.87}$$

For $L - a \leq x \leq L$ $(0 \leq \delta \leq \delta_1)$

$$\delta = \frac{\lambda_1}{\lambda_2} \delta_1 \tanh\left[\frac{1}{2}\lambda_1(L - 2a)\right] \sin[\lambda_2(x - L + a)] + (\delta_1 - \delta_f)\cos[\lambda_2(x - L + a)] + \delta_f \tag{2.88}$$

$$\tau = -\tau_f \frac{\lambda_2}{\lambda_1} \tanh\left[\frac{1}{2}\lambda_1(L - 2a)\right] \sin[\lambda_2(x - L + a)] + \tau_f \cos[\lambda_2(x - L + a)] \tag{2.89}$$

$$\sigma_1 = \frac{\tau_f}{\lambda_1 t_1} \tanh\left[\frac{1}{2}\lambda_1(L - 2a)\right] \cos[\lambda_2(x - L + a)] + \frac{\tau_f}{\lambda_2 t_1} \sin[\lambda_2(x - L + a)]$$
$$+ \frac{\tau_f}{\lambda_1 t_1} \cdot \frac{P}{\lambda_1 \delta_1 b_2 E_2 t_2} \tag{2.90}$$

and

$$P = \frac{\delta_1 \lambda_1^2 b_1 E_1 t_1}{\lambda_2} \times \left\{ \frac{\lambda_2}{\lambda_1} \tanh\left[\frac{1}{2}\lambda_1(L - 2a)\right] \cos(\lambda_2 a) + \sin(\lambda_2 a) \right\} \tag{2.91}$$

It is apparent that P attains its maximum value when $dP/da = 0$. Therefore the a at the maximum load can be determined from the relation

$$\tanh\left[\frac{1}{2}\lambda_1(L - 2a)\right] = \frac{\lambda_2}{\lambda_1} \tan(\lambda_2 a) \tag{2.92}$$

Substituting Eq. (2.92) into Eq. (2.91), we obtain

$$P_{\max} = \delta_f \lambda_2 b_1 E_1 t_1 \sin(\lambda_2 a) \tag{2.93}$$

The problem is that a is defined as an implicit function and can be determined only by iteration. However, it can be illustrated that for large values of L, Eq. (2.93) converges to

$$P_{\max} = \delta_f \lambda b_1 E_1 t_1 \tag{2.94}$$

Crack propagation occurs simultaneously at both ends when $\delta > \delta_f$. This can be studied by decreasing the bond length L. For large values of L, by letting $\delta = \delta_f$ at the left end in Eq. (2.82) or at the right end in Eq. (2.88), we obtain the maximum value of a that is associated with the initiation of macrocrack (debonding):

$$a_{\max} = \frac{1}{\lambda_2} \arctan\left(\frac{\lambda_1}{\lambda_2}\right) \tag{2.95}$$

In case (2), softening appears first at the right end as the load increases. The boundary and continuous conditions can be expressed as

$$\sigma_1 = 0 \quad \text{at} \quad x = 0 \tag{2.96}$$

$$\sigma_1 \text{ is continuous at } x = L - a \tag{2.97}$$

$$\delta = \delta_1 \text{ or } \tau = \tau_f \text{ at } x = L - a \tag{2.98}$$

$$\sigma_1 = \frac{P}{t_1 b_1} \quad \text{at} \quad x = L \tag{2.99}$$

Finally, we obtain the following for $0 \le \delta \le \delta_1$, that is, $0 \le x \le L - a$:

$$\delta = \left\{ \frac{\delta_1}{\cosh[\lambda_1(L-a)]} + \frac{P}{\lambda_1 b_2 E_2 t_2} \tanh[\lambda_1(L-a)] \right\} \cosh(\lambda_1 x) - \frac{P}{\lambda_1 b_2 E_2 t_2} \sinh(\lambda_1 x) \tag{2.100}$$

$$\tau = \left\{ \frac{\tau_f}{\cosh[\lambda_1(L-a)]} + \frac{P\tau_f}{\lambda_1 \delta_1 b_2 E_2 t_2} \tanh[\lambda_1(L-a)] \right\} \cosh(\lambda_1 x) - \frac{P\tau_f}{\lambda_1 \delta_1 b_2 E_2 t_2} \sinh(\lambda_1 x) \tag{2.101}$$

$$\sigma_1 = \frac{\tau_f}{t_1 \lambda_1} \cdot \left\{ \frac{1}{\cosh[\lambda_1(L-a)]} + \frac{P}{\lambda_1 \delta_1 b_2 E_2 t_2} \tanh[\lambda_1(L-a)] \right\} \sinh(\lambda_1 x)$$
$$- \frac{\tau_f}{t_1 \lambda_1} \cdot \frac{P}{\lambda_1 \delta_1 b_2 E_2 t_2} \cosh(\lambda_1 x) + \frac{\tau_f}{t_1 \lambda_1} \cdot \frac{P}{\lambda_1 \delta_1 b_2 E_2 t_2} \tag{2.102}$$

and the following for $\delta_1 < \delta \leq \delta_f$, i.e., $L - a \leq x \leq L$:

$$\delta = \left\{ \frac{\lambda_1}{\lambda_2} \delta_1 \tanh[\lambda_1(L-a)] - \frac{P}{\lambda_1 b_2 E_2 t_2 \cosh[\lambda_1(L-a)]} \right\}$$
$$\times \sin[\lambda_2(x-L+a)] + (\delta_1 - \delta_f)\cos[\lambda_2(x-L+a)] + \delta_f \tag{2.103}$$

$$\tau = \cdot \frac{\tau_f}{(\delta_f - \delta_1)} \left\{ \frac{\lambda_1}{\lambda_2} \delta_1 \tanh[\lambda_1(L-a)] - \frac{P}{\lambda_2 b_2 E_2 t_2 \cosh[\lambda_1(L-a)]} \right\}$$
$$\times \sin[\lambda_2(x-L+a)] + \tau_f \cos[\lambda_2(x-L+a)] \tag{2.104}$$

$$\sigma_1 = \frac{\tau_f}{\lambda_2 t_1} \left\{ \frac{\lambda_2}{\lambda_1} \tanh[\lambda_1(L-a)] - \frac{P}{\lambda_2(\delta_f - \delta_1)b_2 E_2 t_2 \cosh[\lambda_1(L-a)]} \right\}$$
$$\times \cos[\lambda_2(x-L+a)] + \frac{\tau_f}{\lambda_2 t_1} \sin[\lambda_2(x-L+a)] + \frac{\tau_f}{\lambda_2 t_1} \cdot \frac{P}{\lambda_2(\delta_f - \delta_1)b_2 E_2 t_2}$$

$$\tag{2.105}$$

Substituting Eq. (2.99) into Eq. (2.105), we obtain

$$P = \left\{ \frac{\lambda_2}{\lambda_1} \tanh[\lambda_1(L-a)]\cos(\lambda_2 a) + \sin(\lambda_2 a) \right\}$$
$$\times \frac{\tau_f b_1}{\lambda_2} \Bigg/ \left\{ 1 - \frac{\tau_f b_1}{\lambda_1^2 \delta_1 b_2 E_2 t_2} \left[1 - \frac{\cos(\lambda_2 a)}{\cosh(\lambda_1 L - \lambda_1 a)} \right] \right\}$$
$$= \left\{ \frac{\lambda_2}{\lambda_1} \tanh[\lambda_1(L-a)]\cos(\lambda_2 a) + \sin(\lambda_2 a) \right\} \Bigg/$$
$$\left\{ \frac{\lambda_2}{\lambda_1^2 \delta_1 b_1 E_1 t_1} \left[1 + \frac{\cos(\lambda_2 a)}{\beta \cosh(\lambda_1 L - \lambda_1 a)} \right] \right\} \tag{2.106}$$

It is apparent that P in the above equation is larger than that in Eq. (2.35). P attains its maximum value when $dP/da = 0$. For large values of L, the expression of a at the maximum load can be simplified to

$$\tan(\lambda_2 a) = \frac{\lambda_1}{\lambda_2} \tag{2.107}$$

Substituting Eq. (2.107) into (2.106), we obtain the following for large values of L:

$$P_{\max} = \lambda \delta_f b_1 E_1 t_1 = b_1 E_1 t_1 \sqrt{2G_f \left(\frac{1}{E_1 t_1} + \frac{b_1}{b_2 E_2 t_2} \right)} \qquad (2.108)$$

which is identical to Eq. (2.94). For simplicity, we discuss only the case of large values of L, in the following. For the pull—pull joint, Eq. (2.108) can be expressed also as

$$P_{\max} = b_1 \sqrt{2G_f E_1 t_1 \left(1 + \frac{1}{\beta} \right)} \qquad (2.109)$$

P_{max} is observed to be independent of τ_f and δ_f. A similar conclusion can be obtained for the pull—push joint. According to Eqs. (2.20), (2.37), and (2.46), we obtain

$$P_{\max} = b_1 \sqrt{2G_f E_1 t_1 / \left(1 + \frac{1}{\beta} \right)} \qquad (2.110)$$

By comparing Eq. (2.109) with Eq. (2.110), it can be concluded that the P_{max} in a pull—pull joint is twice as that in a pull—push joint when $\beta = 1$. Furthermore, the values of P_{max} in pull—pull and pull—push joints tend to approach each other as β increases.

2.3 Short-term behavior of FRP—concrete interface

2.3.1 Experimental methods and observations

To evaluate the properties of bonds between FRP laminate and concrete, various test methods (shown in Fig. 2.9) have been developed (Wu et al., 2001b). Fig. 2.9A illustrates a direct tensile test specimen. Fig. 2.9B illustrates a bonded pull—push joint that is generally encountered in FRP/steel-strengthened beams. In the single-lap test, the concrete plate is mounted securely to the bottom crosshead of the testing machine. The top of the FRP laminates is clamped in a serrated grip that is free to rotate in all directions. A tensile load is directly applied on the grip. Fig. 2.9C and D depicts the double-lap shear test and inserted shear test, respectively, of FRP laminate-strengthened concrete structures. Fig. 2.9C has been selected by JSCE as a standard test method to assess the bonding and debonding behavior of FRP laminate-strengthened structures. In the double-lap shear test, an FRP laminate is bonded to both sides of a concrete block along the axial direction. A tensile load is applied by pulling both ends of the steel bars embedded in the concrete block on

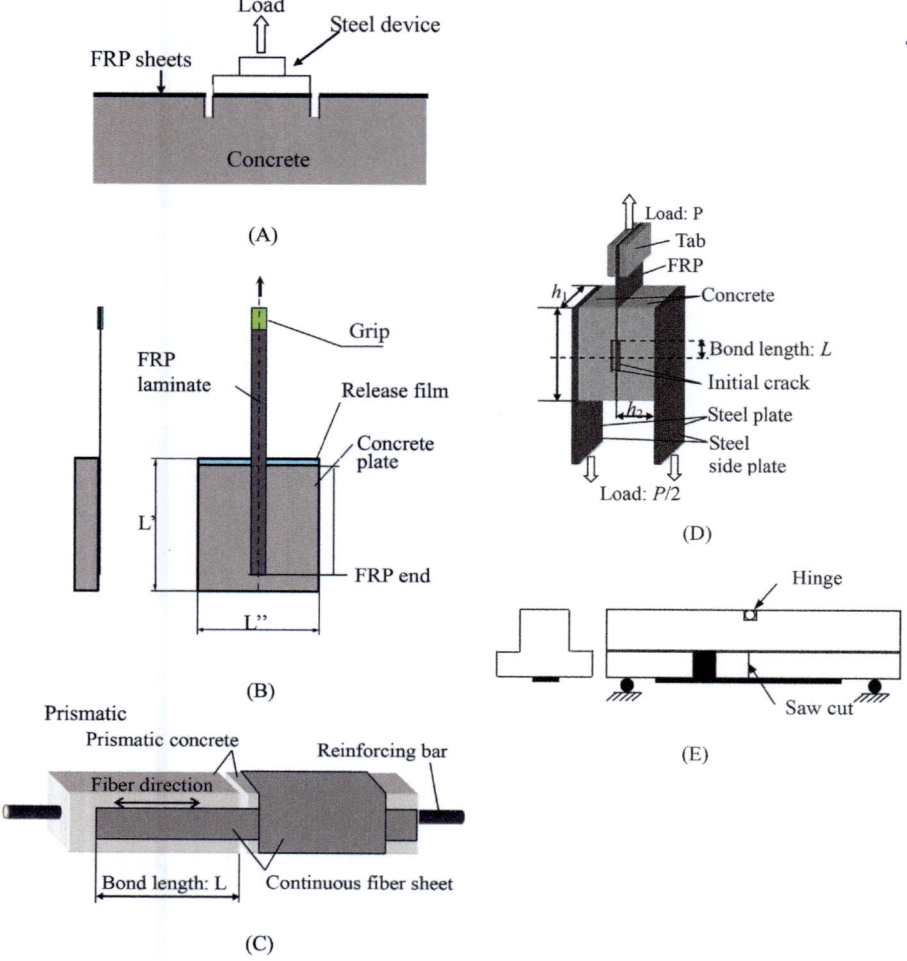

Figure 2.9 Experimental methods (Wu et al., 2001b). (A) Direct tensile test. (B) Single-lap shear test. (C) Double-lap shear test. (D) Inserted shear test. (E) Modified beam test specimen.

which the FRP sheets are bonded. The steel bar is cut off at the position of the notch (preexisting crack). The specimen used for the modified beam test is a plain concrete beam with an inverted-T shape, as shown in Fig. 2.9C. The beam is simply supported. A steel hinge at the top and a saw cut at the bottom, both located at mid-span, are used to control the distribution of the internal forces. During loading, the saw cut causes a crack to develop at the center of the beam and extend up to the hinge. Therefore the compressive force in the beam at midspan is located at the center of the hinge. Moreover, the internal moment arm was known and constant for any specified load level above the stabilized cracking load. This enables an accurate computation of the tensile stress in the FRP.

A series of specimens with adhesively bonded joints both in single-lap and double-lap shear were tested by Wu et al. (2001b) to study the interfacial behavior of FRP—concrete interface. The main test parameter was the type of FRP laminates, that is, high-strength type and high Young's modulus type of carbon fiber sheets, 1- and 2-mm thick carbon fiber plates, and aramid and PBO fiber sheets. Details of the test program and constitutive material properties are available elsewhere (Wu et al., 2001b).

The load—shear displacement relation at the loaded end (as shown in Fig. 2.10) can provide global information on the bond performance and may be measured directly with the aid of a load sensor and displacement gauge. It can be obtained also by measuring the FRP strain distribution for whole loading steps, where the shear displacement can be calculated by integrating the strain distribution along the FRP laminate. Fig. 2.10 shows certain special points such as the occurrence of micro debonding (A), of macro debonding (B), and of final debonding failure (C). Theoretically, there is neither increase nor decrease in loading during the debonding propagation from (B) to (C). However, both case 1 and case 2 (as shown in Fig. 2.10) are observed in the experiment owing to the heterogeneities along the FRP—concrete interface.

Fig. 2.11 shows the strain developed in the bonded FRP laminate versus the distance from free end, for specimens with different FRP stiffness ($E_f t_f$). In both the specimens, debonding fracture was caused by shearing of the concrete close to the bond surface (normal debonding failure). The stress transfer length for the specimen with high FRP stiffness (Fig. 2.11B) is longer than that for the specimen with low FRP stiffness (Fig. 2.11A). Beyond the debonding load, the stress transfer region shifts gradually toward the end of the bonded FRP laminate owing to debonding propagation. The strain distribution becomes horizontal within the range of the debonding zone. This implies that less stress is being transferred to the concrete. The strain distribution of the FRP laminate after interfacial crack indicates a

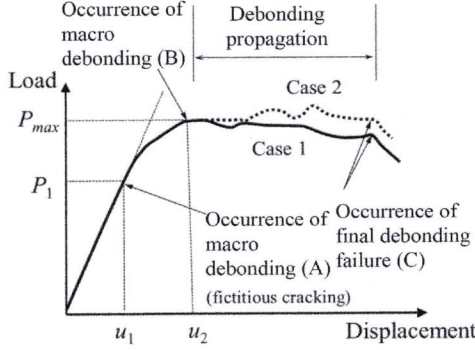

Figure 2.10 Typical load—displacement relations (Wu et al., 2001b).

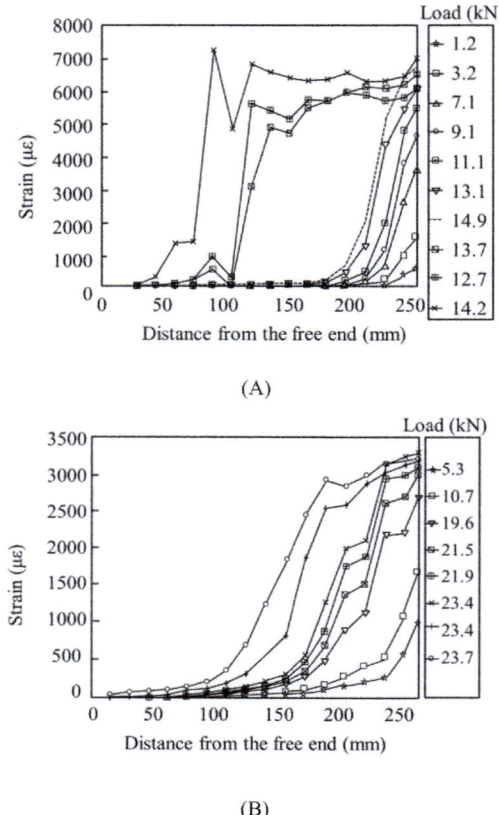

(A)

(B)

Figure 2.11 Examples of strain distribution in FRP laminates (Wu et al., 2001b). (A) E_f $t_f = 51.1$ kN/mm. (B) $E_f t_f = 193.1$ kN/mm.

progressive debonding fracture. Therefore an effective shear stress transfer length exists, as established by previous analytical study.

The mean shear stress between two consecutive gauge positions can be calculated from the measured strain profiles of FRP laminates along the bond line. The local shear stress distribution can be determined when the intervals of two consecutive gauge positions are close enough. The mean shear stress τ_i between two consecutive gauge positions i and $i + 1$ can be expressed based on two strain readings $\varepsilon_{f,i}$ and $\varepsilon_{f,i+1}$, as follows:

$$\tau_i = \frac{E_f t_f (\varepsilon_{f,i+1} - \varepsilon_{f,i})}{\Delta x_i} \tag{2.111}$$

where Δx_i is the distance between gauges i and $i + 1$. Proceeding thus for all the gauge positions, one obtains the trend of the shear stress distribution along the

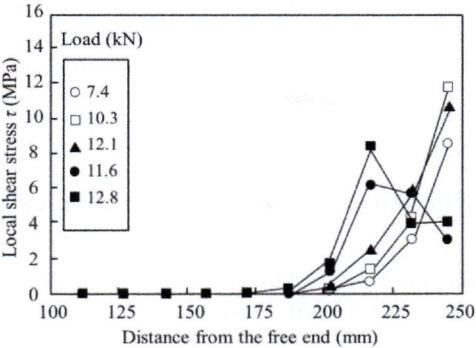

Figure 2.12 Experimental results on local shear stress distribution (Wu et al., 2001b).

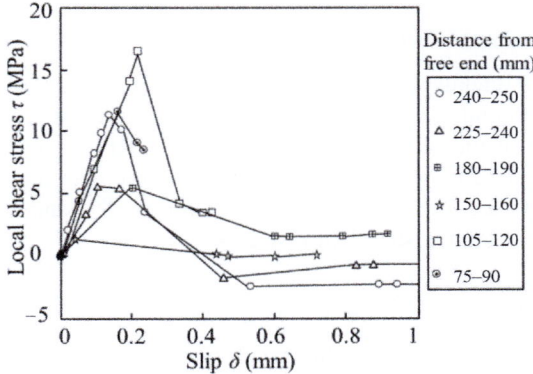

Figure 2.13 Local $\tau-\delta$ relationship (Wu et al., 2001b).

FRP—concrete interface. Fig. 2.12 shows the typical shear stress distribution along the bonded laminate at several load levels. A softening zone is observed to appear at the loading end for large values of load.

For all the measurements of FRP strain distribution, the shear stress can be calculated by Eq. (2.111). The corresponding shear displacement (slip, δ) can be calculated by integrating the strains along the FRP laminate, as follows:

$$s_i = \sum \Delta x_i (\varepsilon_{f,i+1} - \varepsilon_{f,i}) \tag{2.112}$$

Therefore the local $\tau-\delta$ curves for all the positions (consecutive gauge positions) of shear stress transfer including all the interfacial fracturing processes under whole loading can be identified. Moreover, the local $\tau-\delta$ curve can be identified from the variation in shear stress with locations along the FRP—concrete interface at a specified loading step (Fig. 2.12). Fig. 2.13 shows the curves of local shear stress versus interfacial slip at different locations. The identified curves, particularly, the local maximum shear stresses for different locations are largely different owing to the heterogeneities of debonding fracture at different locations. Fig. 2.13 shows that the $\tau-\delta$ curve exhibits both linearly ascending and descending (gradual softening) branches.

2.3.2   Factors influencing FRP–concrete interface

Parameters affecting the interfacial stress transfer and ultimate behavior of FRP–concrete joint include FRP thickness (t_f), width of FRP composites (b_f), width of concrete prism (b_c), modulus of elasticity of FRP laminates (E_f), concrete strength (f_c'), and effective bond length of FRP laminates (L_e). In the following paragraphs, the main parameters influencing the interfacial behavior of FRP–concrete interface are discussed based on an extensive database consisting of available experimental datasets (Sayed, 2014).

2.3.2.1   Influence of FRP thickness

The FRP composite thickness is an important factor that directly affects the strength and stiffness of the bond. According to the experimental results, the FRP thickness causes an increase in the ultimate bond strength of FRP sheets by $t_f^{0.57}$ and $t_f^{0.27}$ when the effective bond length (L_e) is less than and more than the total bond length, respectively. In the case of FRP plates, this increase is $t_f^{0.41}$ and $t_f^{0.32}$, respectively, under a similar condition (as shown in Fig. 2.14).

2.3.2.2   Influence of FRP elastic modulus

The elastic modulus of the FRP laminates (E_f) plays a major role for bond strength, as shown in Fig. 2.15. The bond strength depends on the FRP rigidity $t_f \times E_f$ in all the available models. According to the experimental results, at a certain thickness, the elastic modulus of the FRP composites causes an increase in the ultimate load

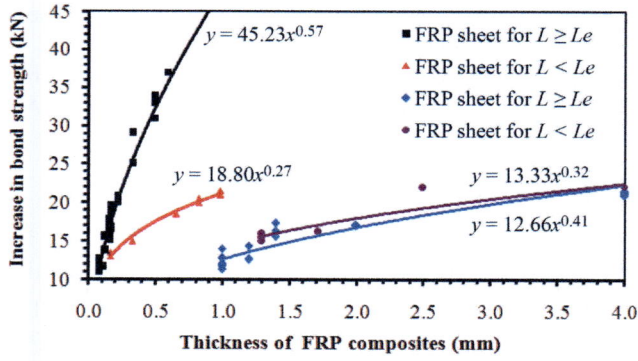

Figure 2.14 Influence of the FRP thickness based on the experimental results (Sayed, 2014).

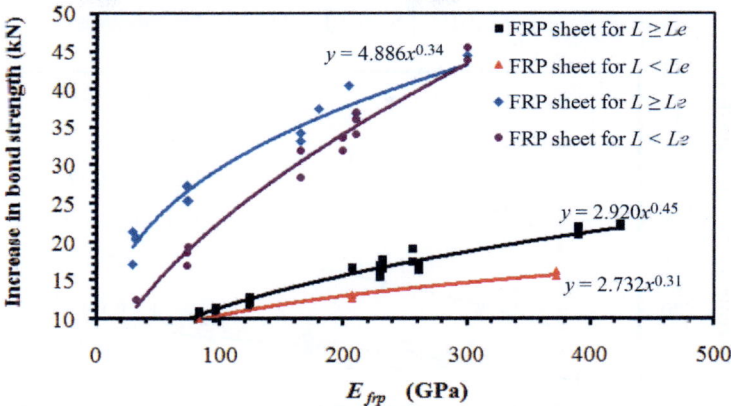

Figure 2.15 Influence of the elastic modulus of FRP laminates based on the experimental results (Sayed, 2014).

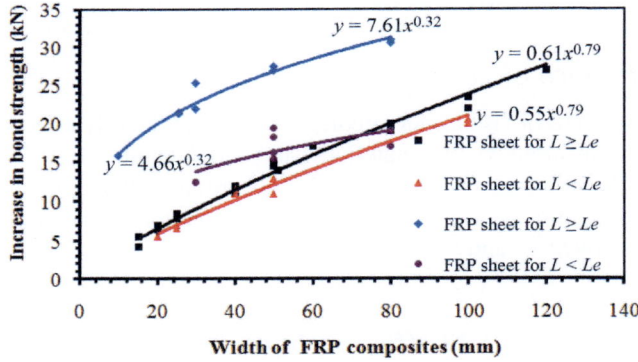

Figure 2.16 Influence of the width of the FRP laminates based on the experimental results (Sayed, 2014).

bond strength of the FRP sheets by $E_f^{0.45}$ and $E_f^{0.31}$ for effective bond lengths less than and more than the total bond length, respectively. In the case of FRP plates, this increase is $E_f^{0.34}$ and $E_f^{0.59}$, respectively, as shown in Fig. 2.15.

2.3.2.3 Influence of width of FRP laminates

The width of the FRP composites plays an important role, and each model considers this influence. The bond strength depends on the width of the plate as a function of b_f and is linear in all the available models. According to the experimental results, the width of the FRP composites causes an increase in the ultimate load bond strength for FRP sheets and plates by $b_f^{0.79}$ and $b_f^{0.32}$, respectively, as shown in Fig. 2.16.

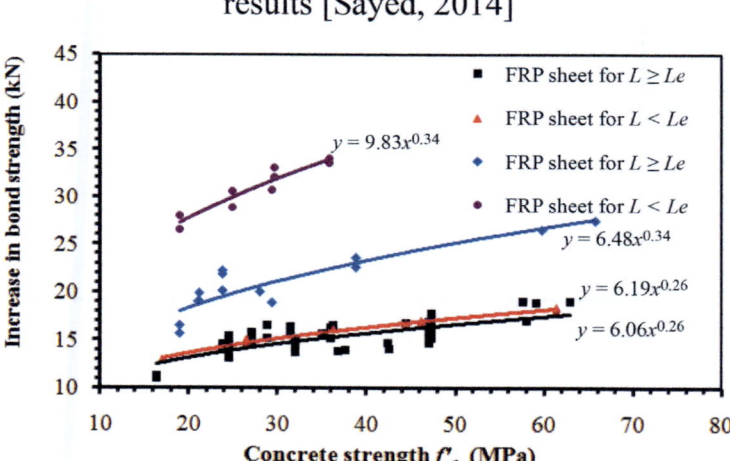

results [Sayed, 2014]

Figure 2.17 Influence of concrete strength based on the experimental results (Sayed, 2014).

2.3.2.4   Influence of concrete strength

The concrete strength (f_c') is also an important factor that directly affects the strength and stiffness of the strengthening material. The bond strength is as a function of the concrete strength for all the available models. According to the experimental results, the concrete strength causes an increase in the ultimate load bond strength for FRP sheets and plates by $f_c'^{0.26}$ and $f_c'^{0.34}$, respectively, as shown in Fig. 2.17.

2.3.2.5   Influence of FRP effective bond length

Various studies have indicated the presence of an effective bond length (L_e) of FRP laminates: an increase in the bond length beyond this does not result in further increase in the anchorage load. According to the experimental results, the effective bond length causes an increase in the ultimate load bond strength of FRP sheets and plates by $(L/L_e)^{0.28}$ and $(L/L_e)^{0.77}$ for effective bond length more than the total bond length, respectively, as shown in Fig. 2.18.

2.3.2.6   Influence of width of concrete prism

The width of the concrete prism influences the bond. The width ratio of bonded FRP to concrete member (b_f/b_c) has significantly affected the ultimate bond strength. If the width of the bonded sheet is smaller than that of the concrete member, the force transfer from the FRP to the concrete results in a nonuniform stress distribution across the width of the concrete member. When the width of the concrete prism increases, the bond strength also increases.

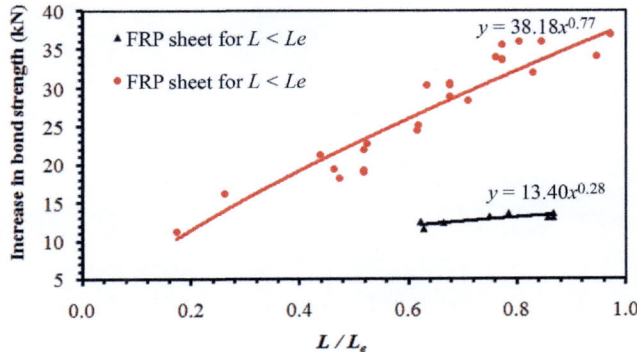

Figure 2.18 Influence of the effective bond length of FRP composites based on the experimental results (Sayed, 2014).

2.3.3 Numerical study of bond behavior of FRP—concrete interface

2.3.3.1 Experimental observation

Unlike the hooked and deformed steel bars in conventional reinforcement concrete, the force transfer between FRP and concrete is achieved by the continuous bond through adhesive resin rather than by the mechanical interlocking between the lugs and the concrete matrix. Owing to the deformation difference between FRP and concrete, the shear stress within the adhesive resin increases and provides stress transfer from FRP to concrete. The debonding failure along FRP—concrete bond interface is generally modeled as Mode II fracture with certain simplified assumptions. This is because the interfacial fractures observed at a macro level in experiments are mostly of shear fracture. However, a closer observation of the fractured interface reveals that concrete of thickness from 2 to 10 mm sticks on the delaminated FRP sheets when debonding failure occurs.

The images in Fig. 2.19 present a few experimental cases of surface crack formed owing to debonding failure. Fig. 2.19A shows delaminated FRP sheets from the bond interface in an FRP-strengthened concrete beam. A close observation of the fractured interface in Fig. 2.19B reveals that it is coarse with part of the concrete broken off the beam body and stuck on the FRP sheets. This is unlike what we would anticipate of a Mode II fracture with a smooth cracked surface. A similar surface crack pattern was observed in a single-lap test by Yoshizawa et al. (2000), as shown in Fig. 2.19C. In addition, finite element simulation of FRP-strengthened concrete beams yielded similar debonding behaviors, wherein a layer of cracked concrete elements in the form of Mode I fracture finally resulted in interfacial debonding failure (Wu and Yin, 2003). Experimental and simulation results reveal that interfacial debonding generally occurs in substrate concrete when the quality of bonding between FRP laminates and concrete is guaranteed.

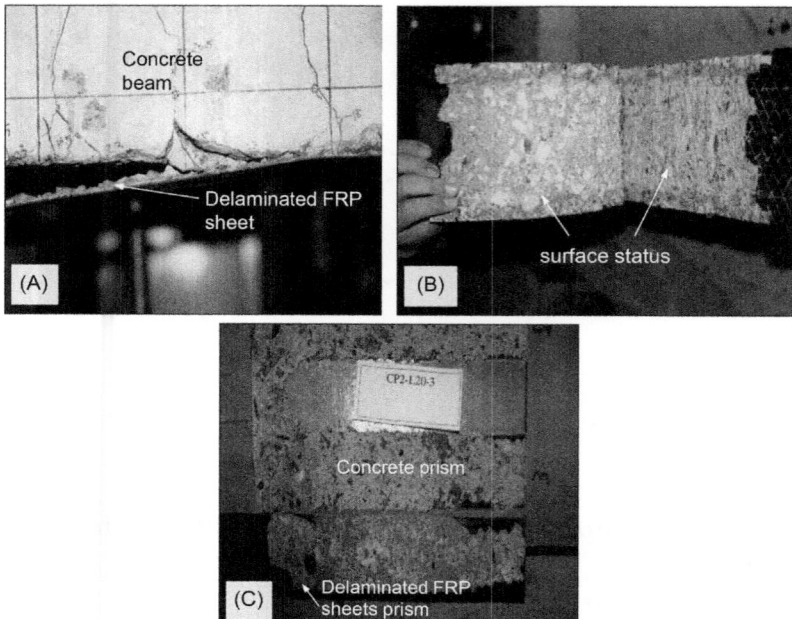

Figure 2.19 Experimental observation of debonding failure in concrete (Yoshizawa et al., 2000).

The stress transfer at the interface heightens the stress level of uncracked concrete adjacent to the bond interface. Once the concrete stress attains its tensile strength, the interfacial concrete starts to crack. In this case, the crack orientation is generally inclined to the bond surface at 45 degrees. With further progress of interfacial stress transfer, more concrete develops cracks. When these concrete cracks join together, macro interfacial debonding occurs. Unlike the Mode II fracture within adhesive resin, crack in concrete is considered to initiate as Mode I fracture, with subsequent aggregate interlocking behavior along the crack surface (representing mixed mode fracture), as shown in Fig. 2.20. Therefore shear-dominated FRP−concrete interfacial debonding could be described by either the macroscopic Mode II fracture that is assumed to occur along the interface bond layer, or the microscopic fracture in concrete adjacent to bond surface that is initiated as Mode I fracture with subsequent transformation to mixed mode fracture.

2.3.3.2 Mode II fracture model (macrorepresentation)

If the interfacial debonding is considered as a Mode II fracture, the constitutive model of the bond interface could be expressed as a relationship of local shear stress, τ, versus relative shear displacement, δ_t, between FRP sheets and concrete matrix, as shown in Fig. 2.21. Here, f_b is the local bond strength. In general,

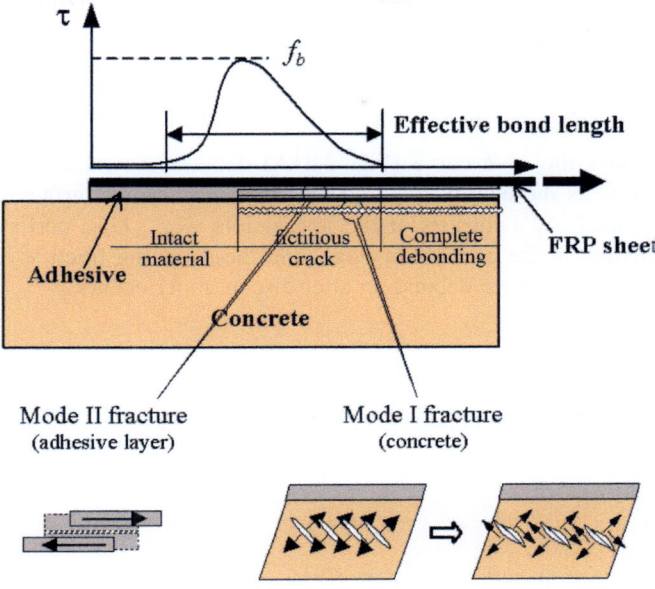

Figure 2.20 Fracture mechanisms of interfacial debonding (Wu and Yin, 2003).

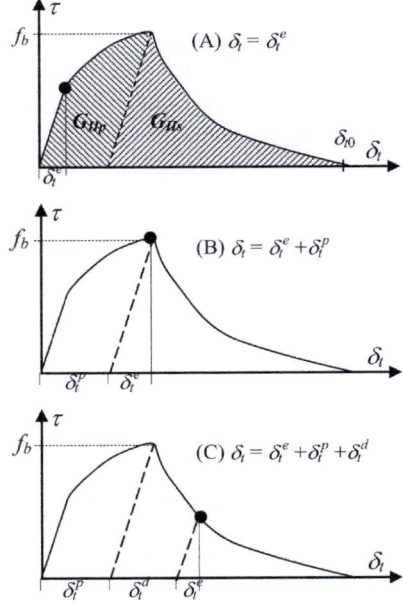

Figure 2.21 A general expression of t–d_t relationship (Niu, 2002).

δ_t consists of an elastic part δ_t^e, a plastic part δ_t^p, and a discontinuous part δ_t^d, yielding the following equation:

$$\delta_t = \delta_t^e + \delta_t^p + \delta_t^d \tag{2.113}$$

The area under the $\tau-\delta$ curve (Fig. 2.21A) denotes the total fracture energy, G_f^{II}. It is defined as the energy required to bring a unit bond surface to complete fracture. In general, G_f^{II}, consists of the irreversible work, G_{IIp}, performed by the plastic deformation in the continuum stage, and the softening fracture energy, G_{IIs}, consumed to cause complete debonding after discontinuity occurs.

$$G_f^{II} = G_{IIp} + G_{IIs} \tag{2.114}$$

The interfacial fracture energy can be computed also by

$$G_f^{II} = \int_0^{\delta_{t0}} \tau \, d\delta_t \tag{2.115}$$

The debonding propagation along the FRP−concrete interface is considered as a fictitious crack. When the bond zone enters nonlinearity, "micro" cracks occur. Nonetheless, it is still capable of transferring forces. Then, when the shear stress decreases to zero, the FRP sheets delaminate from the concrete matrix.

A concrete prism strengthened with FRP sheets is simulated for discussing the shear bond behavior. The finite element analysis is based on the experimental works by Wu et al. (2001b). A schematic sketch is presented in Fig. 2.22. Here, L is the bond length of FRP sheets, $b_c = b_f = 40$ mm is the width of the bonded FRP sheets and concrete prism, $t_c = 60$ mm, and $t_f = 0.22$ mm are the thicknesses of the concrete prism and FRP sheets, respectively.

As stated previously, the $\tau-\delta_t$ relationship can be modeled in various forms. Most of these are composed of the loading stage with increasing stress and the subsequent softening stage, where the shear stress decreases gradually to zero. One is composed of the elastic loading and the subsequent abrupt decrease in shear stress. Six cases are studied, as presented in Fig. 2.23.

The material properties are defined as follows: Young's modulus of concrete $E_c = 3.2 \times 10^4$ MPa, Poisson ratio $\gamma_1 = 0.2$, total interfacial fracture energy $G_f^{II} = 1.2$ N/mm, bond strength $f_b = 8.0$ MPa, $E_{FRP} = 2.3 \times 10^5$ MPa. Except the

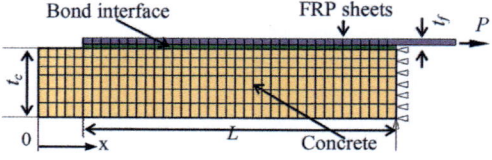

Figure 2.22 Shear bond test of FRP-bonded concrete (Niu, 2002).

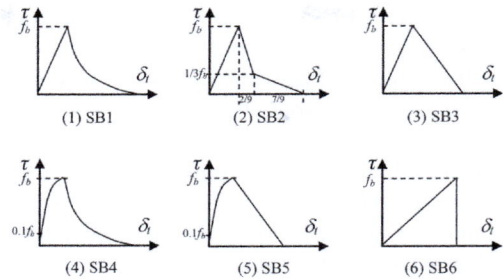

Figure 2.23 Six cases with different of $\tau-\delta_t$ relationships (Niu, 2002).

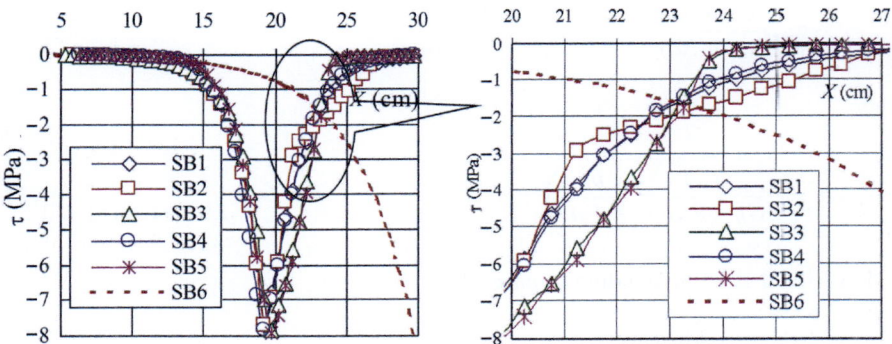

Figure 2.24 Shear stress distribution along bond interface (Niu, 2002).

limit case SB6, all the $\tau-\delta_t$ relationships follow an assumption that the area under the $\tau-\delta_t$ curves before the maximum bond strength is set as 1/6 of G_f^{II}. The initial linear stiffness in the cases SB4 and SB5 is set as five times that of SB1—SB3. Moreover, the linear part terminates when the shear stress attains 1/10 of the bond strength. According to the discussion on fracture energy dissipation in the previous section, G_f^{II} is considered as the sum of G_{IIp} and G_{IIs} in SB4 and SB5 with a linear-—exponential loading curve. However, in SB1—SB3, it is assumed that G_f^{II} is consumed completely in the formation of discontinuous crack with negligible plastic deformation. Therefore only a linear curve is used before the bond strength is attained.

In the finite element simulation, the concrete prism is discretized by four-node plane stress elements. An 8×8-mm plane stress element is used near the bond interface. The occurrence of crack in concrete is simulated by the rotating smeared crack model. The FRP sheets and bond interface are discretized by truss elements and line-to-line interface elements (DIANA), respectively. Here, the FRP sheets are linear elastic until rupture, whereas the debonding behavior of bond interface is modeled by the proposed $\tau-\delta_t$ relationships.

The shear stress distributions, based on the $\tau-\delta_t$ relationships, along the bond interface are shown in Fig. 2.24. Before shear stress attains the bond strength, the results of SB1–SB5 do not display large differences. However, differences are apparent in the softening stage. SB3 and SB5 with linear softening curve display a rapid decrease in the shear stress. Meanwhile, the cases with bilinear and exponential softening curves first display a rapid shear stress degradation. This is followed by a gradual decrease to zero.

In addition, theoretical solutions for two typical models (Model I and Model II presented in Section 2.3) according to the $\tau-\delta_t$ relationships of SB6 and SB3 following Eqs. (2.116) and (2.118), respectively, are compared to the finite element results in Fig. 2.25. The bending effect is omitted. From Fig. 2.25, the finite element results accurately match the theoretical solutions.

For Model I according to case SB6,

$$\tau = \tau_{\max}\frac{\cosh(\lambda x)}{\cosh(\lambda L)} \tag{2.116}$$

where

$$\lambda = \sqrt{\frac{\tau_{\max}^2}{2G_{II}}\left(\frac{1}{E_f t_f} + \frac{b_f}{b_c E_c t_c}\right)} \tag{2.117}$$

For Model II corresponding to case SB3

$$\tau = \tau_{\max}\frac{\cosh(\lambda x)}{\cosh[\lambda(L-a)]} \quad (0<\delta_t<\delta_1) \tag{2.118}$$

$$\tau = -\tau_{\max}\left\{\begin{array}{l}\dfrac{\lambda_1}{\lambda_2}\tanh[\lambda_1(L-a)]\sin[\lambda_2(x-L+a)] \\[2mm] + \cos[\lambda_2(x-L+a)]\end{array}\right\} \quad (\delta_1<\delta_t<\delta_f) \tag{2.119}$$

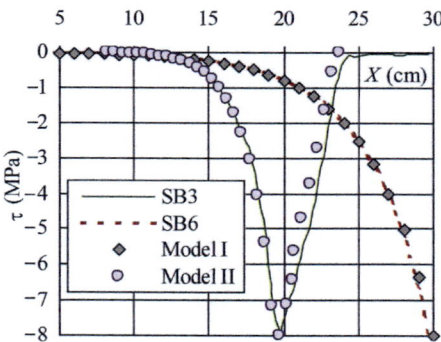

Figure 2.25 Comparison with theoretical results (Niu, 2002).

where

$$\lambda_1 = \sqrt{\frac{\tau_{\max}}{\delta_1}\left(\frac{1}{E_f t_f} + \frac{b_f}{b_c E_c t_c}\right)}$$

(2.120)

$$\lambda_2 = \sqrt{\frac{\tau_{\max}}{\delta_f - \delta_1}\left(\frac{1}{E_{FRP} t_{FRP}} + \frac{b_{FRP}}{b_c E_c t_c}\right)}$$

(2.121)

$$a = \frac{1}{\lambda_2}\arctan\left(\frac{\lambda_1}{\lambda_2}\right)$$

(2.122)

Fig. 2.26 shows the relationship between the load and the pull-out displacement. At the beginning of the loading process, the load–displacement curves of SB1–SB5 are almost identical and match the experimental results. The difference appears only when the load tends to approach the maximum load. Whereas the model with linear softening curve displays a rapid increase to P_{max}, those with bilinear and exponential softening curves display a gradual increase to P_{max}. It is observed that the value of P_{max} from experiments (Yoshizawa et al., 2000) is less than that from FE results. This may be because of the instability near the stage of complete debonding in experiments.

Moreover, Fig. 2.27 shows the FE results of strain distribution in FRP sheets. It is observed that compared to the experiment, the linear softening curve displays better agreement than the other softening curves. Considering the FE results (i.e., the linear curve and linear–exponential curve do not cause apparent difference during the loading stage), a $\tau - \delta_t$ relationship with linear ascending–descending curve (as presented in SB3) is regarded as an applicable model to capture interfacial debonding behavior with adequate accuracy. Assuming that the area under $\tau - \delta_t$ relationships before maximum bond strength is set as 1/6 of G_f^{II}, the initial bond interface stiffness is approximately equal to $K_b = f_b/\gamma_1 = 1.6 \times 10^2$ N/mm^3.

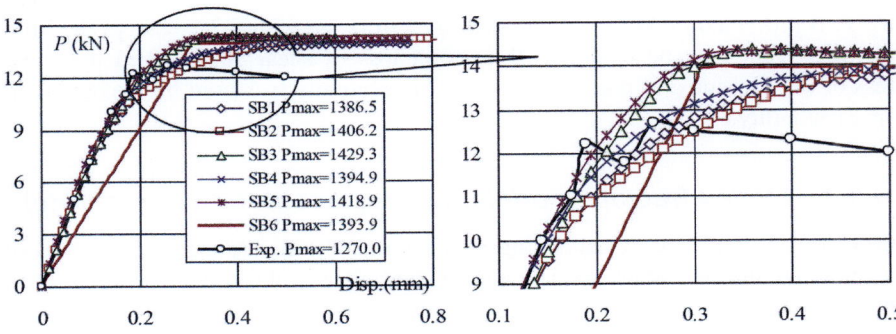

Figure 2.26 Load–displacement curves (Niu, 2002).

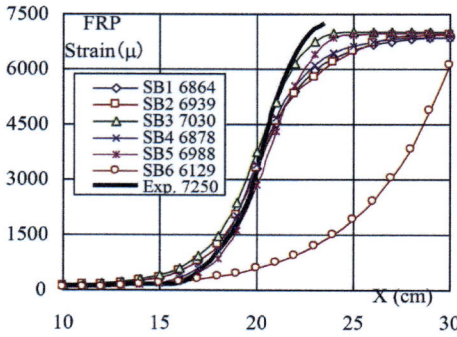

Figure 2.27 FRP strain distribution (Niu, 2002).

Although the initial bond interface stiffness, to an extent, depends on the bonding quality of FRP sheets and the properties of concrete surface, it could be considered as a reference parameter.

From an overall perspective of bond behavior, the selection of different $\tau-\delta_t$ relationships (including the limit case of SB6) does not affect the P_{max} and ε_{max} of FRP sheets. It can be explained that G_f^{II} is an important parameter for determining the stress transfer capability of FRP–concrete bond interface, and P_{max}. This result has been obtained also through the following theoretical solution (Taljsten, 1996; Yuan et al., 2001):

$$P_{\max} = b_f \sqrt{\frac{2G_{II}}{\frac{1}{E_f t_f} + \frac{b_f}{b_c E_c t_c}}} \tag{2.123}$$

$$\varepsilon_{\max} = \frac{P_{\max}}{E_f t_f b_f} \tag{2.124}$$

The theoretical values of P_{max} and ε_{max} of the FRP sheets calculated using Eqs. (2.125) and (2.126) are 13.8664 kN and 6851 $\mu\varepsilon$, respectively. An error comparison is presented in Table 2.1. Most of the errors are within 3%, except the limit case SB6. That may be owing to the abrupt loss of shear stress when the shear stress of the first integral point reaches f_b. This generally is caused by the computational instability.

2.3.3.3 Mode I fracture model (microrepresentation in concrete)

Considering the interfacial debonding in the form of concrete crack, two types of conventional smeared crack models are adopted. One is the rotating crack model, which assumes that the crack continuously rotates with the changing axis of principal stress, as shown in Fig. 2.28A. Because the crack orientation is modified constantly, the shear interlock along the crack surface can be removed. The other is the

Table 2.1 Comparison of FE results and theoretical solutions (Niu, 2002).

	P_{max} (kN)	Error	ε_{max} (μ)	Error
Theory	13.8664	–	6851	–
SB1	13.865	0.01%	6864	0.19%
SB2	14,062	1.41%	6939	1.28%
SB3	14.293	3.07%	7030	2.61%
SB4	13.949	0.6%	6878	0.39%
SB5	14.189	2.32%	6988	2.0%
SB6	13.939	0.46%	6129	10.5%

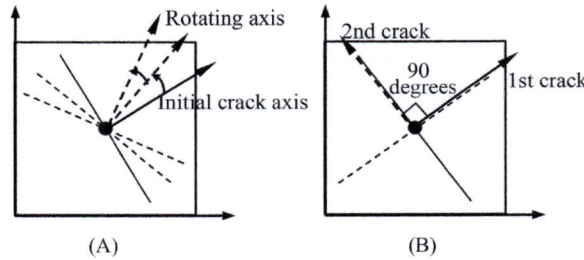

(A)   (B)

Figure 2.28 Two adopted smeared crack models. (A) Rotating crack model. (B) Fixed crack model.

fixed orthogonal crack model, shown in Fig. 2.28B. Here, a shear retention factor τ is introduced to account for the shear transfer effects such as aggregate interlock. In both the crack models, a linear softening curve is used to describe the normal stress degradation on the crack surface. The unloading/reloading behaviors are modeled by a secant path that follows a line back to the origin, as shown in Fig. 2.29.

As discussed in the previous section, many experiments on FRP-strengthened concrete beams (Wu et al., 1997) have revealed that the interfacial debonding generally occurs and propagates in concrete adjacent to the bond interface. Although the debonding owing to the interfacial fracture within the concrete layer can be also addressed as a Mode II fracture because of its macroscopic shear-dominated fracture feature, the cracks in concrete generally initiate as Mode I fracture because concrete is brittle. Therefore the debonding behavior should be related to the property of concrete to a certain extent. Meanwhile, G_f^{II} under the Mode II fracture assumption could be considered only as an equivalent reference rather than a clearly defined material property.

The conventional rotating and fixed smeared crack models are primary approaches for describing crack propagation in concrete. These have been implemented in a

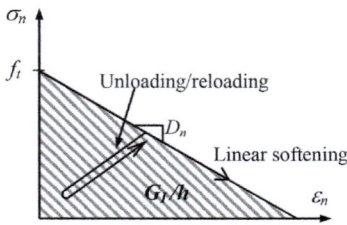

Figure 2.29 Linear tension softening relation.

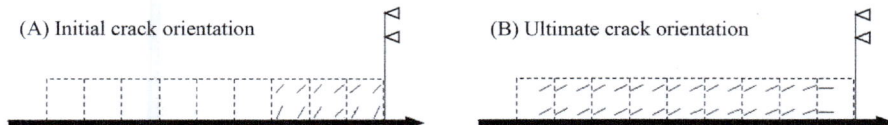

Figure 2.30 Crack evolution along interfacial concrete elements.

significant number of commercial FE codes. Both models consider the concrete crack as Mode I fracture by employing the Mode I fracture energy G_f^I. Herein, we use G_f^I of concrete to address the FRP−concrete interfacial bond capacity.

By using the rotating smeared crack model, the evolution of crack orientations at each cracked element is presented in Fig. 2.30. Interfacial concrete cracks initiate in a direction that is inclined to the loading axis by at least 45 degrees. With continuing load, the crack orientations gradually rotate and ultimately approach the horizontal direction. This implies that the fracture starts under tension and subsequently proceeds under tension−shear. This simulation result implies that the rotating crack model could explain the evolution of interfacial concrete crack that finally results in debonding failure.

Fig. 2.31 presents the load−displacement relationship based on the rotating crack model with $G_f^I = 0.2$ and 1.0 N/mm, and a comparison between this and the Mode II fracture assumption with $G_f^{II} = 1.2$ N/mm. Qualitatively similar to the results obtained using the Mode II fracture model, the load-carrying capacity of the bond interface is enhanced and eventually approaches a constant value as the Mode I fracture of concrete G_f^I increases. Fig. 2.32 shows the shear stress distribution of interface elements along the FRP−concrete interface. It implies that the maximum shear stress and effective bond length are enhanced when the fracture energy G_f^I is increased. However, discrepancy appears when the rationality of the value of mode I fracture energy of concrete is considered. In general, G_f^I is approximately 0.1−0.3 N/mm based on experimental research studies (Lenke and Gerstle, 2001). In accordance with the Mode II fracture assumption, $G_f^{II} = 1.2$ N/mm is also regarded as a general value determined from experimental data (Yoshizawa et al., 2000). Although an identical bond load-carrying capacity should have been

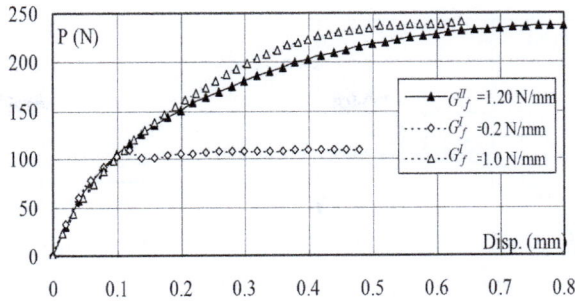

Figure 2.31 Fracture energy release of Mode I and Mode II fracture assumption (Niu, 2002).

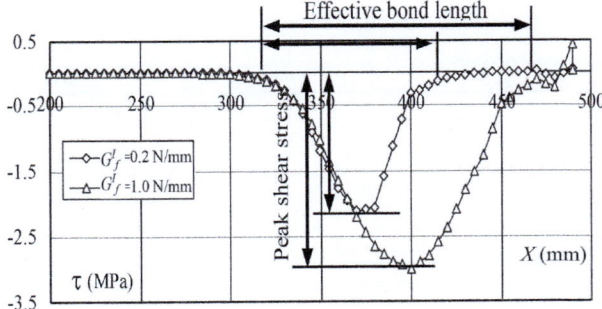

Figure 2.32 Shear stress distribution of interface elements along FRP-to-concrete interface by rotating smeared crack model (Niu, 2002).

obtained irrespective of the interfacial fracture assumption applied, the result reveals a large difference. For the rotating crack model, G_f^I must be increased to 1.0 N/mm (an impracticable value) to obtain an equal load-carrying capacity of bond (Fig. 2.31). It could be inferred that the rotating crack model omits the aggregate interlock on the crack surface, so that the load-carrying capacity of the bond interface is underestimated. Such an aggregate interlock behavior should not be omitted in concrete cracking, particularly when the concrete is subjected to shear loading as in the present case. To consider the aggregate interlock on the concrete crack surface, a fixed smeared crack model with a constant shear retention factor is also used.

Can the fixed smeared crack model obtain a reasonable result by addressing the aggregate interlock through a constant shear retention factor β? By fixing $G_f^I = 0.2$ N/mm and varying β (=0, 0.001, 0.005, and 0.01), Fig. 2.33 shows β's influence on the bond capacity. Here, "Rck" denotes the rotating crack model, and "β" denotes the shear retention factor in the fixed crack model. When β is set to zero or a negligible value, the load-carrying capacity is low and similar to that of the rotating crack model. The aggregate interlock behavior becomes apparent when β increases to 0.005 and 0.01. However, the load intends to increase continuously

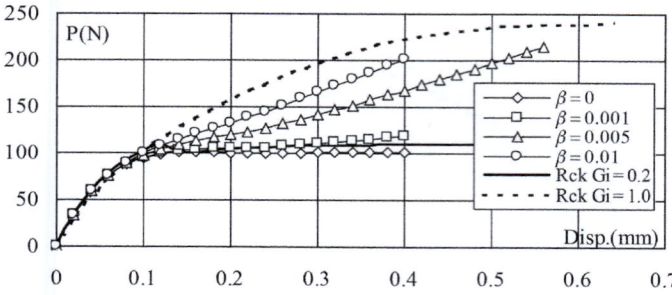

Figure 2.33 Load−displacement curves with different shear retention factor *b*.

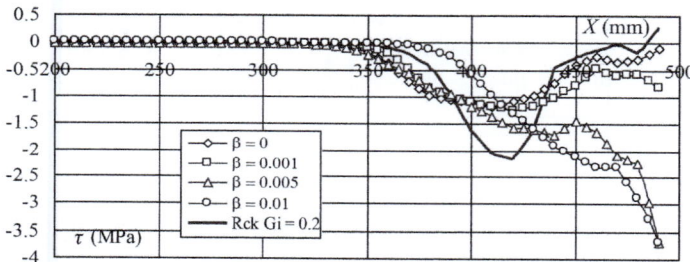

Figure 2.34 Shear stress distribution of interface elements along FRP-to-concrete interface by fixed smeared crack model (Niu, 2002).

but not approaches to an ultimate stable macroscopic debonding propagation. This is apparent also from the shear stress distribution of the interface elements along the FRP−concrete interface in Fig. 2.34. For $\beta = 0.005$ and 0.01, the shear stress at the loading end does not decrease although the interfacial crack has begun to propagate, and the effective bond length is increasing constantly. This is not an objective fracturing behavior of the interface. The aggregate interlock is overestimated owing to the constant accumulation of residual shear stress along the crack surface by a nonnegligible constant shear modulus and is not intended to be released. It is indicated that the shear retention factor has physical significance for describing the aggregate interlock behavior. However, a nonzero constant shear retention factor β tends to cause shear stress accumulation on the crack surface, thereby resulting in incorrect simulation results.

2.3.3.4 Mixed mode fracture and calibration of Gf^{I} and Gf^{II}

Another method is a displacement discontinuity model, which was developed for capturing strong discontinuity in brittle materials. Unlike smeared crack models, the stress degradation on a crack surface is represented by a stress−displacement relationship. Its original version addresses only Mode I fracture. With certain modification by introducing shear fracture energy G_f^{IIs}, the fracturing behavior along the crack surface could also be considered. The detailed finite element formulation is available elsewhere (Wu and Yin, 1998).

When G_f^{IIs} is varied with the concrete fracture energy fixed at $G_f^I = 0.2$ N/mm (as performed previously), the load-carrying capacity of FRP-bonded concrete prism increases, and the ultimate value approaches a constant, as shown in Fig. 2.35.

It is observed that compared to the simulation results using Mode II fracture assumption and Mode I fracture assumption with the rotating smeared crack model, when G_f^{IIs} is set to zero, the load-carrying capacity is at the lower limit (which is similar to the case of the rotating smeared crack model). Meanwhile, an increase in G_f^{IIs} enhances the load-carrying capacity of FRP–concrete interface in the interest of clarity. When $G_f^{IIs} = 0.08$ N/mm, the load-carrying capacity obtained by Mode II fracture assumption with $G_f^{II} = 1.2$ N/mm can be attained without changing the Mode I fracture energy $G_f^I = 0.2$ N/mm. This implies that the shear fracture energy G_f^{IIs} reflects the shear transfer capability of concrete and plays an important role when shear fracture dominates. The shear stress distribution of the interface elements also yields reasonable results (Fig. 2.36). Fig. 2.37 shows the crack pattern in interfacial concrete elements. The crack orientation in each element is fixed as soon as the crack is initiated. This is similar to the crack pattern by the fixed smeared crack model. The difference is that the displacement discontinuity model evaluates the stress state only at the element central point. Unlike the fixed crack model, the irrational shear locking disappears, and the effective bond length is represented well with the increase in G_f^{IIs} at constant G_f^I. A relationship between the results based on the Mode II fracture assumption and Mode I fracture assumption in concrete followed by shear transfer is obtained by numerical fitting using the displacement discontinuity model. It is shown in Fig. 2.38.

It is summarized that when debonding fracture occurs along interfacial concrete, both the Mode II and Mode I fracture assumptions for interfacial debonding should yield an equal load-carrying capacity of the bond interface. G_f^{IIs} is regarded as a specification of both the concrete and the bonding conditions. Therefore the introduction of shear fracture energy to the crack surface after the Mode I crack initiation bridges these two assumptions.

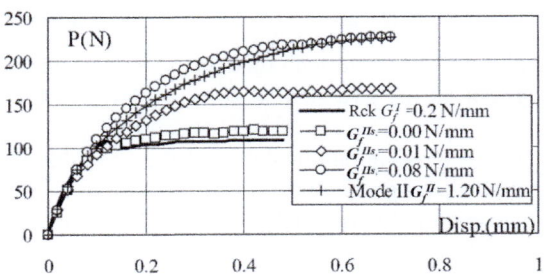

Figure 2.35 Load–displacement curves with varying G_f^{IIs} (Niu, 2002).

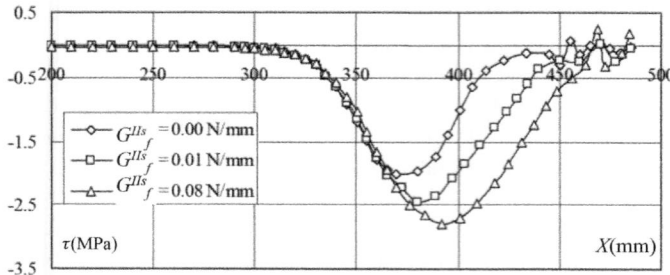

Figure 2.36 Shear stress distribution in interface elements by displacement discontinuity model (Niu, 2002).

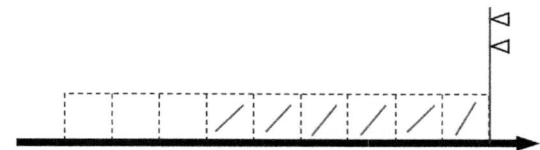

Figure 2.37 Crack pattern along interfacial concrete elements by displacement discontinuity model (Niu, 2002).

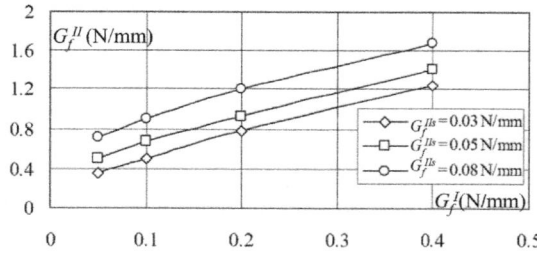

Figure 2.38 $G_f^I - G_f^{II}$ relations with varying G_f^{IIs} (Niu, 2002).

2.3.4   Prediction of bond strength of FRP–concrete interface

Numerous empirical and semiempirical models have been developed to predict interfacial bond strength, in the past several decades. Most of the available models are listed in Table 2.2.

Because of limited success and applicability of the available models, one that is simple to use in practical design has been developed by Wu et al. (2009a,b). The bond strength prediction model is rational. It is capable of capturing the fundamental features of the bond behavior and predicting the maximum bond strength and effective bond length with good accuracy. In the following section, the Wu et al. (2009a,b) model is described in detail and compared with other available models.

Table 2.2 FRP-to-concrete interface bond strength models (Wu and Jiang, 2013).

Models	Bond strength prediction equations
Van Gemert (1980)	$P_u = 0.5 b_f L f_t$
Tanaka (1996)	$P_u = b_f L (6.13 - \ln L)$
Niedermeier (2000)	$L_e = \sqrt{\dfrac{E_f t_f}{4 f_t}},\ G_f = c_f k_b^2 f_t,\ \ c_f = 0.204\text{mm},$ $k_b = \sqrt{1.125 \dfrac{2 - b_f/b_c}{1 + b_f/400}},$ $P_u = \begin{cases} 0.78 b_f \sqrt{G_f E_f t_f} & \text{if } L \geq L_e \\ 0.78 b_f \sqrt{G_f E_f t_f} \dfrac{L}{L_e}(2 - \dfrac{L}{L_e}) & \text{if } L < L_e \end{cases}$
Neubauer and Rostásy (1997)	$k_b = \sqrt{1.125 \dfrac{2 - b_f/b_c}{1 + b_f/400}}\ L_e = \sqrt{\dfrac{E_f t_f}{2 f_t}},$ $P_u = \begin{cases} 0.64 k_b b_f \sqrt{f_t E_f t_f} & \text{if } L \geq L_e \\ 0.64 k_b b_f \sqrt{f_t E_f t_f} \dfrac{L}{L_e}(2 - \dfrac{L}{L_e}) & \text{if } L < L_e \end{cases}$
Maeda et al. (1997)	$L_e = e^{6.13 - 0.58\ln(E_f t_f)}\ P_u = 110.2 \times 10^{-6} E_f t_f b_f L_e$
Khalifa et al. (1998)	$L_e = e^{6.13 - 0.58\ln(E_f t_f)},\ P_u = 110.2 \times 10^{-6} E_f t_f b_f L_e \left(\dfrac{f_c'}{42}\right)^{2/3}$
Chen and Teng (2001)	$L_e = \sqrt{\dfrac{E_f t_f}{\sqrt{f_t}}},\ k_b = \sqrt{\dfrac{2 - b_f/b_c}{1 + b_f/b_c}},$ $P_u = \begin{cases} 0.427 k_b b_f L_e \sqrt{f_c'} & \text{if } L \geq L_e \\ 0.427 k_b b_f L_e \sqrt{f_c'} \sin\dfrac{\pi L}{2 L_e} & \text{if } L < L_e \end{cases}$
Yang et al. (2001)	$L_e = 100\ \text{mm},\ P_u = (0.5 + 0.08\sqrt{0.01 E_f t_f / f_t})\, b_f L_e f_t / 2$
Monti et al. (2003)	$k_b = \sqrt{\dfrac{1.5 \times (2 - b_f/b_c)}{1 + b_f/100\text{mm}}}\ L_e = \sqrt{\dfrac{E_f t_f}{\sqrt{7.2 k_b f_c'}}},$ $P_u = \begin{cases} b_f L \sqrt{\dfrac{1.8 E_f k_b f_c'}{3 t_f}} & \text{if } L \geq L_e \\ b_f L \sqrt{\dfrac{1.8 E_f k_b f_c'}{3 t_f}} \sin\dfrac{\pi L}{2 L_e} & \text{if } L < L_e \end{cases}$
JCI (2003)	$L_e = 0.125(E_f t_f)^{0.57},$ $P_u = 0.93 f_c'^{0.44} b_f L_e,\ \ \text{where } L_e = L\ \ \text{if}\ \ L > L_e$
Dai et al. (2005)	$G_f = 0.514 f_c'^{0.236},\ \Delta b_f = 37\text{mm},$

(Continued)

Table 2.2 (Continued)

Models	Bond strength prediction equations
	$$P_u = \begin{cases} b_f\sqrt{2E_f t_f G_f} & \text{if } b_f < 100\text{mm} \\ (b_f + 2\Delta b_f)\sqrt{2E_f t_f G_f} & \text{if } b_f \geq 100\text{mm} \end{cases}$$
Wu et al. (2009a,b)	$$P_u = \begin{cases} 0.585 k_b b_f f'_c 0.1 (E_f t_f)^{0.54} & \text{if } L \geq L_e \\ 0.585 k_b b_f f'_c 0.1 (E_f t_f)^{0.54} \left(\dfrac{L}{L_e}\right)^{1.2} & \text{if } L < L_e \end{cases},$$ $$k_b = \sqrt{\frac{2.25 - b_f/b_c}{1.25 + b_f/b_c}}$$
Wu and Jiang (2013)	$$P_u = k_L E_f t_f b_f \frac{\alpha}{\beta},\ k_L = \frac{\eta\sqrt{1-\eta^2}\sinh(\sqrt{1-\eta^2}L/\beta)}{1 + \eta\cosh(\sqrt{1-\eta^2}L/\beta)},$$ $$k_b = \lambda + (1-\lambda)b_f/b_c,\ \lambda = 1 + 0.222 f'_c 0.304,$$ $$\alpha = 0.094 f'_c 0.026,\ \beta = \frac{0.134(E_f t_f)^{0.5}}{k_b f'_c 0.082}$$

The theoretical bond strength P_u in simple shear joints can be expressed by

$$P_u = b_f\sqrt{2G_f E_f t_f} \tag{2.125}$$

A three-parameter bond strength model is proposed based on the above equation and using statistical analysis. It also considers the effective bonding length L_e and the width ratio of the bonded sheet to the concrete member (b_f/b_c).

$$P_u = C_1 b_f f'_c C_2 (E_f t_f)^{C_3} \tag{2.126}$$

$$P_u = \begin{cases} 0.585 b_f f'_c 0.1 (E_f t_f)^{0.54} & \text{if } L \geq L_s \\ 0.585 k_b b_f f'_c 0.1 (E_f t_f)^{0.54} \left(\dfrac{L}{L_e}\right)^{1.2} & \text{if } L < L_s \end{cases} \tag{2.127}$$

where k_b is the width factor expressed as follows:

$$k_b = \sqrt{\frac{2.25 - b_f/b_c}{1.25 + b_f/b_c}} \tag{2.128}$$

The proposed model has been evaluated by 311 experimental results, as shown in Fig. 2.39. In particular, the 311 experimental results obtained from the available database

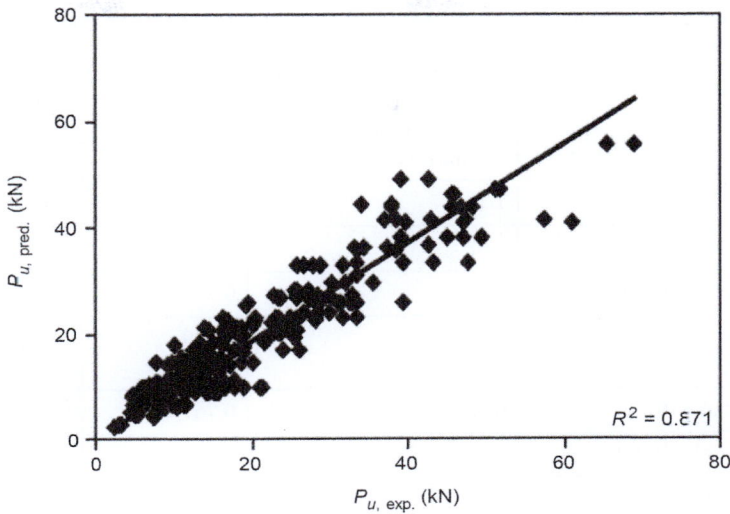

Figure 2.39 P_u predicted versus P_u experiment (Wu et al., 2009a,b).

covers a wide range of parameters. The FRP thickness, width, bond length, and modulus of elasticity vary from 0.083 to 2 mm, 10 to 300 mm, 50 to 700 mm, and 23.9 to 425.1 GPa, respectively. The cylindrical compressive strength and width of concrete prism vary from 16.0 to 75.5 MPa and 100 to 500 mm, respectively. Evidently, the database covers a wide range of each parameter and is likely to provide a reliable benchmark for qualifying different models for predicting debonding behavior and related parameters. As shown in Fig. 2.39, the proposed model provides good predictions of experimental results. It can be adopted for predicting the bond strength of FRP—concrete interface.

A comparison of the model of Wu et al. (2009a,b) with other available models (as presented in Table 2.3 and Fig. 2.40) indicates the superior performance of the new model of bond strength. Table 2.3 illustrates that the Wu et al. (2009a,b) bond strength model exhibits the highest coefficient of correlation as well as the least coefficient of variation. These verify the validity and superiority of the proposed bond strength model over the other available bond strength models.

Recently, Wu and Jiang (2013) proposed a new bond strength model based on the theoretical analysis of a nonlinear continuous bond—slip relationship for FRP—concrete interface (shown in Fig. 2.41A). The coefficients of the bond—slip model are determined by regression analyses based on a comprehensive test database available in Wu and Jiang (2013). The expression of the new model is provided in Table 2.2. The proposed model exhibits higher performance than available models with high accuracy, as shown in Fig. 2.41B.

Table 2.3 Statistic results of the predicted-to-experimental bond strength ratios of different bond strength models (Wu et al., 2009a,b).

Models	Average predicted-to-experimental bond strength ratio	Correlation coefficient (R^2)	Coefficient of variation (CV)
Van Gemert (1980)	1.21	0.471	0.783
Tanaka (1996)	0.68	0.591	0.469
Niedermeier (2000)	0.87	0.706	0.295
Neubauer and Rostásy (1997)	1.10	0.729	0.285
Maeda et al. (1997)	0.95	0.767	0.315
Khalifa et al. (1998)	0.92	0.696	0.376
Chen and Teng (2001)	0.89	0.818	0.254
Monti et al. (2003)	1.19	0.736	0.287
JCI (2003)	1.01	0.716	0.359
Yang et al. (2001)	0.94	0.737	0.335
Wu et al. (2009a,b)	1.00	0.871	0.251

2.4   Long-term behavior of FRP−concrete interface

The long-term behavior of FRP−concrete interface, including under sustained loading and fatigue loading, is essential for extensive promotion of FRP for use in strengthening concrete structures. An improvement in fatigue performance is achievable only when the FRP is adequately bonded to the concrete. This section highlights the bond deterioration mechanism under long-term loading. It includes descriptions of an experimental study of FRP−concrete interface under fatigue and creep loading, and time-dependent modeling of FRP−concrete performance.

2.4.1   Fatigue performance of FRP−concrete interface

Unlike the static behavior of an FRP−concrete interface, the fatigue debond propagates progressively as the number of fatigue cycles increases until the remaining bonded length of the FRP−concrete interfaces can no longer sustain the maximum applied load. In accordance with the experimental method recommended by JSCE for upgrading concrete structures using FRP sheets, Diab et al. (2007) investigated the fatigue behavior of concrete strengthened externally with FRP composites, using

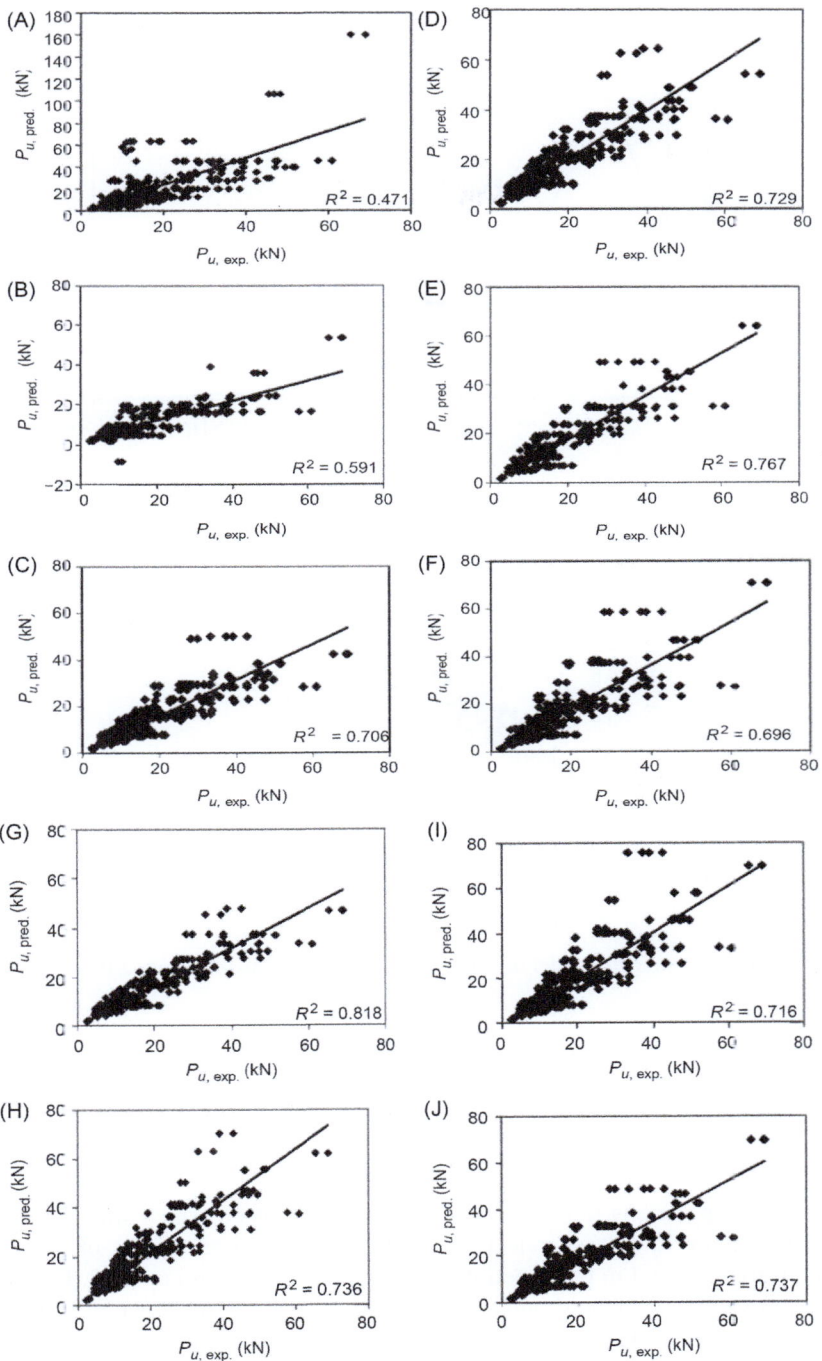

Figure 2.40 Experimental bond strength versus predicted bond strength of existing models (Wu et al., 2009a,b): (A) Van Gemert (1980), (B) Tanaka (1996), (C) Niedermeier (2000), (D) Neubauer and Rostásy (1997), (E) Maeda et al.(1997), (F) Khalifa et al. (1998), (G) Chen and Teng (2001), (H) Monti et al. (2003), (I) JCI (2003), and (J) Yang et al. (2001) (Wu et al., 2009a,b).

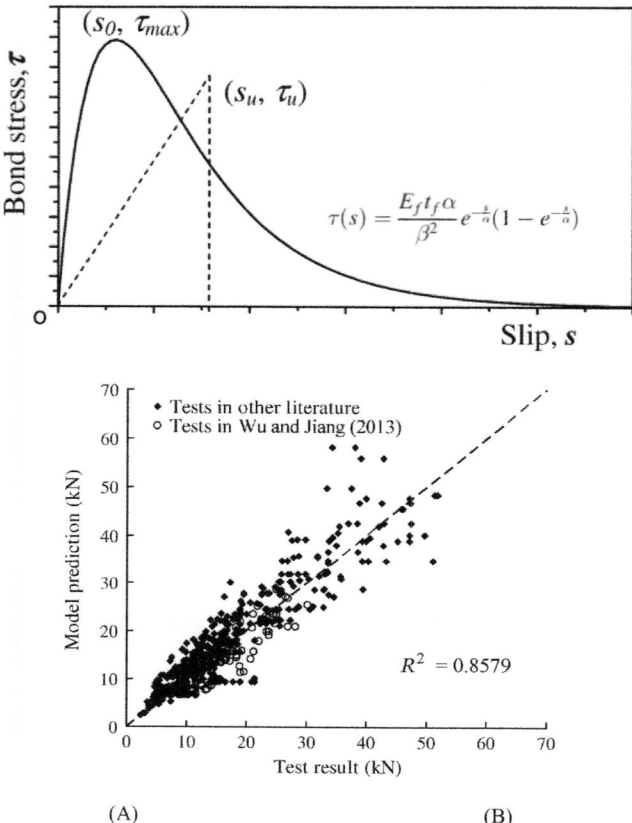

Figure 2.41 Bond strength model of Wu and Jiang (2013) (Wu and Jiang, 2013).

double-lap shear specimens (Fig. 2.9C). The fatigue loading varied from 30% to 80% of the static bond capacity of the FRP−concrete interfaces.

2.4.1.1 Observed fatigue-induced failure modes

Most of the specimens failed by debonding of FRP sheet, which was initiated by the shear failure in the concrete near the loaded end. Three modes of debonding failure were observed in the experimental results, as shown in Fig. 2.42. Debonding Mode A exhibits debonding into coarse aggregate with partial fracture of the FRP fibers accompanied by latitudinal crack between the two sides at a distance of approximately 100 mm from the loaded end. Debonding Mode B exhibits debonding into coarse aggregates, whereas Debonding Mode C exhibits debonding at the interface between the epoxy resin and epoxy primer. Among the three modes, Mode A exhibited the largest load-carrying capacity and fatigue life.

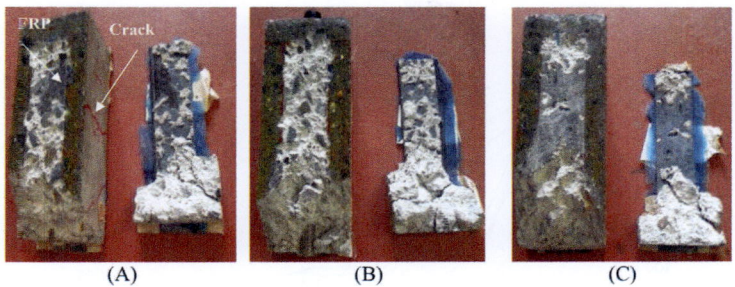

(A) (B) (C)

Figure 2.42 Debonding modes of FRP-to-concrete interface failure due to fatigue loading (Diab, 2008). (A) Mode A. (B) Mode B. (C) Mode C.

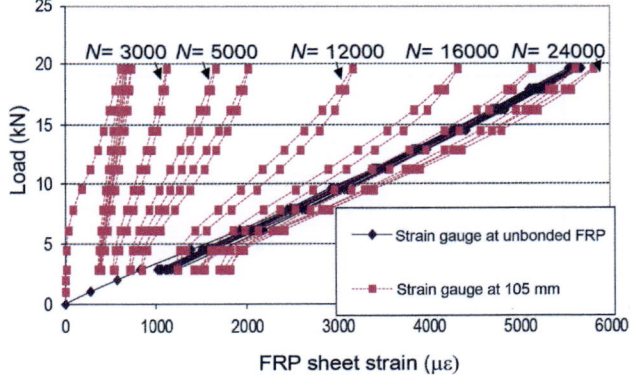

Figure 2.43 Comparison between FRP sheet strain measurements at bonded and unbonded area due to fatigue loading (Diab, 2008).

2.4.1.2 Fatigue damage accumulation

Cyclic loading results in the deterioration of bond between FRP composites and concrete. This causes the redistribution of shear stress along the FRP−concrete interfaces. The bond degradation can be observed from the increase in strain in FRP with an increase in fatigue cycles, as shown in Fig. 2.43. No degradation is observed to be caused by the strain in the FRP attached to an unbonded area and subjected to an identical fatigue loading. Therefore the damage accumulation at FRP−concrete interfaces owing to fatigue loading may be represented conveniently by the stiffness degradation before debonding initiation.

The results of cyclic loading tests on FRP−concrete bonded joints revealed that the interface debonding propagates progressively as the number of fatigue cycles increases, as shown in Fig. 2.44. Apparently, the results revealed that the debonding propagation rate decreases significantly with fatigue cycles. Fatigue failure occurs when the residual bonded length can no longer sustain the maximum fatigue load.

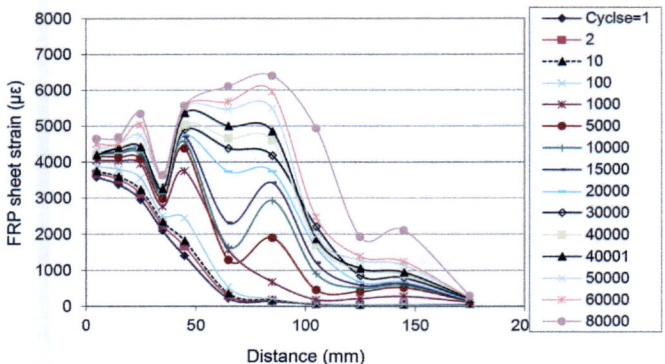

Figure 2.44 FRP time-dependent sheet strain distribution due to fatigue loading (Diab, 2008).

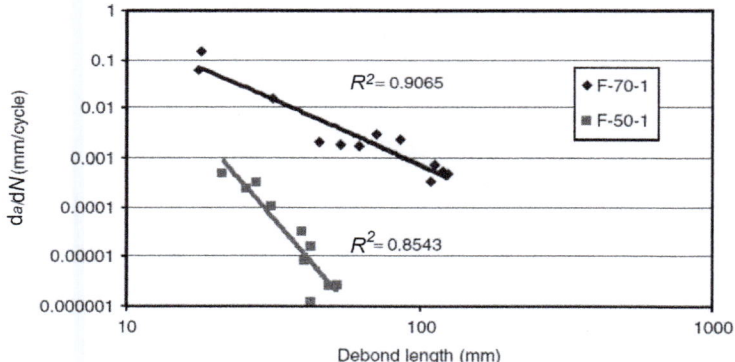

Figure 2.45 Logarithmic debond growth rate versus logarithmic debond length (Diab et al., 2009a).

2.4.1.3 Crack propagation

In contrast to static failure, the formation of macrocrack does not imply failure of fatigue specimens. The debonding propagates progressively with fatigue cycles depending on the stress ratio and the bond capacity of the FRP—concrete interface. Diab et al. (2009a) presented an analytical solution for the evolution and distribution of shear stresses along the entire bond length of FRP—concrete interfaces caused by Mode II fatigue loading. They stated that the debonding length (a) increases with the fatigue cycles (N.) However, the debonding growth rate (da/dN) decreases with an increase in the debond length, as shown in Fig. 2.45. Experimental results revealed that the debond growth begins at a high value. Then, it decreases as the debond length increases until it attains a stable stage, provided that the residual bonded length is adequate to sustain the maximum load Therefore an increase in bonded length is likely to be highly effective for increasing the fatigue life of FRP-strengthened concrete structures.

2.4.1.4 Pre and post fatigue performance of FRP—concrete interfaces

Notwithstanding the extent of debonding along the FRP—concrete interfaces, the post-fatigue bond capacity of FRP—concrete interfaces is not affected by fatigue loading if the remaining bonding length is adequate. Fig. 2.46 shows a comparison between the pre- and postfatigue bond capacity of the specimens. The figure reveals that the ultimate bond capacity of specimens previously subjected to fatigue loading is higher than that of those subjected only to static loading. The likely reason for this phenomenon is fatigue hardening analogous to that in metallic systems. Thus specimens fatigued at a relatively low stress exhibited an increase in the static bond capacity when tested after fatigue.

A comparison between the load—displacement relationships for pre- and post-fatigue loading is presented in Fig. 2.47. Such figure reveals that there is a clear difference between the pre and postfatigue load—displacement relationship by the degradation of the stiffness. Furthermore, the post fatigue load—displacement relationship behaves linearly until failure for all specimens. The increase in compliance is a result of the increase in debonding length with fatigue loading. The fact that bond capacity is not affected by the fatigue loading implies that the bond strength of the FRP—concrete interface is not affected adversely by fatigue loading. Fatigue deterioration by debonding growth is apparent along the FRP—concrete interfaces. Further details are available elsewhere (Diab et al., 2009a,b; Wu and Diab, 2009).

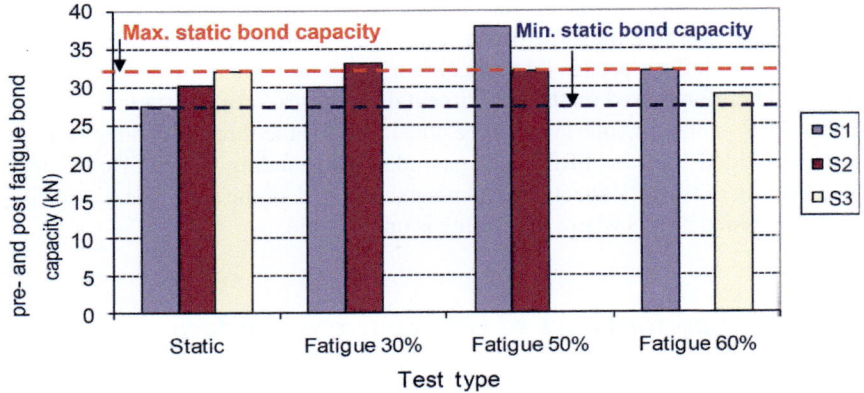

Figure 2.46 A comparison between bond capacity of the FRP-to-concrete interface before and after fatigue loading (Wu and Diab, 2009).

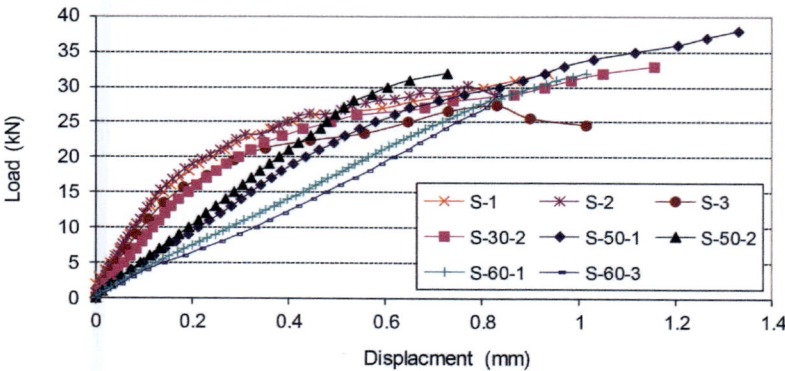

Figure 2.47 A comparison between load—deformation relationships for specimens subjected static loading before fatigue loading (S−1, S−2, and S−3) and after fatigue loading (Wu and Diab, 2009).

2.4.1.5 Fatigue life of FRP—concrete interfaces and evaluation of code provisions

Different entities adopt different approaches to address the behavior of bonded FRP retrofit measures subjected to fatigue loads. The American Concrete Institute (ACI 440.2R-02) and UK Concrete Society (Concrete Society, 2004) direct the designer to verify whether the anticipated FRP stress/strain is lower than an absolute limit associated with the anticipated FRP material performance. These limits are substantially high and are impracticable even under static loading, as shown in Fig. 2.48.

The Japanese Society of Civil Engineers (a pioneer in considering the degradation of interfacial bond strength owing to cyclic loading) (JSCE, 2001) and the Italian National Research Council (CNR, 2004) define the stress in an FRP to be a function of FRP properties and interface characteristics. Whereas the limitation by CNR may be acceptable for direct shear test, the JSCE needs to be revised, as shown in Fig. 2.49. No failure occurred for specimens with maximum fatigue loadings less than 60%. Therefore based on the experimental results, 50% of the static bond capacity of the FRP—concrete interface can be considered as the threshold value of fatigue loading, provided that the FRP sheet has adequate bonding length. However, the fatigue loading of the FRP sheets should not exceed 30% of the static bond capacity of FRP-strengthened beams where debonding is undesirable.

2.4.2 Creep performance of FRP—concrete interface

This section synthesizes the results of a number of creep tests of FRP-strengthened structures. Two types of sustained loading are applied during these tests: direct tensile sustained shear load on the FRP specimens and sustained loading using prestressed FRP sheets.

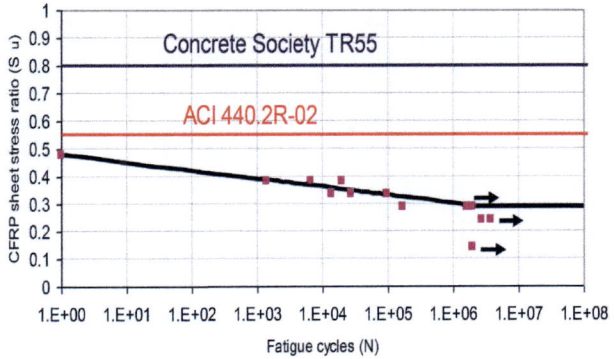

Figure 2.48 S−N relationship of the CFRP composites (Wu and Diab, 2009).

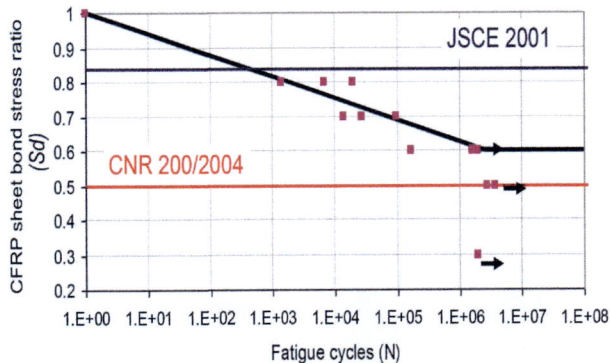

Figure 2.49 S−N relationship of the FRP-to-concrete interface (Wu and Diab, 2009).

2.4.2.1   Pull-out test for double-shear specimens

In accordance with the experimental method recommended by the JSCE guideline for upgrading concrete structures by using continuous fiber sheets, prism specimens of length 450 mm were used in the investigation by Wu and Diab (2007). The details of the prism specimens are schematically shown in Fig. 2.9C. The sustained load levels varied from 35% to 86% of the static bond capacity of the specimens. This static capacity was determined as the minimum of the load-carrying capacities of three static specimens.

The experimental results revealed that the adhesive creep resulted in initiation and increase in the debonding along the FRP—concrete interface. Fig. 2.50 shows the process of time-debonding propagation along the FRP—concrete interface. Debonding failure can be classified into two modes: Debonding Mode A and Debonding Mode B (Fig. 2.51). In general, under high sustained loads, the behavior

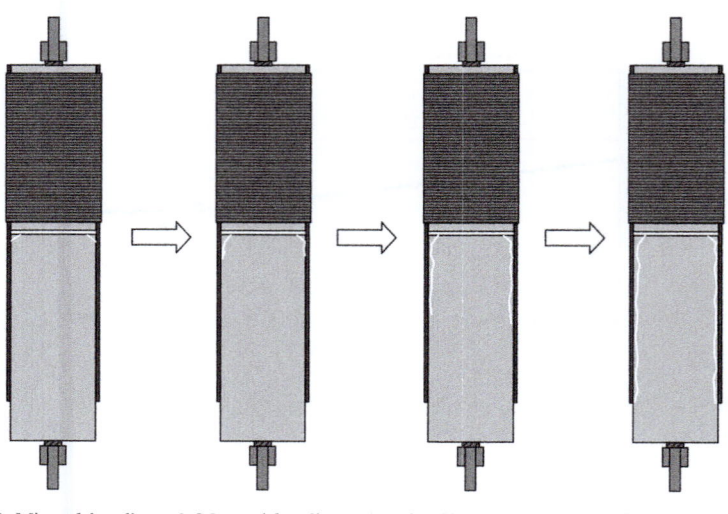

1. Micro-debonding   2. Macro-debonding   3. Debonding propagation   4. Final debonding
occurrence   formation (10-20 mm) (Ultimate state)

Figure 2.50 Process of debonding propagation due to creep (Diab and Wu, 2007).

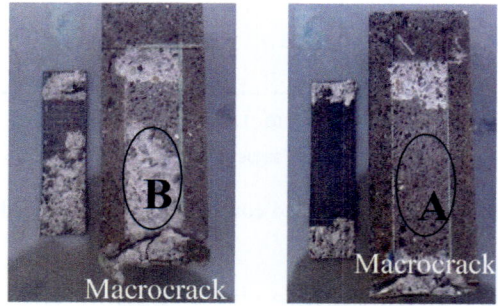

Figure 2.51 Failure modes and concrete cracks for the specimens due to creep (Diab et al., 2009a).

is conditioned more by the properties of concrete. Meanwhile, at low sustained loads, the behavior is conditioned by the properties of epoxy resin. Thus one can conclude that the deterioration effect of creep is higher in epoxy resin than in concrete.

The creep bond strength is lower than the standard static capacity. The relationship between the sustained load and the creep life of specimens is linear on a semilogarithmic scale, as shown in Fig. 2.52. There was no failure until 1000 h when the sustained load was approximately 35% of the static bonding capacity of the

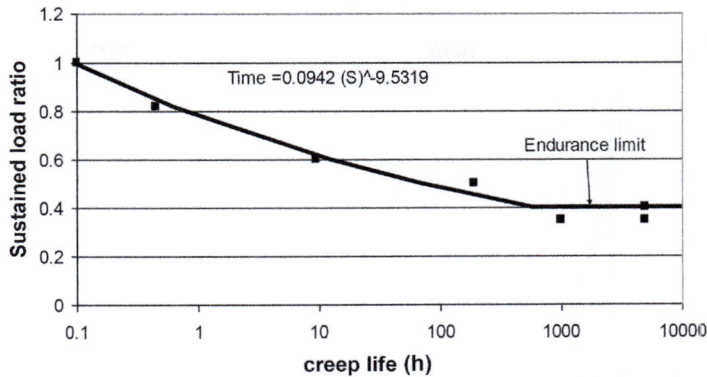

Figure 2.52 Semilogarithmic relation between applied load and creep life (Wu and Diab 2009).

Figure 2.53 Test setup of the concrete beams strengthened with prestressed with FRP sheets (Wu and Diab, 2007).

specimen for which the stress in the FRP sheets was approximately 25% of the tensile strength.

2.4.2.2 Beam strengthened with prestressed FRP sheet

Diab et al. (2009b) presented an experimental campaign concerning the long-term behavior of the anchorage ends of prestressed FRP sheets. The experimental program was conducted to test seven beams prestressed with FRP sheets. Two anchorage ends were created for each beam, as shown in Fig. 2.53. Using

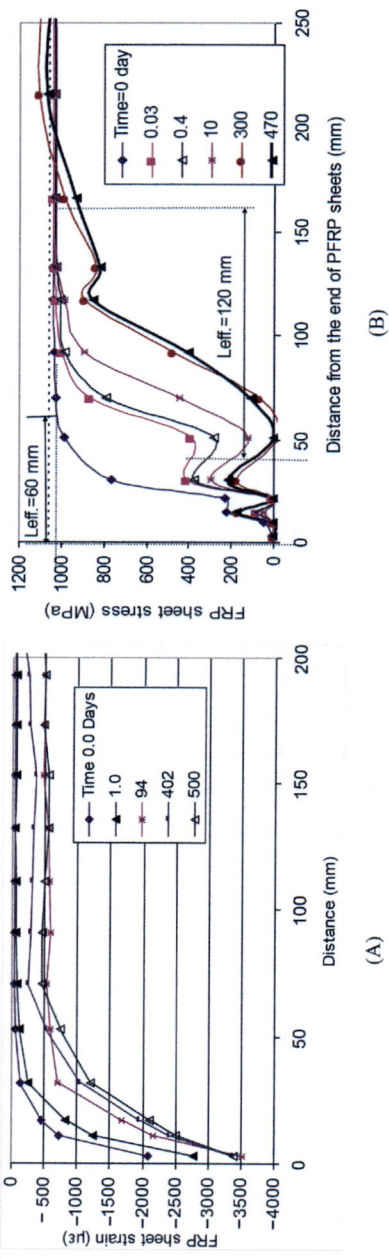

Figure 2.54 Time-dependent strain distribution along the bond length of the prestressed FRP sheet (one layer of FRP sheet): (A) 25% FRP tensile strength; (B) 35% FRP tensile strength (Diab, 2008).

different carbon fiber-reinforced polymer (CFRP) sheet layers, the prestress level was varied from 20% to 40% of the guaranteed tensile strength of FRP sheet ($\sigma\mu = 350C$ MPa and $\varepsilon\mu = 14500\ \mu\varepsilon$). Tests were performed at normal temperature over approximately 20 months.

The experimental results revealed that no debonding occurred along anchorage ends for prestress levels less than 30% of the FRP tensile strength within 500 days, as shown in Fig. 2.54A. It was indicated that the creep of the adhesive layer was mainly responsible for the losses of prestressed FRP sheet, whereas variation in temperature and creep of concrete significantly affected the prestress losses away from the effective bonding length. Fig. 2.54B shows that an increase in the prestress level results in debonding at the anchorage and further prestress loses. According to the results obtained from the debonding and undebonding specimens, the effective bonding length increased with time. The predicted percent of increase at 500 days is approximately 50%. In addition, this study established that a mechanical anchorage system is an effective solution for preventing premature failure at the ends of FRP sheets prestressed at a high load level for short and long-term loading.

The FRP—concrete interface where the load is transferred from the FRP sheet/laminates to the concrete substrate is subjected to creep and an increase with time. Therefore an increase in slip of an FRP sheet with time causes a redistribution of the shear stress along the bonded length: the high shear stresses close to the loaded end decrease, and a longer length of interface is required to carry the applied prestress. In addition, increasing the slip of an FRP sheet beyond the microslip limit results in debonding initiation and propagation, as shown in Fig. 2.55. This propagation occurs rapidly initially (within 10 days), and then gradually reduces with time.

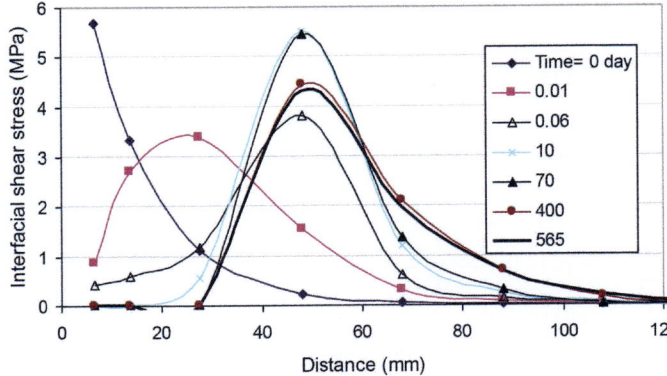

Figure 2.55 Time-dependent interfacial shear stress distributions along the anchorages (Diab et al., 2009b).

2.4.3   Modeling of time-dependent performance of FRP concrete interfaces

This section summarizes the time-dependent FRP—concrete models that imitate the long-term behavior of FRP—concrete interface and can be used for the general purpose of simulating the long-term behavior of FRP-strengthened structures, creep, and fatigue loading, based on the works by Diab and Wu (2007), Diab et al. (2007), and Diab et al. (2009b).

2.4.3.1   Finite element hybrid viscoelastic model of FRP—concrete interface

The time-dependent (fatigue and creep) fracture process of FRP—concrete interfaces can be divided into two stages:

1. A debonding initiation stage, which relies on a linear viscoelastic model controlled by the time of creep or total number of fatigue cycles and the properties of the adhesive layer.
2. A time-dependent debonding propagation stage, which depends on a bond—slip model and a damage accumulation factor.

2.4.3.1.1   Before debonding initiation

The constitutive model describing the nonlinear behavior of FRP—concrete interfaces based on the previous assumptions can be represented by a hybrid viscoelastic model, as shown in Fig. 2.56. A system of Maxwell's chains, which describe the creep—fatigue interaction phenomenon before debonding, is placed in series with an element that schematizes the softening model. The cycle-dependent stiffness and the increment in creep fatigue slip at the first stage can be obtained from Eqs. (2.129) and (2.130):

$$K(t + \Delta t) = \sum_{\mu=0}^{n} \left[\frac{\lambda_\mu}{\Delta t} \left(1 - e^{-\frac{\Delta t}{\lambda_\mu}} \right) K_\mu(t) \right] \tag{2.129}$$

$$\Delta S^c(t + \Delta t) = \frac{1}{K(t + \Delta t)} \sum_{\mu=0}^{n} \left[\tau(t) \left(1 - e^{-\frac{\Delta t}{\lambda_\mu}} \right) \right] \tag{2.130}$$

Meanwhile, the shear stress increment owing to the cycle increment is obtained from Eq. (2.131):

$$\Delta \tau = \sum_{\mu=0}^{n} \left(1 - e^{-\frac{\Delta t}{\lambda_\mu}} \right) \left(\frac{K_\mu(t) \lambda_\mu}{\Delta t} \Delta s - \tau_\mu(t) \right) \tag{2.131}$$

where n is the number of terms of the series, $K_\mu(t)$ are elastic aging coefficients to be determined, $\lambda_\mu = \eta_\mu(t_0)/K_\mu(t_0)$ are the relaxation times, $\eta_\mu(t_0)$ are the viscosities of the dampers, Δs is the total slip increment at the FRP—concrete interface element, $\Delta \tau$ is the shear stress increment to be updated at each time step for each

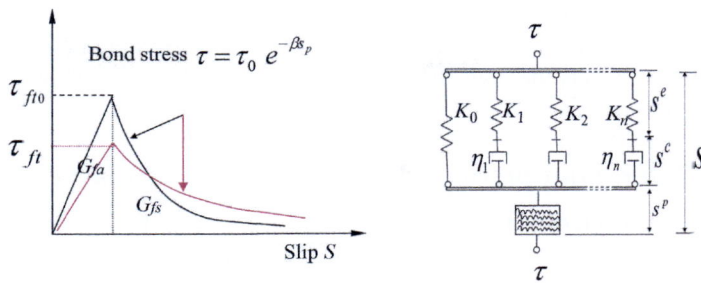

Figure 2.56 A hybrid viscoelastic model and exponential bond—slip model (Diab and Wu, 2007).

chain, and Δt is the time increment (at the fatigue loading time, the increment represents the number of fatigue cycle increments divided by the fatigue load frequency, $\nu(\Delta t = \Delta N/\nu)$). Further details regarding these equations are available in Wu and Diab (2007). The parameters of the viscoelastic model must be determined from experimental tests before debonding of the FRP—concrete interface is initiated under sustained loading or fatigue loading. It should be mentioned that these parameters are different owing to the type of loading (i.e., sustained or fatigue loading).

2.4.3.1.2 After debonding initiation

We assume that debonding will be initiated if the total slip of an intact FRP—concrete interface element, $s^e + s^c$, increases to the microslip deformation, s_p, which is determined from the short time interfacial model. The constitutive equation of slip relationship (as shown in Fig. 2.56) for the shear stress at softening is expressed as

$$f(s_p) = \tau_{ft}\exp(-\beta_f s_p) \tag{2.132}$$

where τ_{ft} is the bonding strength and depends on the time when debonding initiates in the element, s_p is the softening slip, and β_f is a coefficient that depends on the fracture energy of the FRP-interface element and the fatigue damage accumulation in the case of fatigue loading. β_f is obtained from the following equation:

$$\beta_f = \tau_{ft}/(\zeta \cdot G_{fs}) \tag{2.133}$$

$$\zeta = 1 - \frac{K}{(1+t_{ci})^{0.1}}\log_{10}(60\,\nu(t_t - t_{ci})) \tag{2.134}$$

where ζ is a damage accumulation factor; t and t_i are the total fatigue time and debonding initiation time of the element, respectively, in minutes; and K is an experimental coefficient obtained by regression analysis. G_{fs} is the fracture energy and released strain energy for the softening branch of the fatigue and creep loading.

To study the capability of the present models to reproduce the long-term behavior of FRP—concrete interfaces, the viscoelastic parameters of the proposed model were identified based on experimental results before debonding initiation. A comparison of the predicted and experimental results of these viscoelastic parameters of creep or fatigue reveals the high accuracy of the predicted results. This is discussed in the following sections.

2.4.3.2   Determination of parameters of the viscoelastic model

The material parameters of the viscoelastic model ($k\infty$ and the various $k\mu$ and $\lambda\mu$) were obtained based on cyclic-dependent slip measured for creep and fatigue loading. The proposed model consisting of five Maxwell elements is capable of imitating the time-dependent behavior of FRP−concrete interfaces. A comparison between experimental and predicted results of the proposed models is shown in Fig. 2.57. This figure depicts a comparison between normalized slip resulting from sustained loading and fatigue loading, and the predictions using the viscoelastic models.

2.4.3.3   Evaluation of the hybrid viscoelastic model based on the experimental results

2.4.3.3.1   Sustained loading

Using the experiment results of creep specimens presented previously, a comparison between the experimental and predicted time-dependent FRP-sheet strain is shown in Fig. 2.58. This figure demonstrates the capability of the proposed models to predict the time-dependent strain along an FRP sheet under low and high sustained loads. Fig. 2.58A shows that the prestress losses are higher at the effective bonding length. Beyond this length, where the shear stress equals zero, marginal prestress

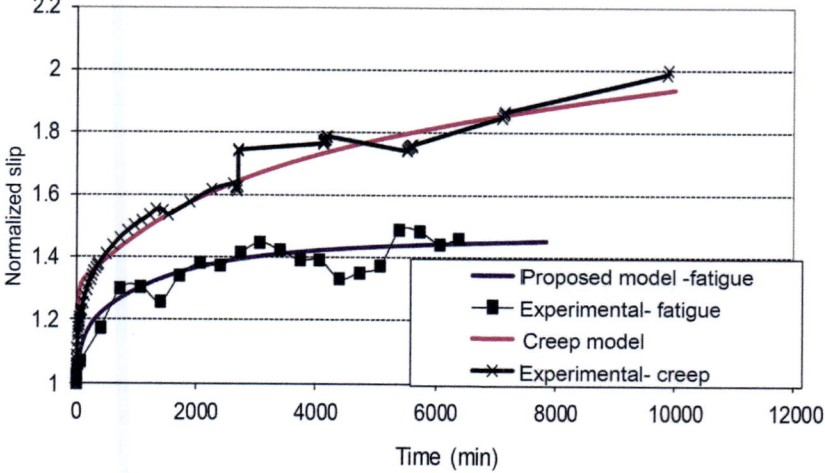

Figure 2.57 Experimental and numerical normalized maximum CFRP sheet slip versus time for sustained and fatigue loading (Wu and Diab 2009).

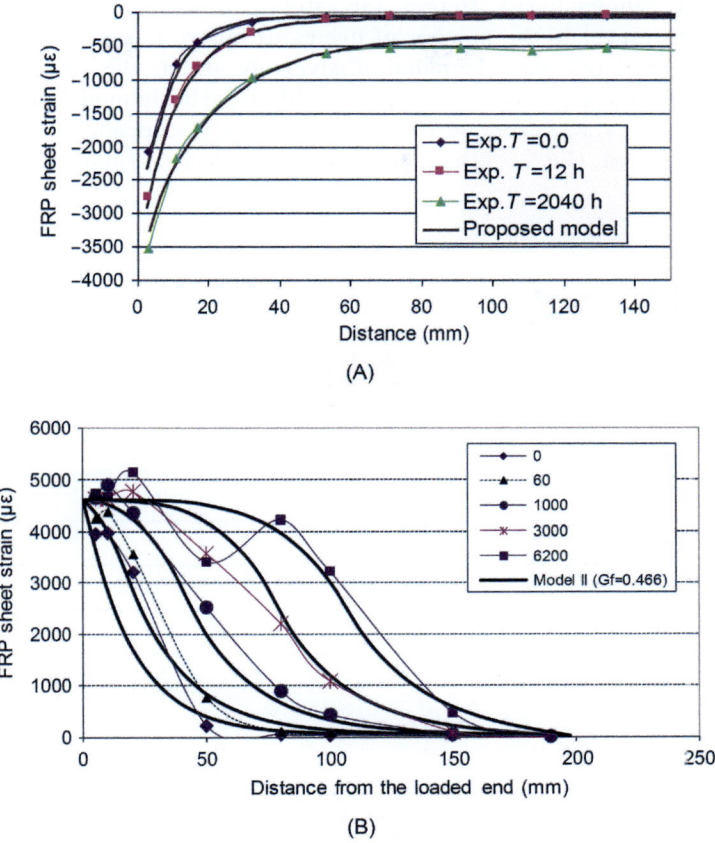

Figure 2.58 Comparison between numerical results with experimental results for FRP shear strain distribution under creep loading: (A) FRP-prestressed beams; (B) double-lap shear specimens (Wu and Diab, 2009).

losses occur. The creep of the adhesive layer is responsible for the losses at the effective bonding length. The actual debonding process is more complicated because real concrete is an inhomogeneous material. However, this model appears to be capable of predicting the debonding and creep fracture propagation along FRP−concrete interfaces, as shown in Fig. 2.58B.

2.4.3.3.2 Fatigue loading

Cycle-dependent degradation of the interfacial stiffness, the damage accumulation factor, and the bond−slip model are the factors governing the debonding propagation along FRP−concrete interfaces. In the available studies, the most common means for representing the fatigue life of FRP−concrete bonded joints is the Whöler curve. Herein, the fatigue strength of a structure (S) is plotted with respect to the number of cycles to failure (N_f). Nevertheless, the progressive local interface degradation adversely affects the serviceability of FRP-strengthened structures. Using the hybrid viscoelastic model, a theoretical S−N diagram can be generated (Fig. 2.59) and cycle-dependent displacement can be simulated (Fig. 2.60). These figures clearly demonstrate that the

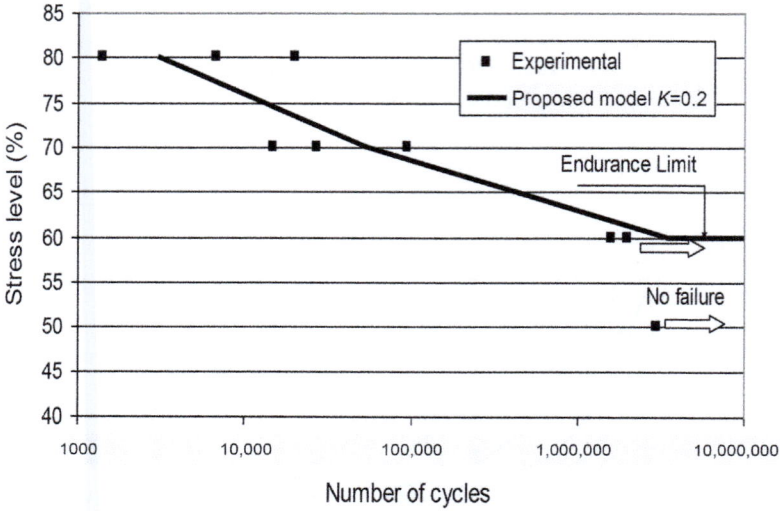

Figure 2.59 Experimental and predicted S−N diagram (Wu and Diab, 2009).

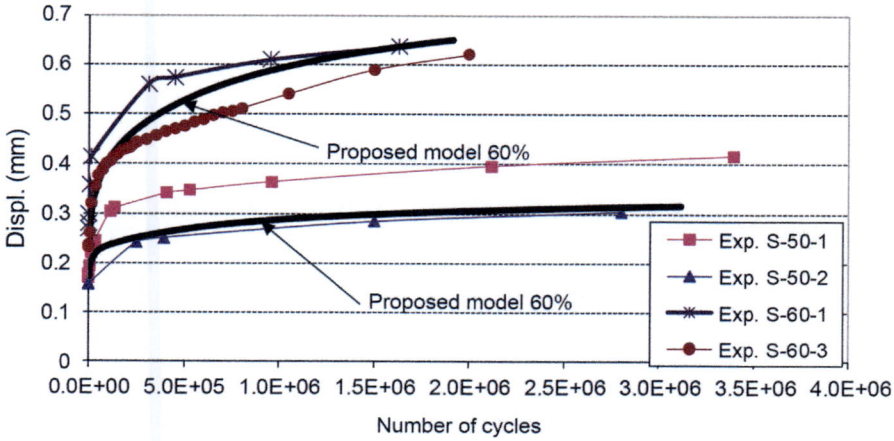

Figure 2.60 Experimental and predicted displacement versus the number of cycles, unfailure specimens (Wu and Diab, 2009).

proposed model is capable of simulating the real behavior under fatigue loading as revealed by the experimental results. Fig. 2.60 shows the predicted and experimental cycle-dependent loaded end displacements for the specimens that did not fail under fatigue loading. It is observed that the proposed model correctly predicts the fatigue displacement and its rate of development for different fatigue loading levels. The predicted results match with the experimental result for each fatigue loading. The predicted results indicate a behavior similar to that revealed by the experimental results: a high increase in displacement during early fatigue cycles and the subsequent steady increase.

2.4.3.4 Comparison between creep and fatigue performance of FRP—concrete interfaces

Fig. 2.61 compares the loading period of test specimens subjected to sustained and fatigue loading with the predictions based on the previously mentioned models. A comparison between the experimental results of creep and fatigue under equal maximum load reveals that specimens display higher damage under a high fatigue loading level, whereas creep specimens display higher damage under low sustained loads. The viscoelastic properties of the FRP—concrete interface are considered to be the dominant factor in the case of sustained loading. Meanwhile, fatigue damage accumulation owing to the fretting or friction heating between CFRP sheet and concrete at the crack tip is the dominant factor in the case of fatigue loading. Evidently, the proposed models provide similar explanation of the failure mechanism of FRP—concrete interfaces.

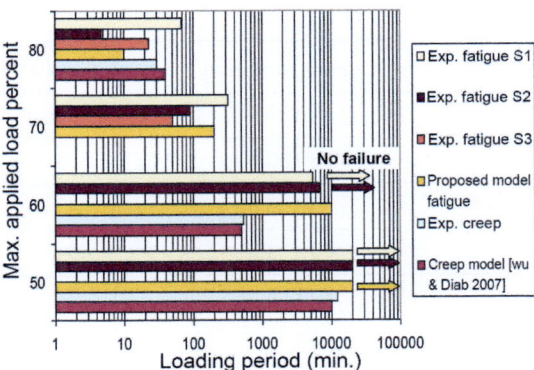

Figure 2.61 Comparison among loading periods of specimens under fatigue, sustained loading and the predicted results (Wu and Diab, 2009).

2.5   Durability of FRP—concrete interface

2.5.1   Temperature effect

As mentioned in Chapter 1, Fundamental Behavior of FRPs and Their Bonding Technique, FRP composites present many advantages such as high tensile strength and high stiffness-to-weight ratio. However, the tensile strength is generally not utilized completely owing to premature debonding between FRP sheet and concrete as discussed previously in this chapter. Moreover, FRP bonding properties under different severe environmental conditions continue to be an important issue that should be addressed. In general, epoxy resin softens and its bond capacity decreases under temperatures higher than the glass transition temperature (T_g) or heat deflection temperature (HDT) under load. Therefore an FRP—concrete interface can also fail under temperature near to or higher than T_g and HDT. The following sections describe an investigation of the bonding behavior of FRP laminates bonded to concrete, under a high temperature ($> 30°C$) environment (Wu et al., 2004).

The thermotolerance capacity of the epoxy is characterized by T_g and HDT. The test values of the bond adhesive properties according to ISO standards are summarized in Table 2.4.

Double-lap shear specimens are used in the investigation, as shown in Fig. 2.9C. Carbon fiber sheets are bonded to both sides of the concrete block along the axial direction. Details of the test specimens are available in Wu et al. (2005). A cup oven is used to maintain the temperatures of prism specimens. The temperatures in the cup oven are increased with a rubber heater and maintained using an aluminum sheet, pasteboard, and sponge rubber (Fig. 2.62). The tests on prism specimens with ordinary epoxy resin were conducted in the cup oven at temperatures ranging from 30°C to 50°C. Meanwhile, the tests on specimens with thermal-resistant epoxy resin were conducted in the cup oven at temperatures ranging from 40°C to 60°C. The

Table 2.4 Summary of bond adhesive properties.

Type of epoxy	Curing condition	T_g (°C)	HDT (°C)
Ordinary epoxy	80°C 3 h	38	48
	60°C 3 h	34	43
Ordinary primer	80°C 3 h	42	NA
	60°C 3 h	34	NA
Thermo-resistant epoxy	80°C 3 h	38	55
	60°C 3 h	40	57
Thermo-resistant primer	80°C 3 h	46	68
	60°C 3 h	55	76

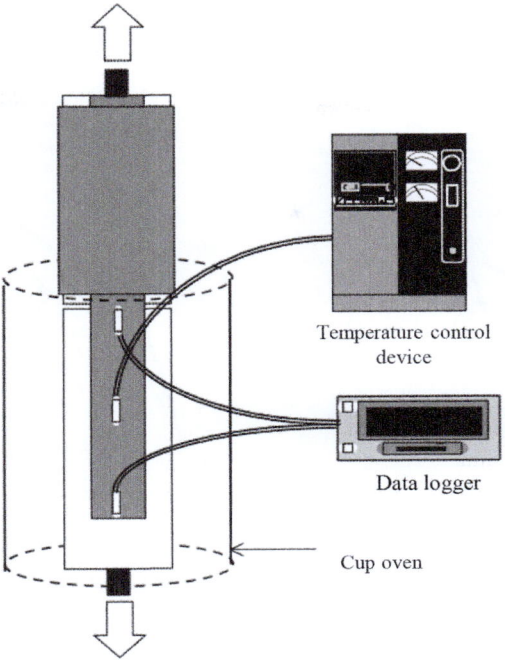

Figure 2.62 Details of the tensile test under different temperatures (Wu et al., 2005).

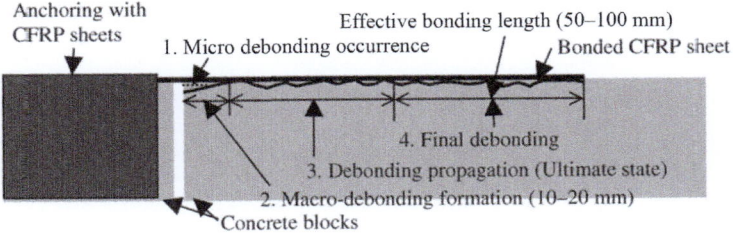

Figure 2.63 Process of debonding propagation (Wu et al., 2005).

reference temperature is maintained at approximately 26°C Each specimen was maintained under the test temperature for at least 6.5 h before the load is applied.

Fig. 2.63 shows the process of debonding propagation. Fig. 2.64 and Fig. 2.65 show the load—displacement curves and bond strength of specimens with both ordinary and thermal-resistant epoxy resins, at different temperatures. First, microdebonding initiates at the tensile end of the bonded FRP sheets. The microdebonding propagates gradually and as a result, macrodebonding occurs. Once a macrodebonding area is formed at approximately 20—30 mm, debonding propagates toward the

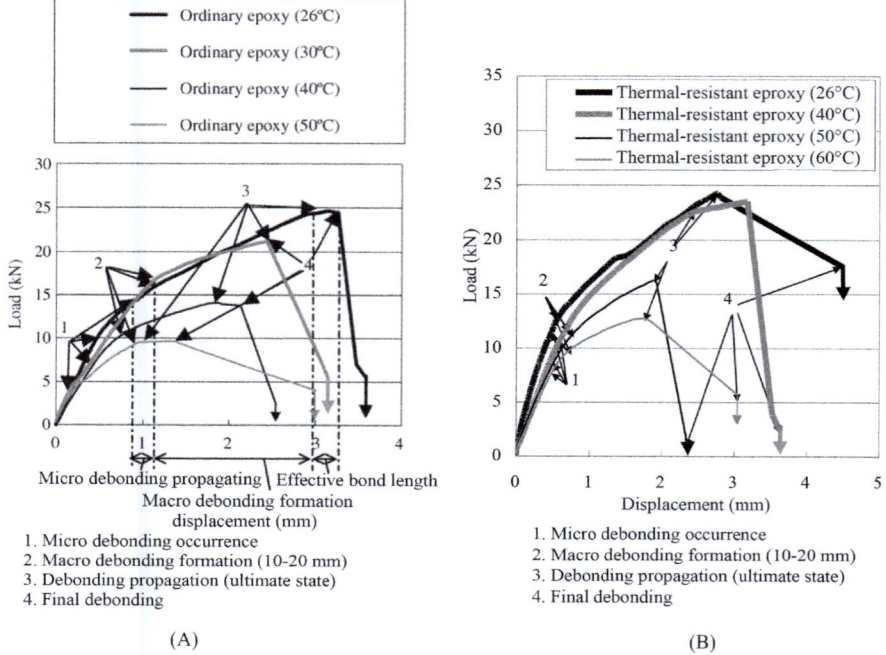

Figure 2.64 Load–displacement curve: (A) ordinary epoxy and (B) thermal-resistant epoxy (Wu et al., 2005).

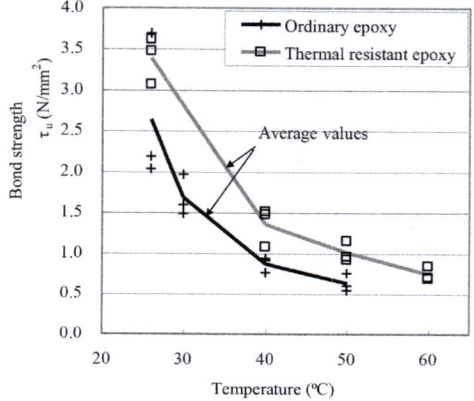

Figure 2.65 Bond strength under different temperatures (Wu et al., 2005).

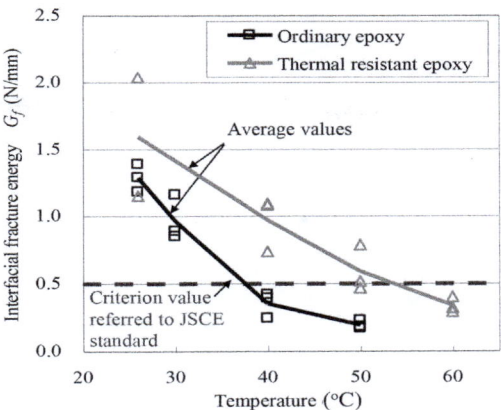

Figure 2.66 Interfacial fracture energy under different temperatures (Wu et al., 2005).

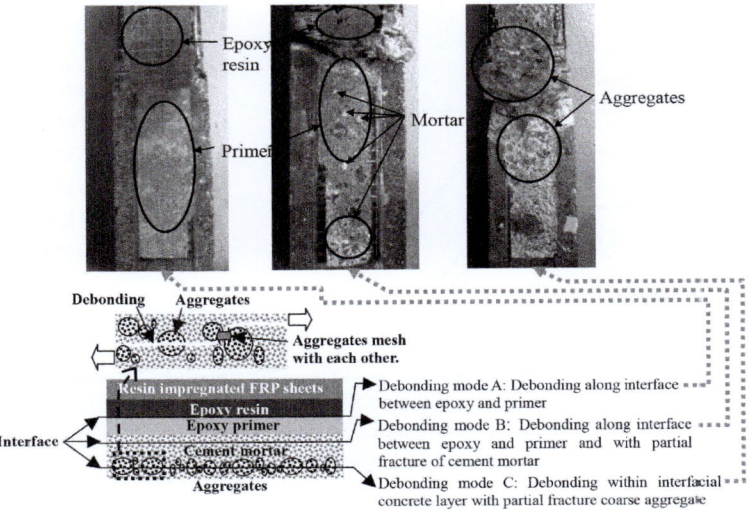

Figure 2.67 Different debonding modes along CFRP-to-concrete interface (Wu et al., 2005).

free end of the bonded sheet. Finally, a debonding crack forms through the entire bond length when debonding propagation reaches a critical location.

The debonding behavior is similar in both specimens (the one using ordinary epoxy resin and that using thermal resistant epoxy resin). The results of the prism tests are analyzed for determining the interfacial fracture energy (G_f), bond strength (τ_u), failure mode, and effective bonding length (L_e). G_f is inversely proportional to temperature (Fig. 2.66). In Fig. 2.66, G_f values decrease when the temperature is near to or higher than T_g. It is also observed that the bonding behavior of CFRP sheets degrades considerably with the increase in the service temperature even

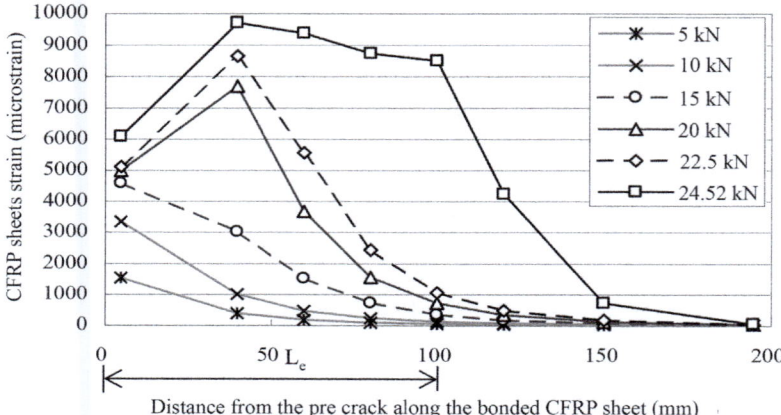

Figure 2.68 Strain distribution along bonded CFRP sheet for a specimen by using ordinary epoxy resin (26°C) (Wu et al., 2005).

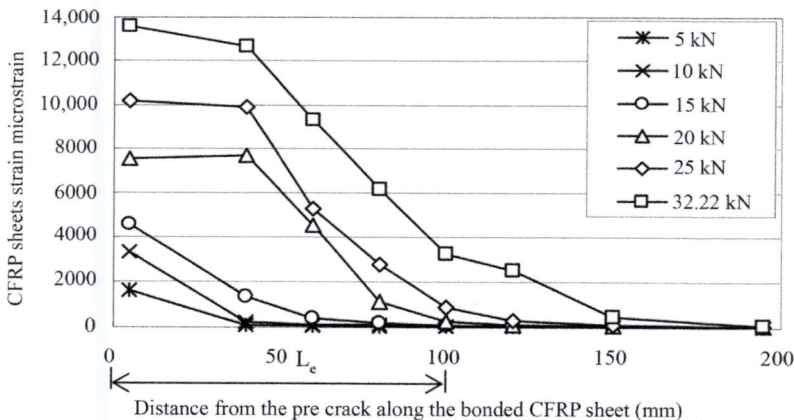

Figure 2.69 Strain distribution along bonded CFRP sheet for a specimen by using thermal resistant epoxy resin (26°C) (Wu et al., 2005).

before the T_g of epoxy resin is attained. The minimum value of G_f for normal bending situations at ambient temperature (26°C) is specified to be 0.5 N/mm. It is observed for both the epoxy resins that G_f reduces below the minimum value when the service temperature exceeds T_g. However, the thermally resistant epoxy resin exhibits higher thermal tolerance for debonding.

Based on direct inspection of debonding location and situations, debonding failure can be classified into three modes: Debonding Mode A exhibits debonding along the interface between the epoxy resin and epoxy primer, Debonding Mode B exhibits

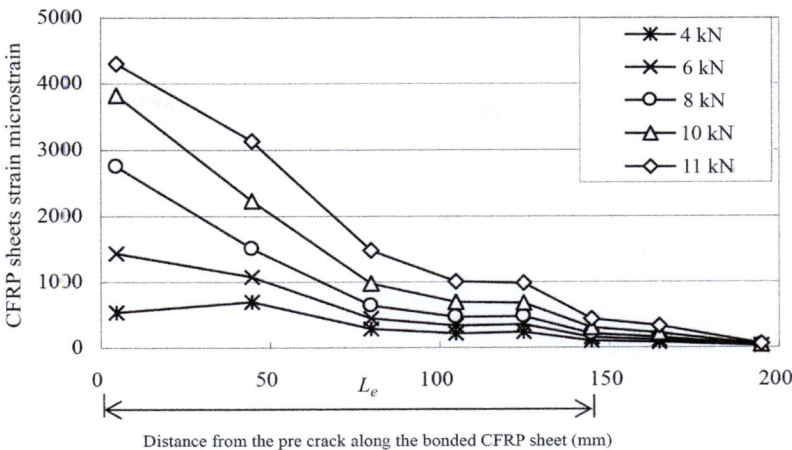

Figure 2.70 Strain distribution along bonded CFRP sheet for a specimen by using thermal-resistant epoxy resin (50°C) (Wu et al., 2005).

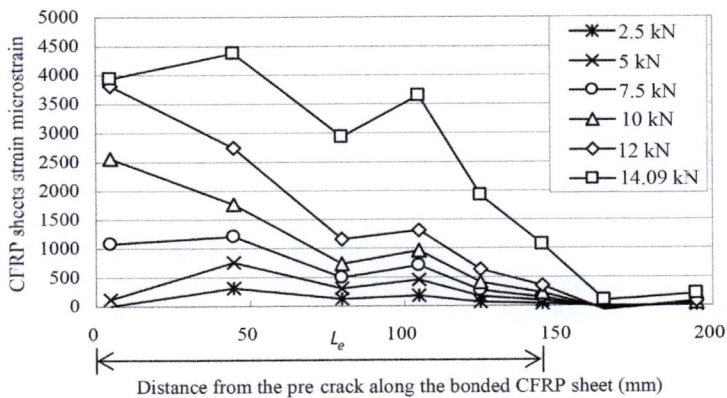

Figure 2.71 Strain distribution along bonded CFRP sheet for a specimen by using ordinary epoxy resin (40°C) (Wu et al., 2005).

debonding along the interface between the epoxy resin and epoxy primer and partial fracture of cement mortar, and Debonding Mode C exhibits crack within the concrete (Fig. 2.67). The debonding mode at temperatures below T_g is either "B" or "C" and that at temperatures above T_g is mostly "A." Fig. 2.68 and Fig. 2.69 show the strain distributions of specimens with ordinary epoxy resin, along the fiber direction at temperatures of approximately 26°C and 40°C.

The strain distributions of specimens with thermal resistant epoxy resin at 26°C and 50°C are shown in Fig. 2.70 and Fig. 2.71. Effective bonding length (L_e) is defined as the distance from the precrack of a prism specimen to the position where

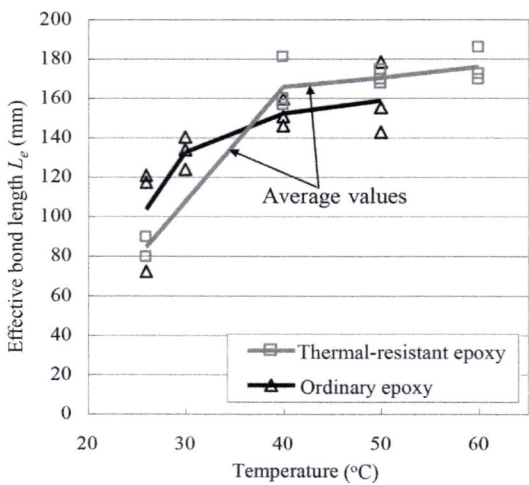

Figure 2.72 Effective bond length under different temperature (Wu et al., 2005).

97% of the strain value at the precrack occurs. Fig. 2.72 shows the effect of temperature on L_e of the specimens. In both the resins, the L_e values appear to increase as the service temperatures increase.

2.5.2 Combined effect of freeze−thaw cycling and sustained load

Freeze−thaw cycling is a primary concern that severely threatens the service life of concrete structures in cold regions. When FRP composites are applied to strengthen concrete structures, the concrete substrate, reinforcing material, adhesive, and FRP−concrete interface are affected by the freeze−thaw environment. Temperature-induced stresses in the adhesive layer caused by differential thermal expansion and contraction of FRP composites and concrete materials as well as the expansion and contraction of water in the freeze−thaw environment may damage the interface and contribute to premature bond failure.

A number of tests were carried out in recent years to observe the durability of FRP-strengthened concrete structures. A few of these were focused on the bond behavior of the FRP−concrete interface after the action of freeze−thaw cycling. Certain researchers (e.g., Davalos et al., 2005; Silva and Biscaia, 2008; Subramaniam et al., 2008; Yun and Wu, 2011) examined the durability of normal concrete−FRP interfaces in freeze−thaw environments. It was observed that the bond capacity decreases remarkably as the number of freeze−thaw cycles increases, and the failure mode in the concrete substrate is debonding (Fig. 2.73). The main reason for this dramatic degradation is that the normal concrete is highly sensitive to freeze−thaw cycles. The bonded joints with normal concrete can sustain only a highly limited number of freeze−thaw cycles. Meanwhile, Mukhopadhyaya et al. (1998)

Figure 2.73 Failure modes of FRP-to-concrete joints subjected to freeze—thaw cycling: normal concrete specimen (Yun and Wu, 2011).

and Green et al. (2000) examined the freeze—thaw response of GFRP—and CFRP—concrete interface, respectively. Air-entrained concrete was used in their studies. Both the study groups indicated that the delamination force of FRP—concrete interfaces decreased negligibly after freeze—thaw exposure. Moreover, shifts in failure plane location from the concrete substrate to the concrete—adhesive interface or the adhesive layer were reported. Fava et al. (2007) also observed that freeze—thaw cycles do not markedly affect the delamination force of CFRP—concrete interfaces although they reduce the bond—slip parameters (local bond shear strength, interfacial fracture energy, and peak slip).

Furthermore, the bond between an FRP laminate and the concrete substrate is typically in a stressed condition in real strengthening applications of externally bonded FRP. During freeze—thaw cycling, the cyclic stress induced by the expansion and shrinkage action of the pore water in the adhesive and concrete are likely to be magnified in the presence of a service load (sustained or fatigue loads). It is essential to consider the coupled effects of the harsh environment and service load to determine the durability characteristics of FRP—concrete interfaces in concrete structures strengthened using externally bonded FRP. The following section introduces an investigation of the coupled effects of freeze—thaw cycling and sustained loading on the bond behavior of FRP—concrete interfaces (Shi et al., 2013).

2.5.2.1 Experimental program

A double-lap shear specimen was used in the test, as shown in Fig. 2.9C. A layer of basalt fiber-reinforced polymer (BFRP) sheet 50-mm wide and 0.156-mm thick was bonded to opposite sides of the concrete blocks along the axial direction. A specially designed spring reaction frame loading system, as shown in Fig. 2.74, was used to apply and maintain the sustained load on the BFRP—concrete interface. Before the application of the sustained load, unconditioned control specimens

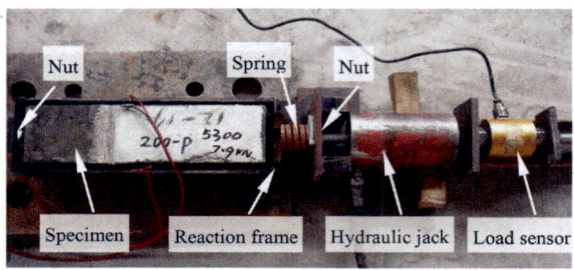

Figure 2.74 Applying sustained load on specimen (Shi et al., 2013).

were tested to determine the ultimate load and load—strain response of the BFRP—concrete interface.

In accordance with the rapid freeze—thaw test method recommended by China National Standard (GB/T 50082-2009), specimens were soaked in water in rubber boxes and subjected to accelerated freeze—thaw cycles each of 3—4 h. The temperature in the concrete prism varied from $+8°C$ to $-17°C$ during the freeze—thaw cycling. The typical temperature fluctuations recorded by the two sensors during the freeze—thaw cycling is shown in Fig. 1.51.

The experimental parameters included the number of freeze—thaw cycles, sustained load level, and adhesive type. As the freeze—thaw cycling was carried out in water, all the specimens except the control specimens were soaked in water for an identical length of time to prevent variation in the material mechanical properties owing to different water soaking times, and to study the effect of water soaking on the bond behavior of the BFRP—concrete interface.

2.5.2.2 Test results

In the control specimens, debonding failure occurred in a thin concrete layer beneath the BFRP. In addition, concrete shear failure occurred at the loaded end (Fig. 2.75A). After soaking in water and the freeze—thaw cycling, the failure modes changed from debonding in the concrete layer to partial or complete debonding in the adhesive layer (Fig. 2.75B). As the number of freeze—thaw cycles increased, less concrete was observed on the debonded BFRP, and fewer concrete shear failures were observed at the loaded end (Fig. 2.75C). This change is mainly because of the degradation of the adhesive after the freeze—thaw cycling, as discussed previously. After 200 freeze—thaw cycles, the specimens subjected to sustained loads failed partially in the adhesive layer. Meanwhile, the specimens not subjected sustained loads failed completely in the adhesive layer. Additional damage in the concrete layer caused by the coupled effects may explain this phenomenon.

Fig. 2.76 shows typical experimental load versus global slip graphs of the BFRP—concrete interface. Fig. 2.76A shows the effect of only freeze—thaw cycling on the load versus global slip response. Freeze—thaw cycling reduced the ultimate load, as well as the ultimate slip and the stiffness before initial debonding. It is

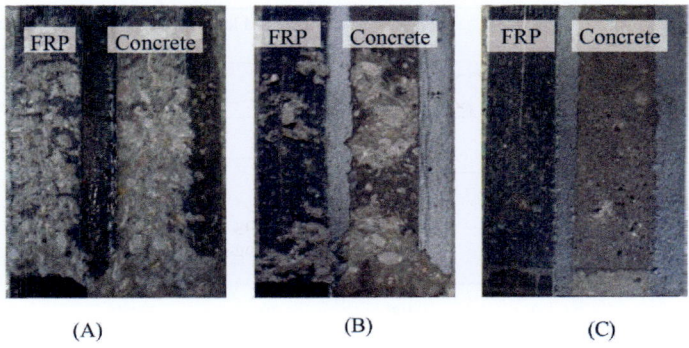

Figure 2.75 Failure modes of FRP-to-concrete interface (Shi, 2014).

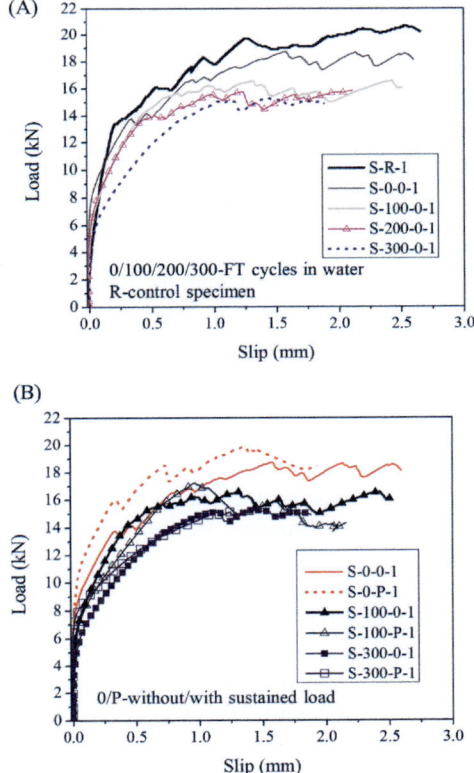

Figure 2.76 Load versus global slip of the BFRP-to-concrete interface (Shi et al., 2013). (A) Effect of freeze—thaw cycling. (B) Effect of sustained load.

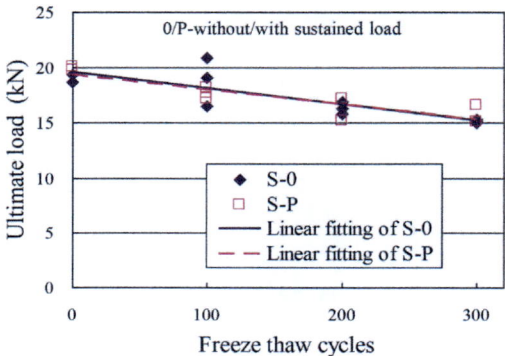

Figure 2.77 Ultimate load of double-lap shear specimens (Shi et al., 2013).

observed that under the sustained loading during freeze−thaw cycling (shown in Fig. 2.76B), the ultimate slip is reduced as is evident in Fig. 2.77, the ultimate load decreased linearly as the number of freeze−thaw cycles increased. Moreover, as the two linear best-fit lines demonstrate, there is no negative effect of the sustained loading on the ultimate load.

The bond−slip relationship is the constitutive model of the FRP−concrete interface and is essential to the study of the behavior of FRP-strengthened concrete structures. A detailed description of a method for determining the bond−slip relationships from test data is available elsewhere (Shi et al., 2013). Fig. 2.78 shows the degradation of the bond−slip relationship after the freeze−thaw cycling. As the number of cycles increased, significant and continuous reductions were observed in the maximum shear stress (τ_0) and interfacial fracture energy (G_f) for specimens not under and under sustained loading (Fig. 2.78A and B), respectively. Fig. 2.78A shows that the water soaking environment adversely influenced the bond−slip relationship of the BFRP−concrete interface (the specimens S−R and S−0−0). The freeze−thaw cycling exerted negligible influence on the slip corresponding to the maximum shear stress (s_0) in the bond−slip relationship.

A detailed comparison of the key parameters of the bond−slip relationship of the BFRP−concrete interface is shown in Fig. 2.79. Linear trend lines are used to illustrate the degradation of the parameters. Continuous degradation is observed in the maximum shear stress (Fig. 2.79A) and interfacial fracture energy (Fig. 2.79B). The specimens under sustained loading tended to exhibit a lower G_f albeit on average a marginally higher τ_0 than the corresponding specimens not under sustained loading. This indicates that sustained loading during freeze−thaw cycles may cause the bond behavior of the BFRP−concrete interface to be more brittle. An increasing trend in the BFRP effective bond length (L_e) was observed as the number of freeze−thaw cycles increased, as shown in Fig. 2.79C. As L_e is inversely proportional to concrete strength according to the available literature (e.g., Wu et al., 2009a,b), it is reasonable to conclude that the increasing trend of L_e is caused by the degradation of the concrete substrate. L_e remained almost unaltered for the specimens under

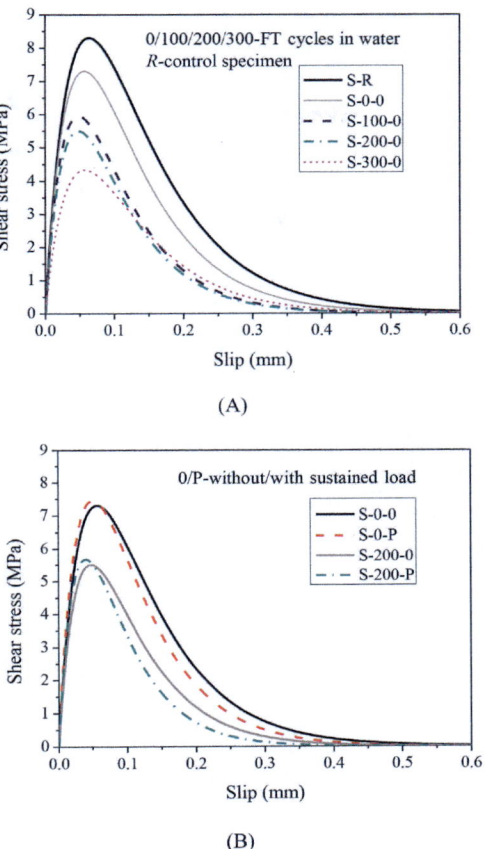

Figure 2.78 Effect of freeze—thaw cycling and sustained load on the bond—slip relationship (Shi et al., 2013). (A) Freeze—thaw cycling. (B) Sustained load.

sustained loading. This could be explained by the additional brittleness of the bond—slip relationship, introduced by the sustained loading during freeze—thaw cycling.

2.5.2.3   Discussion of the coupled effects

The test results presented here reveal that there was negligible additional loss of ultimate load for the specimens under sustained loading compared to those not one sustained loading. However, the fitting results from the analysis of slip data reveal a large influence of the coupled effects on the bond—slip relationship. The coupled effects may magnify the microcracks in the adhesive layer and the adhesi-ve—concrete interaction region beneath the interface, causing additional damage to the BFRP—concrete interface, as illustrated in Fig. 2.80A. However, the sustained

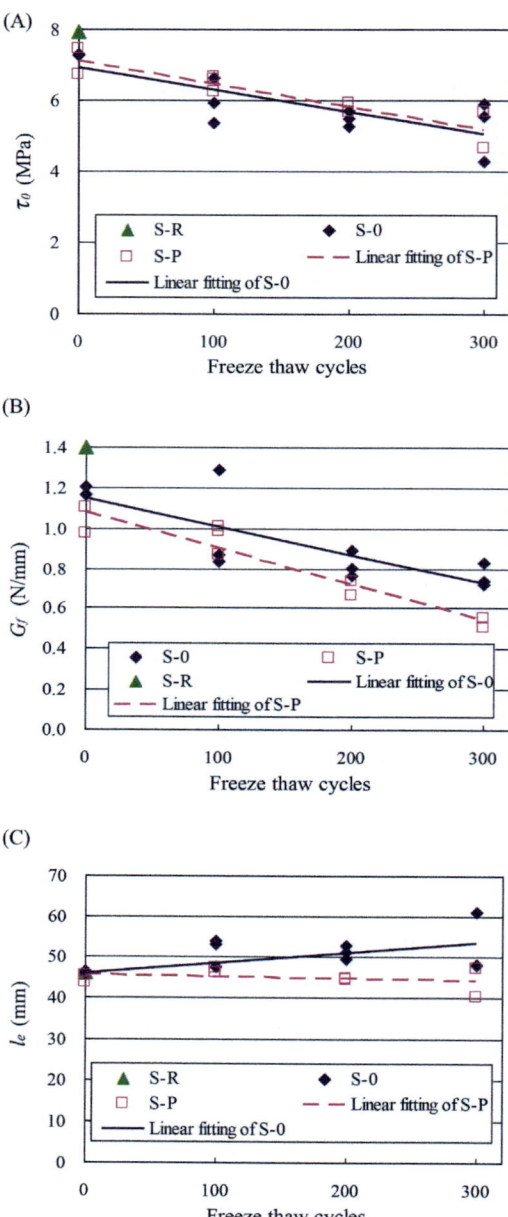

Figure 2.79 Variation of key bonded parameters (Shi et al., 2013). (A) Maximum shear stress. (B) Interfacial fracture energy. (C) Effective bond length.

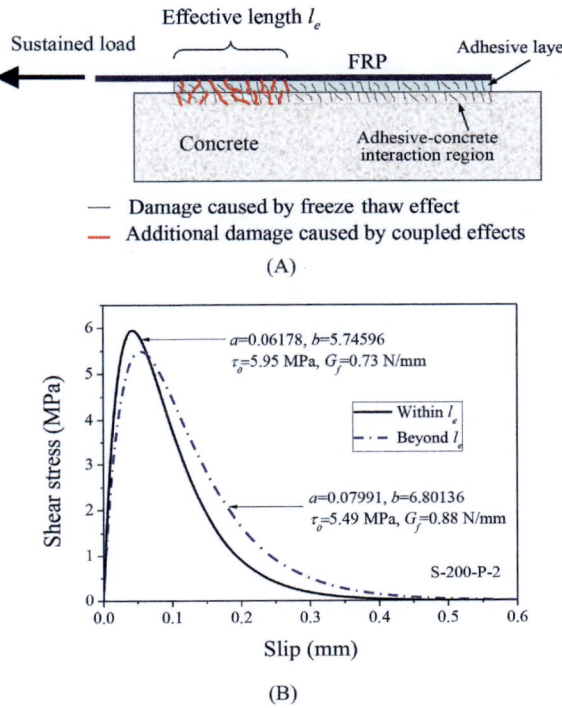

Figure 2.80 Additional damage caused by coupled effects (Shi et al., 2013). (A) Damage of the FRP-to-concrete interface. (B) Bond—slip relationships along bonded length

load is effective only within the limited bond region because of the presence of the effective transfer length l_e. In the bonded region beyond l_e, the bond behavior is influenced only by the freeze—thaw environment. Moreover, the bond—slip relationship is different compared to that within l_e, as shown in Fig. 2.80B. Within the bond length l_e, the debonding resistance of the BFRP—concrete interface was reduced owing to the additional damage caused by the sustained load. Therefore debonding may occur and propagate under the combined action of sustained loading and the cyclic thermal stress generated by freeze—thaw cycling. In summary, if the BFRP bond length is adequately long (i.e., $>2L_e$), although the coupled effects marginally influence the ultimate load of the BFRP—concrete interface, the additional damage caused by sustained loading is likely to increase the likelihood of debonding in freeze—thaw environments.

2.5.3 Moisture effect

FRP-strengthened RC structures in marine environments (e.g., port structures) are generally subjected to adverse environmental conditions such as concrete surface

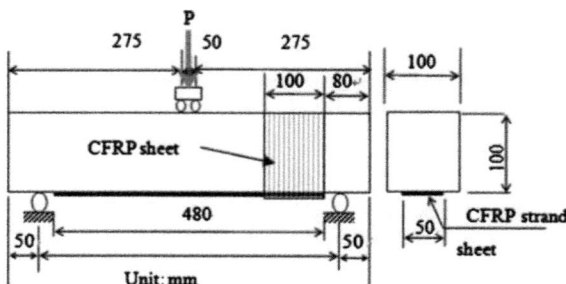

Figure 2.81 Geometry of specimens for bending tests (Dai et al., 2010).

moisture, external moisture in the air during the bonding of FRP, and frequent wet−dry (WD) cycles during the service life. The present understanding of the short- and long-term durabilities of FRP−concrete bond interfaces under moisture is limited. In particular, there are minimal long-term exposure data on the fundamental mechanical properties of FRP−concrete bonded interfaces affected by local climatic conditions. In the following sections, a two-year experimental investigation to examine the effects of moisture on the short- and long-term bond performance of FRP−concrete interfaces is introduced (Dai et al., 2010).

Plain concrete beams strengthened with soffit FRP sheets were prepared for the bending tests (Fig. 2.81). In the test program, coupled thermal and moisture action was applied to accelerate the moisture exposure. The specimens were divided into two series: Group A and Group B. The Group A specimens were tested for short-term moisture effects on bond performance, and the Group B specimens were tested for long-term moisture effects. The test variables included the preconditioned concrete substrate moisture content (WD) at the time of FRP bonding, relative humidity (RH) of air (40% and 90%) during the FRP composite curing, adhesive primer type (normal and hydrophilic), bonding adhesive type (normal and ductile), and exposure duration (unexposed, 8 months, 14 months, and 2 years). Each WD cycle consisted of a 4-day immersion in 60°C sea water and a 3-day exposure to dry air in a laboratory. Further details of the test specimens are available in the literature (Dai et al., 2010).

2.5.3.1 Effect of moisture at the time of FRP installation (construction moisture)

Bending tests were conducted on 21 FRP-bonded preconditioned concrete beams to evaluate the effects of construction moisture (concrete surface moisture and air humidity for curing FRP composite) and the primer and adhesive, on the short-term shear bond performance of FRP−concrete interfaces. All the beams failed owing to debonding between the FRP composite and concrete substrate. Three types of interfacial failure have been identified as shown in Fig. 2.82: (1) failure in the concrete substrate (Mode A), (Fig. 2.82A); (2) failure in a thin mortar layer of the concrete

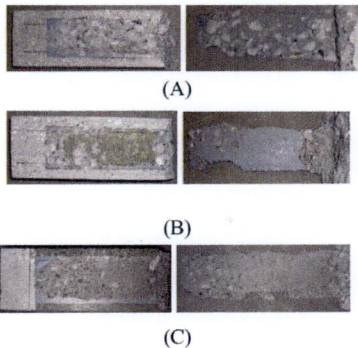

Figure 2.82 Effect of short-term moisture on the shear failure of FRP-to-concrete interface (Dai et al., 2010). (A) Mode A. (B) Mode B. (C) Mode C.

substrate (Mode B), (Fig. 2.82B); and (3) failure at the primer—adhesive interface (Mode C), (Fig. 2.82C). When the concrete substrate was dry, and a normal primer was applied, Failure Mode "A" was observed regardless of the curing air humidity. When the concrete substrates were wet, Failure Mode B was observed for all the specimens. A likely reason for such a failure is that penetration into the wet concrete substrates was superficial. Failure at the primer—adhesive interface (Mode C) was also observed when the concrete substrates were wet. However, this failure mode did not result in decrease in the flexural capacity of the FRP-bonded concrete beams.

Fig. 2.83A—C presents the effects of the concrete surface moisture and air moisture at the time of FRP installation, on the bending load versus midspan deflection curves of the FRP-bonded concrete beams. In the figures, three identical specimens are presented to illustrate the degree of the experimental scatter in each test series. When the concrete substrates were dry, a high RH curing resulted in a marginal decrease in flexural capacity (Fig. 2.83A). However, when the concrete substrate was wet, a high RH curing resulted in a marginal increase in flexural capacity (Fig. 2.83B). Because the flexural response of FRP-bonded plain concrete beams directly reflects the FRP—concrete interface shear bond performance, it can be concluded that variation in RH curing values from 48% to 90% marginally affects the shear bond performance of FRP—concrete interfaces. The application of the normal type of primer onto the wet substrate caused approximately 20% loss of flexural capacity (Fig. 2.83C).

2.5.3.2   Effect of moisture during service life (WD) cycles

Fig. 2.84 presents the bending load versus midspan deflection curves of all the FRP-bonded concrete beams that experienced different periods of WD cycling. The WD cycling did not influence the ascending linear sections of the curves. This is because the strength and stiffness of the concrete are negligibly influenced by a regime of WD cycling. However, following the cracking of concrete, the

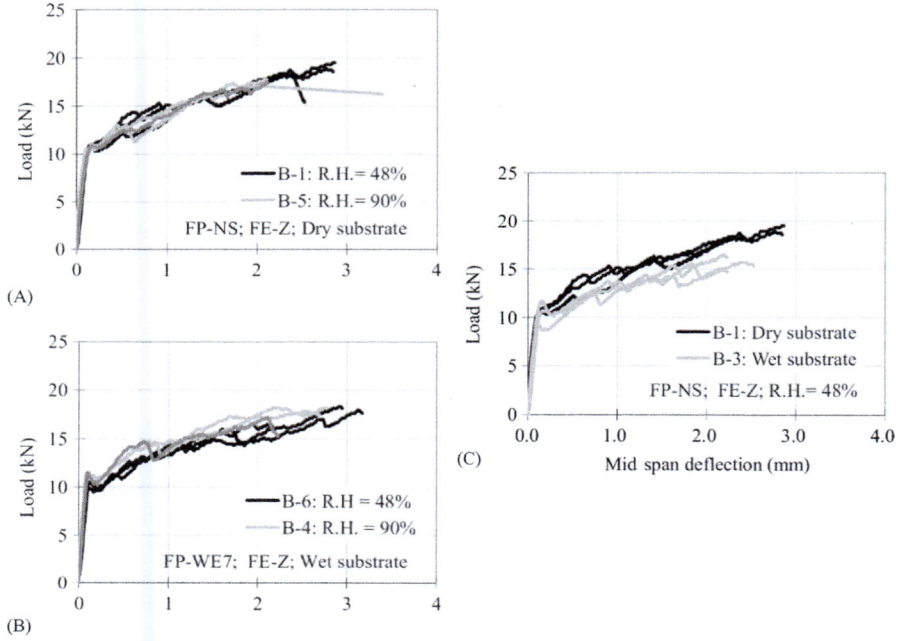

Figure 2.83 Effect of moisture on the short-term performance of FRP bonded concrete beams (Dai et al., 2010).

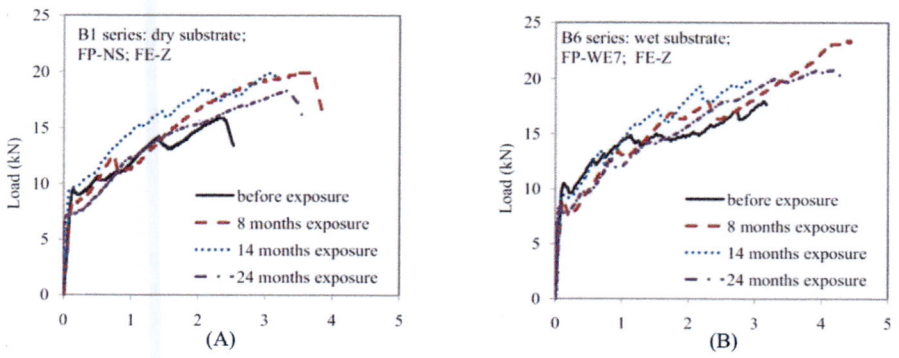

Figure 2.84 Effects of moisture on the long-term performance of FRP bonded concrete beams (Dai et al., 2010).

load−deflection curves directly reflect the shear bond performance of the FRP−concrete interfaces. Both the flexural capacity and ductility increased notice-ably after 8 and 14 months of exposure. Further exposure till 24 months resulted in a marginal decrease in flexural capacity. However, the values were still higher than

Figure 2.85 Failure at the primer-to-concrete interface after WD cycling (Dai et al., 2010).

those of the reference beams unexposed to weathering. All the exposed FRP-bonded concrete beams failed by debonding at the primer—concrete interface, as shown in Fig. 2.85. After the bending tests, it was observed that almost no concrete was attached to the peeled off FRP. There was no decrease in the flexural capacity of the FRP-bonded concrete beams after the long-term accelerated WD cycling.

2.6 Enhanced FRP bonding system

As an adhesive bond is relatively weak, various techniques have been developed to increase the bond strength. Conventional mechanical anchors or bolts generally cannot be used directly for anchoring normal FRP laminates (plates, strips, or sheets). This is because FRP does not exhibit sufficient bearing strength to prevent longitudinal cutting by the anchors. To resolve this problem, various special methods for bond enhancement have been developed and reported in the literature: end anchorage, fiber anchor (or anchor spike), and U-jacketing. End anchorage is effective for constraining debonding initiated at the two ends of an FRP strip or for preventing complete detachment of FRP from the concrete substrate (Garden and Hollaway, 1998; Spadea et al., 1998). However, it cannot effectively block the early development of debonding elsewhere. Hence, it is generally ineffective for IC debonding, which usually exerts the highest influence the bond strength of flexural members with high span/depth ratios. In fiber anchoring systems, anchors made from bundles of fibers are inserted into epoxy-filled holes in the concrete substrate. The free ends of the fibers are then splayed and bonded onto the surface of the FRP laminate with adhesive (Teng et al., 2000, 2003; Ekenel et al., 2006). U-jacketing is a popular and extensively investigated bond enhancement method. Herein, many U-shaped FRP or steel strips are used to wrap the longitudinal FRP strip along the span (Ritchie et al., 1991; Swamy and Mukhopadhyaya, 1999; Smith and Teng, 2001; Brena et al., 2003; Masoud et al., 2005; Kotynia et al., 2008; Wang and Hsu, 2009). This method can delay IC debonding with a limited increase in bond strength, and was included in the former Chinese FRP design code. However, because of the limited strength enhancement and its uncertainty, it was removed from the subsequent Chinese FRP design code.

It has been observed that bond strength can be increased significantly when FRP is embedded in a cover layer of reinforced concrete (De Lorenzis and Teng, 2007) —a technique referred to as near-surface mounting. However, the grooving process may cut the existing transverse bars. Another effective method known as mechanically fastened FRP (MF-FRP) has been demonstrated to be a viable method for flexural strengthening of RC members (Lamanna et al., 2001). However, this method is applicable only to special FRP strips, and the product has a limited lifetime.

Two bond enhancement systems, fiber anchor and hybrid-bonded (HB)-FRP, are introduced in this section.

2.6.1   Fiber anchorage system

Several types of fiber anchorage methods have been developed to achieve higher levels of fiber utilization prior to premature debonding failure. A number of experimental and theoretical research studies have been carried out to validate these anchorage methods and develop optimal design methods. In the following section, two general approaches of FRP anchorage system are briefly introduced.

2.6.1.1   FRP sheet anchorage method

Two types of FRP sheet anchorage methods were developed and applied to FRP—concrete joint specimens by Wu et al. (1999), as shown in Fig. 2.86. FRP sheets with two types of anchorage were bonded to the concrete surface, and FRP

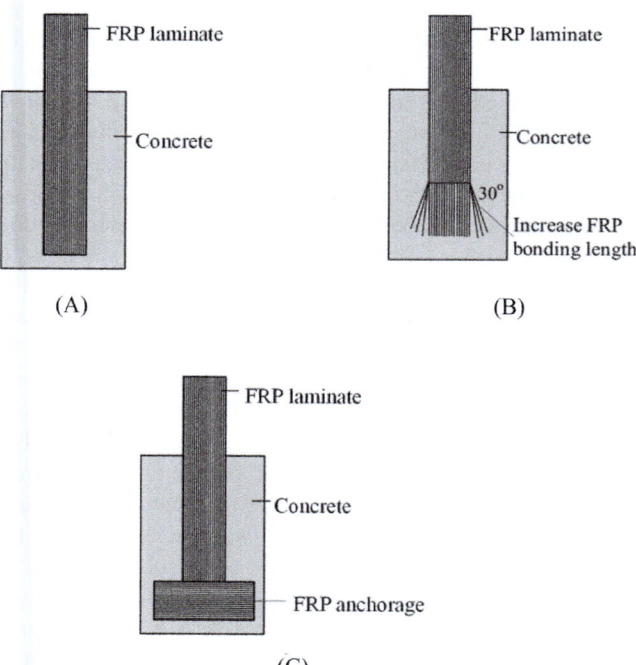

Figure 2.86 FRP sheet anchorages of the test specimens.

Table 2.5 Details of the test specimens.

Specimen No.	FRP bond length L_f (mm)	Bonding area (mm^2)	FRP anchorage	Failure mode
1	15	150	No anchorage	FRP debonding
2	20	200	No anchorage	FRP debonding
3	15	164	Anchorage A-5 cm	FRP debonding
4	15	182.5	Anchorage A-7.5 cm	FRP debonding
5	15	207.7	Anchorage A-10 cm	FRP fracture
6	20	214	Anchorage A-5 cm	FRP debonding
7	20	232.5	Anchorage A-7.5 cm	FRP debonding
8	20	257.7	Anchorage A-10 cm	FRP debonding
9	15	150	Anchorage B-0° fiber	FRP debonding
10	15	150	Anchorage B-90° fiber	FRP debonding

anchorages were applied with different geometrical parameters, as presented in Table 2.5.

After tensile loading of the FRP laminates, all the specimens except No. 5 failed by FRP debonding from the concrete prism. Moreover, FRP fracture was observed in Specimen No. 5, as shown in Fig. 2.87. For the specimens with FRP anchorage A, the maximum load and interfacial fracture energy (G_f) increased apparently with the applied FRP anchorage length. The specimens with longer FRP bond length exhibited higher increasing trend, particularly for the specimen with an FRP anchorage length of 10 cm, as shown in Fig. 2.88 and Fig. 2.89. Both the specimens with FRP anchorage B exhibited higher bond capacity than the corresponding unanchored ones. The specimen with the FRP anchorage sheet bonded parallel to the fiber direction of the bonded FRP laminates (0-degree fiber anchorage) displayed dramatically higher maximum load and G_f. This indicated that the anchorage efficiency of the 0-degree fiber anchorage is significantly higher than that of the 90-degree fiber anchorage.

2.6.1.2 Fiber anchors

Inspired by the research of FRP sheet anchorages, carbon fiber anchors have been widely applied in the FRP-strengthened structures in Japan (Tsukagoshi et al., 1999). The commonly used carbon fiber anchor is made from carbon fiber sheets/strands manufactured in a factory or hand-made in a field. There are two types of fiber anchors depending on the mounting method: through type and inserted type.

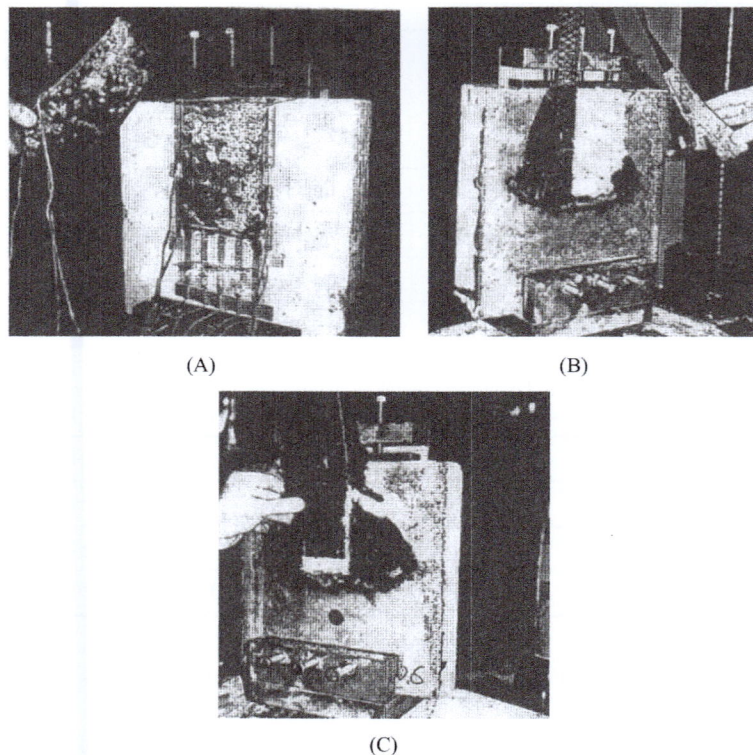

(A)   (B)

(C)

Figure 2.87 Typical failure modes. (A) Specimen No. 1. (B) Specimen No. 3. (C) Specimen No. 5.

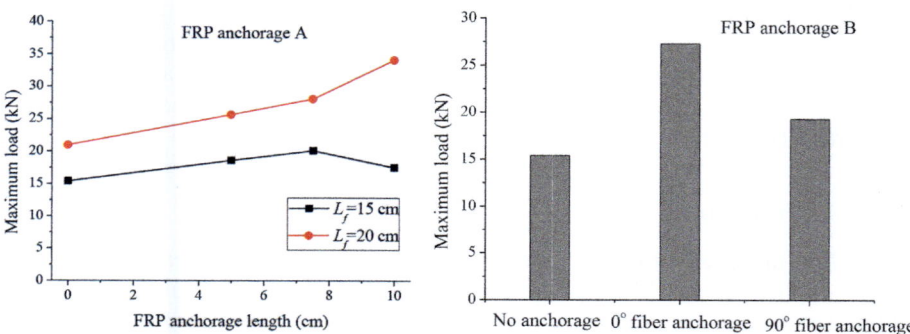

Figure 2.88 Maximum load of the test specimens.

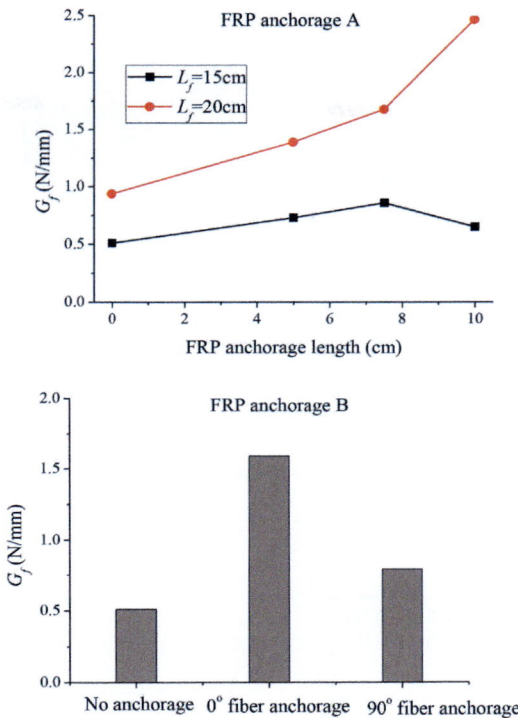

Figure 2.89 Interfacial fracture energy (G_f) of the test specimens.

The through type is used to connect carbon fiber sheets separated by an obstacle. The inserted type is used when the obstacle is thick, and it is challenging to open the through- or when the end of the carbon fiber sheet is directly fixed to the concrete.

Another form of fiber anchor is made from rolled fiber sheets or bundled loose fibers (Kalfat et al., 2013), as shown in Fig. 2.90. One end of the anchor (anchor dowel) is inserted into a predrilled hole in the concrete substrate, and the dowel length can be confined to the cover region of the member. The other end of the anchor is epoxied onto the surface of the FRP plate. The ends of the fibers, which are splayed and epoxied onto the surface of the laminates to disperse local stress concentrations, are hereinafter referred to as anchor fan.

To determine the characteristics of the FRP anchor, Zhang and Smith (2012) investigated FRP–concrete joints with similar FRP anchors. A generic load–slip response of single fan and bow-tie anchors is shown in Fig. 2.91. The three main stages of the load–slip response are denoted by A (debonding and activation of FRP anchor), B (postpeak reserve of strength offered by completely intact FRP anchor and frictional resistance of debonded plate), and C (postpeak reserve of strength offered by partially intact FRP anchor and frictional resistance of debonded

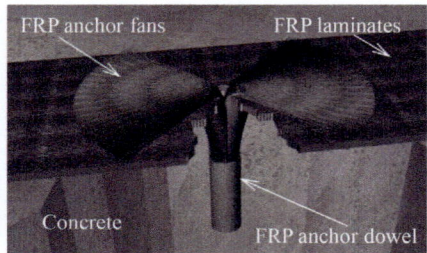

Figure 2.90 FRP anchor (Kalfat et al., 2013).

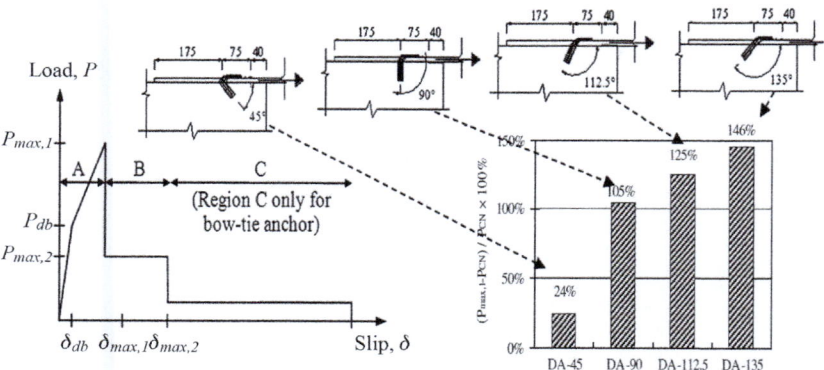

Figure 2.91 Load−slip response and joint strength enhancement of the FRP anchor (Kalfat et al., 2013).

laminates). A review by Smith (2010) reported that FRP anchors with a single fan component increase the shear strength and slip capacity of FRP−concrete joints by up to 70% and 800%, respectively, over those of unanchored control joints. Of particular interest in Fig. 2.91 is the significant effect of the dowel angle on the enhancement of the joint strength over that of the unanchored control joint (Zhang and Smith, 2012).

2.6.2 Hybrid-bonded FRP system

Although the previously described methods can improve FRP−concrete bonding strength, the magnitude of increase is generally limited and insufficient. A simple mechanical fastener (Fig. 2.92) has been developed to enhance bond strength (Wu and Huang, 2008). This mechanical fastener is composed of a thin capping plate that applies normal pressure on the FRP sheet to constrain the vertical separation of FRP from the concrete substrate. In addition, two small anchors are used to fasten

(A)   (B)

Figure 2.92 HB-FRP system: (A) mechanical fastener; (B) HB-FRP retrofitted beam
(bottom view) (Wu and Huang, 2008).

the plate onto the concrete. After the FRP is adhesively mounted, fasteners are
used to "staple" the FRP strip onto the concrete substrate, as shown in Fig. 2.92B.
This system is called the HB-FRP because it uses both an adhesive bond and a
mechanical fastening. The major difference between mechanical fasteners used in
the HB-FRP system and those used in the MF-FRP system (Lamanna et al., 2001)
is that no bearing resistance is required between the anchors and FRP in HB-FRP,
and the fasteners increase the bond by suppressing the interfacial relative slip and
inducing frictional bond (Wu and Huang, 2008; Wu et al., 2010). Consequently,
the system is applicable to all commercially available FRP laminates. Experimental
tests and theoretical studies (Wu and Huang, 2008; Wu et al., 2009a,b; Wu et al.,
2010; Wu and Liu, 2013) revealed that the use of HB-FRP can result in a several-
fold increase in the bond strength that is generally sufficient to prevent the critical
IC debonding.

2.6.2.1   Mechanism of HB-FRP

There are three mechanisms of interfacial shear transfer between two mechanically
connected bodies: adhesion, interlocking (or dowel action), and friction. All the
three mechanisms exist in an HB-FRP system (Fig. 2.93).

When the slip at the loaded end increases from zero to s_u, as shown by the distribution $s_1(x)$ in Fig. 2.93B, the adhesive bond stress develops at a_1 and shifts gradually to a_2 and then to a_3, as shown in Fig. 2.93C. Here, a_3 is a full bond stress
block with $\tau = 0$ at the loaded end. A further pull after a_3, for example, from $s_1(x)$
to $s_2(x)$, shifts the bond stress block backward to the position indicated by a_4. The
area in front of a_4 has attained zero bond stress or has been completely debonded.
Therefore the total adhesive bond, F_a, which is the area enclosed by the bond stress
block, attains its peak at a_3 and remains constant thereafter. The length of the bond
stress block corresponding to a_3, L_e, as shown in Fig. 2.93C, is the effective bond
length for the adhesive bond. It is evident that an increase in the bond length
beyond L_e does not result in further increases in F_a. L_e is generally in the range of

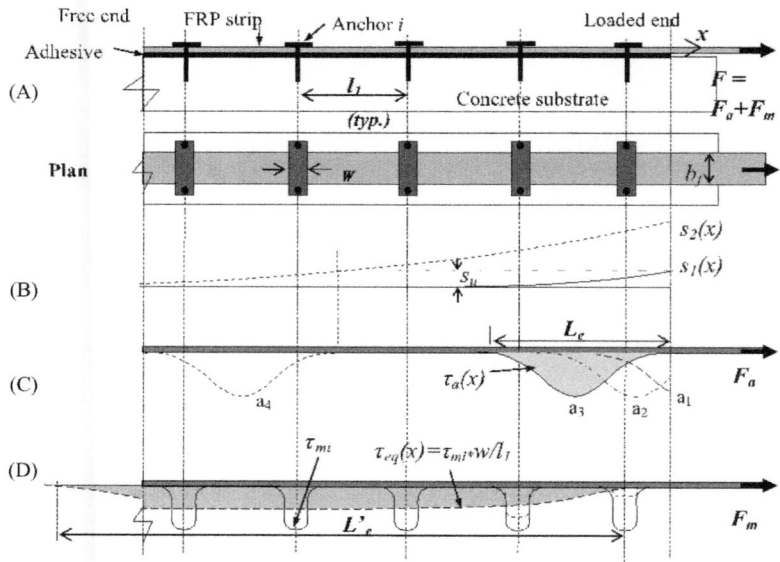

Figure 2.93 Interfacial bonds of HB-FRP system: (A) section; (B) slip; (C) adhesion; (D) mechanical bond (Wu and Liu, 2013).

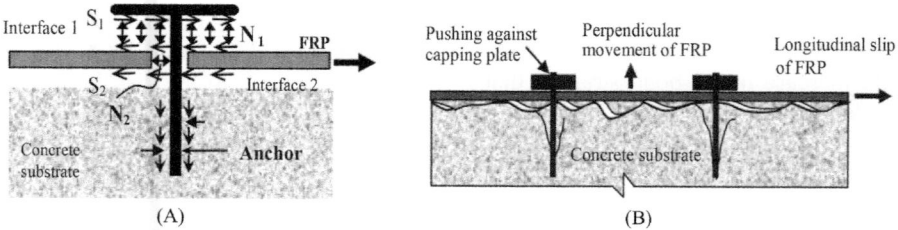

Figure 2.94 Mechanisms of HB-FRP system: (A) internal actions; (B) friction (Wu and Liu, 2013).

50−200 mm, depending on properties of the joint. Because of this limited effective bond length, the capacity of the adhesive bond is highly limited.

The dowel force of the anchor is transmitted through two paths (Fig. 2.94A): (1) bearing force N_2 and (2) adhesion and friction S_1 at Interface 1. The MF-FRP system operates largely through Path 1, whereas fiber anchors operate through Path 2. For the HB-FRP system, there is no direct contact between the anchors and FRP. Therefore Path 1 does not exist. The dowel action through Path 2 is marginal.

In general, when adequate adhesive bonding is used, debonding of FRP occurs in concrete with a rough surface, as shown in Fig. 2.94B. Hence, a longitudinal slip

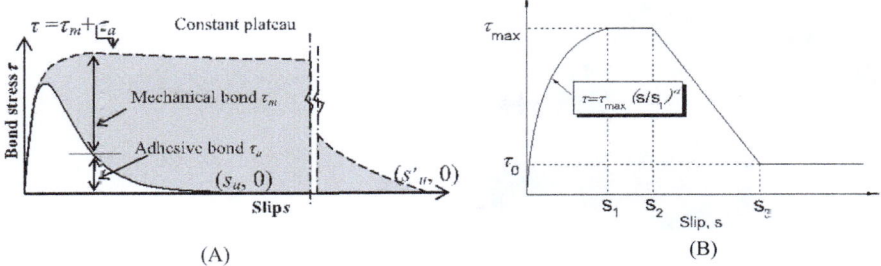

Figure 2.95 Typical bond—slip relationships: (A) HB-FRP system; (B) CEB-FIP model for internal bars (Wu and Liu, 2013).

of the FRP against the concrete substrate induces a vertical movement of the FRP strip. This perpendicular movement of the FRP pushes it against the fasteners, which are firmly embedded in the concrete, result in passive pressure on the FRP and hence, frictional resistance at the interface. This mechanism is identical to the highly popular shear—friction theory for joints in reinforced concrete.

Available tests have revealed that the additional bond strength induced by mechanical fasteners, including friction and dowel action, is directly proportional to the number of fasteners employed (Wu et al., 2010; Wu and Liu, 2013). This indicates that the bond resistance contributed by different individual fasteners is identical, as depicted by the solid lines in Fig. 2.93D. As the slip reduces monotonically from the loaded end to the free end (see $s_2(x)$ in Fig. 2.93B), this necessitates a (nearly) constant bond stress at different slip values. This implies the presence of a large plateau in the bond—slip relationship of the mechanical bond (Fig. 2.95A).

The maximum bond strength is attained when the bond resistance provided by the first mechanical fastener reduces to zero (Fig. 2.93D). This occurs because a further pull at the loaded end can only shift the zero-bond position (and hence, the effective bond area) backward in a manner similar to the shifting of an adhesive bond stress block. The distance between the two zero-bond-resistance positions at the loaded end and the free end is the effective bond length, as shown by L_e' in Fig. 2.93D. This is the case when small concrete nails or screws are used as anchors, as observed in tests where nails were pushed out at a certain slip (Wu and Huang, 2008). However, complete failure of anchor was not observed when larger concrete screws were used. A residual bond stress is present in this case even at a large slip, with a value similar to τ_0 in CEB-FIP (1993) (see Fig. 2.95B). In this case, an effective bond length does not exist in HB-FRP, and the bond strength always increases when the bond length is increased.

2.6.2.2 Experimental tests

Flexural tests of RC beams strengthened with HB-FRP were conducted to evaluate the effectiveness of the system (Wu et al., 2010). The test program and results are shown in Fig. 2.96. The externally bonded (EB) specimen in the

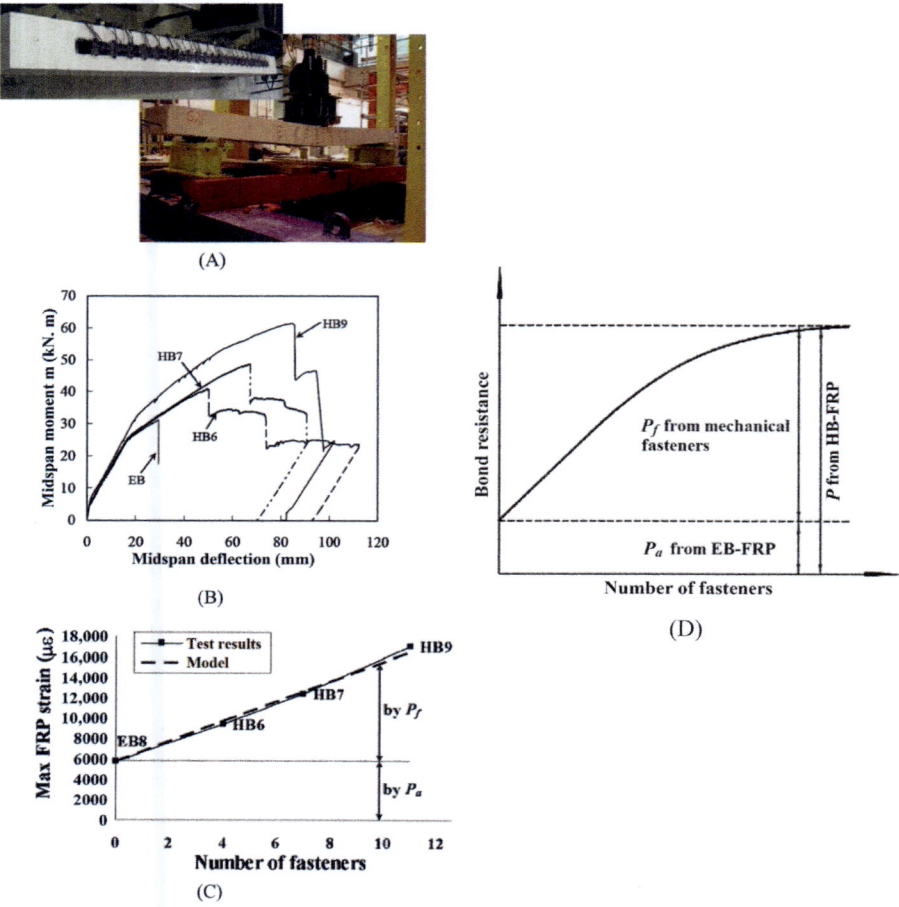

Figure 2.96 HB-FRP-strengthened beam tests (Wu et al., 2010). (A) Test setup. (B) Load−displacement curves. (C) Enhanced bond. (D) Total bond strength.

figure represents normal EB-FRP without fastener. HB6, HB7, and HB9 are specimens with different numbers of fasteners. The test results in Fig. 2.96B and C reveal the increase in the load-carrying capacity and maximum force, respectively, of FRP. Here, P, P_a, and P_f are the total bond strength and bond strength contributions from adhesion and mechanical fasteners, respectively. Fig. 2.96D shows the theoretical relationship between the bond strength and number of fasteners. For the bond−slip relationship that approaches zero at a large slip (Fig. 2.95A), the total bond strength approaches an asymptotic value when the number of fastener increases. For a bond−slip relationship with residual bond, the total bond increases continuously when the number of fasteners increases.

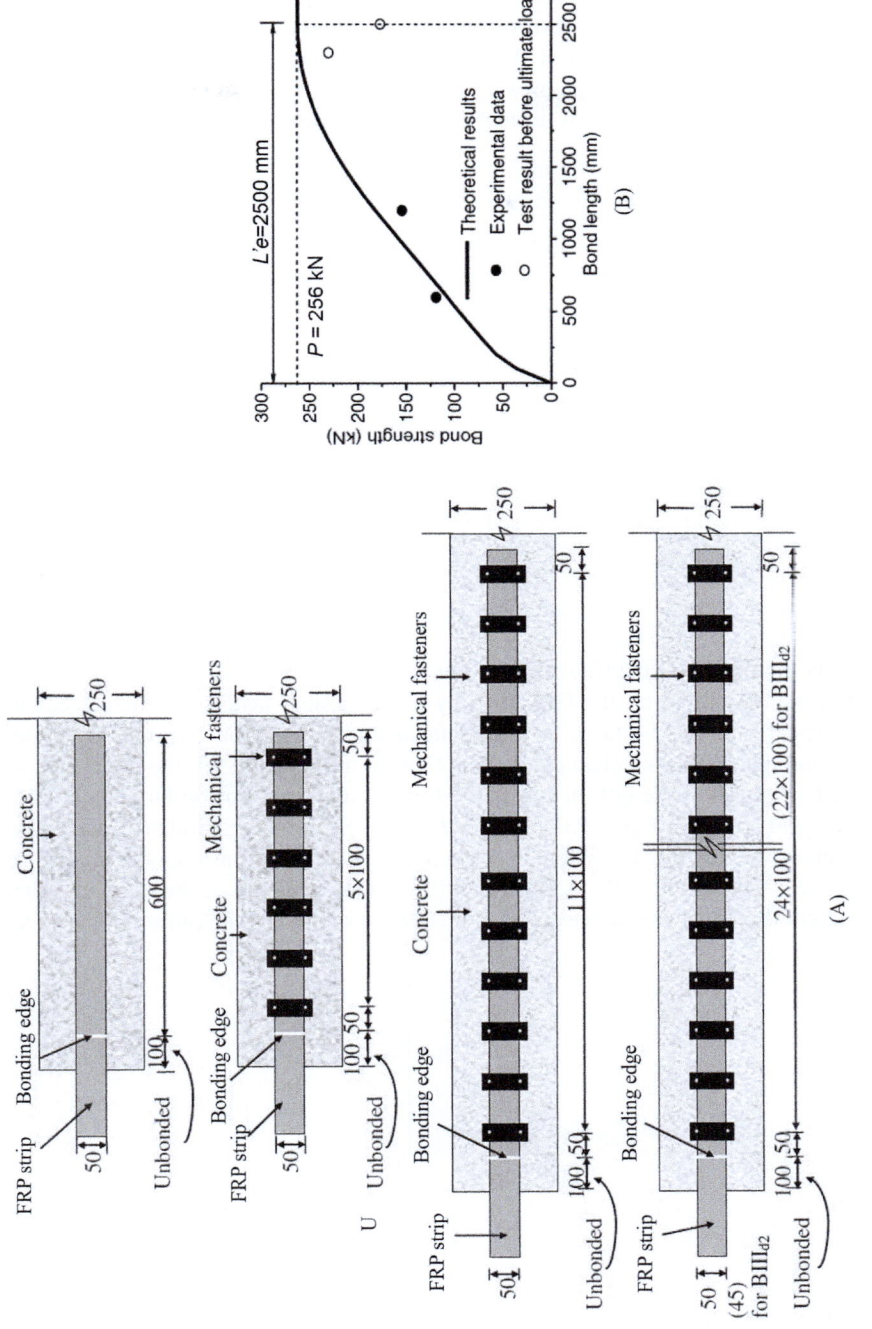

Figure 2.97 Single pull-off tests (Wu and Liu, 2013). (A) Test specimens. (B) Test results.

Single pull-off tests were conducted to characterize the bond of the HB-FRP system (Wu and Liu, 2013). The test specimens are shown in Fig. 2.97A. The test results are illustrated in Fig. 2.97B. For the particular test specimens, the effective bond length of the HB-FRP system was determined to be 2.5 m (Fig. 2.97B). This is significantly larger than that of normal EB-FRP systems.

2.6.2.3 Bond models

Based on experimental results and analyses of the failure mechanisms, the total bond strength P of an HB-FRP system can be determined using the following equation:

$$P = P_a + P_f + P_d \tag{2.135}$$

where P_d is the bond strength contribution from dowel action, which is negligible in the HB-FRP system. The contribution from adhesive bond P_a is similar to that in normal EB-FRP systems and can be calculated using an available model. P_f is the frictional bond strength and is expressed as

$$P_f = \mu.N_a.\frac{L_f}{s} \tag{2.136}$$

where μ is the frictional coefficient between two rough concrete surfaces. N_a is the normal pressure applied on the FRP strip by a mechanical fastener, which is equal to the vertical pullout resistance of the fastener and known for a certain concrete strength and particular mechanical fastener. s is the spacing of the mechanical anchors. L_f is the bond length of the FRP strip, which must not be larger than the effective bond length in Fig. 2.93D, or $L_f \leq L_e'$. L_f/s specifies the number of mechanical fasteners within the bond length. When the number of fasteners is equal to zero, $P_f = P_d = 0$, and $P = P_a$ in Eq. (2.135), as shown in Fig. 2.96C and D. The frictional coefficient μ was determined to be 0.96 in the tests. The pullout resistance

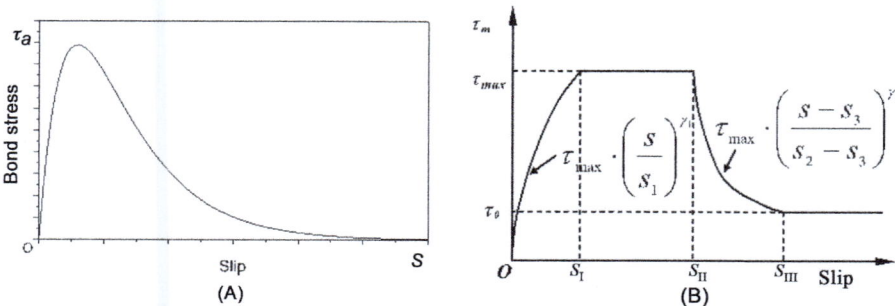

Figure 2.98 Bond—slip models: (A) adhesive bond—slip model; (B) bond—slip model for mechanical mechanism (Wu and Liu, 2013).

for the fastener shown in Fig. 2.92A was determined to be $N_a = 6$ kN. When different fasteners are used, tests should be conducted to evaluate N_a.

Conventional adhesive bond models, such as that in Fig. 2.98A, can be used for the bond stress τ_a outside the fasteners. The bond—slip relationship for the local areas under fasteners is different from that outside it. The model shown in Fig. 2.98B and expressed by Eq. (2.137) is suitable for simulating the mechanical bond stress:

$$\tau_m(s) = \begin{cases} \tau_{max} \cdot \left(\frac{s}{s_1}\right)^{\gamma_1} & s \leq s_1 \\ \tau_{max}, & s_1 < s \leq s_2 \\ \tau_{max} \cdot \left(\frac{s-s_3}{s_2-s_3}\right)^{\gamma_2}, & s_2 < s \leq s_3 \\ \tau_o, & s > s_3 \end{cases} \tag{2.137}$$

where τ_{max}, τ_0, γ_1, γ_2, s_1, s_2, and s_3 are parameters of the model. The overall bond—slip model for HB-FRP system is expressed as

$$\tau(x) = E_f \cdot t_f \cdot \frac{d^2 s}{dx^2} = \begin{cases} \tau_a, & \text{for } x \text{ outside fasteners} \\ \tau_a + \tau_m, & \text{for } x \text{ under fasteners} \end{cases} \tag{2.138}$$

For the particular anchor shown in Fig. 2.92, γ_1 and γ_2 are equal to 1.0. Other parameters were identified to be $\tau_{max} = 4.4$ MPa, $\tau_0 = 0$, $s_1 = 0.2$ mm, $s_2 = 8.0$ mm, and $s_3 = 30$ mm. Further details of the work are available in Wu and Liu (2013). When different anchors are used, tests should be conducted to determine the parameters for Eq. (2.137). The test method can be available in Wu et al. (2016).

References

Brena, S.F., Bramblett, R.M., Wood, S.L., Kreger, M.E., 2003. Increasing flexural capacity of reinforced concrete beams using carbon fiber-reinforced polymer composites. ACI Struct. J. 100 (1), 36—46.

Chen, J.F., Teng, J.G., 2001. Anchorage strength models for FRP and steel plates bonded to concrete. J. Struct. Eng. 127 (7), 784—791.

Concrete Society, 2004. Design guidance for strengthening concrete structures using fiber composite materials, Technical Report 55, Camberly, UK.

Dai, J.G., Ueda, T., Sato, Y., 2005. Development of the nonlinear bond stress—slip model of fiber reinforced plastics sheet—concrete interfaces with a simple method. J. Compos. Constr. 9 (1), 52—62.

Dai, J.G., Yokota, H., Iwanami, M., Kato, E., 2010. Experimental investigation of the influence of moisture on the bond behavior of FRP to concrete interfaces. J. Compos. Constr. 14 (6), 834—844.

Davalos, J.F., Kodkani, S.S., Ray, I., Boyajian, D.M., 2005. A fracture mechanic approach for interface durability of bonded FRP to concrete. In: Proc. of the 7th Int. Symposium on FRP Reinforcement for Concrete Structures (FRPRCS-7), Kansas City, USA, pp. 1465−1479.

De Lorenzis, L., Teng, J.G., 2007. Near-surface mounted FRP reinforcement: an emerging technique for strengthening structures. Compos. Part B: Eng. 38 (2), 119−143.

Diab, H., 2008. Time-dependent and long-term performance of FRP-strengthened concrete structures. Doctoral thesis, Ibaraki University, Japan.

Diab, H.M., Wu, Z.S., 2007. Nonlinear constitutive model for time-dependent behavior of FRP−concrete interface. Compos. Sci. Technol. 67, 2323−2333.

Diab, H.M., Wu, Z.S., Iwashita, K., 2007. Experimental and numerical investigation on the fatigue behavior of FRP−concrete interface. In: Proceeding of FRPRCS-8, Greece.

Diab, H.M., Wu, Z.S., Iwashita, K., 2009a. Theoretical solution for fatigue debonding growth and fatigue life prediction of FRP−concrete interfaces. Adv. Struct. Eng. 12 (6), 781−792.

Diab, H.M., Wu, Z.S., Iwashita, K., 2009b. Short and long-term bond performance of pre-stressed FRP sheet anchorages. Eng. Struct. 31 (5), 1241−1249.

Ekenel, M., Rizzo, A., Myers, J.J., Nanni, A., 2006. Flexural fatigue behavior of reinforced concrete beams strengthened with FRP fabric and precured laminate systems. J. Compos. Constr. 10 (5), 433−442.

Fava, G., Mazzotti, C., Poggi, C., Savoia, M., 2007. Durability of FRP−concrete bonding exposed to aggressive environment. In: Proc. of the 8th Int. Symposium on FRP Reinforcement for Concrete Structures (FRPRCS-8), Patras, Greece.

Garden, H.N., Hollaway, L.C., 1998. An experimental study of the influence of plate end anchorage of carbon fiber composite plates used to strengthen reinforced concrete beams. Compos. Struct. 42 (2), 175−188.

Green, M.F., Bisby, L.A., Beaudoin, Y., Labossière, P., 2000. Effect of freeze−thaw cycles on the bond durability between fiber reinforced polymer plate reinforcement and con-crete. Can. J. Civ. Eng. 27 (5), 949−959.

Japan Society of Civil Engineers (JSCE), 2001. Recommendation for upgrading of concrete structures with use of continuous fiber sheets. In: Concrete Engineering Series 41, JSCE, Japan.

JCI, 2003. Technical report of technical committee on retrofit technology. In: Proceedings of International Symposium on Latest Achievement of Technology and Research on Retrofitting Concrete Structures, Kyoto, Japan.

Kalfat, R., Al-Mahaidi, R., Smith, S.T., 2013. Anchorage devices used to improve the perfor-mance of reinforced concrete beams retrofitted with FRP composites: a-state-of-the-art-review. J. Compos. Constr. 17 (1), 14−33.

Khalifa, A., Gold, W.J., Nanni, A., Aziz, A., 1998. Contribution of externally bonded FRP to shear capacity of RC flexural members. J. Compos. Constr. 2 (4), 195−203.

Kotynia, R., Baky, H.A., Neale, K.W., Ebead, U.A., 2008. Flexural strengthening of RC beams with externally bonded CFRP systems: test results and 3D nonlinear FE analysis. J. Compos. Constr. 12 (2), 190−201.

Lamanna, A.J., Bank, L.C., Scott, D.W., 2001. Flexural strengthening of reinforced concrete beams using fasteners and fiber-reinforced polymer strips. ACI Struct. J. 98 (3), 368−376.

Lenke, L., Gerstle, W., 2001. Tension test of stress versus crack opening displacement using cylindrical concrete specimens. ACI SP. 201, 189−206.

Maeda, T., Asano, Y., Sato, Y., Ueda, T., Kakuta, Y., 1997. A study on bond by mechanism of carbon fiber sheet. In: Proc. of 3rd International Symposium on Non-metallic (FRP) Reinforcement for Concrete Structures, JCI, Sapporo, Japan, vol. 1, pp. 279−285.

Masoud, S., Soudki, K., Topper, T., 2005. Post repair fatigue performance of FRP-repaired corroded RC beams: experimental and analytical investigation. J. Compos. Constr. 9 (5), 441—449.

Monti, G., Renzelli, M., Luciani, P., 2003. FRP adhesion in uncracked and cracked concrete zones. In: Proceedings of the Sixth International Symposium on FRP Reinforcement for Concrete Structures (FRPRCS-6), vol. 1. World Scientific Publishing Co. Pte. Ltd., Singapore, pp. 183—192.

Mukhopadhyaya, P., Swamy, R.N., Lynsdale, C., 1998. Influence of aggressive exposure conditions on the behaviour of adhesive bonded concrete—GFRP joints. Constr. Build. Mater. 12 (8), 427—446.

National Research Council (CNR), 2004. Guidelines for design, execution and control of strengthening interventions by means of fiber-reinforced composites, Advisory Committee on Technical Regulations for Construction, National Research Council, Rome, Italy.

Neubauer, U., Rostásy, F.S., 1997. Design aspects of concrete structures strengthened with externally bonded CFRP plates. In: Proceedings of the 7th International Conference on Structural Faults and Repairs, ECS Publications, Edinburgh, Scotland, vol. 2, pp. 109—118.

Niedermeier, R., 2000. Zugkraftdeckung bei klebearmierten bauteilen (Envelope line of tensile forces while using externally bonded reinforcement). Doctoral Dissertation, TU München, German.

Niu H.D. Interfacial fracture mechanisms and performance evaluation of RC structural system strengthened with FRP sheets. Doctoral thesis, Ibaraki University, Japan, 2002.

Niu, H.D., Wu, Z.S., 2005. Debonding and failure mechanism of FRP-strengthened R/C beams influenced by flexural cracks. Comput. Aided Civ. Infrastruct. Eng. 20 (5), 354—368.

Oehlers, D.J., 1992. Reinforced concrete beams with plates glued to their soffits. J. Struct. Eng. 118 (8), 2023—2038.

Ritchie, P.A., Thomas, D.A., Lu, L.W., Connelly, G.M., 1991. External reinforcement of concrete beams using fiber reinforced-plastics. ACI Struct. J. 88 (4), 490—500.

Roberts, T.M., 1989. Approximate analysis of shear and normal stress concentrations in the adhesive layer of plated RC beams. Struct. Eng. 67 (12), 229—233.

Sayed, A.M., 2014. Modeling advancement for RC beams strengthened with FRP composites. Doctoral thesis, Southeast University, Nanjing, China.

Shi, J.W., 2014. Durability and reliability design of FRP strengthened concrete structures under coupled effects of multi-factors. Doctoral thesis, Southeast University.

Shi, J.W., Zhu, H., Wu, Z.S., Seracino, R., Wu, G., 2013. Bond behavior between basalt fiber-reinforced polymer sheet and concrete substrate under the coupled effects of freeze—thaw cycling and sustained load. J. Compos. Constr. 17 (4), 530—542.

Silva, M.A.G., Biscaia, H., 2008. Degradation of bond between FRP and RC Beams. Compos. Struct. 85 (2), 166—174.

Smith, S.T., Teng, J.G., 2001. Debonding failures in FRP-plated RC beams with or without U strip end anchorage, FRP Compos. Civ. Eng., vols. I and II. Elsevier, New York, pp. 607—615, Proc.

Smith, S.T., 2011, Strengthening of concrete, metallic and timber construction materials with FRP composites. In Advances in FRP Composites in Civil Engineering (pp. 13—19), Springer, Berlin, Heidelberg.

Spadea, G., Bencardino, F., Swamy, R.N., 1998. Structural behaviour of composite beams with externally bonded CFRP. J. Compos. Constr. 2 (3), 132—137.

Subramaniam, K.V., Ali-Ahmad, M., Ghosn, M., 2008. Freeze—thaw degradation of FRP—concrete interface: impact on cohesive fracture response. Eng. Fract. Mech. 75 (13), 3924—3940.

Swamy, R.N., Mukhopadhyaya, P., 1999. Debonding of carbon fiber- reinforced polymer plate from concrete beams. Struct. Build. 134 (4), 301—317.

Taljsten, B., 1996. Strengthening of concrete prisms using the plate-debonding technique. Int. J. Fract. 81, 253—266.

Tanaka, T., 1996. Shear resisting mechanism of reinforced concrete beams with CFS as shear reinforcement. Graduation thesis, Hokkaido University, Japan.

Teng, J.G., Chen, J.F., Smith, S.T., Lam, L., Jessop, T., 2003. Behaviour and strength of FRP-strengthened RC structures: A state-of-the-art review. Struct. Build. 156 (3), 334—335.

Teng, J.G., Lam, L., Chan, W., Wang, J., 2000. Retrofitting of deficient RC cantilever slabs using GFRP strips. J. Compos. Constr. 4 (2), 75—84.

Tsukagoshi, H., Jinno, Y., Iketani, J., et al., 1999. A study on shear reinforcement of T-shape beams by carbon fiber sheets and strands. Proc. Jpn. Concr. Inst. 21 (3), 1531—1536.

Van Gemert, D., 1980. Force transfer in epoxy-bonded steel—concrete joints. Int. J. Adhes. Adhes. 1, 67—72.

Wang, Y.C., Hsu, K., 2009. Design recommendations for the strengthening of reinforced concrete beams with externally bonded composite plates. Compos. Struct. 88 (2), 323—332.

Wu, Z.S., Diab, H.M., 2007. Constitutive model for time-dependent behavior of FRP—concrete interface. J. Compos. Constr. 11 (5), 477—486.

Wu, Z.S., Diab, H.M., 2009. Modeling of time-dependent bonding and debonding in structures externally strengthened with fiber reinforced polymer sheets. Int. J. Model. Ident. Contr. 7 (2), 199—208.

Wu, Y.F., Huang, Y., 2008. Hybrid bonding of FRP to reinforced concrete structures. J. Compos. Constr. 12 (3), 266—273.

Wu, Y.F., Jiang, C., 2013. Quantification of bond-slip relationship for externally bonded FRP-to-concrete joints. J. Compos. Constr. 17 (5), 673—686.

Wu, Y.F., Liu, K., 2013. Characterization of mechanically enhanced FRP bonding system. J. Compos. Constr. 17 (1), 34—49.

Wu, Z.S., Niu, H.D., 2007. Prediction of crack-induced debonding failure in R/C structures flexurally strengthened with externally bonded FRP composites. JSCE J. Mater. Concr. Struct. Pavements 63 (4), 620—639. 2007.

Wu, Z.S., Yin, J., 1998. A hybrid finite shell element with internal displacement discontinuities. Proc. EASEC-6 1, 231—237. Taipei.

Wu, Z.S., Yin, J., 2003. Fracturing behaviors of FRP-strengthened concrete structures. Eng. Fract. Mech. 70 (10), 1339—1355.

Wu, Z.S., Matsuzaki, T., Tanabe, K., 1997. Interface crack propagation in FRP strengthened concrete structures. Proc. FRPRCS-3 319—326.

Wu, Z.S., Murayama, D., Yoshizawa, H., 1999. Experimental study on bonding behavior and its improvement approach of CFRP sheets in anchorage zone. Proc. Jpn. Concr. Inst. 21 (2), 211—216 (in Japanese).

Wu, Z.S., Yuan, H., Yoshizawa, H., Kanakubo, T., 2001b. Experimental /analytical study on interfacial fracture energy and fracture propagation along FRP-to-concrete interface. ACI SP. 201-8, 133—152.

Wu, Z.S., Yuan, H., Niu, H.D., 2002. Stress transfer and fracture propagation in different kinds of adhesive joints. J. Eng. Mech. 128 (5), 562—573.

Wu, Z.S, Iwashita, K., Yagashiro, S., Ishikawa, T., Hamaguchi, Y., 2004. Temperature effect on bonding and debonding behavior between FRP sheets and concrete. J. Soc. Mater. Sci. 54 (5), 474—480.

Wu, Z.S., Islam, S.M., Said, H., 2009a. A three-parameter bond strength model for FRP—concrete interface. J. Reinf. Plast. Compos. 28 (19), 2309—2323.

Wu, Y.F., Wang, Z., Liu, K., He, W., 2009b. Numerical analyses of hybrid-bonded FRP strengthened concrete beams. Comput. Aided Civ. Infrastruct. Eng. 24 (5), 371—384.

Wu, Y.F., Yan, J.H., Zhou, Y.W., Xiao, Y., 2010. Ultimate strength of reinforced concrete beams retrofitted with hybrid bonded fiber reinforced polymer. ACI Struct. J. 107 (4), 451—460.

Wu, Y.F., He, L., Bank, L.C., 2016. Bond test protocol for plate-to-concrete interface involving all mechanisms. J. Compos. Constr. 20 (1), 04015022.

Yang, Y.X., Yue, Q.R., Hu, Y.C., 2001. Experimental study on bond performance between carbon fiber sheets and concrete. J. Build. Struct. 22 (3), 36—42 (in Chinese).

Yoshizawa, H., Wu, Z.S., Yuan, H., 2000. Study on FRP-to-concrete interface bond performance. J. Mater. Concr. Struct. Pavements, JSCE 49 (662), 105—119 (in Japanese).

Yuan, H., Wu, Z.S., Yoshizawa, H., 2001. Theoretical solutions on interfacial stress transfer of externally bonded steel/composite laminates. J. Struct. Mech. Earthq. Eng. JSCE 18 (1), 27—39. Tokyo.

Yun, Y.C., Wu, Y.F., 2011. Durability of CFRP—concrete joints under freeze—thaw cycling. Cold Reg. Sci. Technol. 65 (3), 401—412.

Zhang, H.W., Smith, S.T., 2012. FRP-to-concrete joint assemblages anchored with multiple FRP anchors. Compos. Struct. 94 (2), 403—414.

Fiber-reinforced polymer-strengthened tensile members

3

3.1 Experimental investigations of FRP-strengthened tensile members

3.1.1 Experimental observation

Concrete specimens of length 500 mm and two cross-sections (100 mm × 100 mm and 67 mm × 67 mm) were used in the experiment, as shown in Fig. 3.1. These were selected to ensure an adequate distribution of cracking. The concrete specimens were designed with a notch of width 1 mm and depth 5 mm at the specimen center to control the formation of the first crack. The width of the first crack, axial strains in the fiber-reinforced polymer fiber (FRP) sheet, and debonding behavior around the first crack were monitored by strain gauges, as shown in Fig. 3.2. The test variables included two types of carbon FRP (CFRP) materials (high tensile (HT) strength and high modulus (HM)), concrete strength, steel reinforcing ratio, number of FRP layers, and depths of concrete cover, as listed in Tables 3.1 and 3.2. The FRP sheets were applied according to the manufacturer's specifications (Tonen). Steel grips were used to mount both ends of the steel bar and the connection bolt to investigate the composite behavior of the strengthened member after the steel bars yielded. In addition, steel jigs were used at the two ends of the concrete specimens to stabilize the connection bolt during loading and to transfer the tensile loading to the ends of the FRP sheets. The tests were conducted under load control with a constant loading rate of 2.5 kN/min. The total elongation of the test specimens was measured as the relative displacement of the two steel jigs using two linear variable displacement transducers. In addition to the measurement of the first crack that occurred at the center of the test specimens, the crack widths of the second and third main cracks were measured using clip gauges mounted on the opposite sides of each of the test specimens immediately after the crack initiation. During testing, the debonding formation of FRP sheets was observed.

For the experimental details, readers may refer to Yoshizawa and Wu (1999). Table 3.3 shows a summary of the test results. Load−elongation curves were recorded for each specimen, as illustrated in Fig. 3.3. Fig. 3.4 shows the average crack spacing versus the average strain relationships. The average crack spacing for different cases before and after the yielding of the steel bars is shown in Figs. 3.5 and 3.6, respectively, for two typical average strains. After a certain average strain (0.001−0.0015), the minimum average crack spacing was attained and it was not affected by the variables, such as the ratios of the steel bar and continuous fiber sheet (CFS), concrete strength, and depth of concrete cover. Fig. 3.7 shows the

Structures Strengthened with Bonded Composites. DOI: https://doi.org/10.1016/B978-0-12-821088-8.00003-5

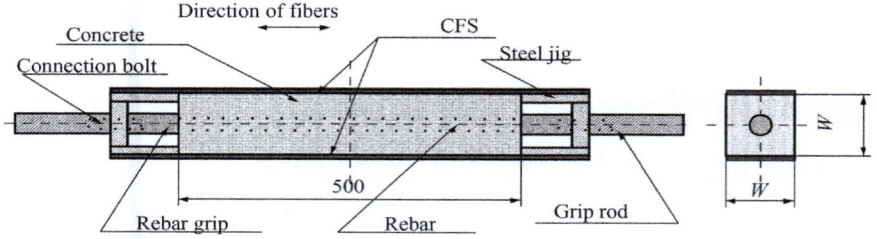

Figure 3.1 Details of specimen (Wu and Yoshizawa, 1999).

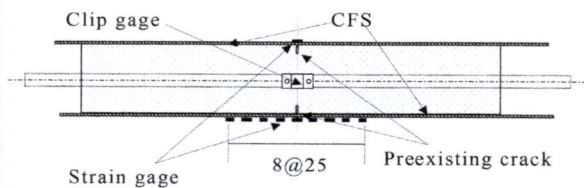

Figure 3.2 Instrumentation (Wu and Yoshizawa, 1999).

Table 3.1 Summary of the material properties (Wu and Yoshizawa, 1999).

Materials			Properties	
CFS	High tensile strength (HT)	FAW 200 g/m²	Design thickness	0.111 mm
			Young's modulus	230 GPa
			Tensile strength	4200 MPa
		FAW 300 g/m²	Design thickness	0.167 mm
			Young's modulus	230 GPa
			Tensile strength	4200 MPa
	High modulus (HM)	FAW 300 g/m²	Design thickness	0.165 mm
			Young's modulus	390 GPa
			Tensile strength	4000 MPa
Concrete	Design strength 22.5	Cement	Normal Portland cement	
		Aggregate size	$G_{max} = 20$ mm	
		Strength	Comp. strength f_c'	22.4 MPa
	Design strength 10.0	Cement	Normal Portland cement	
		Aggregate size	$G_{max} = 20$ mm	
		Strength	Comp. strength f_c'	11.2 MPa
Rebars SD295A	D13	①	f_y: 395 MPa	f_u: 585 MPa
		②	f_y: 370 MPa	f_u: 580 MPa
	D19	①	f_y: 385 MPa	f_u: 580 MPa
		②	f_y: 365 MPa	f_u: 560 MPa
	D22	①	f_y: 385 MPa	f_u: 585 MPa
		②	f_y: 365 MPa	f_u: 570 MPa

Note: FAW is the fabric weight per square meter.

Table 3.2 Experimental variables (Wu and Yoshizawa, 1999).

	No.		Yielding load of rebars (kN)	Max. load (kN)	Max. strain	Number of cracks (before yielding of rebar)	Number of cracks (after yielding of rebar)
Without CFS	Nr19	1	101.4	130.2	0.024	1	5
		2	102.3	129.1	0.0236	1	8
	Nr19-Lc	1	99.8	128.9	0.0169	1	5
	Nr13-W100	1	45.4	62.2	0.0218	1	2
		2	46.3	61.6	0.0204	1	2
	Nr13-W67	1	48.5	49.6	0.0106	3	4
		2	48.5	55.3	0.021	3	7
	Nr22	1	143.9	188.2	0.0193	3	5
		2	143.4	173.9	0.0143	2	3
With CFS	S19	1	132.8	225.4	0.0135	4	9
		2	128.7	201	0.0115	4	7
	S19-Lc	1	126.5	211.1	0.0112	5	6
		2	127.5	199.7	0.0122	3	7
	S19-HM	1	135.3	215.7	0.0074	3	5
		2	137.3	181.4	0.0053	4	6
	S19-0.67P	1	131	179	0.0074	4	6
	S19-3P	1	150.7	248.3	0.0073	4	7
		2	157.6	251.3	0.006	4	5
	S13-W100	1	67.8	100.1	0.0058	4	5
		2	71.6	117.4	0.0072	4	4
	S13-W67	1	58.1	86.3	0.0118	4	12
	S22	1	165.9	243.6	0.0164	3	8
		2	162.5	245.3	0.0126	3	7

crack patterns of typical specimens at the final loading state. Fig. 3.8 shows different curves of the first crack width with the increase in the average strain of the reinforcing bar. The larger constraining effect owing to the higher modulus and larger ratio of the FRP sheets can be observed from the figure. Fig. 3.9 shows the width values of the first crack at an average strain of 0.0015. It is evident that after a certain load, the crack width is insensitive to the ratio of the steel rebar and depth of concrete cover at the same average strain. Fig. 3.10 shows the typical relationships of the load versus the crack width for the first crack to demonstrate the constraining effect to the crack width. The maximum crack spacing l_{max} and the minimum crack spacing l_{min} relative to the average crack spacing l_{av} before steel yielding for the CFS-strengthened specimens are shown in Fig. 3.11. The average values of 0.67 for l_{min}/l_{av} and 1.49 for l_{max}/l_{av} were identified in the experimental series.

Table 3.3 Summary of test results (Wu and Yoshizawa, 1999).

	No.	Rebar	Width, height of specimen (mm)	Compressive strength of concrete (MPa)	Ratio of rebar p_s (%)	CFS type	CFS fiber arial weight (g/m²)	Ratio of CFS p_{cf} (%)	Number of specimens	Comments
Without CFS	Nr19	D19②	100	22.4	2.95				2	Reference specimen without CFS
	Nr19-Lc	D19②	100	11.2	2.95				1	Influence of concrete strength
	Nr13-W100	D13②	100	22.4	1.28				2	Influence of steel ratio
	Nr13-W67	D13②	66.5	22.4	2.95				2	Influence of depth of concrete cover
	Nr22	D22②	100	22.4	4.03				2	Influence of steel ratio
With CFS	S19	D19②	100	22.4	2.95	HT	300	0.34	2	Reference specimen with CFS
	S19-Lc	D19①	100	11.2	2.95	HT	300	0.34	2	Influence of concrete strength
	S19-HM	D19①	100	22.4	2.95	HM	300	0.34	2	Influence of Young's modulus of CFS
	S19-0.67P	D19①	100	22.4	2.95	HT	200	0.22	1	Influence of CFS ratio
	S19-3P	D19①	100	22.4	2.95	HT	300 × 3	1.03	2	Influence of CFS ratio
	S13-W100	D13①	100	22.4	1.28	HT	300	0.34	2	Influence of steel ratio
	S13-W67	D13①	66.5	22.4	2.95	HT	200	0.34	2	Influence of steel ratio
	S22	D22①	100	22.4	4.03	HT	300	0.35	2	Influence of steel ratio

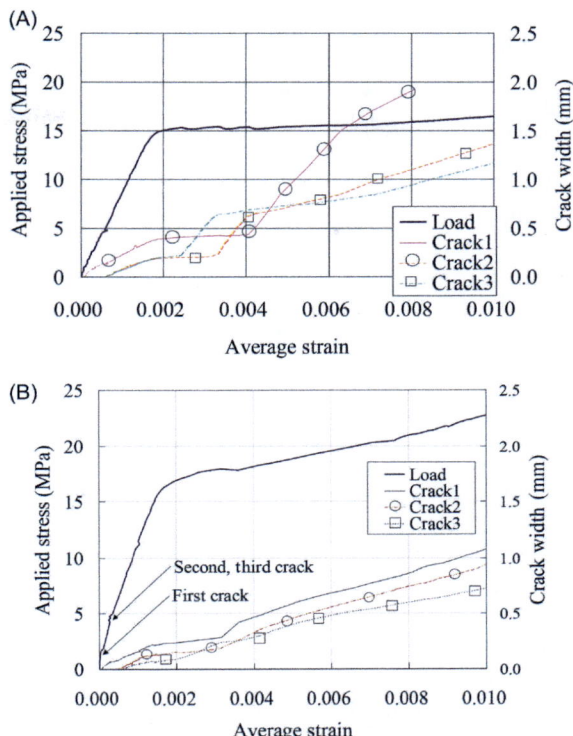

Figure 3.3 Average stress versus average strain relationship (Wu and Yoshizawa, 1999). (A) Without CFS (Nr22). (B) With CFS (S22).

3.1.2 Debonding mechanism

Fig. 3.12 shows the crack front propagation behavior and the formation processes observed from the experiment. Owing to the constraining effect of the FRP sheets, the tensile concrete cracking first occurred from the belly of the specimen and propagated perpendicularly toward the FRP sheet. When the concrete crack propagated to the vicinity of the FRP sheet, one to three diagonal shear cracks occurred gradually at the front of the tensile crack and extended toward the FRP sheet. These shear cracks would deteriorate the shear stiffness of the interfacial layer. It was observed that the front of the diagonal crack was along the FRP–concrete interface and resulted in a debonding of the FRP sheet. Therefore although the debonding along the FRP sheet–concrete interface could be attributed to many reasons, the debonding was found to be initiated from the tensile crack front of concrete. Based on the previous argument, the constitutive relationship along the bond surface before the initiation of the interfacial debonding may be expressed as shown in Fig. 3.13A. The length of the fracture area with diagonal microcracks can be considered the initial debonding length (Fig. 3.12D). Moreover, if debonding propagation occurs

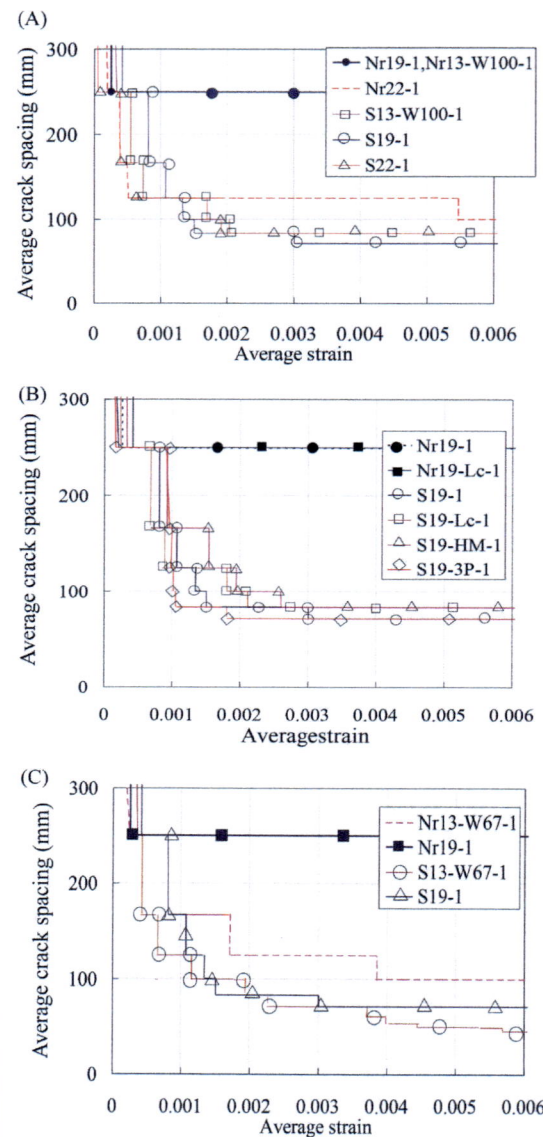

Figure 3.4 Average crack spacing−average strain relationships (Wu and Yoshizawa, 1999). (A) Effect of steel ratio. (B) Effect of CFS ratio. (C) Effect of depth of concrete cover.

along the FRP−concrete interface, the shear transfer ability would drop suddenly, similar to the mode I fracture of brittle materials. When the bond fracture propagates along the interface or through the concrete adjacent to the adhesive layer, the descending portion of the interface shear could be caused by the softening behavior

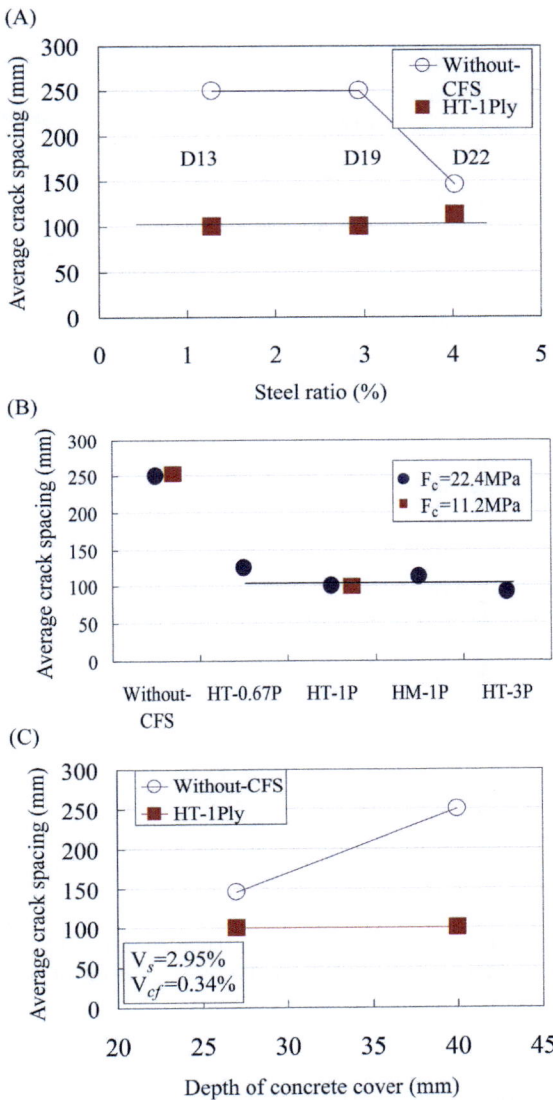

Figure 3.5 Average crack spacing at average strain 0.015 (Wu and Yoshizawa, 1999). (A) Effect of steel ratio. (B) Effect of CFS ratio. (C) Effect of depth of concrete cover.

of concrete in shear, as shown in Fig. 3.13B. The softening branch in Fig. 3.13B is characterized by the peak strength and the characteristic slip, after which the shear resistance decreases gradually to a value near zero. The area G_f in Fig. 3.13B can be defined as the fracture energy based on nonlinear fracture energy concept. Therefore the shear stress−debonding displacement relationship is considered as a

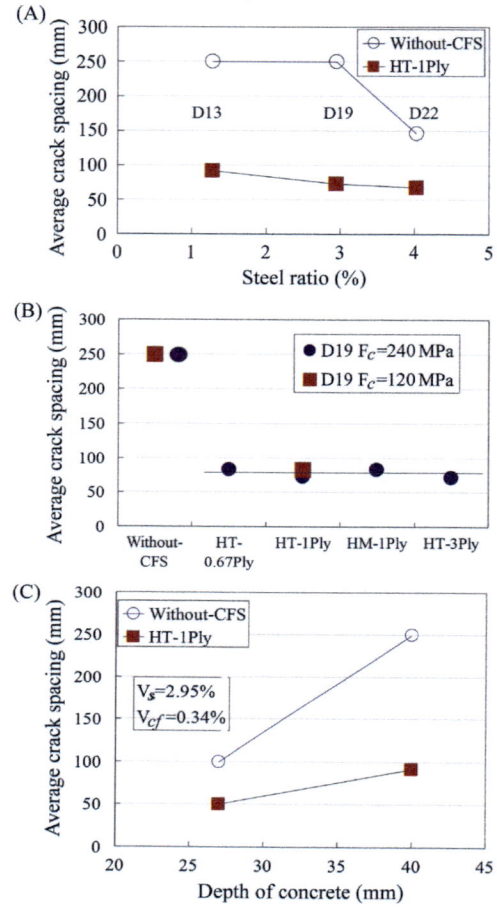

Figure 3.6 Average crack spacing at average strain 0.05 (Wu and Yoshizawa, 1999). (A) Effect of steel ratio. (B) Effect of CFS ratio. (C) Effect of depth of concrete cover.

constitutive law in front of the interfacial crack during its propagation. Meanwhile, the bond between the composite sheet and concrete may fracture abruptly as a result of a catastrophic propagation of the crack along the FRP–concrete interface. Therefore the debonding (interface crack) propagation may be addressed as a fracture mechanics problem, in which fracture occurs when the strain energy release rate equals the critical strain energy release rate of the interface G_{IIC} (G_f may be considered as its approximate value) (Traiantafillou and Plevris, 1992). To measure the critical strain energy release rate for the FRP–concrete interface, a double-shear specimen pulled in tension was used in a study by Traiantafillou and Plevris (1992).

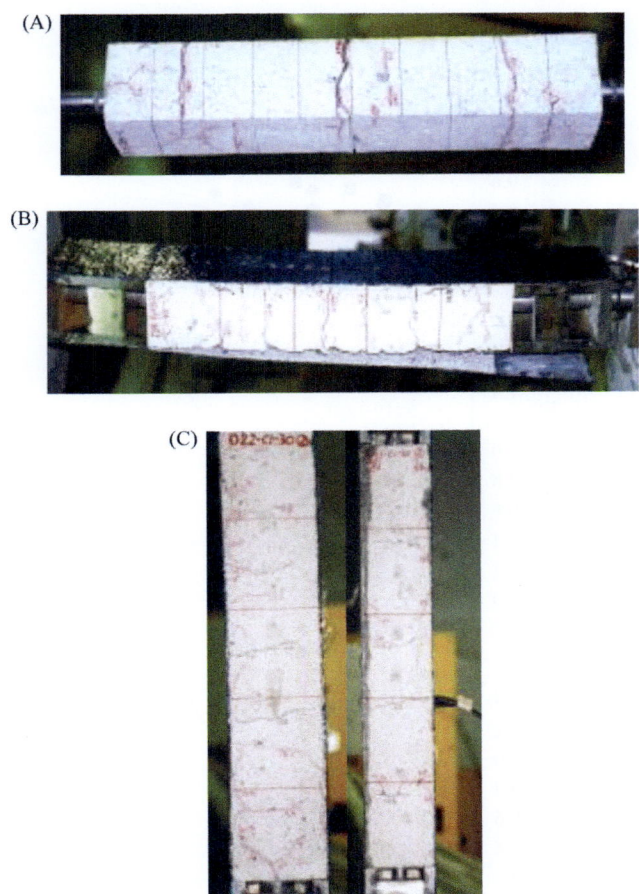

Figure 3.7 Failure patterns at final loading states. (A) Nr 19. (B) S19. (C) S22.

In conclusion, the bond—debonding mechanism on the FRP—concrete interface is different from the bond—slip behavior between a reinforcing bar and the surrounding concrete, which results from the friction effect and aggregate interlocking.

3.2 Fracture energy approach for analyzing tensile properties of FRP-strengthened tensile members

An analytical model for determining the tensile properties of FRP-reinforced concrete (RC) members was developed using the fracture energy approach of Ouyang and Shah (1994), which is based on fracture mechanics. The outline of the analytical model is as follows:

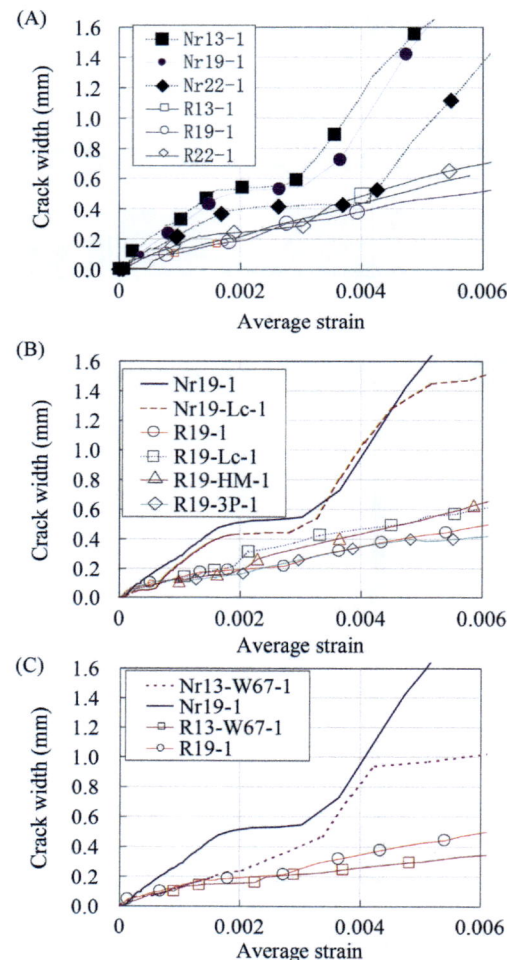

Figure 3.8 Crack width–average strain (Wu and Yoshizawa, 1999). (A) Effect of steel ratio. (B) Effect of CFS ratio. (C) Effect of depth of concrete cover.

3.2.1 Energy balance and strain formulation in reinforced concrete members

Assume that N cracks occurred in an RC member with externally bonded FRP sheet/sheets, as shown in Fig. 3.14, when a tensile stress σ was applied to the end of the member. Here, the average strains in the concrete, reinforcing bars, and FRP sheet/sheets are denoted as ε_c, ε_s, and ε_{cf}, respectively. When crack number is N and assuming that the crack opening width is w_o, the average strain occurring in an RC member can be expressed as

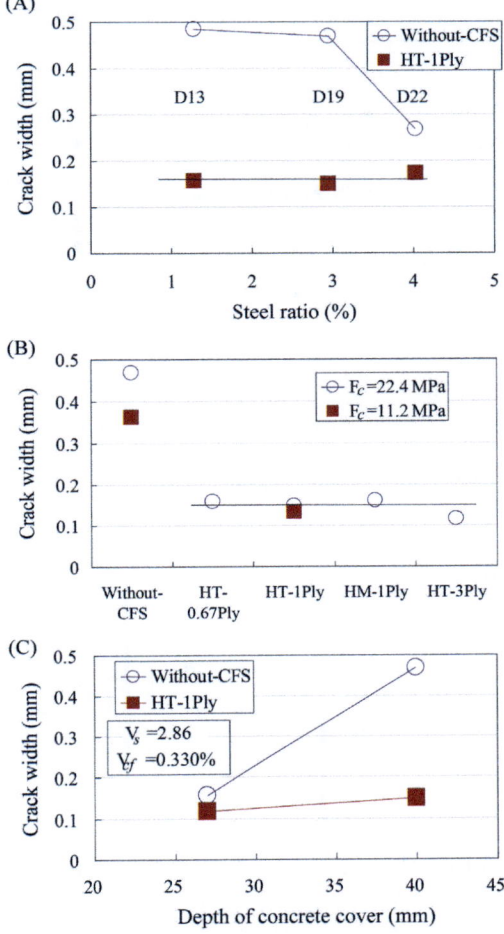

Figure 3.9 Width of the first crack at average strain 0.0015 (Wu and Yoshizawa, 1999). (A) Effect of steel ratio. (B) Effect of CFS ratio. (C) Effect of depth of concrete cover.

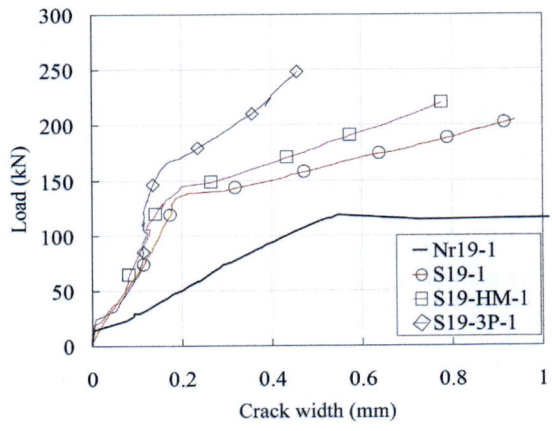

Figure 3.10 Load versus crack width relationships (Wu and Yoshizawa, 1999).

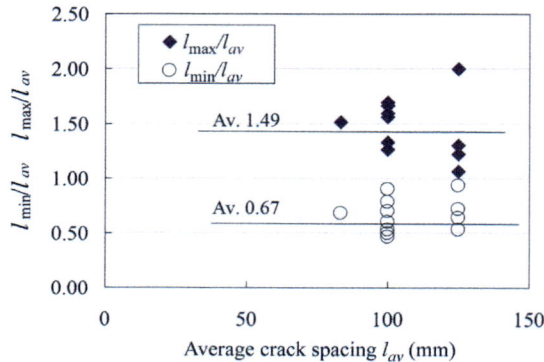

Figure 3.11 Minimum and maximum crack spacing (With CFS) (Wu and Yoshizawa, 1999).

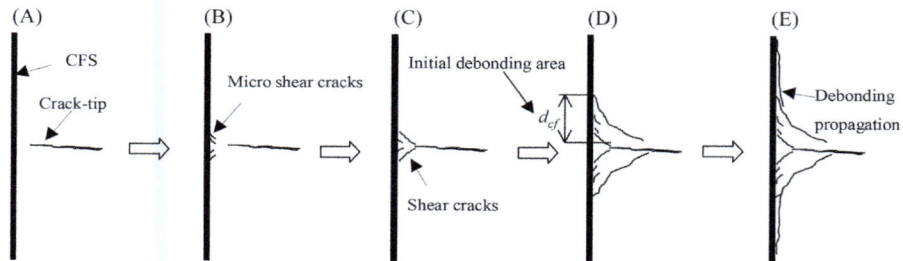

Figure 3.12 Cracking behavior at crack tip and formation of debonding (Wu and Yoshizawa, 1999). (A) Occurrence of a tension crack in concrete. (B) Occurrence of microshear cracks. (C) Occurrence of shear cracks. (D) Occurrence of initial debonding area. (E) Occurrence of debonding.

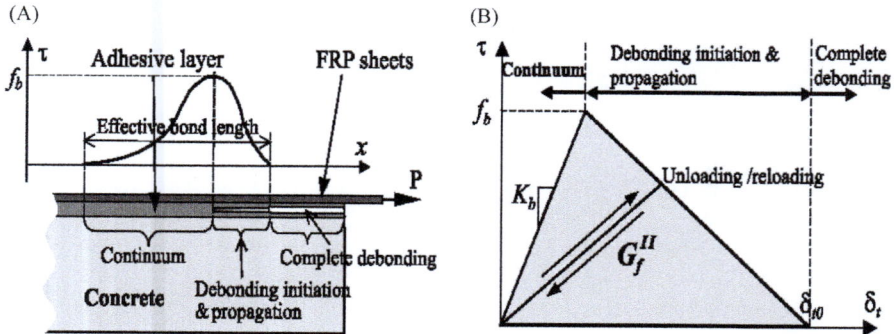

Figure 3.13 Interfacial debonding fracture within adhesive layer (Wu and Yin, 2003). (A) Debonding propagation procedure. (B) A $\tau - \delta_t$ relationship of adhesive layer.

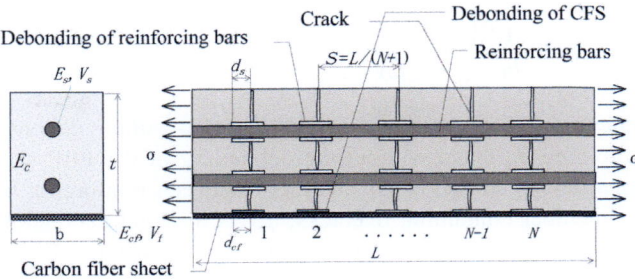

Figure 3.14 Cracking of FRP sheets-reinforced RC member (Wu and Yoshizawa, 1999).

$$\varepsilon_s = \varepsilon_{cf} = \varepsilon_c + \frac{Nw_o}{L} \tag{3.1}$$

Here, it is assumed that ε_c is not affected by w_o; however, the strains in the reinforcing bars and FRP sheet are affected by w_o. The symbols used in Fig. 3.14 are denoted as follows:

L: length of an RC member;
t: thickness of an RC member;
b: width of an RC member;
$S = l_{av}$: average crack spacing;
d_s: debonding length of reinforcing bars;
d_{cf}: debonding length of FRP sheets;
E_s: Young's modulus of reinforcing bars;
E_c: Young's modulus of concrete;
E_{cf}: Young's modulus of FRP sheets;
V_s: volume ratio of reinforcing bars;
V_{cf}: volume ratio of FRP sheets

The average crack spacing S can be expressed as follows:

$$S = \frac{L}{N+1} \tag{3.2}$$

In FRP-reinforced RC members with cracks occurring under a tensile load, the debonding fracture of the reinforcing bars at the cracked sections, bond sliding at the interface between the reinforcing bars and concrete, and debonding fracture of FRP sheet−concrete interfaces occur. Simultaneously, the number of cracks and the crack opening width increase gradually.

Based on fracture mechanics, energy equilibrium during cracking requires the strain energy release rate to be equal to the sum of the rates of the debonding energy and the sliding energy at the interface between the reinforcing bars and concrete, debonding energy along the FRP−concrete interface, and concrete fracture resistance, as shown in Eq. (3.3):

$$-\frac{1}{bt}\frac{\partial \varphi_c}{\partial N} = R_{lcf} + \frac{1}{bt}\frac{\partial \left(\varphi_s + \varphi_{ds} + \varphi_{dcf}\right)}{\partial N} \qquad (3.3)$$

φ_c is the strain energy of concrete containing N cracks, without debonding and sliding of the reinforcing bars and without the debonding of the FRP sheet; φ_s is the total sliding energies of all debonded interfaces between reinforcing bars and concrete of an RC member containing N cracks; φ_{ds} is the total debonding energies of all debonded interfaces between reinforcing bars and concrete of an RC member containing N cracks; φ_{dcf} is the total debonding energies of all debonded interfaces between CFS and concrete of an RC member containing N cracks; R_{lcf} is the fracture resistance of a plain concrete member.

Detailed formulations for calculating each type of strain energy and the stress intensity of the FRP sheet around the cracks are described in JCI Technical Committee on Continuous Fiber Reinforced Concrete (1997, 1998).

3.2.2   Debonding fracture energy

The debonding fracture energy φ_{dcf} of an FRP sheet with the number of cracks N in concrete is calculated by the following equation ($G_{IIc} = \gamma_{cf} = G_f$):

$$\varphi_{dcf} = 2Nbd_{cf}\gamma_{cf}, \qquad (3.4)$$

where γ_{cf} denotes the debonding fracture energy per unit area of an FRP sheet−concrete crack surface. Experimentally, it has demonstrated that the debonding length of CFS before the yielding of reinforcing bars is small. Therefore in this study, $d_{cf} \leqq d_s$ was assumed up to the maximum number of crack N developed. The debonding fracture energy G_{IIc} was measured using a double-shear test, as shown in Fig. 3.15. An average value of 6.0 N/m ($=0.006$ N/mm) was used in the calculation.

3.2.3   Calculation of stress in reinforced concrete members

From Eq. (3.3), ε_c for a given number of cracks N is calculated. The main condition, under which Eq. (3.3) applies, is expressed as follows:

$$2d_s \leq S \qquad (3.5)$$

First, the concrete stress is calculated as

$$\sigma_c = \frac{\varepsilon_c}{M_c} = E_c\varepsilon_c e^{-\xi N}, \qquad (3.6)$$

where M_c is the elastic compliance of concrete and ξ is a coefficient for expressing the effect of concrete containing N cracks.

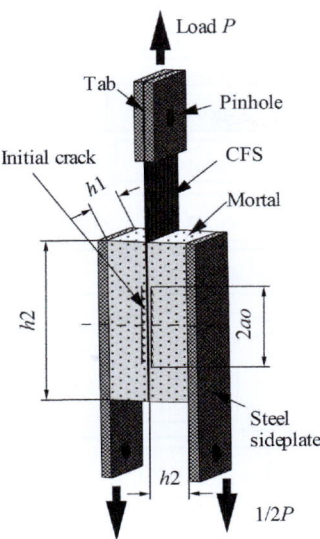

Figure 3.15 Double-side shear specimen (Wu and Yoshizawa, 1999).

Denoting the yielding stress of the reinforcing bars as f_y and the corresponding strain ε_y, the average stress of the reinforcing bars before and after yielding can be expressed by the following equations:

$$\sigma_s = E_s \varepsilon_s = E_s \left(\varepsilon_c + \frac{Nw_o}{L} \right) : \text{if } \varepsilon_s < \varepsilon_y \tag{3.7}$$

$$\sigma_s = f_y : \text{if } \varepsilon_s \geq \varepsilon_y \tag{3.8}$$

$$\sigma_{cf} = E_{cf} \varepsilon_{cf} = E_{cf} \left(\varepsilon_c + \frac{Nw_o}{L} \right) \tag{3.9}$$

$$\sigma = \sigma_s V_s + \sigma_c \left(1 - V_s - V_f \right) + \sigma_{cf} V_f \tag{3.10}$$

The computational flow for predicting the tensile properties of the FRP-sheet-reinforced RC members based on the above is shown in Fig. 3.16.

When the number of cracks N becomes maximum under the condition of Eq. (3.5) and does not increase further, the strain in the reinforcing bars and FRP sheets is increased by the widening of the crack opening w_o. At this moment, most of the concrete separates from the reinforcing bars. Therefore the concrete strain increases slightly, and only the strains of the reinforcing bars and FRP sheets at the cracked sections of the concrete are assumed to increase.

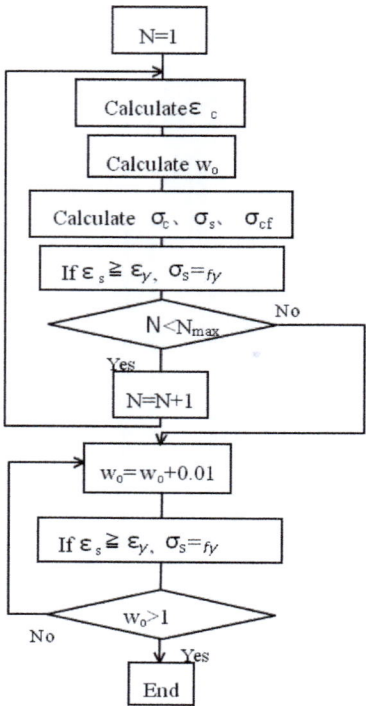

Figure 3.16 Computational flow (Wu and Yoshizawa, 1999).

3.2.4 Comparison of analytical/experimental results

The usefulness and applicability of the fracture energy approach introduced earlier were demonstrated numerically by Wu (1997) and Wu et al. (1997). In this study, it was used to analyze the test specimens. In the calculations, the value of fracture resistance of plain concrete R_{Icf} estimated by Ouyang and Shah (1994) was used. To obtain the debonded length of a reinforcing bar, the following equation proposed by Gilbert (1992) was adopted, in which r is the radius of the reinforcing bar:

$$d_s = \frac{r}{20V_s} \qquad (3.11)$$

The initial debonded length of the FRP sheets is based on the experimental results. The value was observed to be equal to approximately 15 mm. The values of the parameters above and that of the concrete tensile strength are shown in Table 3.4. Fig. 3.17A shows a comparison between the predicted and experimental results for both the load−average strain and the crack width relationships of a No. S19 specimen with a D19 reinforcing bar and a layer of high tensile strength FRP

Table 3.4 Mechanical properties of materials (Wu and Yoshizawa, 1999).

Members width b (mm)	Members height t (mm)	Member length L (mm)	Coarse aggregate size a_0 (mm)	Tensile Strength of concrete f_t (MPa)	Young Modulus of concrete E_c (GPa)	R_{lcf} (Nmm)	Yielding stress of rebar σ_y (MPa)	Debonding length of FRP sheet d_{cf} (mm)
100	100	500	20	2.30	30	0.009	370	15

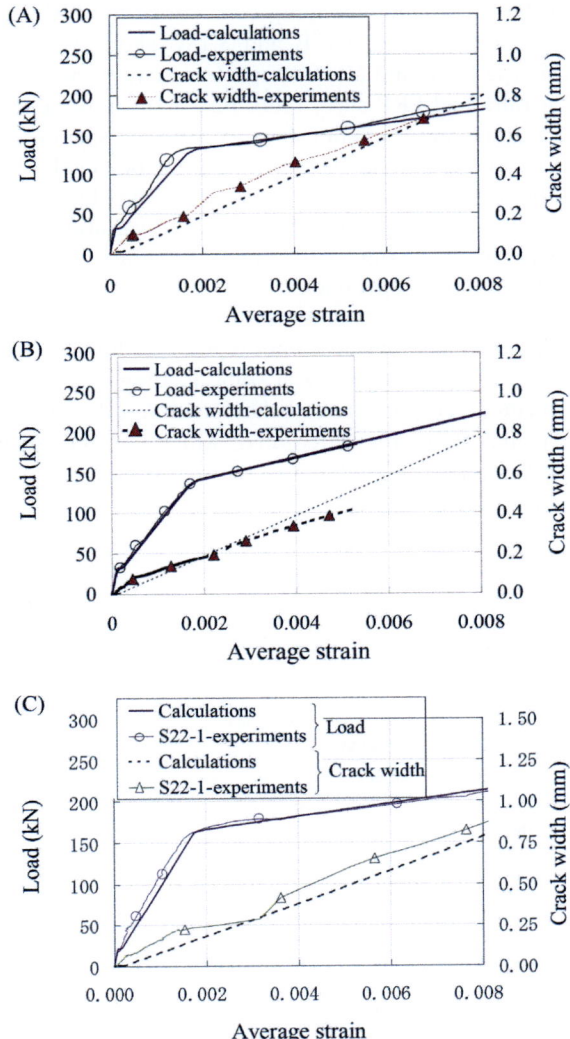

Figure 3.17 Comparison of experimental and calculated results (Wu and Yoshizawa, 1999). (A) S19. (B) S19-HM. (C) S22.

(HT type CFS) sheet. As shown, the behavior of the test specimen was represented well by the analytical technique. The load capacity of the RC members increased owing to the reinforcement effect of the FRP sheets, and even after the reinforcing bars had yielded, the load could be increased with the increase in deformation owing to the FRP sheets. The debonding length d_{cf} varied with the stress of the FRP sheets at a cracked section; however, because the dominant debonding propagation was not observed in this experimental series, the initial value of the debonding length was set as a constant. The initial debonding length of the FRP sheets minimally affected the stress−strain relationship of the structures. However, it significantly affected the stress and debonding behavior of the FRP sheets around the cracked section of concrete. Figs. 3.17B and 3.18C show the results for the No. S19-HM (with HM type CFS) and No. S22 specimens, respectively, to illustrate the effects of the rebar ratio and FRP sheet ratio. Fig. 3.18 shows a comparison of the calculated stress−average strain relationships of different specimens. The different reinforcing effects on the tension stiffness owing to the bond−slip between the reinforcing bars and the surrounding concrete and the debonding behavior along the FRP−concrete interface are shown in the figure. The variation in CFS stress with the average strain is shown in Fig. 3.19. Fig. 3.20 shows the relationship of the load−crack width for different specimens. The larger constraining effect to cracking by the HM sheet at the same average strain was observed in the experiments; however, the numerical simulations could not reproduce this effect well. In fact, the different constraining effects to cracking at some loading states could be predicted by the analytical method. To predict the behavior more effectively, an assumption of average strain in the structures is required. Moreover, the prediction of crack spacing is inappropriate after the yielding of the reinforcing bars. All of these problems must be resolved in the future.

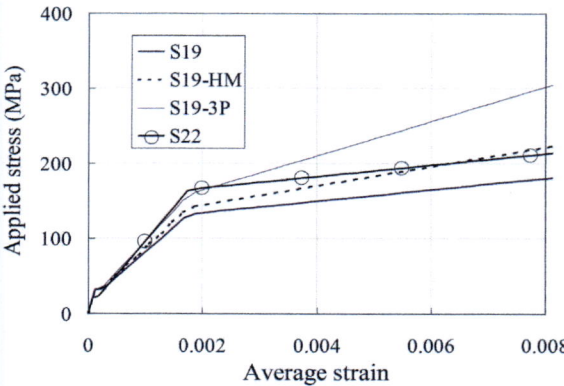

Figure 3.18 Calculated stress−average strain relationships (Wu and Yoshizawa, 1999).

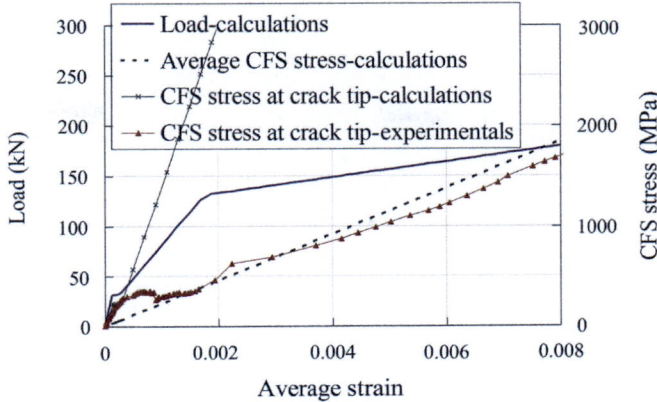

Figure 3.19 Stress of CFS (Wu and Yoshizawa, 1999).

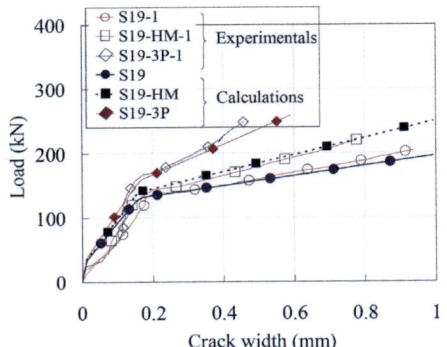

Figure 3.20 Load versus crack width relationships (Wu and Yoshizawa, 1999).

3.3   Modeling of tension stiffening effect

3.3.1   Yoshizawa—Wu model

Compared with plain concrete, the tension stiffness of cracked reinforced concrete improved owing to the bond—slip between the rebars and concrete. This tension—stiffness effect is also present in FRP-bonded concrete members.

Fig. 3.21 shows the pattern of load versus average strain for an RC member reinforced with CFRP sheets. In this figure, point A is the position where the initial cracking occurs, point B corresponds to the position until subsequent cracking, and

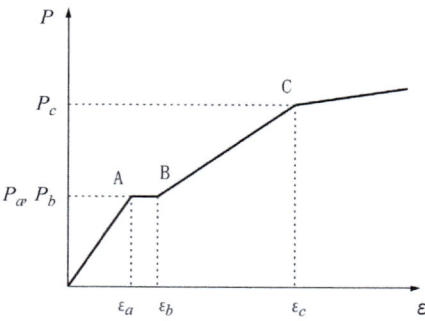

Figure 3.21 Load—average strain relationships.

point C is the position of the rebar yielding. Moreover, the shared load of the RC member is decreased by reducing its stiffness.

The strain at point A is calculated by the following equation:

$$\varepsilon_a = \frac{f_t}{E_c} = \varepsilon_{ct}, P_a = A_c f_t \left(1 + n\rho + n_{cf}\rho_{cf}\right), \tag{3.12}$$

where n and n_{cf} are Young's modulus ratios of the steel rebars and CFRP sheets to that of concrete, respectively; ρ and ρ_{cf} are the cross-sectional area ratios of the steel rebars and CFRP sheets to concrete, respectively; f_t is the cracking stress of concrete; ε_{ct} is the cracking strain of concrete; A_c is the cross-sectional area of concrete; E_c is Young's modulus of concrete.

Following ACI-318, Young's modulus of concrete can be calculated by the following equation:

$$E_c = 5000\sqrt{f'_c}(f':\text{MPa}) \tag{3.13}$$

The cracking stress of concrete f_t is calculated using Eq. (3.14):

$$f_t = 0.09f'_c \tag{3.14}$$

where f'_c is the concrete compressive strength.

In concrete without FRP sheet, the strains of points B and C calculated from Eqs. (3.15) and (3.16) by Gupta and Maestrini (1990) agree well with the experimental results, in which the considered Young's modulus ratio and the volume ratio of the CFRP sheets should be transformed:

$$\varepsilon_b = \frac{f_t}{E_c}\left[1 + \frac{1 + \left(n\rho + n_{cf}\rho_{cf}\right)}{10\left(n\rho + n_{cf}\rho_{cf}\right)}\right], P_b = A_c f_c\left(1 + n\rho + n_{cf}\rho_{cf}\right) \tag{3.15}$$

$$\varepsilon_c = \frac{f_y}{E_s}\left[1 - \frac{1}{2\left(n\rho + n_{cf}\rho_{cf}\right)}\frac{f_t}{E_c}\right], P_c = A_c f_y\left(n\rho + n_{cf}\rho_{cf}\right) \tag{3.16}$$

where f_y is the yielding stress of the reinforcing bars.

Fig. 3.22 shows the load versus average strain obtained from the experiment, and the loads and strains of points A, B, and C calculated by Eqs. (3.12), (3.15), and (3.16), respectively. Moreover, the load versus strain calculated by ignoring the loading capacity of concrete is shown in Fig. 3.22, in which WOCCE represents "without consideration of the concrete effect."

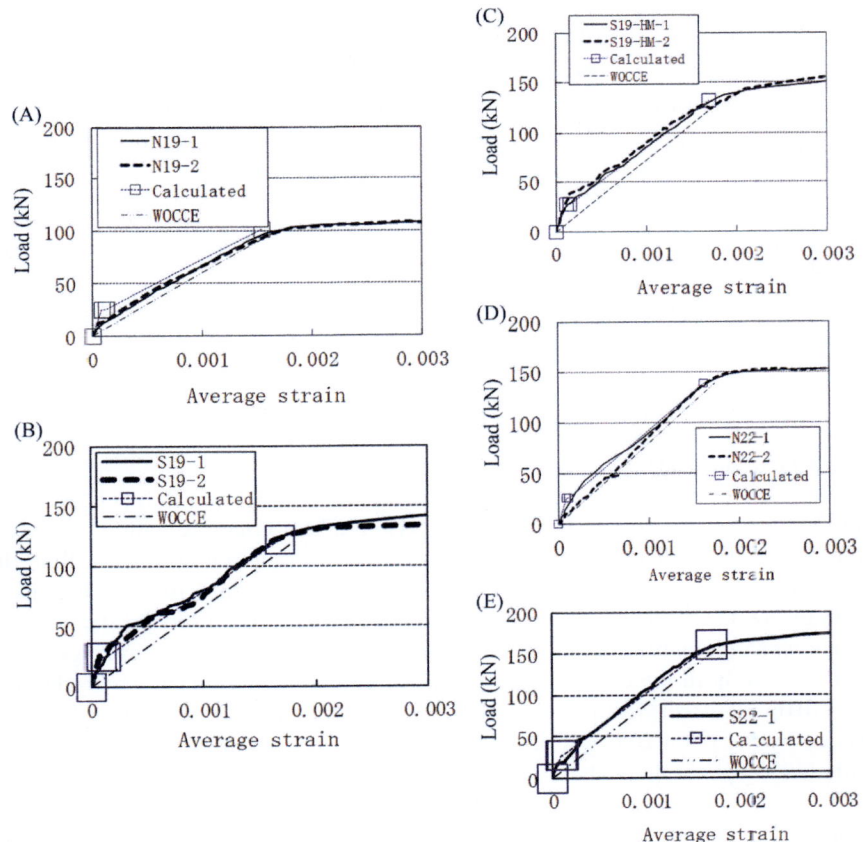

Figure 3.22 Experimental and calculated load—average strain relationships (WOCCE represents without consideration of the concrete effect). (A) N19. (B) S19. (C) S19-HM. (D) N22. (E) S22. (F) N13-W100. (G) S13-W100. (H) N13-W67. (I) S13-W67.

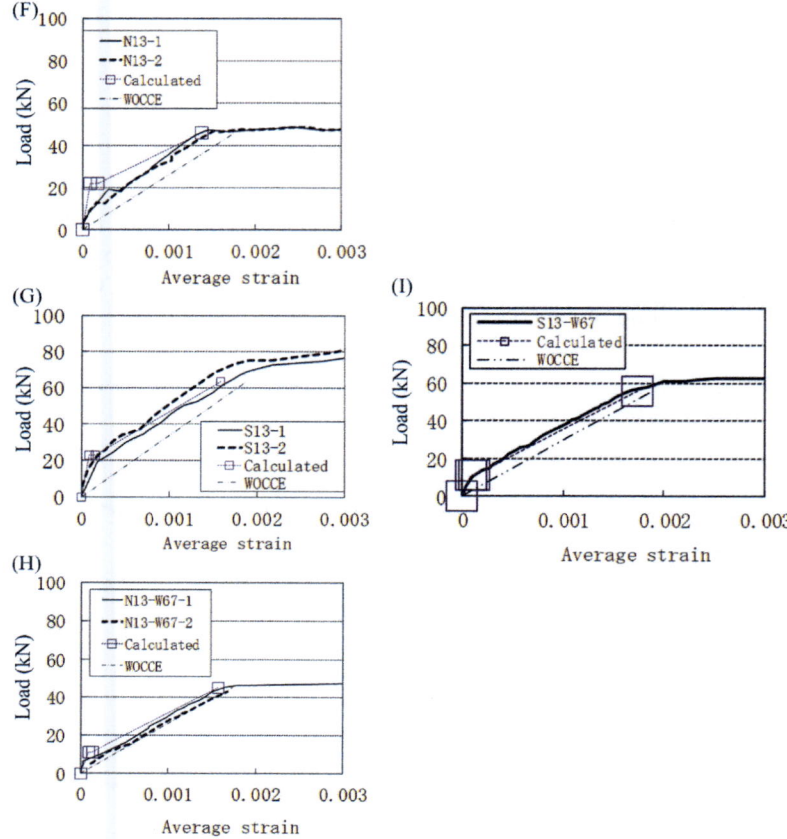

Figure 3.22 (Continued).

In the cases of Nr19, Nr13-W100, and N13-W67 for concrete without the reinforcing effect by CFRP sheets, the cracking loads of concrete are lower than the calculated values because the cracking load reduced owing to the significant effects of cracks in the RC members when little rebars were used. In the cases of N22 and specimens reinforced by the CFRP sheet, the load versus average strain obtained from both the experiment and calculation agreed well, in which the tension—stiffness effect from the CFRP sheet before the yielding of rebar was replaced by the rebar in terms of the CFRP's Young's modulus. Additionally, the detailed study of the load shared by the rebars should be further directly gauged.

3.3.2 Experimental studies

Fundamental data are required for developing a reliable constitutive model in pure tension of concrete elements with internal steel reinforcement and external FRP sheets, such as the effect of FRP sheets on the average bond stress—average strain relationship of steel reinforcement, the average stress—average strain relationship of concrete and steel reinforcement, and crack width. In this section, two types of specimens were studied for experimental modeling (Ueda et al., 2002). Type A specimens were tested for investigating the relationship between the average strain and average concrete stress, and between the average strain and average bond stresses of steel reinforcement and CFS. Type A specimens were concrete prisms with a deformed bar embedded at its center of the cross-section and CFS externally bonded to two opposite sides, as shown in Fig. 3.23.

The experimental parameters were the steel reinforcement ratio (ρ) and sheet ratio (ρ_{cf}) (see Table 3.5). The width of the CFS was 120 mm, which is narrower than the prism width. Strain gauges were mounted on the steel bar at 40 mm spacing and on the top layer of the CFS at a 20 mm spacing in the test zone measuring 1200 mm long. Contact chips were mounted on the concrete surface at a 60 mm spacing to measure the crack widths. The center of the prism contained a notch for

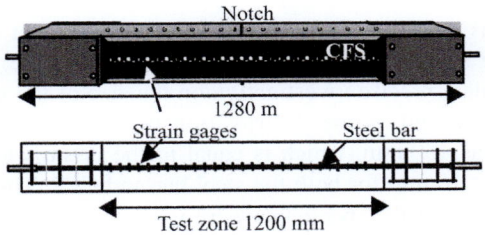

Figure 3.23 Type A specimen (Ueda et al., 2002).

Table 3.5 Experimental specimens (Ueda et al., 2002).

Type of specimen	Specimen	Type of Re. Bar	Cover (mm)	A_c (mm × mm)	ρ (%)	ρ_{cf} (%)
A	S-1-1″	D16	92	200 × 200	0.5	0.07
	S-1-2″	D16	92	200 × 200	0.5	0.13
	S-1-3″	D16	92	200 × 200	0.5	0.20
	S-2-0	D16	67	150 × 150	0.89	0
	S-2-3	D16	67	150 × 150	0.89	0.40
	S-3-0	D19	65.5	150 × 150	1.27	0
	S-3-1	D19	65.5	150 × 150	1.27	0.12
	S-3-2	D19	65.5	150 × 150	1.27	0.23
B	S-4-1′	D10	70	150 × 150	0.32	0.15
	S-4-2′	D10	70	150 × 150	0.32	0.29
	S-4-3′	D10	70	150 × 150	0.32	0.44
	S-2-2′	D16	67	150 × 150	0.89	0.29
	S-5-0	D22	64	150 × 150	1.72	0
	S-5-2′	D22	64	150 × 150	1.72	0.29

Table 3.6 Material properties of steel bars (Ueda et al., 2002).

Steel bar	Diameter (mm)	Young's modulus (GPa)	Yielding strain (%)
D10	9.53	188	0.178
D16	15.9	177	0.223
D19	19.1	170	0.230
D22	22.2	178	0.220

Table 3.7 Material properties of CFS (Ueda et al., 2002).

Thickness (mm)	Fiber density (g/m²)	Tensile strength (MPa)	Young's modulus (GPa)	Fracturing strain (%)
0.11	200	3479	230	1.5

initial cracking. Both ends of the prism were reinforced by steel plates and lateral reinforcement.

Type B specimens were similar to Type A specimens except for the width of the CFS, spacing of the strain gauges, and length of the test zone. A smaller spacing was adopted to enable a precise investigation on the size of the yielding zone of the steel reinforcement as well as the deterioration of the CFS bond properties. The width of the CFS was 150 mm, which is equal to the prism width. The spacing of the strain gauges was 10 mm for both the steel bar and CFS. The test zone size was 600 mm long. The ends of the specimens were reinforced by wrapping the CFS laterally over the longitudinal CFS. Table 3.5 shows details of all Type B specimens.

The concrete compressive strength was ~ 30 MPa for all specimens. The material properties of the deformed bars and CFS are shown in Tables 3.6 and 3.7, respectively.

3.3.3 Numerical studies

3.3.3.1 Outline of analysis

In this section, an analytical method based on force equilibrium shown in Fig. 3.24 is introduced to evaluate the tension stiffening effect of reinforced concrete members with CFS strengthening (RCC). The flow chart of the program is shown in Fig. 3.25. In the analysis, the Shima model (Shima, 1987) shown in Eq. (3.17) was adapted to a bond stress—slip—strain relationship for steel bar, while Sato's model (Sato et al., 2001) shown in Eqs. (3.18)—(3.21) was used as that for the CFS.

The Shima model

$$\tau_s = 0.73 \frac{[\ln(1+5s)]^3}{1+10^6 \varepsilon_s} f_c', \tag{3.17}$$

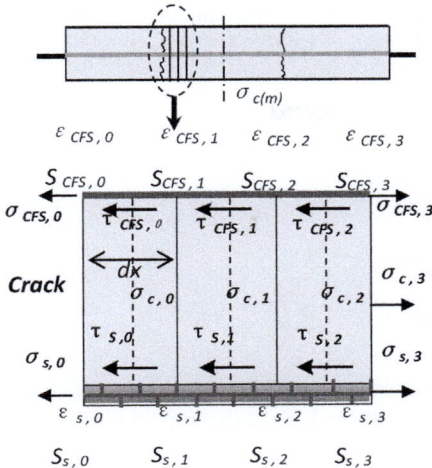

Figure 3.24 Stress components in a small element.

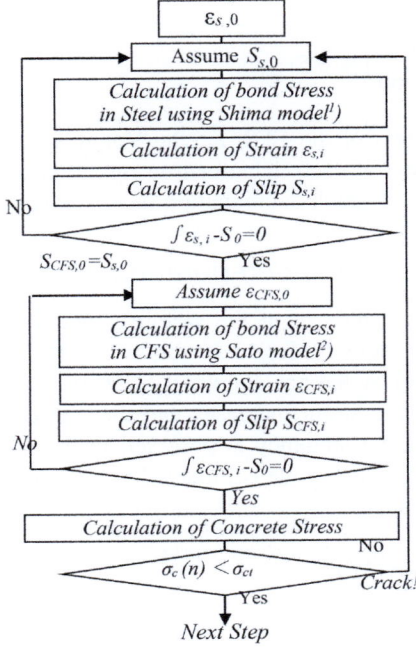

Figure 3.25 Program flow.

where τ_s is the bond stress and s is the slip of the bar (S_s) normalized by the bar diameter (D) $(s = 1000S_s/D)$.

The Sato model

For an ascending branch,

$$\tau_{CFS} = 0.2 \frac{S_{CFS}}{1 + 10^3 \varepsilon_{CFS}} f_c^{\prime 0.2} \tag{3.18}$$

For a descending branch,

$$\tau_{CFS} = \tau_{max-i} \, e^{-10(S-S_0)}, \tag{3.19}$$

where τ_{CFS} is the bond stress of the CFS, ε_{CFS} the CFS strain, τ_{max-i} the maximum bond stress at the point concerned, S_{CFS} the slip of CFS, and S_0 the slip at the maximum bond stress.

The maximum bond stress can be calculated as follows:

$$\tau_{max} = 9.1 f_c^{\prime 0.2} t_{CFS} \cdot E_{cf} \, 10^{-5} \le 3.49 f_c^{\prime 0.2} \tag{3.20}$$

with

$$\tau_{max-i} = \tau_{max} \left(1 - \frac{x}{60}\right) \le 0.5\tau_{max}, \tag{3.21}$$

where t_{CFS} is the thickness of the CFS, E_{cf} the Young's modulus of the CFS, τ_{max} the maximum bond stress at each delamination initiation point, and x the distance from the point concerned to each delamination initiation point.

3.3.3.2 Modeling of bond deterioration of steel bars

In the postyielding stage, the size of the yielding zone must be evaluated for the tension stiffening of the member. Fig. 3.26 shows that the relationship between the yielding zone normalized by the bar diameter, L_p/D, and the bar strain at crack. L_p/D increases linearly as the strain increases and then becomes constant at a value between 3 and 4. The same tendency is observed in the other specimens. Finally, the development of the yielding zone is modeled as shown in Fig. 3.27, and the following equations are obtained:

$$L_p = k(\varepsilon_{max} - \varepsilon_y)D \le 3.5D \tag{3.22}$$

with

$$k = \frac{\left(100\rho_{cf} + 1\right)^{0.46}}{500} - 0.0005, \tag{3.23}$$

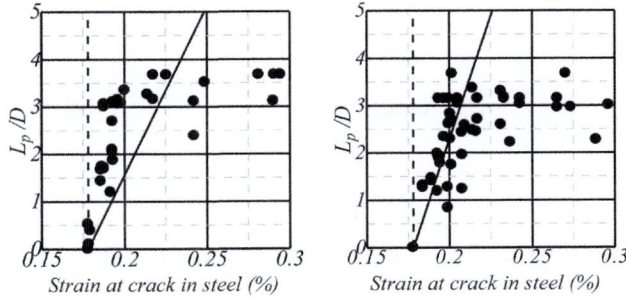

Figure 3.26 Variation of plastic length. (A) S-4-1′ (B) S-4-3′.

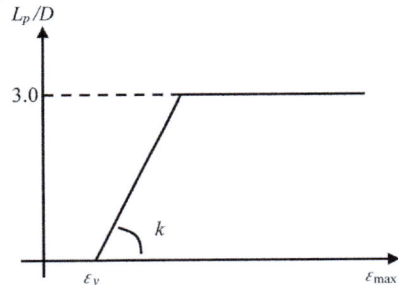

Figure 3.27 Modeling of L_p–ε_{max} relationship.

where k is the initial slope, and ε_{max} is the steel strain at the crack.

It is widely known that the bond deterioration of a steel bar around a crack must be considered. Hence, the length of the bond deterioration zone is assumed to be the same as the length of the plastic yielding zone. That is, the bond stress decreases linearly to zero at a distance of $3.5D$ from the crack surface.

3.3.3.3 Modeling of bond deterioration in fiber-reinforced polymer sheets

Bond deterioration in steel bars occurs near cracks. Similarly, bond deterioration in CFSs occurs near cracks. The deterioration mechanism is considered as follows: when a crack occurs, concrete near the crack moves suddenly owing to the opening at the crack. This movement causes some slip between concrete and CFS. Sudden cracking damages the bond near the crack. If a bond-damaged zone exists, concrete near the crack may be deformed, as shown in Fig. 3.28.

Based on Fig. 3.28 and sensitivity analysis, Eqs. (3.24), (3.25), and (3.26) are assumed to represent the bond deterioration of CFSs near a crack:

$$\tau_{max-i} = \frac{x}{60} \tau_{max} \tag{3.24}$$

$$\tau_{\max-i} = \left[1 - \left(\frac{x-60}{60}\right)\right]\tau_{\max} \tag{3.25}$$

$$\tau_{\max-i} = 0.5\tau_{\max} \tag{3.26}$$

3.3.3.4 Modeling of bond deterioration owing to decrease in crack spacing

As shown in Fig. 3.29, the direction of the bond stress can change when a second crack occurs. Therefore unloading/reloading paths must be prepared in the bond stress—slip—strain model. Because a small residual slip occurred in the experiment,

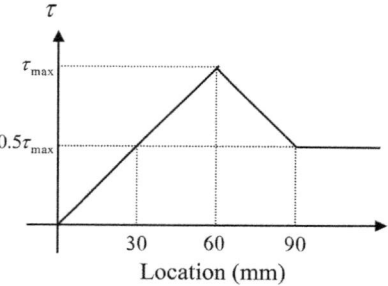

Figure 3.28 Bond deterioration model.

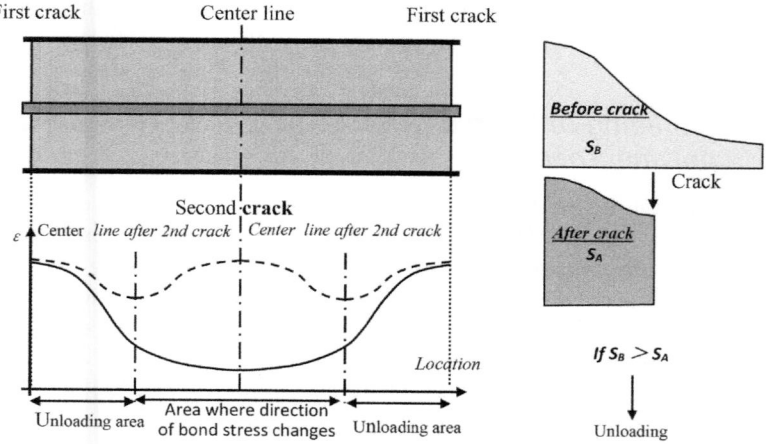

Figure 3.29 Variation of strain distribution due to second cracking.

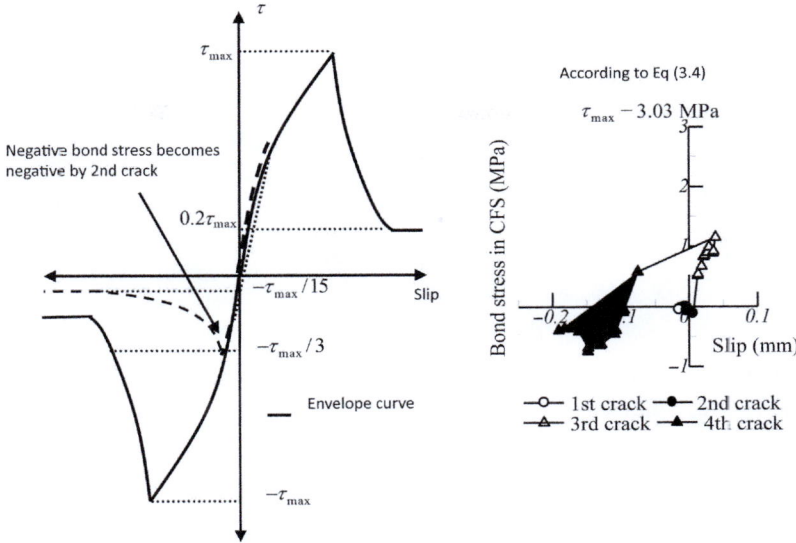

Figure 3.30 Model of bond deterioration in CFS caused by decease of crack spacing (Ueda et al., 2002).

linear curves connecting the origin and the stresses on the envelope curve were assumed for the unloading and reloading paths.

Fig. 3.30 shows the relationships between the local bond stress and local slip in the CFS. As shown, the slip became negative after the fourth cracking, and the peak bond stress after the fourth cracking became lower than the original one (Ueda et al., 2002). In this study, the maximum bond stress was assumed to be one-third of the original one when the slip was greater than 0.025 (before the cracking changed its direction owing to a decrease in crack spacing).

3.3.4 Comparison of analytical and experimental studies

3.3.4.1 Load–average strain relationship

Fig. 3.31 shows the experimental and analytical relationships between the load and the average strain of the specimens with steel reinforcement ratios of 0.5% and 1.27%. It can be concluded that the CFS significantly affected the stiffness at the postyielding stage (yielding strain was $\sim 2000\mu$), and the analytical results agreed well with the experimental results.

3.3.4.2 Average bond stress in steel and continuous fiber sheets

The average bond stress and strain curves of steel and CFS in specimens with steel reinforcement ratios of 0.5% and 1.27% are shown in Figs. 3.32 and 3.33, respectively. The average bond stress in the steel bars decreased as the number of CFSs

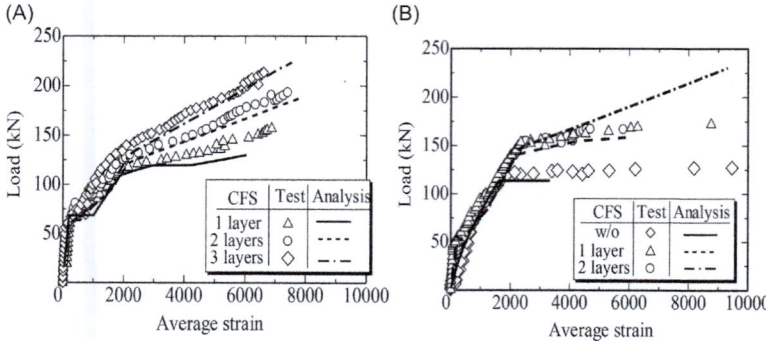

Figure 3.31 Relationships between load and average strain. (A) Series S-1 ($p_s = 0.5\%$).
(B) Series S-3 ($p_s = 1.27\%$).

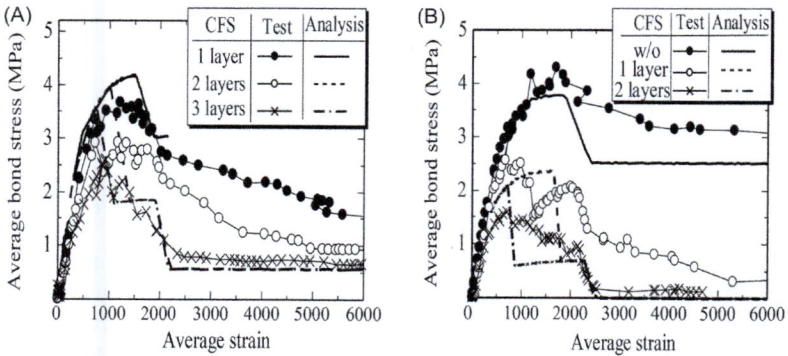

Figure 3.32 Relationships between average bond stress in steel and average strain.
(A) Series S-1 ($p_s = 0.5\%$). (B) Series S-3 ($p_s = 1.27\%$).

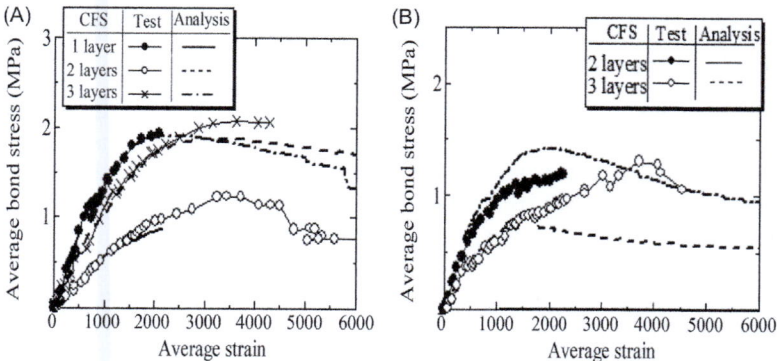

Figure 3.33 Relationships between average bond stress in CFS and average strain.
(A) Series S-1 ($p_s = 0.5\%$). (B) Series S-3 ($p_s = 1.27\%$).

increased, while the bond stress in the CFS increased. The analysis could simulate the experimental results well. In the analysis, the bond stress in steel and CFS decreased rapidly with the decrease in the crack spacing and at the yielding of the steel reinforcement.

3.3.4.3 Tension stiffening effect

Figs. 3.34 and 3.35 show the experimental and analytical relationships between the average stress and strain of concrete in specimens with steel reinforcement ratios of 0.5% and 1.27%, respectively. The solid line in Fig. 3.35 shows the tension stiffening curve for reinforced concrete predicted by Eq. (3.27) (Okamura and Maekawa, 1991):

$$\sigma_c = f_t \left(\frac{\varepsilon_{tu}}{\overline{\varepsilon}}\right)^{0.4}, \tag{3.27}$$

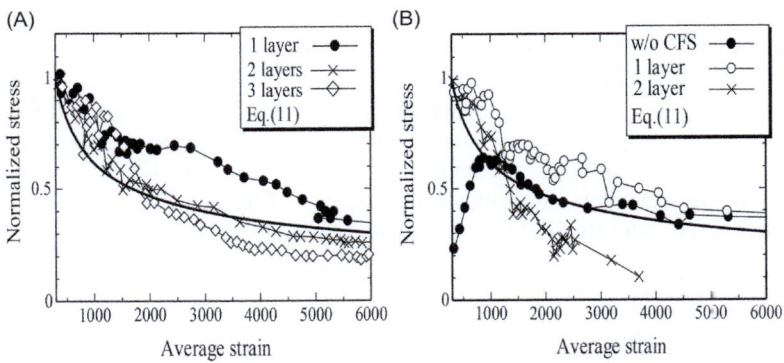

Figure 3.34 Tension stiffening curves observed in the experiment. (A) Series S-1 ($p_s = 0.5\%$). (B) Series S-3 ($p_s = 1.27\%$).

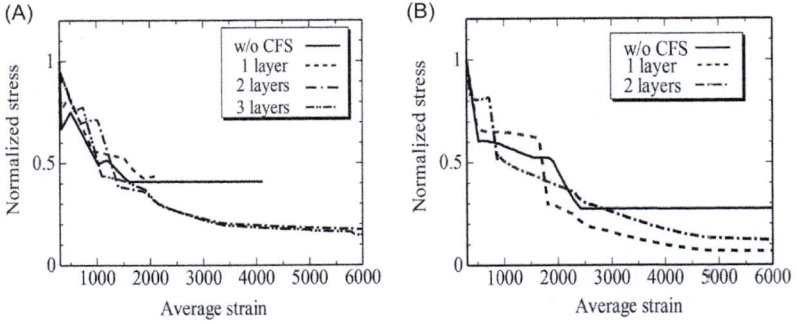

Figure 3.35 Tension stiffening curves observed in the analysis. (A) Series S-1 ($p_s = 0.5\%$). (B) Series S-3 ($p_s = 1.27\%$).

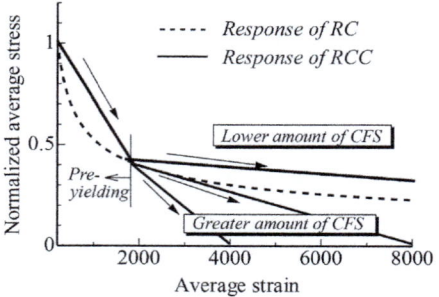

Figure 3.36 Characteristics of tension stiffening in RCC.

where σ_c is the average stress in concrete, f_t the tensile strength of concrete, ε_{tu} the strain at cracking, and $\bar{\varepsilon}$ the average strain.

As shown, the tension stiffening in RCC with one layer of CFS becomes greater than that in RC. In the case of two or three layers, however, the tension stiffness of concrete in RCC decreases more than that in RC after the yielding of steel.

Fig. 3.36 illustrates the general tension difference in stiffening curves between RCC and RC cases. The tension stiffening curve in RCC can be represented by two parts: the part in the preyielding range where the tensile stress is greater than that without the CFS but no effect on the amount of CFS is observed, and the part in the postyielding range where the greater number of CFSs generates a lower tensile stress in concrete after the yielding of the steel bar.

3.4 Formulation of crack spacing

3.4.1 Wu–Yoshizawa model

If the crack interval of members strengthened with continuous fiber sheets is the same as those for reinforced concrete members, the safe crack width can be determined by substituting the stress level of the reinforcement derived by considering the effect of continuous fiber sheets in Eq. (3.28). This is extracted from the Standard Specifications for Design and Construction of Concrete Structures (Design) (JSCE Guidelines for Concrete, 2007). In general, cracks in members to which continuous fiber sheets have been bonded are dispersed; accordingly, the crack width is reduced. In pull-out tensile strength tests of members bonded with carbon fiber sheets, the crack width is almost proportional to the average strain of the sheet and reinforcement, and it is almost independent of the concrete cover, steel diameter, rigidity of continuous fiber sheets, and compressive strength of concrete. At the level immediately before the yield point of the reinforcement, the crack width is approximately 0.3–0.7 times the width of cracks in members not bonded with sheets. It is noteworthy that when cracks have already occurred in a structure governed by dead loads, it is unclear whether the effect of the distribution of further cracking can be anticipated even if an upgrade is conducted with continuous fiber sheets attached to the underside of the girders. However, Eq. (3.28) extracted from the Standard Specifications for Design and Construction of Concrete Structures (Design) may be used for estimating the flexural crack width. In other words, the crack width after upgrading should be the width of cracks produced in the existing structure by drying shrinkage and dead loads, added to the additional crack width caused by the additional load (live loads, etc.) after upgrading with continuous fiber sheets. To calculate the additional crack width, the reinforcement strain caused by the additional load considering the continuous fiber sheets should be used. Even when the dead load is large, for structures with no cracking or those governed by live loads, the flexural crack width may be calculated using Eq. (3.29), in which the crack width calculated using Eq. (3.28) is multiplied by the maximum crack width ratio of 0.7:

$$w = k[4c + 0.7(C_s - \Phi)]\left[\frac{\sigma_{se}}{E_s}\left(or\,\frac{\sigma_{pe}}{E_p}\right) + \varepsilon'_{cs}\right] \tag{3.28}$$

$$w = 0.7k[4c + 0.7(C_s - \Phi)]\left[\frac{\sigma_{se}}{E_s}\left(or\,\frac{\sigma_{pe}}{E_p}\right) + \varepsilon'_{cs}\right] \tag{3.29}$$

Here, c is the concrete cover; c_s is the distance between the reinforcing bars and the centroid of a member; Φ is the diameter of tensile reinforcement; ε'_{cs} is the compressive strain for the evaluation of crack width increment owing to shrinkage and creep of concrete; σ_{se} and σ_{pe} are stress increments of reinforcement and prestressing steel, respectively, from the state in which concrete stress at the portion of reinforcement is zero.

For shear cracks, the mechanisms of the initiation and propagation of cracks are different from those of flexural cracks. Therefore they should be studied using appropriate methods. The examination of crack widths is not necessary for concrete structures upgraded with continuous fiber sheets because the surface is protected.

3.4.2 Sato's model for crack spacing estimation

Because cracking is inevitable in reinforced concrete, a number of investigations have been conducted to quantify the relationship between bar arrangement and crack spacing. The CEB-FIP model code is a popular model for crack spacing estimation using Eq. (3.30):

$$s_r = (c_s + s_b/10) + 0.1D/\rho, \tag{3.30}$$

where s_r is the crack spacing; s_b is the bar spacing; ρ is reinforcement ratio.

Crack spacing is governed by the concrete tensile strength and the bond strength between a bar and concrete. Therefore the deviation of Eq. (3.28) ranges from 50% to 200%. However, a previous study indicated that FRP sheets reduced the crack spacing by 50% in certain conditions. Therefore an alternative method is required for to estimate the crack spacing of RC with FRP sheets. The calculation of equilibrium and compatibility conditions can be used to determine the crack spacing using interfacial bond stress—slip relations along a reinforcement. Sato and Vecchio established a model equation to estimate the crack spacing of RC with FRP sheets and implemented it into a FEM algorithm (Sato and Vecchio, 2003). Eq. (3.31) is a fundamental differential equation for the bond between reinforcement and concrete:

$$\frac{d^2 S_{s,i}(u_1)}{du_1^2} = \frac{1}{\cos\theta_{s,i}} \left\{ \frac{\psi_{s,i}\, \tau_{s,i}(u_1)}{A_{s,i}E_{s,i}'} + \sum_{i'=1}^{m} \left(\frac{\psi_{s,i'}\cos\theta_{s,i'}\, \tau_{s,i'}(u_1)}{A_{ce}E_c} \right) \right\}, \tag{3.31}$$

where u_1 is the coordinate along the principal tensile concrete stress direction, θ_s the angle between the reinforcement and the principal tensile concrete stress direction, A_s the bonded area per unit reinforcement length, and A_{ce} the effective cross-sectional area of concrete.

Eq. (3.31) presents a similar formulation to that for steel bar but can be extended to general conditions where different types of reinforcements exist and where the fiber direction does not coincide with the principal tensile concrete stress direction. Substituting the bond stress—slip relation into Eq. (3.31) allows the distributions of bond stress, slip, sheet stress, and concrete stress to be computed, in which crack occurs when the tensile stress induced by the bond reaches the concrete tensile strength. The number of the cracks increases as the tensile strain increases and finally converges to a constant crack spacing. Hence, a parametric study was conducted to estimate the final crack spacing s_r using the tension chord model shown in Fig. 3.37.

Eqs. (3.32) and (3.33) express a crack spacing model, which was developed based on parametric computations.

$$s_r = \frac{f_t}{2 \sum_{i=1}^{m} \frac{\rho_i\, \tau_{b0,i}\, \cos\theta_{s,i}}{D_i} + \frac{1}{220} \sum_{j=1}^{n} \frac{\rho_{cf,j}\, c_{3,j}\cos^2\theta_{F,j}}{t_{CFS,j}}} \quad (s_r \le 220 \text{ mm}) \tag{3.32}$$

$$s_r = \frac{f_t - \sum_{j=1}^{m} \frac{\rho_{cf,j}\, c_{3,j}\cos^2\theta_{F,j}}{t_{CFS,j}}}{2 \sum_{i=1}^{n} \frac{\rho_i\, \tau_{b0,i}\, \cos\theta_{s,i}}{D_i}} \quad (s_r > 220 \text{ mm}) \tag{3.33}$$

$$c_3 = \left(15.8 + 1.34\sqrt{t_{CFS}E_{cf}} \right)\sqrt{G_f} \quad (\text{N}/\text{mm}), \tag{3.34}$$

where τ_{b0} is the average bond stress along the steel bar (twice the tensile concrete strength), θ_F the angle between the FRP sheet and the principal tensile concrete stress direction, and G_f the area enveloped by the bond stress—slip curve.

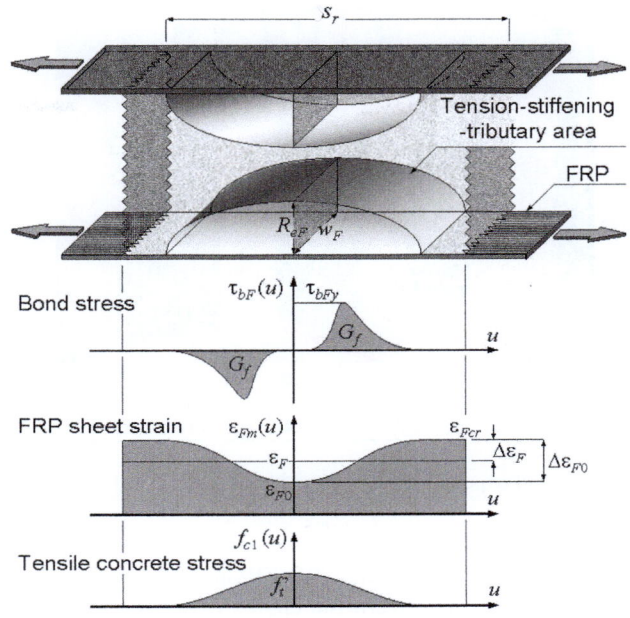

Figure 3.37 Tension chord model (Sato and Vecchio, 2003).

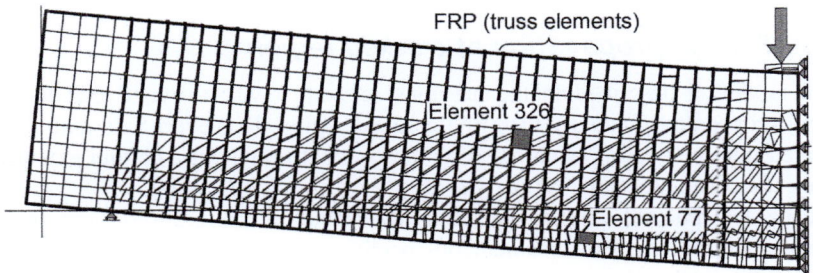

Figure 3.38 FE model (Sato and Vecchio, 2003).

Based on Eqs. (3.32)−(3.34), the model equations can be implemented into an FEM algorithm to analyze the crack spacing of RC beams strengthened with FRP sheets, the analysis model of which is presented in Fig. 3.38 (Sato and Vecchio, 2003). The analysis results indicated that the contribution of the FRP to the crack spacing induced a slight difference in the macro load−deflection behavior of the beams. On the contrary, differences in the calculated crack spacing were remarkable. Fig. 3.39 shows the crack pattern of the beam close to the loading point. The analyzed and experimental crack widths are indicated in the figure. The black areas are carbon fiber sheets, which were provided for shear strengthening. Because the

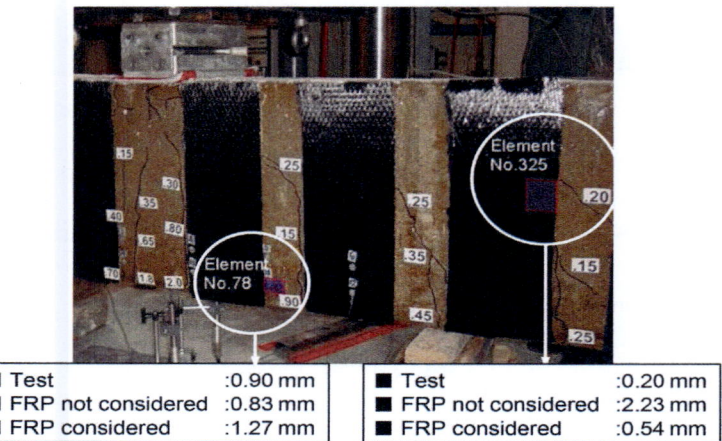

■ Test	:0.90 mm		■ Test	:0.20 mm
■ FRP not considered	:0.83 mm		■ FRP not considered	:2.23 mm
■ FRP considered	:1.27 mm		■ FRP considered	:0.54 mm

Figure 3.39 Analyzed and experimental crack widths.

bond along the flexural reinforcing bars governs the flexural cracks on the bottom of the beam, the difference between the CEB-FIP and the proposed models is insignificant. Meanwhile, the sheets govern the cracks in the mid-depth region, and the difference between the models becomes remarkable. The proposed model provides a closer estimate of crack width for the mid-depth region.

3.4.3   Peeling propagation at cracked region

In the modeling described in the previous section, it was assumed that the sheet peeling propagated gradually from the cracked region. However, Sato and Vecchio indicated that the peeling occurred instantaneously throughout a certain extent from a crack (Sato and Vecchio, 2003).

Hence, six uniaxial tension specimens (Sato and Vecchio, 2003) were analyzed to investigate the effect of instantaneous peeling at the crack region. The specimen was a 1200-mm-long concrete block with a 150 mm^2 cross section. A steel bar of diameter 16 or 19 mm was placed longitudinally along the block. Table 3.8 presents the reinforcing conditions and fracture modes of the specimens.

Fig. 3.40 illustrates the bond strength reduction model proposed by Sato and Vecchio (2003). The x-coordinate represents the distance from the crack. b_{max} is the maximum local bond stress and b_y is the bond strength obtained by material test. This model represents the assumption that the instantaneous peeling reduces the bond strength linearly up to a distance of 60 mm from a crack (original paper assumes that the bond strength further reduces at $x > 60$ mm; however, this is neglected herein). The analysis conducted in this study assumed an interfacial fracture energy $G_f = 0.48$ N/mm.

Fig. 3.41 shows the relationship between the average cross-sectional stress and axial strain. The bond strength is reduced based on the model shown in Fig. 3.41.

Table 3.8 Uniaxial tension specimens.

Specimen	Cross-sectional area ratio of reinforcement (%)		Fracture mode
	Steel bar	**Carbon fiber sheet**	
S20	0.89	0.00	Steel bar yield
S21	0.89	0.12	Sheet rupture
S23	0.89	0.36	Sheet rupture
S30	1.26	0.00	Steel bar yield
S31	1.26	0.12	Sheet rupture
S32	1.26	0.36	Sheet rupture

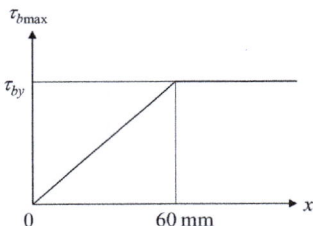

Figure 3.40 Bond strength reduction model.

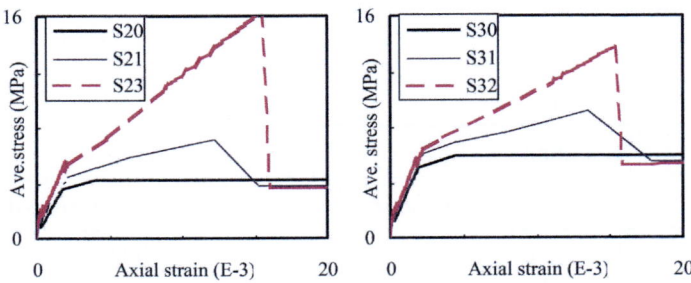

Figure 3.41 Relationships between the average cross-sectional stress and the axial strain.

Table 3.9 compares the experimental and analyzed maximum average stresses. The analysis overestimated the maximum stress considerably. Further consideration is required for the estimation of the FRP sheet's local stress.

Fig. 3.42 illustrates the relationship between the crack spacing and axial strain of Specimens S20, S21, and S23. Table 3.10 compares the experimental and analyzed crack spacing. The bond strength reduction model resulted in a considerable difference in the analyzed crack spacing of Specimen S21. The analysis without the

Table 3.9 Maximum average stress.

Specimen	Test (MPa)	Analysis (MPa)	
		Bond strength reduced	**Not reduced**
S20	3.8	4.7	4.7
S21	4.9	7.2	7.8
S23	8.1	16.3	16.2
S30	5.8	6.7	6.7
S31	7.5	9.2	9.5
S32	7.3	13.7	13.7

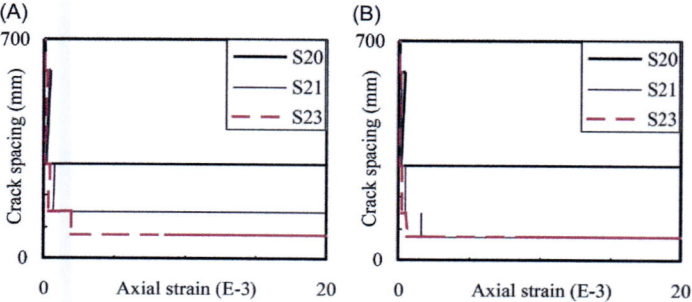

Figure 3.42 Relationships between crack spacing and axial strain. (A) Bond strength reduced. (B) Not reduced.

Table 3.10 Crack spacing.

Specimen	Test (mm)	Analysis (mm)	
		Bond strength reduced	**Bond strength reduced**
S20	400	300	300
S21	200	150	75
S23	125	75	75

strength reduction overestimated the tensile concrete stress induced by the sheet bond and resulted in an underestimation of the crack spacing.

Fig. 3.43 shows the relationships between the average rebar bond stress and the axial strain of Specimens S30, S31, and S32. The average rebar bond stress is defined as the average of bond stress between adjacent cracks.

The experimentally observed maximum values of the average bond stresses (before sheet rupture) were 4.3 MPa for specimen S30, 3.1 MPa for S31, and

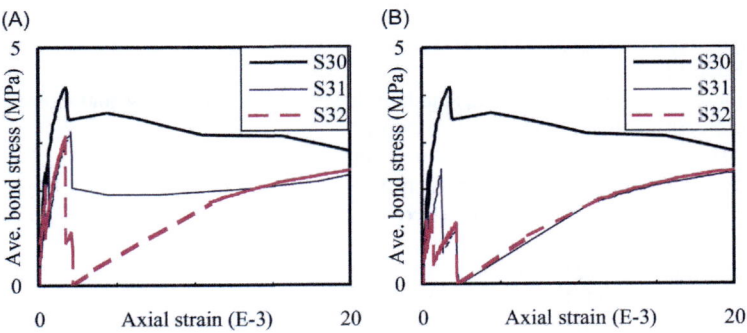

Figure 3.43 Relationships between average bond stress along steel bar and axial strain. (A) Bond strength reduced. (B) Not reduced.

Table 3.11 Maximum value of average bond stress along steel bar.

Specimen	Test (MPa)	Analysis (MPa)	
		Bond strength reduced	**Not reduced**
S30	4.3	4.2	4.2
S31	3.1	3.2	2.4
S32	2.5	3.1	1.5

2.5 MPa for S32. Table 3.11 indicates that the analysis with the strength reduction provided closer estimates to the test. Meanwhile, the analysis without the strength reduction resulted in lower average bond stresses; 2.4 MPa for specimen S31 and 1.5 MPa for S32. These results were caused by the larger bond stresses of the sheets, which increased the number of cracks and accelerated bond deterioration along the steel bar.

In this analysis, the bond stress reduced linearly up to a distance of 60 mm from a crack. This assumption should be modified in different testing conditions. Therefore further consideration is required with respect to the relationship among crack size, loading rate, and peeling propagation.

References

Gilbert, R.I., 1992. Shrinkage cracking in fully restrained concrete members. ACI Struct. J. 89 (2), 141−149.

Gupta, A.K., Maestrini, S.R., 1990. Tension-stiffness model for reinforced concrete bars. J. Struct. Eng. 116 (3), 769−790. ASCE.

JCI Technical Committee on Continuous Fiber Reinforced Concrete, 1997−1998. A State-of-the-Art Report on Repair and Strengthening Technique with Continuous Fiber Sheet, JCI Report I, II.

JSCE Guidelines for Concrete, 2007. Standard Specifications for Design and Construction of Concrete Structures (Design), Tokyo, Japan.

Okamura, H., Maekawa, K., 1991. Nonlinear analysis and constitutive models of reinforced concrete. Gihodo Shuppan Company, p. 182.

Ouyang, C., Shah, S.P., 1994. Fracture energy approach for predicting cracking of reinforced concrete tensile members. ACI Struct. J. 91 (1), 69−78.

Sato, Y., Vecchio, F.J., 2003. Tension stiffening and crack formation in RC members with FRP sheets. J. Struct. Eng. 29 (6), 717−724. ASCE.

Sato, Y., Asano, Y., Ueda, T., 2001. Fundamental Study on Bond Mechanism of Carbon Fiber Sheet, vol. 37. Concrete Library International, JSCE, pp. 97−115.

Shima, H., 1987. Micro and macro models for bond in reinforced concrete. J. Faculty Eng. XXXIX (2), 1987. The University of Tokyo (B).

Traiantafillou, T.C., Plevris, N., 1992. Strengthening of RC beams with epoxy-bonded fiber-composite materials. Mater. Struct. 25, 201−211.

Ueda, T., Yamaguchi, R., Shoji, K., Sato, Y., 2002. Study on behavior in tension of reinforced concrete members strengthened by carbon fiber sheet. J. Compos. Constr. 6 (3), 168−174. ASCE.

Wu, Z.S., 1997. Research trends on retrofitting and strengthening concrete structures by FRP sheets and plates. In: Proceedings of the International Conference on Fibre Reinforced Concrete, Guangzhou, pp. 8−20.

Wu, Z.S., Yin, J., 2003. Fracturing behaviors of FRP-strengthened concrete structures. Eng. Fract. Mech. 70, 1339−1355.

Wu, Z.S., Yoshizawa, H., 1999. Analytical/experimental study on composite behavior in strengthening structures with bonded carbon fiber sheets. J. Reinf. Plast. Compos. 18 (12), 1131−1155.

Wu, Z.S., Matsuzaki, T., Tanabe, K., 1997. Interface crack propagation in FRP-strengthened concrete structures. Proc. Third Int. Symp. Non-Metallic (FRP) Reinf. Concr. Struct. 1, 319−326.

Yoshizawa, H., Wu, Z.S., 1999. Cracking behavior of plain concrete and reinforced concrete members strengthened with carbon fiber sheets. In: Fourth International Symposium on Fiber Reinforced Polymer Reinforcement for Reinforced Concrete Structures, ACI International SP-188, pp. 767-779.

Further reading

CEB-FIP, 1978. Model Code for Concrete Structures: CEB-FIP International Recommendations, third ed Comité Euro-International du Béton, Paris, p. 348.

Japan Concrete Institute, 2003. Proceedings and Technical Report on JCI Technical Committee, International Symposium on Latest Achievement of Technology and Research on Retrofitting Concrete Structures, Kyoto, Japan.

JSCE, 2001. Recommendations for Upgrading of Concrete Structures with Use of Continuous Fiber Sheets, Tokyo, Japan.

Suga, M., Nakamura, H., Higai, T., Saito, S., 2001. Effect of bond properties on the mechanical behavior of RC beam. Proc. Jpn. Concr. Inst. 23 (3), 295−300.

Flexural strengthening of structures

4

4.1 Introduction

Traditional methods that are commonly used for the flexural strengthening of reinforced concrete (RC) members include the external cable method, overlaying and jacketing method, and external steel plate bonding. These methods suffer from inherent disadvantages ranging from difficult application procedures to a lack of durability. Using fiber-reinforced polymer (FRP) composites to increase the flexural strength of existing structures has gained wide acceptance in recent years. Using high-strength and high-quality adhesives, FRP laminates are externally bonded to the structure to be strengthened. Although the cost of the materials, including fibers and resins, is higher than that of traditional materials, the total life-cycle cost may be lower owing to the lower installation costs, higher performance, and lower maintenance costs (Meier and Winistorfer, 1995; Teng et al., 2001).

Many experimental investigations and practical applications have been performed, which have shown that the strengthening effects can be reflected in a wide range of aspects, such as the enhancement of the structural stiffness, load-carrying capacity, ductility, and corrosion resistance (Wu and Kurokawa, 2002). As discussed previously, FRP composites remain elastic until failure and fail in a noticeably brittle manner. Moreover, the limited ability of the FRP−concrete interface to transfer stresses from the concrete substrate to FRP laminates often greatly reduces the expected gains and results in a catastrophic, brittle failure, such as delamination at the anchorage zone, peeling-off caused by shear cracks, and debonding initiated from intermediate flexural cracks (Wu et al., 1997). Generally, the strengthening effect of FRP cannot be fully utilized owing to the debonding of FRP laminates and sufficient ductility may not be gained owing to the debonding or brittle rupture of FRP composites. Moreover, the strengthening effects with aramid, glass, or basalt fiber on both the flexural stiffness before yielding of steel reinforcement and the yielding load are not significant. Even for the case of carbon fiber with a regular Young's modulus equivalent to steel reinforcement, the effects are still insufficient because the reinforcing ratio of FRP is relatively low. Hence, a significant increase in the service load could hardly be achieved. To improve the performance of FRP-strengthened structures and efficiently utilize the strengthening effect of FRP, it is essential to develop effective strengthening techniques such as an FRP prestressing method and a hybrid FRP strengthening technique.

Structures Strengthened with Bonded Composites. DOI: https://doi.org/10.1016/B978-0-12-821088-8.00004-7

4.2   Flexural strengthening methods for structural members

Existing infrastructure, such as bridges, buildings, tunnels, and chimneys with insufficient flexural resistance can be conveniently strengthened with externally bonded FRP laminates. Flexural strengthening with FRP laminates can be applied to enhance the structural performance in different ways, as shown in Fig. 4.1. For the

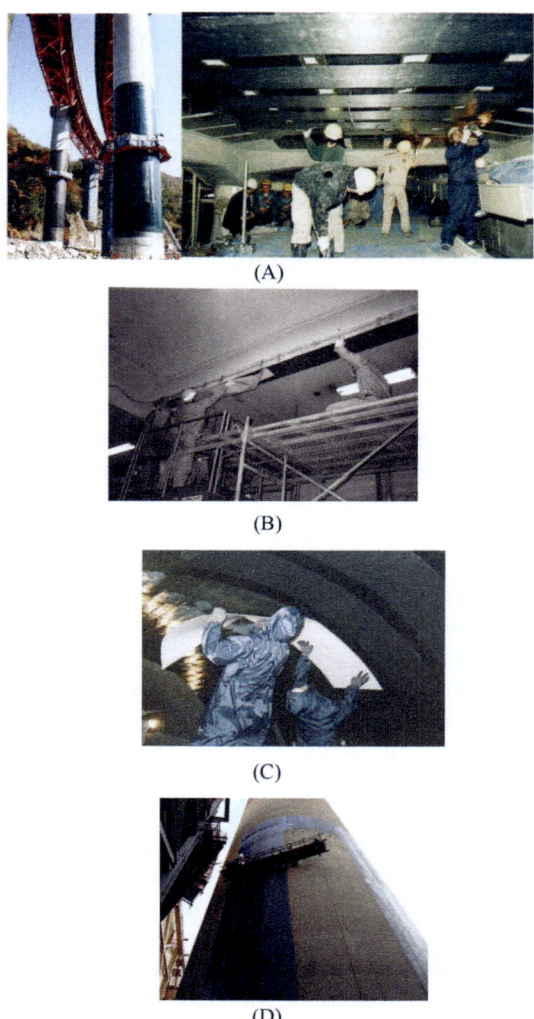

(A)

(B)

(C)

(D)

Figure 4.1 Flexural strengthening method for different types of structures. (A) Bridge. (B) Building. (C) Tunnel. (D) Chimney.

strengthening of bridge structures, FRP laminates can be bonded to the bottom of girders or decks for the flexural capacity enhancement; FRP laminates can also be
bonded to the side of bridge piers to enhance the flexural resistance capacity and to
prevent the flexural-shear failure of piers or lap splice failure of columns with longitudinal reinforcement cut-offs, as shown in Fig. 4.1A. Similar strengthening methods
can be found in Fig. 4.1B−D for buildings, tunnels, and chimneys, respectively.

4.3 Effect of flexural strengthening on the performance of structural members

4.3.1 Strengthening effect under monotonic load

4.3.1.1 Failure modes

Previous research has demonstrated that significant improvements in FRP-
strengthened structures can be achieved in terms of stiffness, strength, crack resistance, fatigue resistance, and durability. However, a wide variety of premature failure modes may limit these theoretical gains, such as crushing of concrete, shear
failure, FRP rupture, and bond failures, as depicted in Fig. 4.2. Among them, the

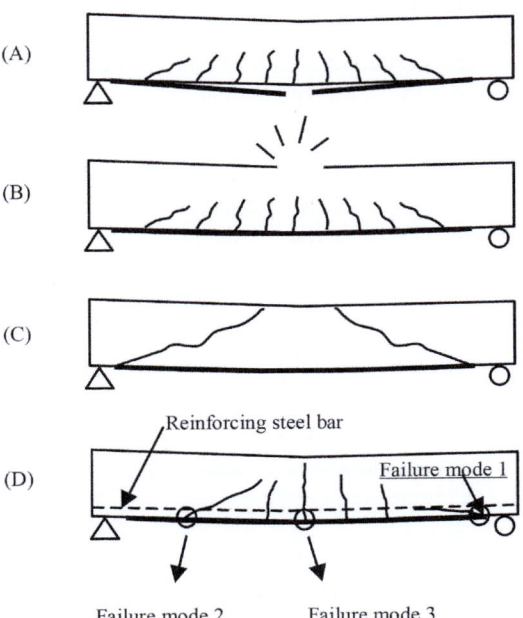

Figure 4.2 Failure modes in FRP-strengthened RC beams: (A) FRP rupture; (B) concrete
crushing failure; (C) concrete shear failure; (D) bond failure modes (Failure mode 1:
delamination at the cut-off point; Failure mode 2: peeling-off due to shear cracks; Failure
mode 3: intermediate crack-induced debonding failure).

first three failure modes can be predicted by using conventional or modified RC flexural theory. Bond failures (Fig. 4.2D), including delamination at the cut-off point, peeling-off due to shear cracks, and debonding failure due to flexural cracks, are new failure modes, which invalidate the composite action between concrete and FRP and often significantly reduce the expected load-carrying capacity of the retro-fitted system. To improve the FRP-to-concrete bond and take full advantage of the strength of FRP composites, sound modeling, understanding, and evaluation of the interfacial fracture behavior are essential for the development of the FRP bonding technique.

4.3.1.2   Experimental observations

1. *FRP-strengthened plain concrete beams*

A series of plain concrete beam tests were carried out by Wu et al. (1997) to investi-gate the debonding behavior of FRP-strengthened concrete structures. The primary aim was the observation of the debonding occurrence, crack propagation, and failure mode of the strengthened beams. A total of eight concrete beams were cast, as shown in Fig. 4.3. Half of the concrete beams were designed with a notch located in the bottom of the beams at the midspan, as shown in Fig. 4.3. To observe the debonding occurrences along the FRP-to-concrete interface, piezoelectric sensors were bonded to the bottom of the carbon FRP sheet at four points, as shown in Fig. 4.4. Small vibrations due to debonding along the FRP-to-concrete interface were transformed into electrical signals and their wave shapes were shown on a synchroscope.

Plots of the load versus deflection of both notched and unnotched beams are shown in Fig. 4.5. Fig. 4.6 shows the crack propagation in the concrete of both notched and unnotched beams, and Fig. 4.7 shows the debonding appearances. Only one flexural crack initiated nearly at the midspan and propagated from the bottom to the top of the beams. Around this flexural crack, one or two diagonal cracks initiated from the beam base and propagated through to the flexural crack, especially for the notched beams. The interfacial crack along the FRP-to-concrete interface developed from the bottom tip of the flexural crack. The diagonal cracks were found to accelerate the interfacial crack propagation. A debonding region (approximately 100 mm) between the tips of the flexural and first

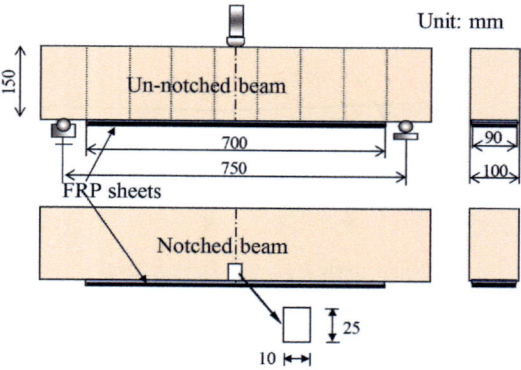

Figure 4.3 Details of the test beams.

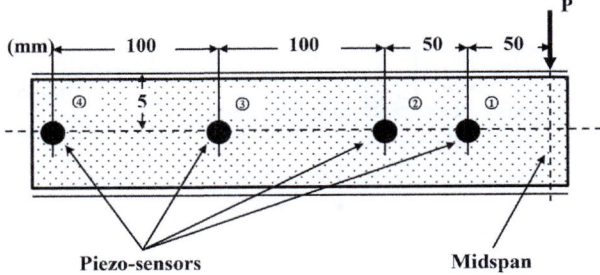

Figure 4.4 Locations of mounted piezo-sensors for measuring debonding.

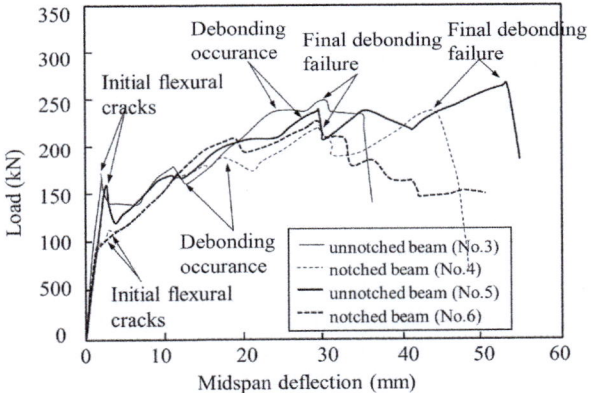

Figure 4.5 Load versus deflection relationship of the test beams.

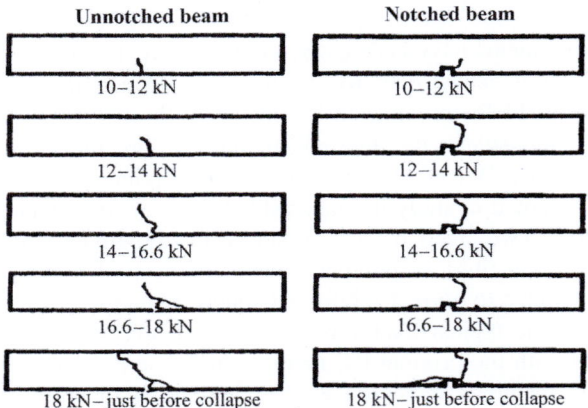

Figure 4.6 Crack propagation in concrete beams.

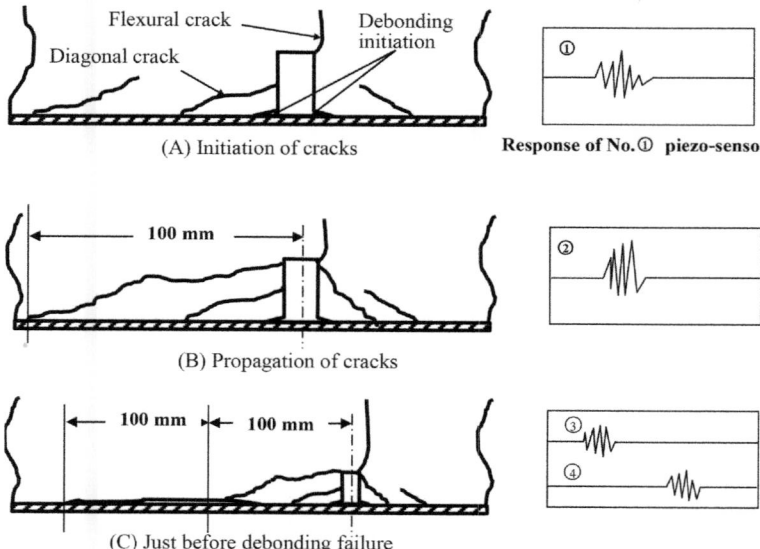

(A) Initiation of cracks   **Response of No.① piezo-sensor**

(B) Propagation of cracks

(C) Just before debonding failure

Figure 4.7 Fracturing process.

diagonal cracks formed with the propagation of the concrete cracks and gradually extended to the tip of the second diagonal crack with a relatively high speed. The interfacial cracks continuously extended toward the end of the FRP sheet to a length of 200 mm, approximately half of the total bonding length. Then, a sudden collapse occurred. All of these observations were confirmed by the sequence of responses of the synchroscope initiated from the piezoelectric sensor No. 1 (Fig. 4.4) and the visual observations as shown in Figs. 4.6 and 4.7. A strong peeling phenomenon of the FRP sheet at the bottom tip of the main diagonal crack was observed during the tests.

2. *FRP-strengthened RC beams*

Extensive experiments have been carried out to study the effect of FRP strengthening on the flexural behavior of RC beams. This section introduces the typical flexural behavior of FRP-strengthened RC members through experimental observations. Fig. 4.8 demonstrates the dimensions and reinforcement details of the test beams carried out by Wu et al. (1999a). The four-point or three-point bending method is often adopted to study the flexural behavior of strengthened beams under monotonic loading. Two layers of high-tensile (HT) strength carbon fiber (C), high-modulus (HM) carbon fiber, or polypara-phenylene-benzo-bis-oxazole (PBO) fiber (P) sheets were bonded to normal-reinforced concrete beams.

Fig. 4.9 shows the typical load versus midspan displacement of the test specimens. Compared with the original RC member, the load-carrying capacity of FRP-strengthened RC members increased at the yielding of the reinforcing steel and afterward, as shown in Fig. 4.9A. Increasing the number of FRP layers improved the ultimate load capacity of the strengthened beams but reduced the ductility, as

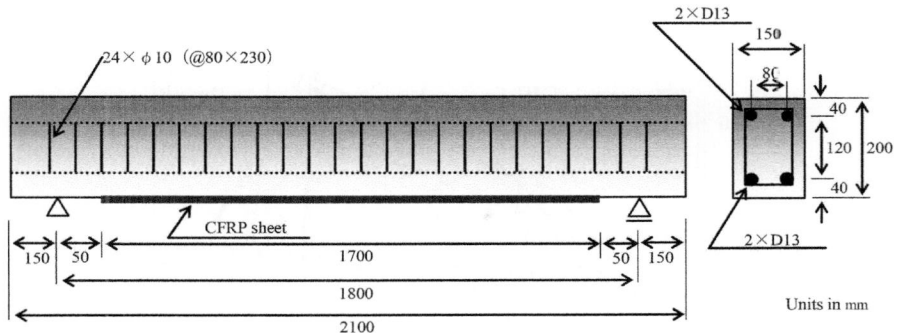

Figure 4.8 Details of the experimental beams.

shown in Fig. 4.9B. This is mainly due to the increased brittleness of the premature FRP debonding failure when more layers of FRP sheets were applied.

From the continuous monitoring of strain gauges attached to the FRP laminate, the strain profile along the length of the beam can be established for various load levels. Fig. 4.10A demonstrates the typical stress/strain distributions of the FRP. Increasing the number of FRP layers leads to the decreased ultimate FRP stress/strain, as shown in Fig. 4.10B–D.

Fig. 4.11 shows the crack pattern of the tested specimens, in which the solid and dashed lines represent cracks before and after yielding of the reinforcements, respectively. It can be seen that the cracks of carbon fiber-reinforced polymer (CFRP)-strengthened RC members tended to be distributed more uniformly and closely, compared with those of the unstrengthened RC beam.

Fig. 4.12 demonstrates the relationship between the measured crack width and strains of the rebar and FRP. The strain of the FRP is the average strain of 15 strain gauges pasted in a pure bending span. The crack width and strain of the CFRP are approximately proportioned before and after yielding of the rebar.

Fig. 4.13 shows the applied load versus the average crack width of FRP-strengthened and unstrengthened beams. The average crack width of the FRP-strengthened beam is smaller than that of the unstrengthened beams. Moreover, the average crack width of the 2C-HM specimen with high FRP stiffness tends to be smaller than that of the 2C-HT specimen with low FRP stiffness under the same load.

4.3.2 Strengthening effect under fatigue load

Fatigue damage can be observed in various types of structures, especially bridges, under cyclic loading, which is often characterized by multiple cycles and low load amplitudes. Many concrete bridges in use today have exceeded their original design life. In addition, allowable traffic loads have increased over the years. Owing to these combined factors, many existing bridges are

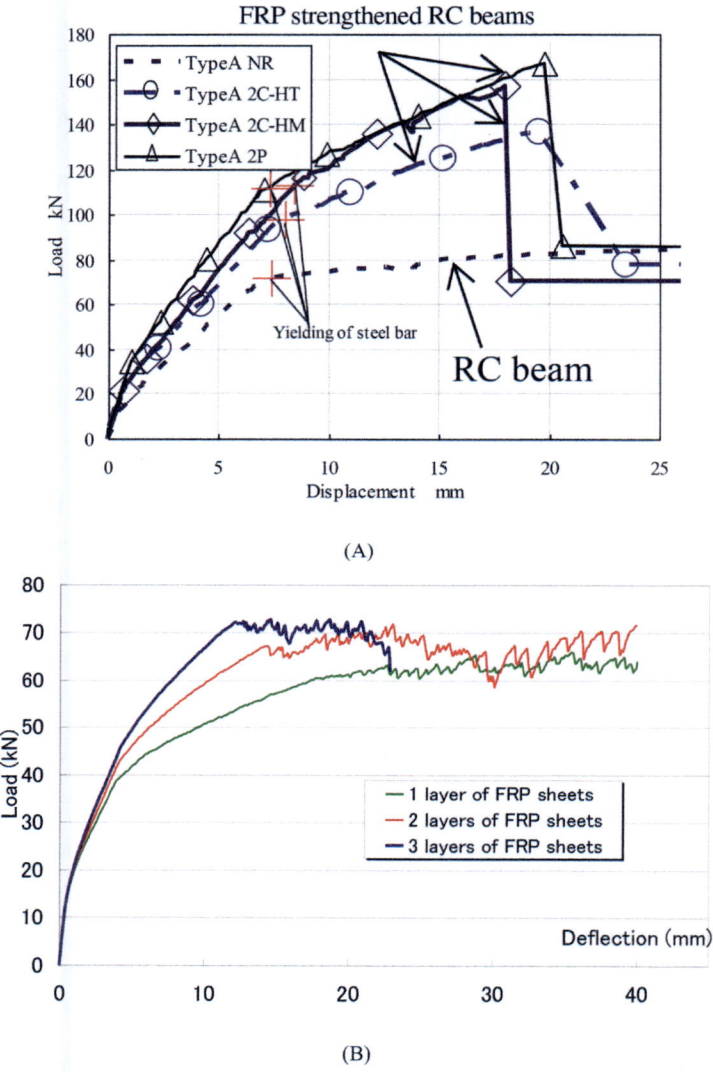

Figure 4.9 Load−displacement response.

structurally deficient. As discussed previously, FRP composites have been successfully used to strengthen or retrofit bridge girders and decks. Given that bridges are subjected to fatigue loads, research on flexural members strengthened with FRP composites under fatigue loading is required. The following sections introduce the strengthening effect of FRP-flexural-strengthened RC members under fatigue loading.

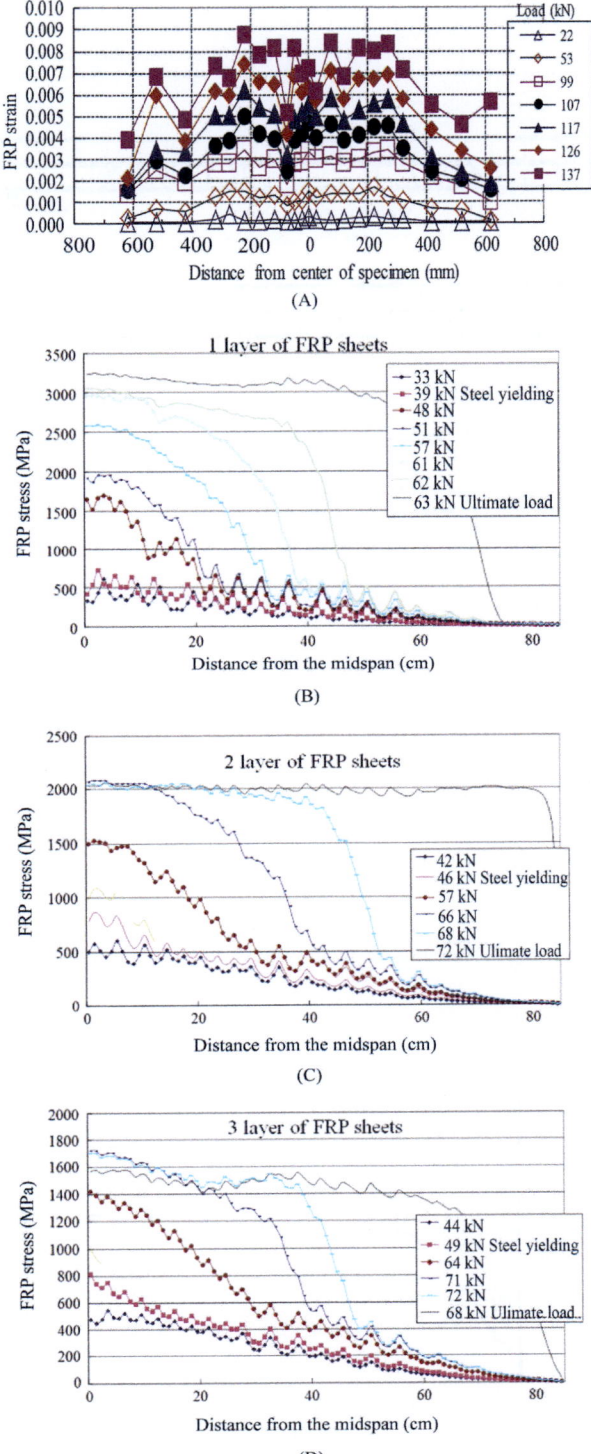

Figure 4.10 FRP strain/stress distribution of FRP-strengthened RC beams.

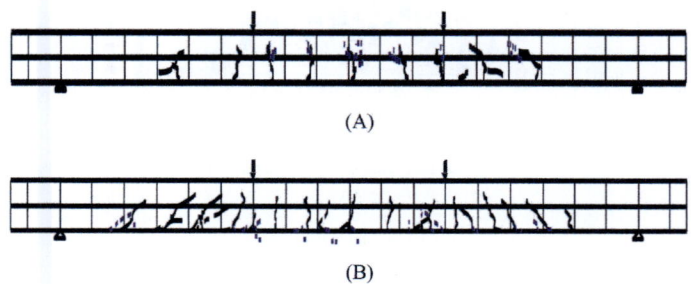

Figure 4.11 Crack pattern. (A) RC beam. (B) FRP-strengthened RC beam.

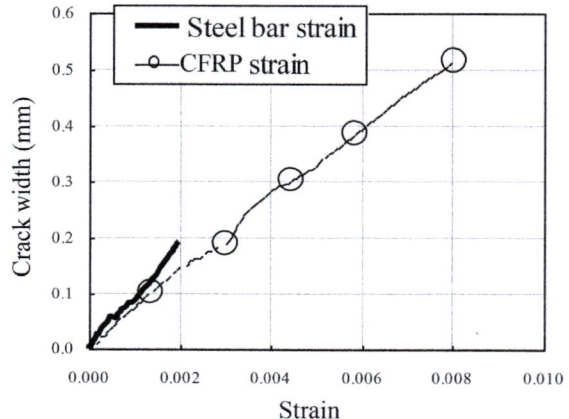

Figure 4.12 Relationship between crack width, rebar, and strain of FRP.

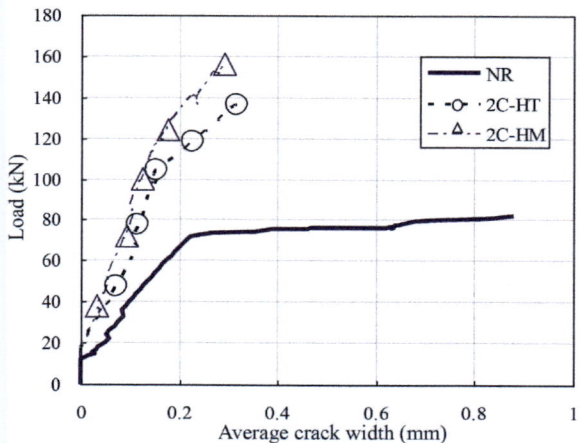

Figure 4.13 Load versus average crack width.

4.3.2.1   Failure mode under fatigue loading

The most commonly observed fatigue failure mode of FRP-strengthened beams is the rupture of tensile reinforcing steel followed by FRP failure (Kim and Heffernan, 2008). Thus the fatigue failure of the strengthened beams is primarily governed by the fatigue life of the steel reinforcement, rather than that of the concrete or the FRP. Nevertheless, the FRP significantly contributes to extending the fatigue life of strengthened beams by reducing stresses in the steel reinforcement. A schematic description of the fatigue failure sequence of FRP-strengthened RC beams is shown in Fig. 4.14. The steel reinforcement is susceptible to fatigue failure at stress amplitudes far lower than those that would cause the FRP reinforcement to fail. The secondary failure of the strengthened beams is generally due to the delamination of FRP bonded on the substrate of the beam. This secondary failure often occurs immediately following the primary failure. Other secondary failure modes are possible; for example, a shear failure mode induced by diagonal shear cracks may occur before the delamination of FRP in the case where a strengthened

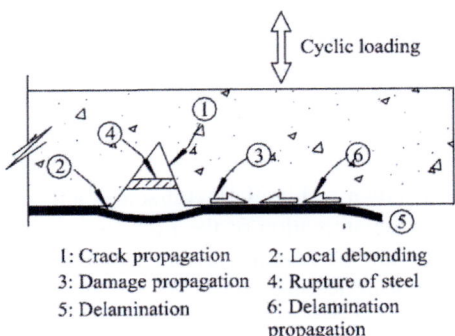

1: Crack propagation   2: Local debonding
3: Damage propagation   4: Rupture of steel
5: Delamination   6: Delamination
                         propagation

(A)

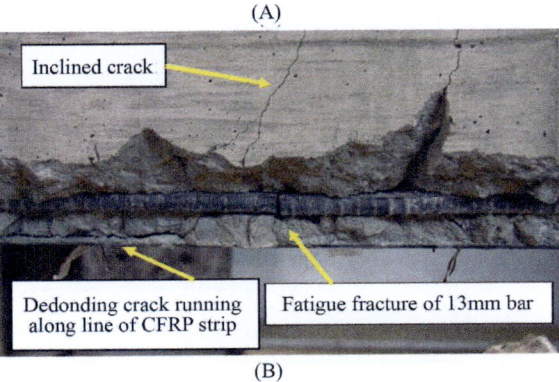

(B)

Figure 4.14 Typical failure sequence of an RC beam strengthened with FRP subjected to fatigue loading: (A) Kim and Heffernan (2008); (B) Quattlebaum et al. (2005).

beam is subjected to high levels of fatigue load. This possible failure indicates the importance of the applied fatigue load levels and the intrinsic soundness of the beam.

4.3.2.2   Experimental observations

As presented previously, RC beams strengthened with FRP composites exhibit better structural performance than unstrengthened beams under monotonic loads, including higher flexural capacity, delayed cracking, and reduced crack widths. These advantages are also observed when strengthened members are subjected to fatigue loading. The improved cracking behavior may reduce the local stresses in the steel reinforcement that is likely to fail under fatigue loading. The enhanced resistance to deflection may also preclude the gradual degradation of the beam's flexural stiffness commonly observed in RC flexural members subjected to fatigue loading (Kim and Heffernan, 2008).

Notable fatigue damage in FRP-strengthened beams, generally, accumulates rapidly during the early cycles. The rate of damage accumulation then slows considerably until just before the imminent fatigue failure of the beams, as shown in Fig. 4.15. The flexural crack propagation and the development of the maximum crack width follow the same trend. It should be noted that once the cracks form during the early cycles, no significant changes in the crack patterns are noted thereafter, despite an increased number of fatigue cycles.

As the number of cycle increases, the loading and unloading paths in a load−FRP strain response approach each other owing to less friction between the concrete and the FRP, as shown in Fig. 4.16 (Larson et al., 2005). This observation may indicate a progressive delamination of the FRP sheets, caused by gradual bond failure due to accumulated fatigue damage, even though no other indications of this delamination were noted during testing.

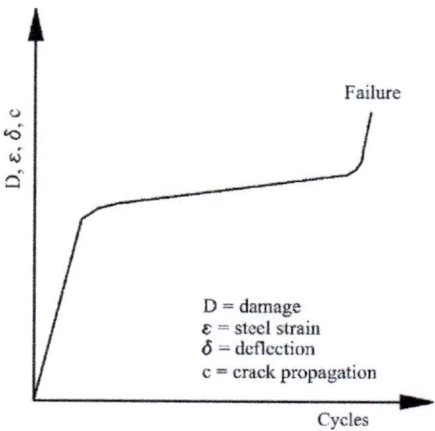

Figure 4.15 Fatigue performance parameters of RC beams strengthened with FRP (Kim and Heffernan, 2008).

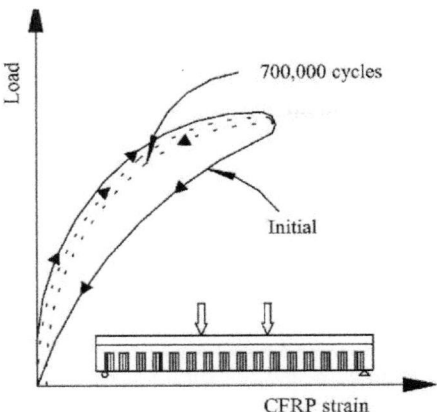

Figure 4.16 Load—FRP strain response under fatigue loads (Larson et al., 2005).

Existing studies have concluded that the fatigue life of unstrengthened beams could be increased after strengthening, taking into account the stress redistributions between the steel and the FRP, resulting in lower stresses in the steel reinforcement. It should be noted that the increase in fatigue life depends on the quantity of FRP relative to that of steel reinforcement and also on the bond quality between concrete members and FRP reinforcement.

4.3.2.3 Influence factors of the fatigue behavior of FRP-strengthened RC beams

Many experimental studies have been conducted and confirmed the effect of cyclic loading on the fatigue responses of RC structures strengthened with FRP. However, it is still difficult to accurately predict the flexural fatigue behavior of an RC beam strengthened with FRP sheets, owing to the complex mechanisms induced by the heterogeneity of the materials and various parameters that influence the flexural fatigue life. The main factors include the fatigue characteristics of the steel reinforcement, FRP sheets, and FRP-to-concrete bonding. In addition, minor factors include the cross-sectional area and yield strength of the longitudinal steel reinforcement, beam dimension, concrete strength, shear span, thickness, width, and tensile strength of the FRP, and the maximum load applied to the beam. In the ACI design guideline (ACI 440.2R, 2008), the designed static bending moment, M_n, of an FRP-strengthened RC beam is calculated as follows:

$$M_n = A_{st}f_{yst}\left(d - \frac{\beta_1 C}{2}\right) + \psi_f A_f f_f\left(h - \frac{\beta_1 C}{2}\right) \tag{4.1}$$

$$C = \frac{A_{st}f_{yst} + A_f f_f}{\alpha_1 f_c' \beta_1 b} \tag{4.2}$$

where $M_n = P_n a$; P_n is the ultimate load and a is the shear span. A reduction factor, $\psi_f = 0.85$, is applied to the flexural strength contribution of the FRP reinforcement. The terms α_1 and β_1 in Eq. (4.2) are parameters that define a rectangular stress block in the concrete equivalent to the nonlinear distribution of stress (ACI 318, 2011). Eq. (4.1) can be simplified by defining variables K_1 and K_2 as follows:

$$K_1 = d - \frac{\beta_1 C}{2} \tag{4.3}$$

$$K_2 = h - \frac{\beta_1 C}{2} \tag{4.4}$$

The parameters associated with static flexural load are comprehensively considered in the model, and all parameters have linear effects under monotonic loading. Focusing on this model, the following sections address the influences of various parameters on the flexural fatigue capacity based on the results of finite-element analysis (FEA).

1. *Finite-element modeling*
    RC beams of various cross-sections and lengths were studied by FE modeling, as shown in Fig. 4.17. Details of the modeling and properties of the materials can be found in Wang et al. (2015). The minimum fatigue load was equal to 10% of the ultimate load. In addition to the numerical values of the maximum fatigue loads, the corresponding number of cycles, steel stress at failure, and various possible failure modes were considered. The $S-N$ relationships of the steel bar, FRP, and FRP-to-concrete interface were evaluated using existing models. The fatigue limit of RC members was determined by the minimum numbers of load cycles of the three aforementioned $S-N$ relationships.
2. *Parametric study and prediction of the flexural fatigue load of RC beams*

A parametric study was conducted to evaluate the effects of several parameters on the flexural fatigue behavior. The parameters included the area of the tensile longitudinal reinforcement bars, yield strength of the steel reinforcement, beam width, concrete strength, beam depth, shear span, FRP thickness, width of the FRP sheet, tensile strength of the FRP sheet, and maximum load applied to the beam.

- Influence of area of steel reinforcement and yield strength
    The cross-sectional area of the tensile longitudinal reinforcement bars is an important factor that directly affects the strength and stiffness of the beam under monotonic and

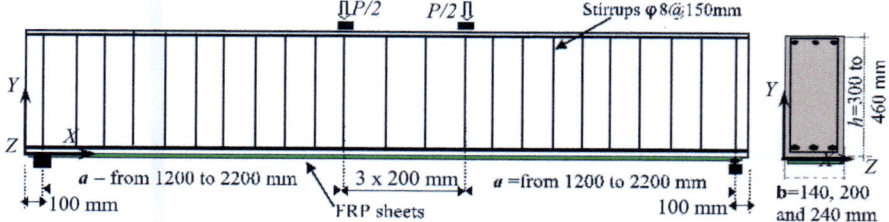

Figure 4.17 Geometrical details of the RC beams (Wang et al., 2015).

cyclic loading. Fig. 4.18 shows the influence of the area of the steel reinforcement on the allowable number of flexural fatigue load cycles, according to the FEA simulations. The FEA simulation results indicate that the effect of the area of the steel reinforcement on the allowable number of fatigue cycles can be expressed as $A_{st}^{0.846}$. The yield strength of the steel reinforcement has a major effect on the allowable number of flexural fatigue load cycles, as shown in Fig. 4.19. The FEA simulation results indicate that the allowable number of load cycles increases nonlinearly with the yield strength of the steel reinforcement, according to the function $f_{yst}^{0.595}$. These values were obtained by using a power regression equation to determine the influence of the area and the yield strength of the steel reinforcement. The power regression analysis resulted in higher R^2 values than other types of regression.

- Influence of concrete strength and beam width and depth

In Eq. (4.1), for beams without FRP sheet strengthening, the influences of the concrete strength, beam depth, and beam width are reflected in the parameter K_1. This parameter is an important factor that directly affects the strength and stiffness of beams subjected to cyclic flexural fatigue loading. In Eq. (4.1), for beams strengthened with FRP sheets, the influences of the concrete strength, beam height, and beam width are reflected in the parameter K_2. This parameter is another important factor that directly affects the strength

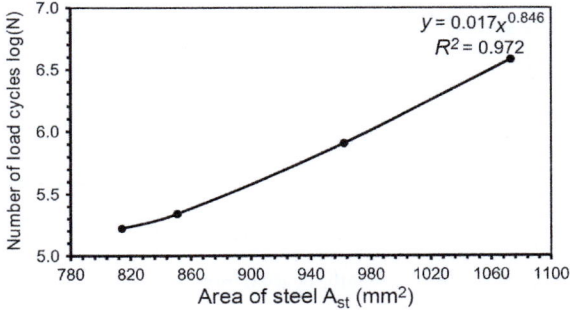

Figure 4.18 Influence of area of steel reinforcement (Wang et al., 2015).

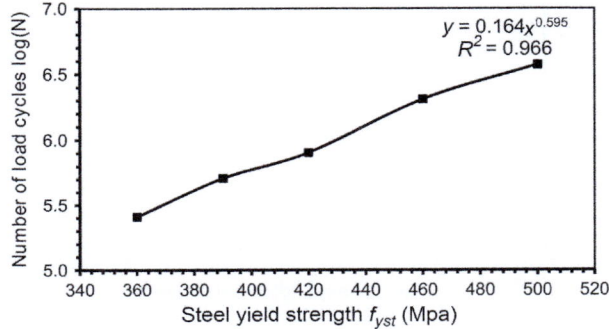

Figure 4.19 Influence of steel yield strength (Wang et al., 2015).

and stiffness of strengthened beams and the number of load cycles to failure. For a beam strengthened with an FRP sheet, the total number of load cycles must be increased compared to a beam without strengthening under the same conditions. Thus the difference between the number of load cycles with and without strengthening is represented by K_2. Figs. 4.20 and 4.21 show the influences of K_1 and K_2 on the number of load cycles, according to the FEA simulation results. The FEA simulation results indicate that the number of load cycles to failure changes with the values of K_1 and K_2, according to functions $K_{0.94\ 1}$ and $K_{1.07\ 2}$.

- Influence of thickness and width of FRP sheet

  The thickness and width of the FRP sheet are the primary factors affecting the strength and stiffness of the strengthening material. As Figs. 4.22 and 4.23 show, the number of load cycles to failure increases nonlinearly with the thickness and width of the FRP. The FEA simulation results indicate that the number of load cycles to failure changes with the thickness and width according to the functions $t_{0.922f}$ and $w_{0.717f}$, respectively.

- Influence of ultimate strength of FRP sheet

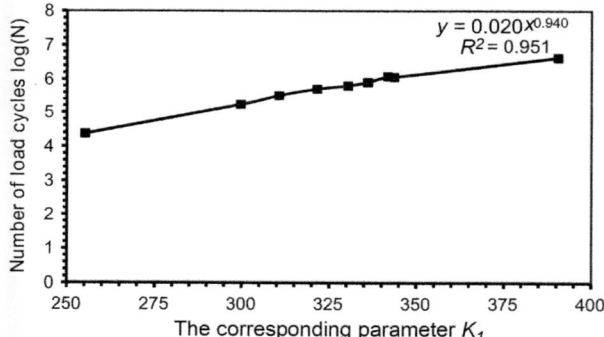

Figure 4.20 Influence of concrete strength, beam depth, and beam width for unstrengthened beam (Wang et al., 2015).

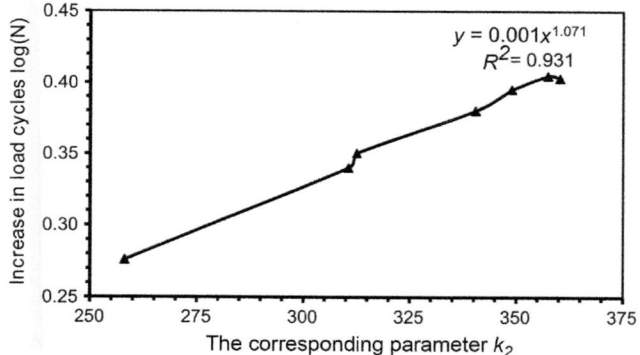

Figure 4.21 Influence of concrete strength, beam depth, and beam width for FRP-strengthened beam (Wang et al., 2015).

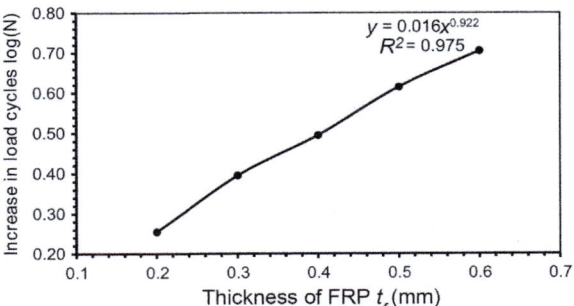

Figure 4.22 Influence of thickness of FRP (Wang et al., 2015).

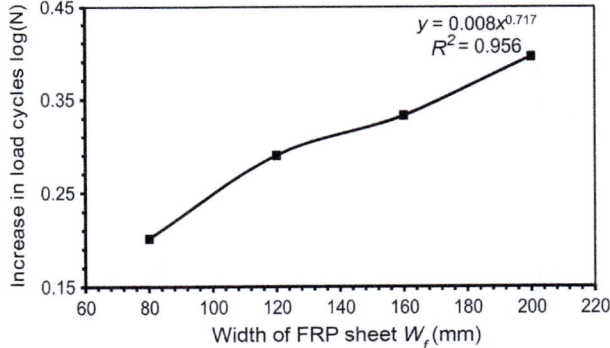

Figure 4.23 Influence of width of FRP (Wang et al., 2015).

The influence of the ultimate strength of the FRP sheet plays a major role in design guidelines and the number of load cycles to fatigue failure, as shown in Fig. 4.24. The ultimate strength of an FRP sheet is a nearly linear function of FRP strengthening. The FEA simulation results show that the number of load cycles to failure increases almost linearly with the ultimate strength of the FRP sheet, according to the function $f^{0.976}_{fu}$, as shown in Fig. 4.24.

- Influence of shear span

The shear span has a direct influence on the ultimate flexural load capacity of beams. The influence of the shear span is considered in the ACI 440 (2008) model, as shown in Eq. (4.1). The influence of the shear span is analyzed for beams with and without FRP strengthening. The FEA simulation results indicate that the number of load cycles to failure decreases almost linearly with the shear span, according to the functions $a^{-0.997}$ and $a^{-1.004}$ for beams with and without strengthening, respectively, as shown in Fig. 4.25.

- Influence of applied load range

The applied load range has a major influence on the number of load cycles to failure, as reflected in existing design models. The influence of the applied load range was analyzed for beams with and without strengthening. The slope of the curve in the FEA simulation results indicates that the number of load cycles to failure decreases almost linearly

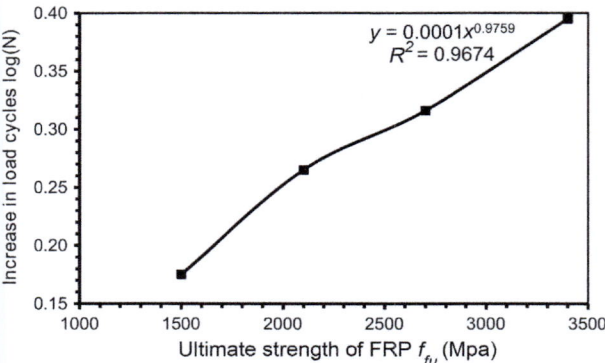

Figure 4.24 Influence of ultimate strength of FRP sheet (Wang et al., 2015).

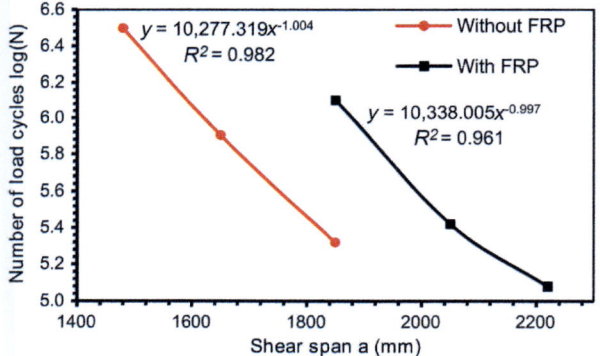

Figure 4.25 Influence of shear span (Wang et al., 2015).

with the applied load range, according to the functions $P_{-0.995 \; range}$ and $P^{-0.993}_{range}$ for beams with and without strengthening, respectively, as shown in Fig. 4.26.
- Correction of the influence of parameters included in K_1 and K_2

The factors K_1 and K_2 are functions of other parameters apart from the concrete strength and beam depth and width. These parameters are the area and yield strength of the steel reinforcement for K_1; and the thickness, width, and tensile strength of the FRP sheet for K_2. These parameters cannot be separated from K_1 and K_2. The correct values of the area and yield strength of the steel reinforcement are given by $A^{-0.09}_{st}$ and $f_{-0.10y,st}$, respectively. The correct values of the thickness, width, and tensile strength of the FRP sheet are given by $t_{-0.06f}$, $w_{-0.03f}$, and $f_{-0.03f, u}$, respectively. As Eqs. (4.1) and (4.2) show, a decrease in the power of the variable C has an increasing effect on the values of these parameters. The number of load cycles to failure varies with the area and yield strength of the steel reinforcement and the thickness, width, and tensile strength of the FRP sheet according to the functions $A^{0.95}_{st}$, $f_{0.71y,st}$, $t_{0.97f}$, $w_{0.74f}$, and $f_{1.00f,u}$, respectively.

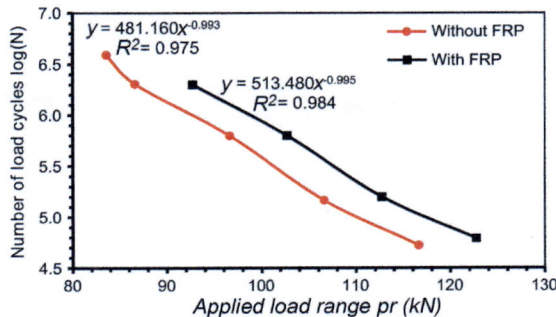

Figure 4.26 Influence of the applied load range (Wang et al., 2015).

4.4 Bonding and debonding mechanisms in FRP-strengthened RC members

In this section, particular emphasis is placed on clarifying the failure mechanisms of flexural crack-induced debonding and shear crack-induced peeling-off failure. By investigating the experimental observations, fracture models are proposed for different interfacial failure modes and nonlinear fracture mechanics-based FEAs are performed to investigate the influencing factors on debonding and peeling-off fracture behavior.

4.4.1 Flexural crack-induced debonding in FRP-strengthened RC beams

Experiments were conducted on beams retrofitted with FRP sheets (Wu et al., 1997, 1999a; Wu and Kurokawa, 2002) to investigate the strengthening performance, failure modes, bond mechanism, and effects of important design parameters such as concrete strength, surface preparation, number of plies and types of FRP sheets, prestress level, anchorage treatment, and diameter of reinforcing steel bar. Two typical flexural crack patterns that accompanied the flexural crack-induced debonding failure mode were observed, as shown in Fig. 4.27. These were a single localized crack near the maximum-moment region in the strengthened plain concrete beams and multiple or distributed cracks in the strengthened RC beams. Studies have found that the number of flexural cracks may have a considerable effect on the debonding load (Bizindavyi and Neale, 1999; Wu et al., 1997; Ueda et al., 2002). In what follows, an FRP-strengthened RC beam subjected to three-point bending load (Niu, 2002) is used to conduct numerical simulations of the effects of distributed flexural cracks on the debonding behavior and failure mechanism.

The FE model of the FRP-strengthened RC beams was simulated using the FE program DIANA, as shown in Fig. 4.28. The retrofitted beam was loaded by an increasing displacement in the Y-direction at the top of the midspan. The thick lines

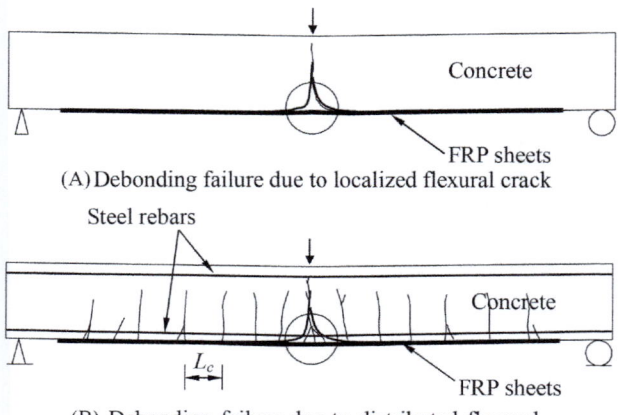

(A) Debonding failure due to localized flexural crack

(B) Debonding failure due to distributed flexural

Figure 4.27 Typical failure modes of the flexural crack-induced debonding (Niu, 2002).

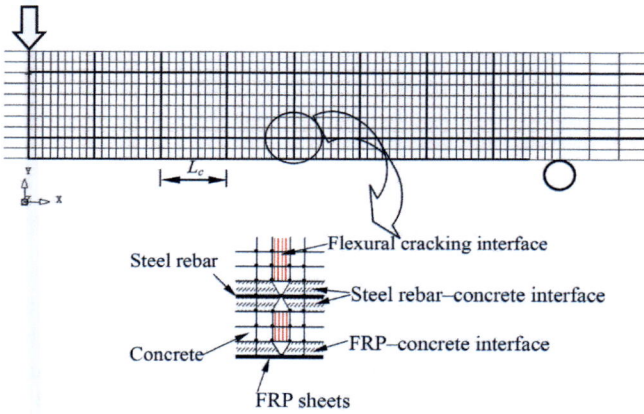

Figure 4.28 Discretization model for FRP-strengthened RC beam (Niu and Wu, 2005).

represent FRP sheets, steel rebar, flexural crack planes, and FRP-to-concrete and steel rebar-to-concrete interfaces. The vertical flexural cracks are modeled by zero-thickness interface elements with a spacing of L_c from the midspan to the supports of the beam. Details of the FE modeling can be found in Niu and Wu (2005).

Through varying the crack spacing to simulate the different crack patterns encountered in experiments, the effect of the crack spacing L_c on the structural performance is shown in Fig. 4.29. It can be seen that the rigidity of the strengthened structure decreases with decreasing crack spacing. One localized crack and a large crack spacing give almost the same ultimate load (Fig. 4.29A), which is lower than those with smaller crack spacings. It is noted in Fig. 4.29B that a similar crack pattern makes the structural response of the case with a crack spacing of 56.25 mm

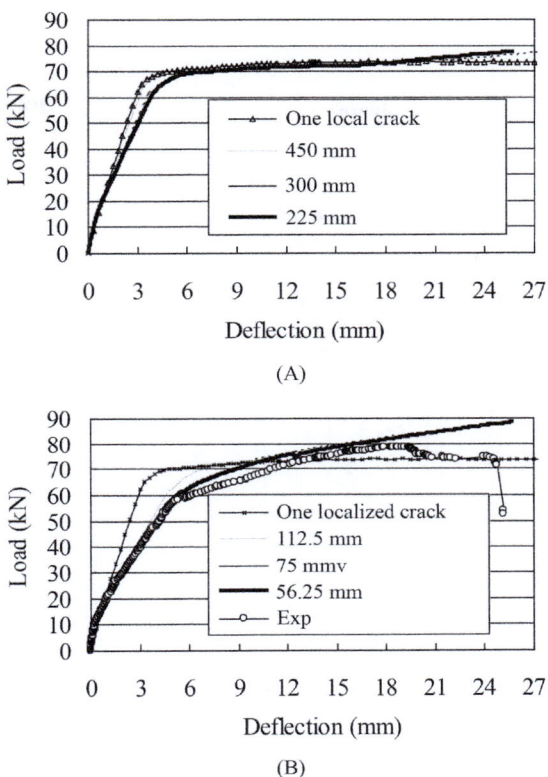

Figure 4.29 Load versus deflection for various crack spacings (Niu and Wu, 2005). (A) Large crack spacings. (B) Short crack spacings.

very close to that of the experiment before the deflection of 20 mm, and the difference afterwards may be dependent on the chosen parameters. In Fig. 4.30A, the FRP exhibited approximately linear behavior in the four stages corresponding to pre- and postcracking, yield of rebar, and interface debonding. Once the rebar increases to its yield strength, the strengthening effect of the FRP sheet is triggered and the FRP stress significantly increases. For the cases with large crack spacings, FRP stresses remain almost constant after the occurrence of the macrodebonding owing to the debonding propagation along the FRP–concrete interface, which can explain why the load remains nearly constant. However, in the cases with small crack spacings, the FRP stress increases continuously and contributes to a sustained increase in the load-carrying capacity. Therefore the decrease in crack spacing may be helpful in enhancing the strengthening effect of both the internal rebar (Fig. 4.30B) and the external FRP sheet.

The above result can be clearly explained by considering three cases: one localized crack and crack spacings of 300 and 75 mm. For the case of one localized

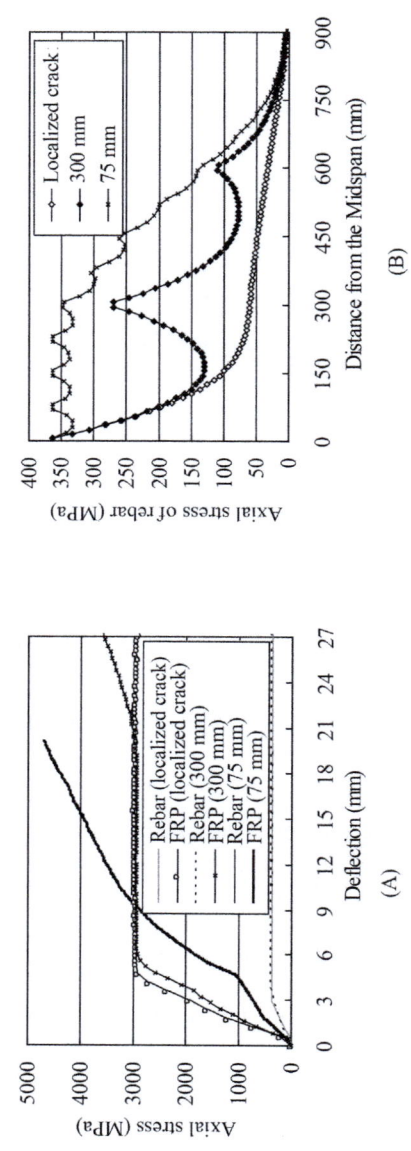

Figure 4.30 Reinforcement stresses for various crack spacings (Niu and Wu, 2005). (A) Reinforcement stresses versus deflection (midspan). (B) Rebar stress distributions (deflection = 20 mm).

crack (Fig. 4.31), once the macrodebonding is initiated at the midspan, the FRP stress attains the maximum value and then remains constant during the debonding propagation from the midspan to the end of the FRP sheet. Either from the interfacial shear stress distribution or the FRP stress distribution shown in Fig. 4.31, the effective shear transfer length required to attain the ultimate load-carrying capacity is approximately 100 mm. For the case of a crack spacing of 300 mm (Fig. 4.32),

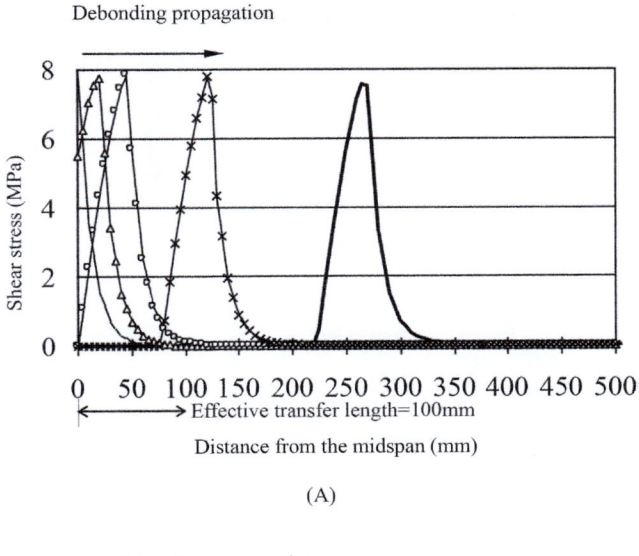

(A)

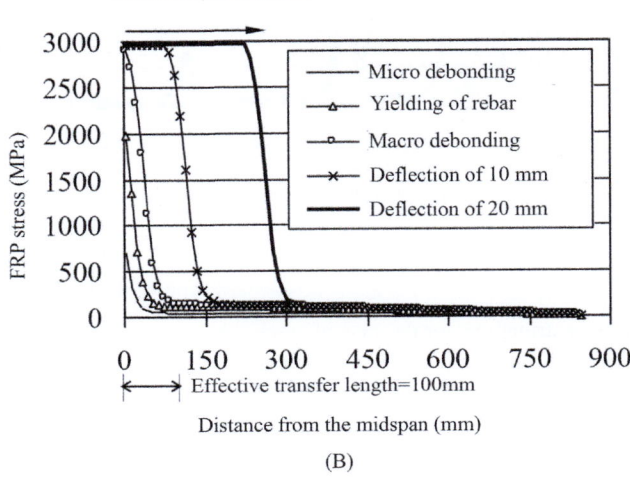

(B)

Figure 4.31 Stress distributions for the case of one localized crack (Niu and Wu, 2005). (A) Interfacial shear stresses. (B) FRP stresses.

Debonding propagation

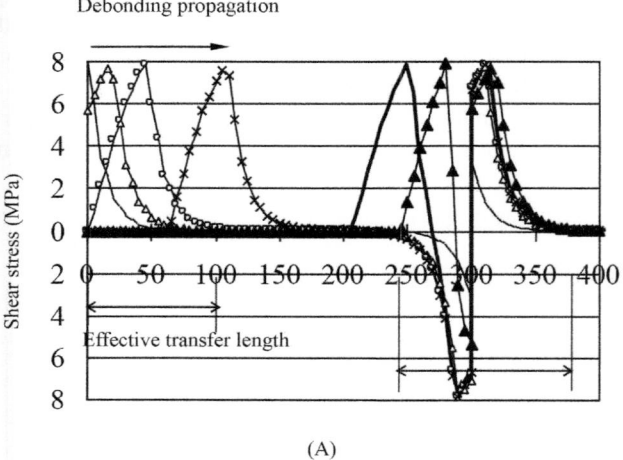

(A)

Debonding propagation

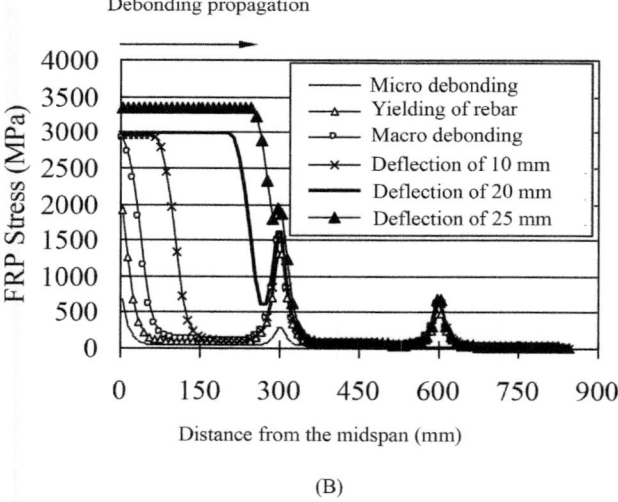

Distance from the midspan (mm)

(B)

Figure 4.32 Stress distributions for the case of crack spacing = 300 mm (Niu and Wu, 2005). (A) Interfacial shear stresses. (B) FRP stresses.

similar debonding behavior is observed during the debonding propagation from the midspan to the nearby crack. It should be noted that the nearby crack may be helpful to increase the shear transfer length and additional work is required for the debonding to propagate through the crack, which increases the stress in the FRP sheet (Fig. 4.32B) and yields a slight increase in the ultimate load as shown in Fig. 4.29A. In practice, such a phenomenon may not occur because the impact effect of a certain length of debonding may accelerate the debonding propagation and cause the structure to fail without a further increase in load-carrying capacity. Different from the cases of one localized crack and large crack spacing, the macrodebonding at the midspan crack does not result in final debonding failure in the case of a crack spacing of 75 mm. Although the macrodebonding occurs at some locations, the FRP sheet can still carry the additional load and this yields a continuous increase in load. As shown in Fig. 4.33, the macrodebonding first occurs at the midspan, but because the crack spacing is less than the effective transfer length of approximately 100 mm, there is not enough length for the debonding to propagate and the propagation encounters resistance from the opposite direction at the location of the adjacent crack, which leads to an increased effective transfer length as shown in Figure Fig. 4.33B. The debonding propagation in this case does not have a very smooth appearance as in the cases of one localized crack and large crack spacing, and more energy is required for the debonding to propagate through the cracks, which contributes to the increase in external load.

To summarize, the flexural crack spacing has a significant effect on the interfacial debonding mechanism and the ultimate load-carrying capacity. For the case of the crack spacing larger than the effective transfer length of FRP sheets, the debonding mechanism and the structural performance are similar to that of a case with a single localized crack. Once macrodebonding is initiated, the debonding would propagate toward the end of the FRP sheets and the load would remain constant until the final debonding failure. However, for the case of a crack spacing less than or close to the effective transfer length of the FRP sheets, the initiation of macrodebonding would not lead to the complete debonding failure of the retrofitted structure. Because more external work is needed to redistribute the stress between adjacent cracks, this contributes to an increase in the load-carrying capacity.

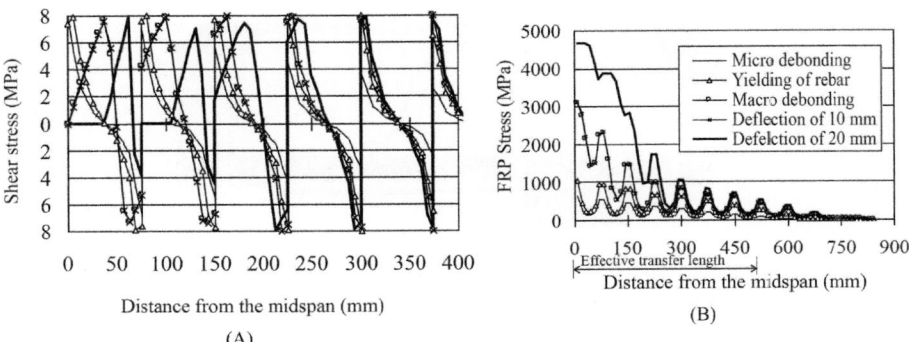

Figure 4.33 Stress distributions for the case of crack spacing = 75 mm (Niu and Wu, 2005). (A) Interfacial shear stresses. (B) FRP stresses.

4.4.2 Shear crack-induced debonding failure in FRP-strengthened RC beams

A series of three-point loading tests were conducted on notched plain concrete beams and RC beams by Wu et al. (2000) and Wu and Kurokawa (2002). The final failure modes are illustrated in Fig. 4.34. For the PS-2-5 and CS1-a5 specimens, the retrofitted structures failed due to debonding caused by flexural cracks, whereas for the PS-4-5 specimen, peeling-off failure was caused in by shear cracks.

Without loss of generality, two simplified crack patterns are adopted in the following analysis to give a clear insight into the debonding/peeling-off failure mechanism caused by flexural/shear cracks (Niu et al., 2006), as shown in Fig. 4.35. Flexural cracks usually yield a high shear stress concentration over a short distance at the interface and may lead to debonding of FRP sheets (mode II), whereas diagonal cracks induce high normal and shear stress concentrations and may lead to peeling-off of FRP sheets due to tension—shear action (mixed mode).

With the discrete crack concept, the position and direction of cracks in an RC beam are predefined within an FE structural model, as shown in Fig. 4.36. The retrofitted beam is loaded by increasing the displacement in the Y-direction at the top of the midspan. The interfacial fracture energy at the diagonal flexural/shear cracks should be separated owing to the mixed-mode fracture behavior. In the following, G_{i1} denotes the interfacial fracture energy for the mode II fracture behavior and G_{i2} denotes the interfacial fracture energy for the mixed-mode fracture behavior.

1. *Notched plain beams strengthened with PBO sheets*

Local interfacial stresses can be obtained at some Gauss integration points: point Int1 at midspan and both sides of shear cracks Int2 and Int3, as shown in Fig. 4.37, which can be used to verify the incorporated models. Fig. 4.37A shows that the peeling stress is

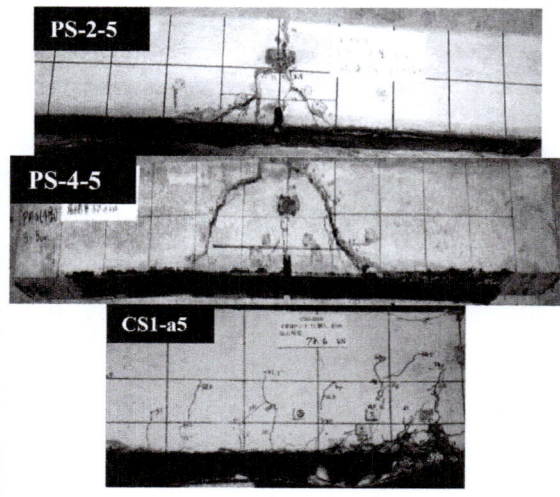

Figure 4.34 Debonding/peeling-off failure (Niu et al., 2006).

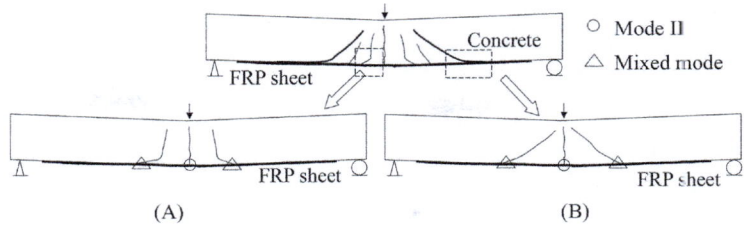

Figure 4.35 Simplified crack patterns for possible peeling-off failure (Niu et al., 2006). (A) Diagonal flexural crack. (B) Shear cracks.

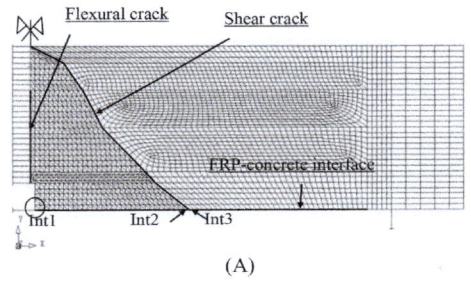

(A)

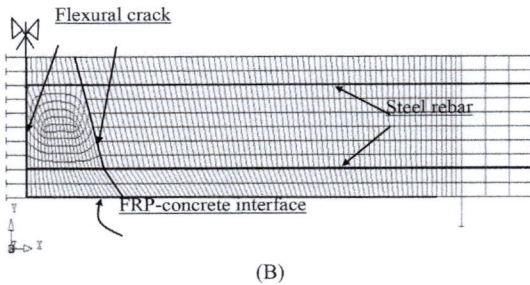

(B)

Figure 4.36 Finite-element discretization model (Niu et al., 2006). (A) FRP-strengthened plain beam with notch. (B) FRP-strengthened RC beam.

under tension and compression at the opposite interfaces of the shear crack. Owing to this peeling stress, a peeling-off local fracture can be observed in Fig. 4.38.

Local peeling-off fracture only occurs within one interface element at the end of the shear crack, and final peeling-off failure is caused by mode II fracture, which may be due to the fact that the peeling angle is very small for this structure. Accordingly, it can be concluded that the load decreases with the decrease in interfacial fracture energies G_{t2}, as shown in Fig. 4.39A. However, for the PS-2-5 specimen, the shear crack does not evolve and the structure fails by debonding caused by flexural cracks for different interfacial fracture energies G_{i2}. The experimental values are not provided in Fig. 4.39B due to a problem with the linear variable differential transformers in the tests. The experimental loads of the crack initiation and failure were 8 and 21.7 kN for the PS-2-5 specimen, respectively.

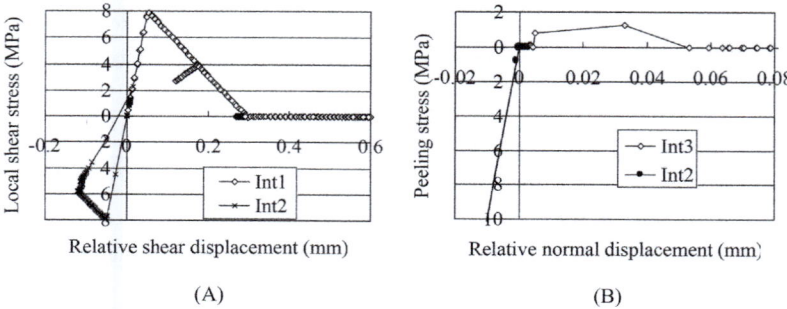

Figure 4.37 Interfacial stresses ($G_{i2} = G_{i1}$) for PS-4-5 (Niu, 2002). (A) Output for local shear stress. (B) Output for local peeling stress.

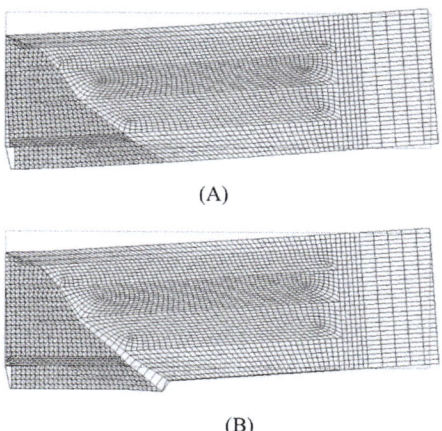

Figure 4.38 Structural deformation of PS-4-5 (Niu, 2002). (A) Deflection = 0.2 mm. (B) Deflection = 4 mm.

2. RC beam strengthened with CFRP sheet

In practice, final debonding failure is accompanied by many distributed flexural cracks in concrete. This is the reason the predicted curves have a larger stiffness than that of the experimental curve, as shown in Fig. 4.40A. Although local peeling-off fracture was found within one interface element at the end of a diagonal flexural-shear crack, the debonding propagation path remained unchanged from the end of the central flexural crack to the end of the plate, which is illustrated in Fig. 4.40B.

To summarize, diagonal cracks induce local bond fracture (mode I debonding) due to high peeling stress concentrations and provide a possible failure path. The final debonding failure is mainly governed by mode II fracture behavior rather than the mode I component because mode I debonding propagates very slowly as

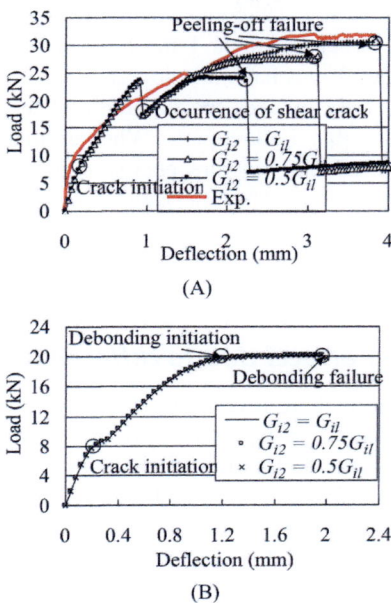

(A)

(B)

Figure 4.39 Load versus deflection relationships (Niu, 2002). (A) PS-4-5. (B) PS-2-5.

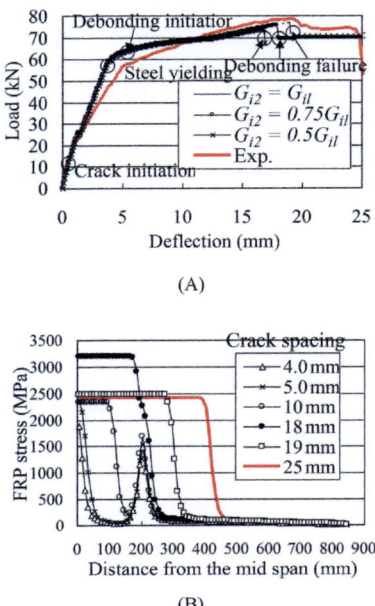

(A)

(B)

Figure 4.40 Structural performance for CS1-a5 (Niu, 2002). (A) Load versus deflection relationships. (B) FRP stress distribution ($G_{i2} = G_{i1}$).

compared to mode II debonding. Diagonal macrocrack-inducing debonding results in mixed-mode fracture. Whereas mode I mechanisms are responsible for initial local bond failures, mode II mechanisms govern the subsequent propagation of debonding. Peeling stress only induces local fracture at the diagonal flexural-shear cracks and provides a possible failure path. Whether this peeling-off propagates mainly depends on the interfacial fracture energy consumed for mode II fracture behavior between crack locations. The interfacial fracture energy has a significant effect on the fracture behavior and the load-carrying capacity of the strengthened RC beam.

4.4.3   Strengthening effects due to different interfacial behavior

In this section, an effort is made to investigate the effect of interfacial properties on the debonding behavior and the strengthening effect of FRP in the strengthened RC beam with multiple or distributed flexural cracks. Numerical simulations were conducted to investigate the effect of interfacial parameters such as the stiffness k_s, local bond strength τ_f, fracture energy G_f^b, and local bond curves on the strengthening performance (Niu and Wu, 2006). Details of the parameters can be found in Table 4.1.

4.4.3.1. *Interfacial stiffness k_s*
As shown in Fig. 4.41, in the case of a localized crack pattern, the interfacial stiffness only has an effect on the structural response from the yielding point to the macrodebonding point. The yield load can be enhanced to some extent with the increase in the interfacial stiffness, but the debonding point (load and deflection of macrodebonding) remains the same, regardless of the interfacial stiffness. As shown in Fig. 4.41B, low stiffness results in a low stress transmission capacity from concrete to FRPs and causes reinforcing bars to yield early. Only after the yielding of reinforcing bars, FRP stresses increase quickly, which means that low stiffness may delay the occurrence of the microdebonding. It should be noted that hardly any effect is observed on the yield load and structural stiffness if the interfacial stiffness is large enough to cause microdebonding to occur prior to the steel yielding (for example, $k_s = 25$ and 125 MPa/mm). Although low stiffness may yield a large transfer length (Fig. 4.42A), it has no effect on the load transfer capacity and this explains why all cases give the same ultimate load-carrying capacity. Fig. 4.42B illustrates debonding propagation for the cases of $k_s = 2-125$ MPa/mm. Once macrodebonding initiates, the FRP stress remains constant until the final debonding.
In the case of a distributed crack pattern, cracks cause the stresses to uniformly distribute in the reinforcing bars and FRPs and thus greatly reduce the effect of the interfacial stiffness. As shown in Fig. 4.43A, the interfacial stiffness has a similar but smaller effect on the structural performance as compared to that of the localized crack pattern. The results demonstrate that occurs more easily microdebonding with the increased interfacial stiffness. Fig. 4.43B shows similar FRP stress distributions and debonding propagation for different interfacial stiffnesses. In practical applications, the choice of adhesive thickness can be equivalent to the choice of the interfacial stiffness. Therefore this shows that adhesive thickness has no significant effect on the structural strength.

Table 4.1 Interfacial properties investigated in numerical simulations (Niu and Wu, 2002).

FRP–concrete bond behavior	Crack patterns	Interfacial mechanical parameters		
		k_s (MPa/mm)	τ_f (MPa)	G_f^b (N/mm)
k_s (MPa/mm) τ_f (MPa) G_f^b (N/mm)	One localized crack (midspan) / Uniformly distributed cracks $L_c = 56.3$ mm	2	2	1.2
		5		
		25		
		125		
		5	0.5	
			1	
			2	
			3	
		5	2	0.6
				0.8
				1.2
				1.8
	One localized crack (midspan) / Uniformly distributed cracks $L_c = 75$ mm	10	—	—
		100		
		1000		
		1.0E6		

(Continued)

Table 4.1 (Continued)

FRP–concrete bond behavior	Crack patterns	Interfacial mechanical parameters		
		k_s (MPa/mm)	τ_f (MPa)	G_f^b (N/mm)
	One crack (midspan)	Curve 1	2	1.2
	Uniformly distributed cracks $L_c = 56.3$ mm	Curve 2		
		Curve 3		
		Curve 4		

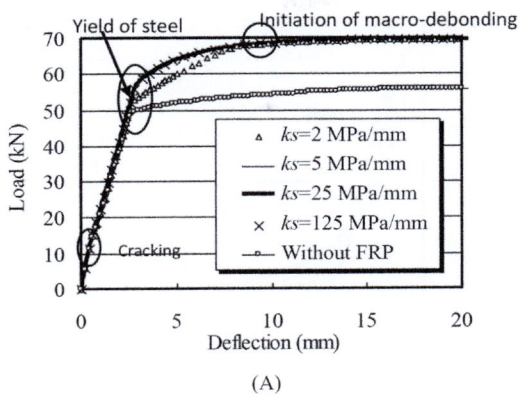

(A)

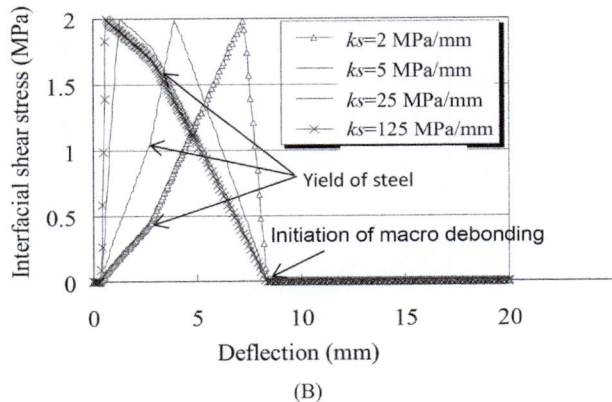

(B)

Figure 4.41 Effect of FRP–concrete interfacial stiffness (one localized crack at the beam midspan) (Niu and Wu, 2002). (A) Load versus deflection curves. (B) Interfacial shear stress at the midspan crack.

To further investigate the effect of stiffness on the strengthened RC beams with distributed cracks, a linear interfacial model without softening behavior is used. From Fig. 4.44, it is confirmed that interfacial stiffness only affects the yield load and has almost no effect on the load-carrying capacity. For this case, the final failure mode is the crushing of concrete.

4.4.3.2. *Local bond strength τ_f*

Fig. 4.45 illustrates the effect of interfacial bond strength on the strengthening performance of structures. From both cases, it can be concluded that interfacial bond strength has nearly no effect on yield load, but it does affect the structural behavior after the yielding of reinforcing bars. Low bond strength results in the early occurrence of microdebonding and may slightly decrease the stress transmission capacity, but this has nearly no effect on yield load (Fig. 4.45). Only after the steel yields, more stress is transferred to the FRPs and this process is governed by the following interfacial transmission behavior. For a higher bond strength, as shown in

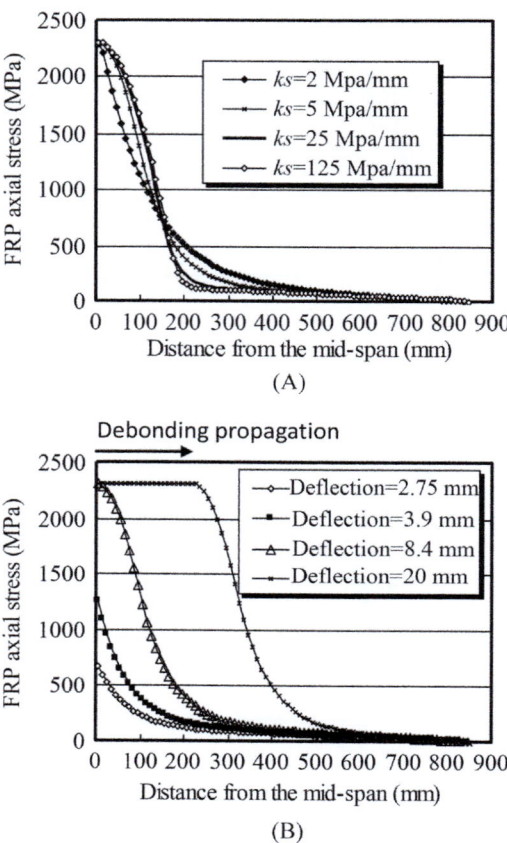

Figure 4.42 FRP stress distributions in the beam with one localized crack at midspan (Niu and Wu, 2002). (A) Initiation of macrodebonding. (B) Debonding propagation (k_s = 5 MPa/mm).

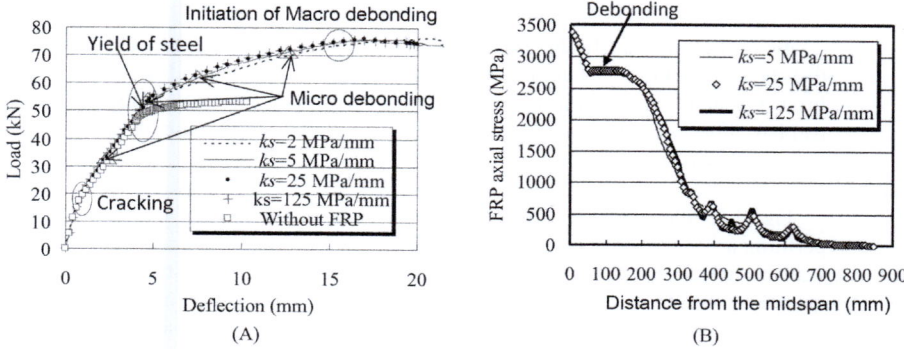

Figure 4.43 Effect of FRP—concrete interfacial stiffness (distributed cracks) (Niu and Wu, 2002). (A) Load versus deflection curves. (B) FRP stress distributions (deflection = 20 mm).

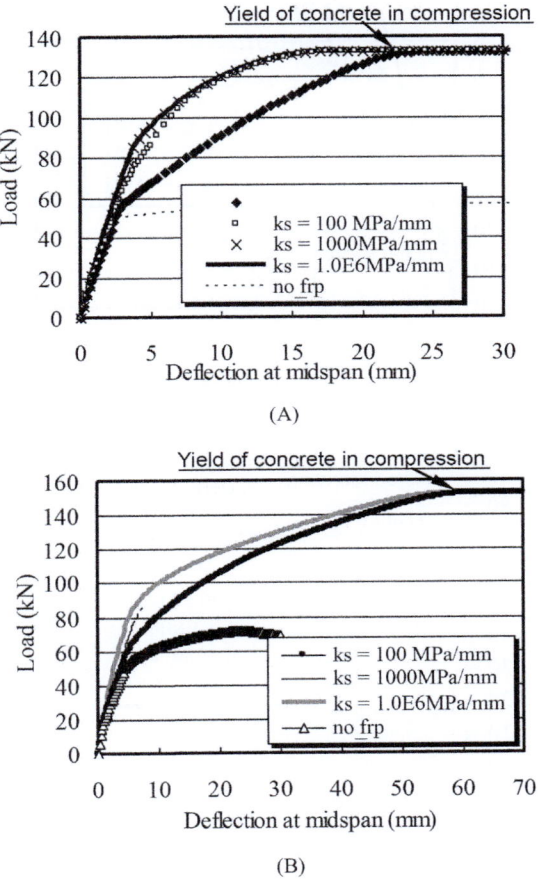

Figure 4.44 Load versus deflection curves (linear bond-slip model without softening) (Niu and Wu, 2002). (A) One crack at the midspan. (B) Distributed cracks with $L_c = 75$ mm.

Fig. 4.46A, interfacial shear stress increases more rapidly and this correspondingly makes macrodebonding occur earlier. This suggests that a low bond strength may be helpful to improve ductility (defined as the ratio of the deflections corresponding to ultimate and yield loads) of a structure. The interfacial bond strength has no effect on the maximum transferable load, as shown in Fig. 4.46B, and a low bond strength can increase the FRP stress transfer length and take full advantage of FRPs.

Unlike the case of a localized crack pattern, the ultimate load may be slightly increased with the increase in the interfacial bond strength in the case of a distributed crack pattern. This may be explained by the fact that a higher bond strength yields higher FRP stresses (Fig. 4.47). Different bond strengths may affect the debonding propagation paths. A low bond strength yields debonding failure initiated from the midspan crack, whereas a high bond strength results in debonding failure initiated from the crack adjacent the midspan one.

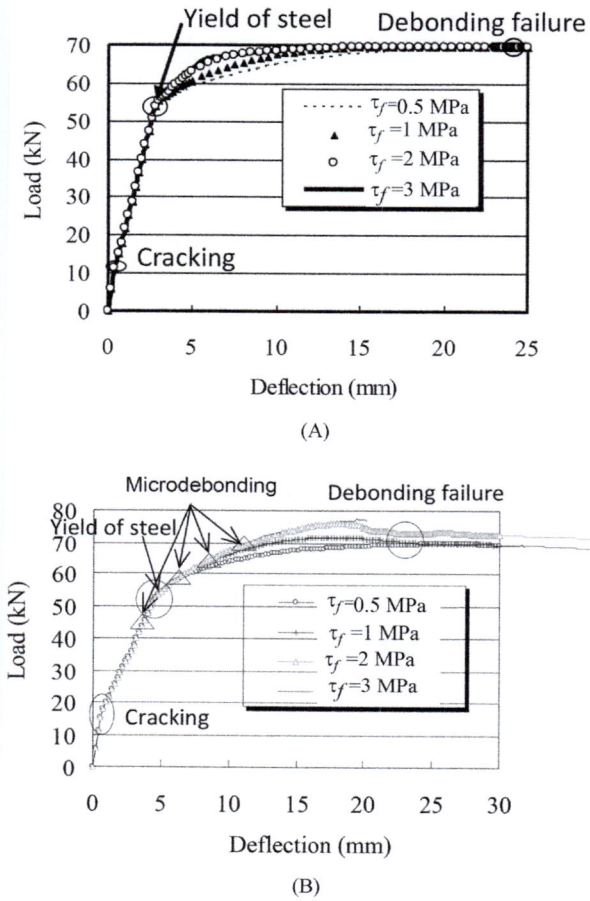

Figure 4.45 Effect of FRP–concrete interfacial bond strength (Niu and Wu, 2002). (A) Localized crack. (B) Distributed cracks.

4.4.3.3. *Interfacial fracture energy G_f^b*

As expected, the interfacial fracture energy or toughness has a significant effect on the load-carrying capacity (Fig. 4.48). A higher interfacial fracture energy results in more difficult debonding. With the increase in the interfacial fracture energy, the ultimate load and the ductility can be enhanced. The interfacial fracture energy itself does not affect the yield load, but it determines the maximum achievable FRP stress, which can be concluded from Fig. 4.48 together with the previous results.

4.4.3.4. *Effect of bond curve shape*

As reviewed in Niu and Wu (2005), two typical curves are observed in either simple shear tests or flexural tests: linear softening curve between FRP and concrete or a plastic curve in adhesives. Herein, four different bond curves (Table 4.1) are used to investigate the effect of the curve shape on the structural performance. As shown in Fig. 4.49, it

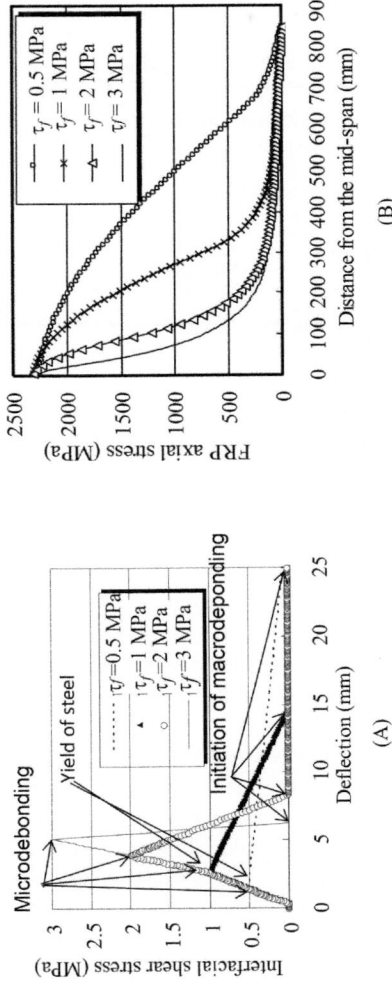

Figure 4.46 Interfacial shear and FRP stresses (one localized crack at midspan) (Niu and Wu, 2002). (A) Interfacial shear stress at midspan crack. (B) FRP stress distribution.

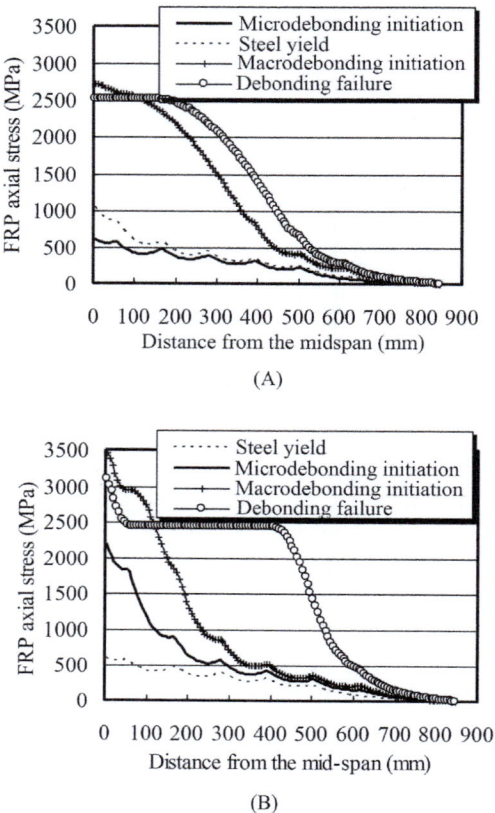

Figure 4.47 FRP stress distributions (distributed cracks) (Niu and Wu, 2002). (A) Local bond strength = 1 MPa. (B) Local bond strength = 2 MPa.

can be concluded that the bond curve shape has nearly no effect on the yield load and the ultimate load for the same fracture energy. This also confirms that the real bond behavior can be simplified by a linear softening curve or other simple models. Actually, the effect of the bond shape may be equivalent to the effect of stiffness. Fig. 4.49 shows that the descending branch may affect the macrodebonding point. The more slowly the local bond stress decreases after microdebonding, the later macrodebonding occurs.

4.5   Design of flexural strengthening

4.5.1   Design flexural load-carrying capacity

The design of concrete members strengthened with externally bonded FRP laminates shall be determined by an appropriate method concerned with the failure

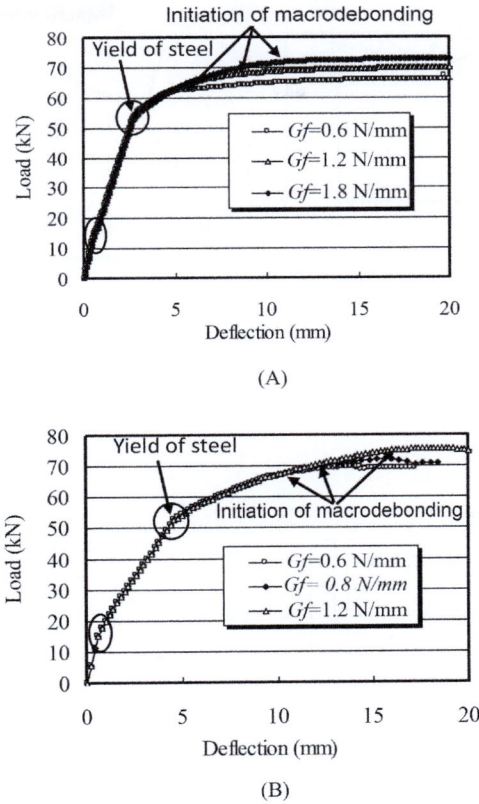

Figure 4.48 Effect of interfacial fracture energy (Niu and Wu, 2002). (A) Localized crack at midspan. (B) Distributed cracks.

mode, for example, considering whether the debonding failure of the FRP laminates occurs. As illustrated previously, three main debonding modes may occur in FRP-strengthened flexural members (Fig. 4.2). Among these, the failure modes of delamination at the cut-off point and peeling-off due to shear cracks can be avoided by the application of different anchorage schemes and rational shear design/strengthening. In practice, the failure mode of intermediate crack-induced (IC) debonding is more critical and sometimes unpreventable, even with anchor devices, as compared to cut-off point delamination failure. For the flexural design, the FRP IC debonding failure stress/strain may be determined using predictive models (details in Section 4.5.2). In other words, no FRP debonding occurs when the FRP stress/strain at the location of flexural cracking caused by the maximum bending moment in the member satisfies the IC debonding criterion. If debonding failure of the FRP laminates does not occur, the design flexural capacity of the member may be determined using the same method as for RC members.

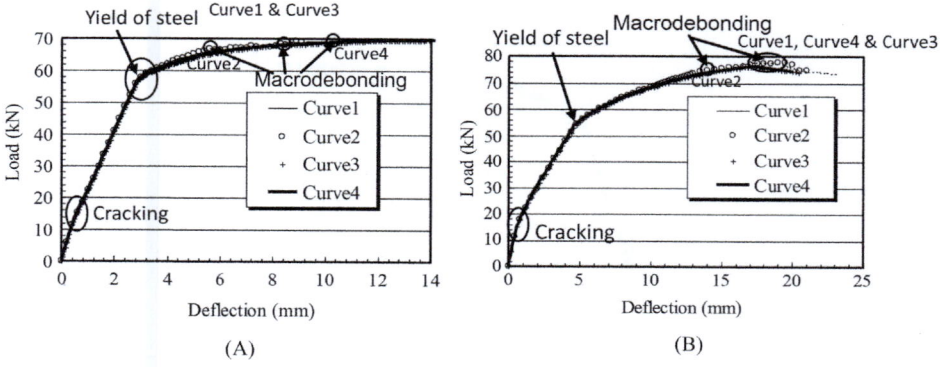

Figure 4.49 Effect of local bond curve shape (Niu, 2002). (A) Localized crack at midspan. (B) Distributed cracks.

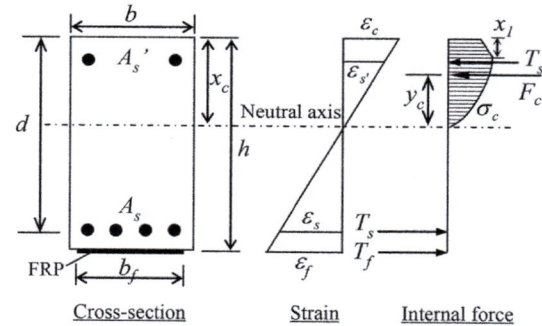

Figure 4.50 Strain compatibility-based flexural sectional analysis.

To calculate the flexural capacity of FRP-strengthened concrete members, a strain compatibility-based model is often adopted in the current design codes/guide-lines to calculate the external moment acting on a section and the FRP stress/strain on the same section, as shown in Fig. 4.50. When performing the sectional analysis, the following assumptions are often made: (A) a plane section is assumed to remain planar during bending, (B) no slip develops between the steel (or externally bonded FRP) reinforcement and the concrete substrate, (C) no premature FRP separation failure is considered, and (D) the contribution of the tensile stress of concrete is negligible.

It is assumed that failure is reached when the FRP strain or variation in the stress satisfies the criteria of IC debonding. The bending capacity of FRP-strengthened RC members at IC debonding can be divided into two parts: the contribution of the steel reinforcement and that of the externally bonded FRP composites, which can be expressed by the following equation:

$$M = \varepsilon_s E_s A_s (d - x_c + y_c) + \varepsilon_{fd} E_f t_f b_f (h - x_c + y_c) \tag{4.5}$$

in which ε_s, E_s, and A_s are the strain, elastic modulus, and area of the tensile steel reinforcement; d is the effective depth of the concrete section; x_c is the depth of the concrete compression zone; y_c is the distance between the neutral axis and the compressive resultant force of the concrete and the compressive steel reinforcement (F_c); and h is the height of the concrete section.

4.5.2   Existing models for the prediction of IC debonding failure

For the flexural design method, the main difference among the existing design codes is in the calculation of the FRP debonding strain, ε_{fd}, in the ultimate state. The flowing sections present an overview of several typical IC debonding models in the existing design codes and also list some common models' reference.

1. *Japan Society of Civil Engineers JSCE (2001) recommendations and Wu and Niu (2007) model*

Recommendations by the *Japan Society of Civil Engineers (JSCE) (2001)* on the prediction of IC debonding failure, which are based on the study of Wu and Niu (2000), suggest that no IC debonding failure occurs when the maximum value $\Delta\sigma_f$ of the difference in the FRP tensile stress satisfies Eq. (4.6):

$$\Delta\sigma_f = \sigma_{f2} - \sigma_{f1} \leq \sqrt{2G_f E_f / t_f} \tag{4.6}$$

where σ_{f2} is the FRP tensile stress at the maximum-bending-moment critical section and σ_{f1} is the FRP tensile stress at a section located at a distance L from the section of the maximum moment, as shown in Fig. 4.51A. This approach is based on fracture mechanics analysis and employs the concept of a smeared crack to consider the effect of distributed cracks on the bond capacity of the FRP−concrete interface, as shown in Fig. 4.51B. The application of this approach is controlled by two main parameters: the interfacial fracture energy G_f and the distance L along which $\Delta\sigma_f$ is calculated. Japan Society of Civil Engineers JSCE (2001) suggests that $G_f = 0.5$ N/mm and L is in the range of 150−250 mm. Wu and Niu (2007) developed the following equations to accurately calculate G_f and L:

$$G_f = 0.644 f_c'^{0.19}, L = Max.(L_e', l_y) \tag{4.7}$$

in which L_e' is the FRP effective bond length and can be approximately computed as

$$L_e' = 1.3 \sqrt{E_f t_f} / f_c'^{0.095} \tag{4.8}$$

and l_y is the distance between the maximum-bending-moment critical section and the end of the steel yield region, as shown in Fig. 4.51C. $\Delta\sigma_f$ and l_y may be calculated using an iterative trial-and-error method. For convenience, in the following analysis, the recommendations for G_f and L by Wu and Niu (2007) are referred to as the Wu and Niu (2007) model.

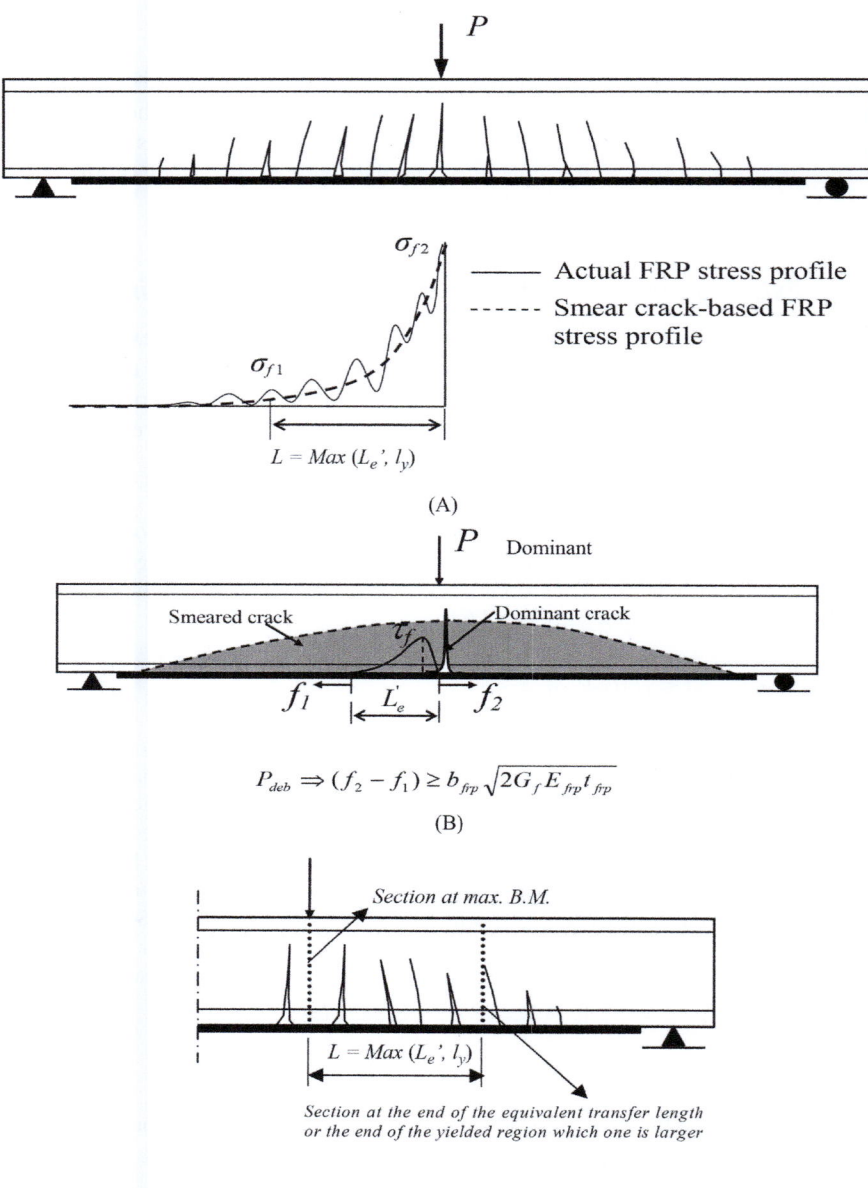

Figure 4.51 Prediction for debonding failure in FRP-strengthened RC beams (Wu and Niu, 2007).

2. *fib (2001) recommendations*

The *fib* Bulletion 14 (fib 2001) is the FRP design guideline used in many European countries. This approach employs a methodology similar to the Wu and Niu (2007) model and states that the FRP stress check against debonding between two adjacent cracks can be conducted by applying the following equation:

$$\Delta\sigma_f \le \Delta\sigma_{f,\max,fib} \tag{4.9}$$

Three main steps are included in the prediction approach: (1) determination of the most unfavorable spacing of flexural cracks, (2) determination of the difference between the FRP tensile stresses of two consecutive cracks $\Delta\sigma_f$, and (3) determination of the maximum possible increase in tensile stress in the FRP $\Delta\sigma_{f,max,fib}$. These three steps are elaborated in detail elsewhere (e.g., fib, 2001).

3. *Teng et al. (2003) model*

Based on a study of FRP—concrete bonded joints, Teng et al. (2003) suggested an empirical method to avoid IC debonding by proposing a certain limit on the allowable FRP strain. This approach facilitates the calculation of FRP debonding strain; it has been adopted by the Australia guideline and has been cited numerous times in the literature. In addition to f_c' and $E_f t_f$, the model considers the effect of the FRP-to-concrete width ratio by introducing the width factor β_w for the debonding failure. The model suggests that the FRP strain at debonding can be expressed by the following equations:

$$\varepsilon_{fd} = \begin{cases} \alpha\beta_w\sqrt{\sqrt{f_c'}/(E_f t_f)} & \text{if } l_f \ge \sqrt{E_f t_f/\sqrt{f_c'}} \\ \alpha\sin\left(\frac{1}{2}\pi l_f/\sqrt{E_f t_f/\sqrt{f_c'}}\right)\beta_w\sqrt{\sqrt{f_c'}/(E_f t_f)} & \text{if } l_f < \sqrt{E_f t_f/\sqrt{f_c'}} \end{cases} \tag{4.10}$$

in which

$$\beta_w = \sqrt{(2 - b_f/b)/(1 + b_f/b)} \tag{4.11}$$

where α is a factor determined by regression analysis of the test data. In the following analysis, $\alpha = 1.1$ was used to calculate the average prediction results according to the suggestion by Teng et al. (2003). l_f is the distance from the FRP cutoff to the nearest applied load. b_f and b are the widths of the FRP composites and the concrete member, respectively.

4. *ACI 440 (2008) guide*

Based on a best-fit analysis of average measured FRP strains at debonding from a database for flexural tests, the ACI 440 (2008) guide also suggests a simple method to prevent IC debonding by limiting the effective strain in FRP reinforcement, as defined by Eq. (4.12).

$$\varepsilon_{fd} = 0.41\sqrt{f_c'/(E_f t_f)} \le 0.9\varepsilon_{fu} \tag{4.12}$$

where ε_{fu} is the rupture strain of the FRP reinforcement.

5. *Said and Wu (2008) model*

To simplify the problem and comprehensively reflect the effect of various conditions, Said and Wu (2008) collected more than 200 flexural specimens with IC debonding failures and proposed a simple and effective model to predict FRP debonding strain based on a statistical analysis. The model can be expressed by the following relation:

$$\varepsilon_{fd} = 0.23 f_c'^{0.2} / (E_f t_f)^{0.35} \tag{4.13}$$

6. *GB (2010) recommendations*

To prevent IC debonding failure, the Chinese code GB 50608-2010 (GB, 2010) suggests that the effective strain in FRP reinforcement should be limited to the strain level at which debonding may occur, as defined in the following equation:

$$\varepsilon_{fd} = (1.1 / \sqrt{E_f t_f} - 0.2 / l_f) \beta_w f_t \tag{4.14}$$

The width ratio β_w was considered in this model and defined by the following equation:

$$\beta_w = \sqrt{(2.25 - b_f/b)/(1.25 + b_f/b)} \tag{4.15}$$

where f_t is the concrete tensile strength and can be estimated by the following equation according to American Concrete Institute ACI (2011):

$$f_t = 0.53 \sqrt{f_c'} \tag{4.16}$$

4.5.3   Reliability-based approach for flexural design of IC debonding failure

As introduced previously, IC debonding is one of the most common failure modes in FRP-strengthened flexural members and controls their ultimate capacities. Although the application of additional anchorage may increase the bond strength and ductility, it cannot always circumvent IC debonding failure. Considerable research has been conducted to investigate the effects of different parameters on IC debonding failure and several models for predicting IC debonding failure have been proposed and developed. However, owing to the complexity and the brittle characteristics of the failure mechanism, large dispersions have been observed in the predicted-to-experimental ratios of all typical IC debonding models (Said and Wu, 2008). Although current FRP guidelines consider the strength reduction of FRP materials, the design methods in these guidelines are mostly based on the empirical revisions of those in the contemporary codes for RC structures and do not consider the uncertainties introduced by the brittle characteristics of FRP composites, especially the debonding failure between FRP and concrete members. Reliability-based limit-state design techniques have been extensively used in the research carried out on civil infrastructure (Nowak and Collins, 2013). This approach allows designers to rationally assess the potential for structural failure and achieve a balance between safety and cost. The application of reliability analysis in FRP-strengthened concrete structures can advance the acceptance of FRP composites in civil infrastructure applications.

The following sections will introduce a reliability approach to investigate the statistics of IC debonding model uncertainties and to recommend appropriate reduction factors to be used in the design based on reliability analysis (Shi et al., 2015). The results presented herein demonstrate the viability of, and may provide, a reliability-based analysis for the development of current FRP guidelines.

4.5.3.1. *Evaluation of model uncertainties*

Model uncertainties in the calculation of resistance are primarily caused by the approximations and assumptions when utilizing different IC debonding models. To achieve a rational evaluation of the model uncertainties, only IC debonding failure modes were considered during the calculation of the failure load. Therefore the model uncertainty factor can be treated as a random variable K_p and expressed by the ratio of the experimental results P_{exp} divided by the model-predicted results P_{pred}, as defined in Eq. (4.17).

$$K_p = P_{exp}/P_{pred} \qquad (4.17)$$

Based on a test database that contains 217 FRP-strengthened RC specimens with IC debonding failure, the model uncertainty for the aforementioned IC debonding models and the main statistical parameters are presented in Fig. 4.52.

To determine the probabilistic characteristics of the model uncertainties, a hypothesis testing method was applied that assumed several types of possible probability distributions (e.g., normal, log-normal, and Weibull distributions). Taking the ACI 440 (2008) model as an example, Fig. 4.53 compares the probability density and the cumulative probability of K_p with the assumed probabilistic distribution fitting results. As shown in the figure, the fitting results of the normal and log-normal distributions conform comparatively well to the distribution of K_p calculated from the test database for ACI 440 (2008) model. The Kolmogorov–Smirnov test was performed with a significance level of 0.05 to determine the distribution of K_p. The models of Teng et al. (2003), Wu and Niu (2007), ACI 440 (2008), and Said and Wu (2008) can be taken as obeying the normal distribution, whereas the K_p distributions of fib (2001) and model are closer to the log-normal and Weibull distributions, respectively. In the following reliability evaluation, the probabilistic distribution and the statistical parameters of the model uncertainties were determined according to the hypothesis testing results.

4.5.3.2. *Calculation of reliability index*

The reliability analysis requires probabilistic models of the design variables and supporting databases to characterize the uncertainties in these variables. In the following analysis, the adopted probabilistic characteristics of the material properties and geometric dimensions are based on a statistical survey of conventional RC structures (Nowak and Szerszen, 2003). Given that dead (*D*) and live (*L*) loads are the most dominant in the design of RC flexural members, only these loads are considered. In the load and resistance factor design for buildings (ASCE, 2010), the most commonly used load combination is

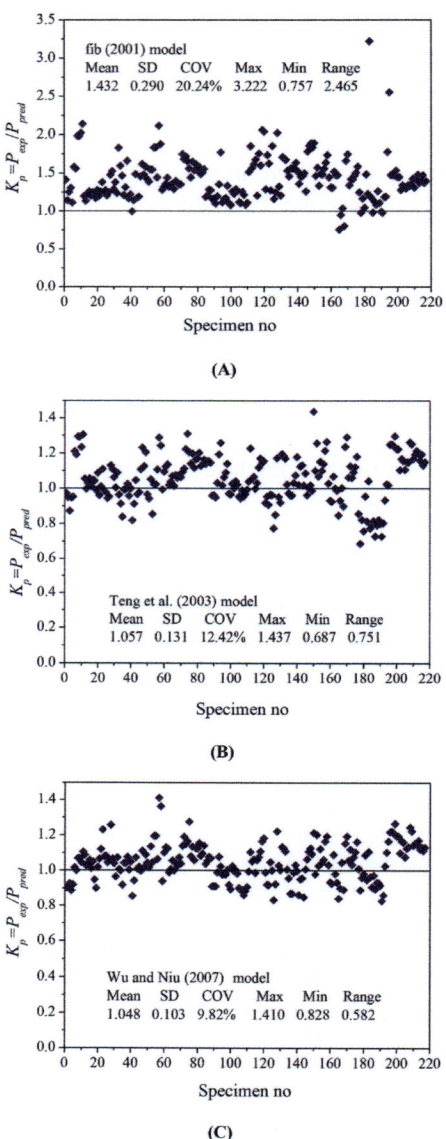

Figure 4.52 Model uncertainties for IC debonding models. (A) fib (2001); (B) Teng et al. (2003); (C) Wu and Niu (2007); (D) American Concrete Institute ACI (2008); (E) Said and Wu (2008); (F) GB (2010) (Shi et al., 2015).

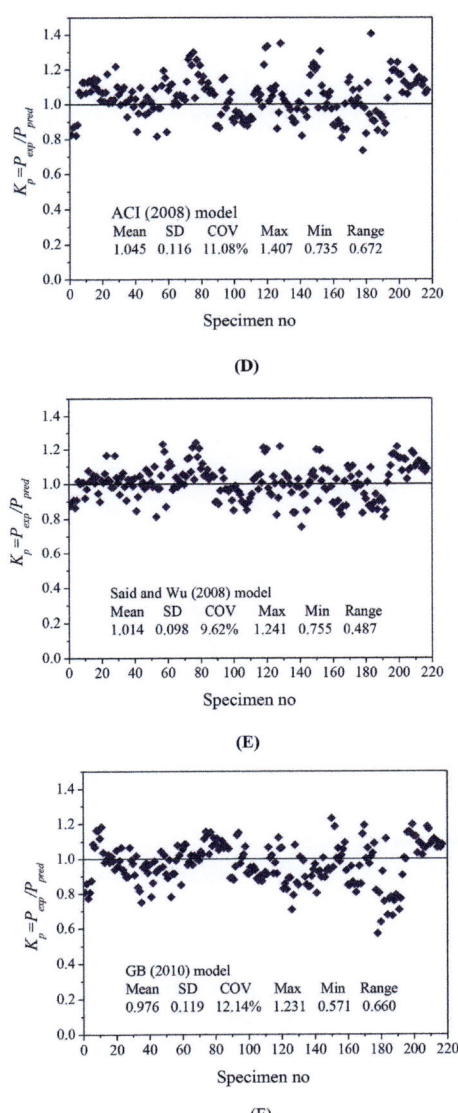

Figure 4.52 Continued

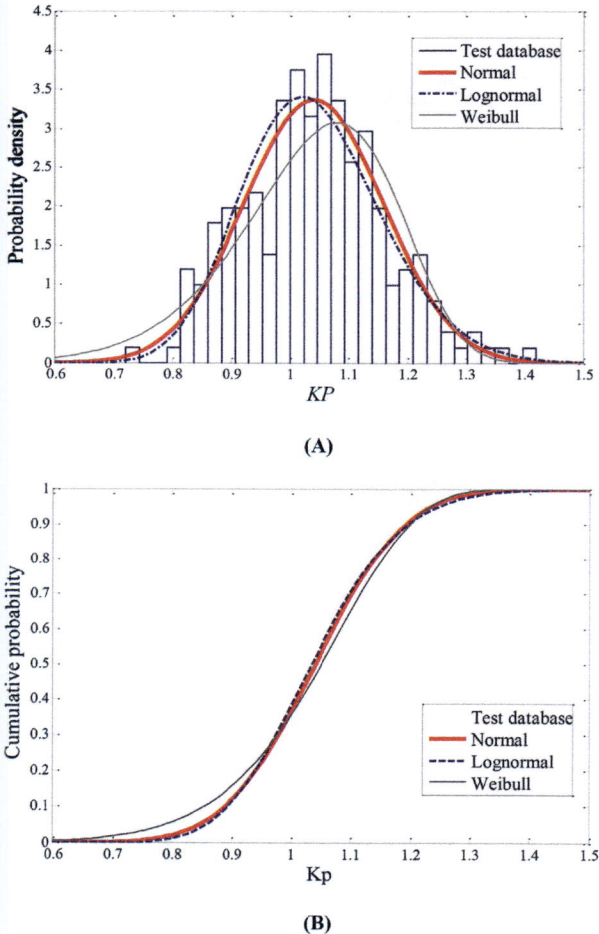

Figure 4.53 Determination of the probabilistic distribution of model uncertainty factor (ACI 440 (2008) model): (A) probability density; (B) cumulative probability (Shi et al., 2015).

$$S_d = 1.2D_n + 1.6L_n \tag{4.18}$$

in which S_d is the design load, D_n is the nominal dead (permanent) load, and L_n is the nominal occupancy live load. The uncertainties associated with the load statistics were determined according to Galambos et al. (1982).

The limit state function of the FRP-strengthened RC member can be represented by the following equation:

$$Z = R - D - L \tag{4.19}$$

in which R is a random resistance and can be expressed by the predicted resistance value [Eq. (4.5)] multiplied by the model uncertainty factor K_p:

$$R = K_p M = K_p \left(\varepsilon_s E_s A_s (d - x_c + y_c) + \varepsilon_{fd} E_f t_f b_f (h - x_c + y_c) \right) \qquad (4.20)$$

The mean values of the design variables were applied in this equation for the calculation of the mean resistance M. D and L are the random load effects caused by the dead and live loads, respectively. The design expression under the ultimate limit state can be expressed as

$$R_d = S_d \qquad (4.21)$$

in which R_d is the design resistance of the FRP-strengthened concrete member and can be expressed by the following equation:

$$R_d = \varphi M_n = \varphi \left(\varepsilon_s E_s A_s (d - x_c + y_c) + \psi_f \varepsilon_{fd} E_f t_f b_f (h - x_c + y_c) \right) \qquad (4.22)$$

Here, $\phi = 0.9$ is the resistance factor applied to the total flexural resistance (American Concrete Institute ACI, 2011; Szerszen and Nowak, 2003), M_n is the nominal resistance calculated by the nominal value of the design variables, and ψ_f is the reduction factor for the FRP debonding strain that needs to be determined. By substituting Eq. (4.18) into Eq. (4.21), we obtain the expressions for D_n and L_n, respectively, as follows:

$$D_n = \frac{R_d}{1.2 + 1.6\eta}, L_n = \frac{\eta R_d}{1.2 + 1.6\eta} \qquad (4.23)$$

in which η is the nominal live-to-dead load ratio L_n/D_n. The three most probable load ratios were selected in the following reliability analysis, namely 0.5, 1.0, and 1.5.

In practice, the probability of failure can be evaluated approximately in terms of the reliability index β. Owing to the nonlinear form and the complexity of the resistance equations, ensuring convergence is challenging when using the theoretical method to calculate the reliability index directly. To overcome this problem, the reliability index was calculated by a hybrid approach. The resistance R is treated as a single random variable and determined using a Monte Carlo simulation (MCS). The probability distribution of R is determined by a hypothesis testing method based on the MCS results, which is similar to the determination of the model uncertainty distribution. Then, the Rackwitz−Fiessler first-order reliability method (FORM), along with the distributions for the dead and live loads, was applied to compute the reliability index. When the random variables did not follow a normal distribution, the equivalent normalization transformation process was performed to calculate the equivalent normal values for the mean and standard deviation for each nonnormal random variable. FORM was selected owing to its appealing trade-off between accuracy and computational efficiency. The main steps for the reliability analysis implemented in MATLAB. According to the study of Atadero and Karbhari (2008), the number of trials to include in each MCS was selected as $N = 100,000$ in the reliability analysis of FRP-strengthened RC beams. To generalize the reliability assessment approach, two rather extreme nominal values were selected for each design variable.

4.5.3.3. Calibration of the reduction factor

The objective for the reliability analysis is to determine an overall value of the FRP reduction factor Ψ_f that yields a target reliability index β_T. To be consistent with design recommendations for beams and slabs under flexural loading, for probability-based limit state design (American Concrete Institute ACI, 2011, 2008), $\beta_T = 3.5$ is selected as the target. To accomplish this goal, Ψ_f is reduced from 1.0 by a step size of 0.05 and minimized with respect to Ψ_f using the following least-squares average:

$$H_{\psi f} = \frac{1}{n_c} \sum_{i=1}^{n_c} (\beta_i - \beta_T)^2 \tag{4.24}$$

The behavior of $H_{\psi f}$ as a function of ψ_f for each IC debonding model is illustrated in Fig. 4.54. In all cases, $H_{\psi f}$ initially decreases and subsequently increases with decreasing ψ_f. The calibrated FRP reduction factor is the value of ψ_f required to meet the target reliability and can be achieved when $H_{\psi f}$ reaches the minimum value. Detailed results of the calibrated reduction factors for each IC debonding model are 0.2, 0.65, 0.75, 0.75, 0.75, and 0.6 for the models proposed by fib (2001), Teng et al. (2003), Wu and Niu (2007), ACI 440 (2008), Said and Wu (2008), and GB (2010), respectively. The calibrated reduction factors are mainly influenced by the model uncertainties, and a model with a lower dispersion model uncertainty possesses a relatively higher reduction factor. Owing to the extremely high dispersion of the model uncertainty, the $H_{\psi f}$ of the fib (2001) model could only attain the minimum value when ψ_f was reduced to 0.20, indicating that this model is not suitable to be used as the FRP debonding criterion in the design guideline. The collaborated FRP reduction factor for the

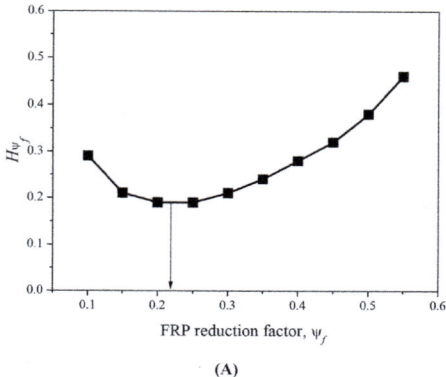

(A)

Figure 4.54 Least-squares average analysis to determine the FRP reduction factor. (A) fib (2001); (B) Teng et al. (2003); (C) Wu and Niu (2007); (D) American Concrete Institute ACI (2008); (E) Said and Wu (2008); (F) GB (2010) (Shi et al., 2015).

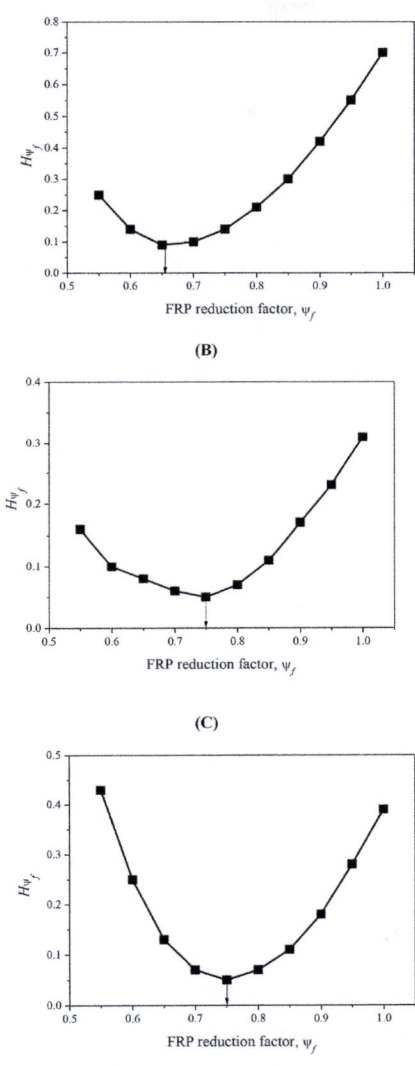

Figure 4.54 Continued

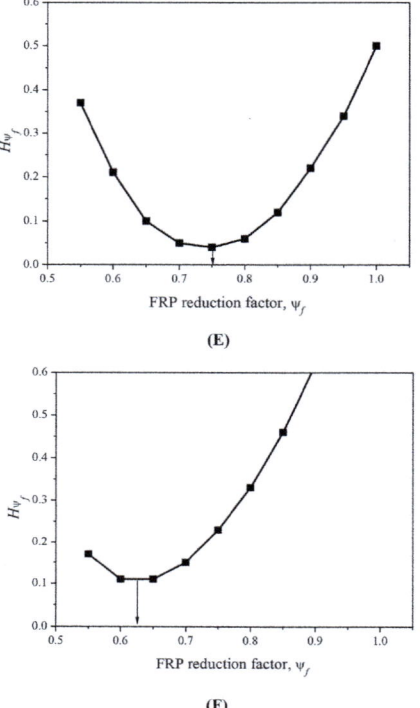

Figure 4.54 Continued

ACI 440 (2008) model is $\psi_f = 0.75$, which is lower than that specified in the ACI 440 (2008) guideline for FRP flexural strengthening ($\psi_{f\_ACI} = 0.85$). The higher scattering of the IC debonding failure mode, compared with the others, that is, the concrete crushing and the FRP rupture, could potentially be the main reason for this difference. To achieve adequate reliability, it is suggested to consider the combined FRP reduction factors when these models are applied in the FRP design guideline.

4.5.4 Flexural design recommendations under fatigue loads

When an existing RC structure strengthened with FRP is subjected to a fatigue load, engineers should consider several factors in designing an FRP system. The stress conditions in both the steel and FRP reinforcements should be considered when designing RC structures strengthened using FRP and subjected to fatigue loading (Harries and Aidoo, 2006). Whether for strengthened or unstrengthened RC beams, the most likely failure mode resulting from fatigue is fracture of the reinforcing steel, which is principally determined by the applied stress range. Thus structures will likely fail at similar fatigue lives, regardless of whether they are strengthened or not, when the stress range applied to the reinforcing steel is the same (Oudah and El-Hacha, 2012). The usual approach in design codes is to limit the stress range in the steel reinforcement while maintaining the maximum stress in the FRP reinforcement below certain limits.

It should be noted that the fatigue life of an FRP-strengthened structure would increase if the structure were subjected to the same fatigue load before and after the strengthening by reducing the applied stress in the steel reinforcement, considering the stress sharing mechanism (Heffernan and Erki, 2004; Brena et al., 2005). The predicted model for the fatigue life of steel reinforcement has been studied widely and proposed in many design codes. The conventional method to predict and prevent steel fatigue failure for unstrengthened RC beams may be reasonably applicable to FRP-strengthened structures. With regard to the allowable cyclic stress limits in FRP reinforcements, the ACI Committee 440 guideline limits both the sustained and the cyclic stress limits in the glass fiber-reinforced polymer, aramid fiber-reinforced polymer, and CFRP to $0.2f_{fu}$, $0.3f_{fu}$, and $0.55f_{fu}$, respectively, where f_{fu} is the ultimate FRP strength. The JSCE limits the tensile stress to 80% of the characteristic creep limit of the FRP, whereas the stress should not be greater than $0.7f_{fu}$. However, current design recommendations for mitigating debonding failures of externally bonded FRP are inadequate to effectively mitigate intermediate crack-induced debonding in flexural members (Kim and Heffernan, 2008). For example, the ACI 440.2R recommendation for fatigue is based on FRP behavior and does not address bond degradation that may accompany fatigue loading.

Studies indicate that many factors may influence the fatigue life of FRP-strengthened concrete members. It is not practical to conduct experimental studies that consider all of the parameters that affect the fatigue capacity of such members. However, it is feasible to use FEA to simulate the behavior of RC members strengthened with FRP under flexural fatigue loading and to use this simulation to investigate the effect of each parameter. Based on an FE investigation on the fatigue behavior of FRP-strengthened concrete members (Wang et al., 2015), a fatigue capacity prediction model was developed to reflect the influences of the major and minor parameters. The major parameters include the fatigue behavior of steel reinforcement, FRP sheets, and FRP-to-concrete bonding, and the minor parameters include the yield strength of steel reinforcement, concrete strength, width and thickness of the FRP sheet, and others.

As stated in the previous section, the flexural resistance of an FRP-strengthened beam can be calculated as the sum of the flexural resistances of the FRP and the steel reinforcement. As a result, the flexural strength of an FRP-strengthened RC beam is expressed as the sum of these two flexural strength components. Based on FEA simulations of 47 beams (as discussed in Section 4.3.2.3), the flexural fatigue capacity is found to vary with A_{st}, f_{yst}, d, b, f_{0c}, t_f, w_f, f_{fu}, a, and P_r. All of these parameters have linear effects on monotonic loading. Some of these parameters have linear effects on fatigue loading, namely, the applied loading range, tensile strength of FRP sheets, shear span, and concrete dimensions; this means that these parameters do not affect the number of cycles to failure. On the other hand, certain parameters have a nonlinear effect on fatigue loading, namely, the area of the steel reinforcement, yield strength of the steel, factors K_1 and K_2, width of the FRP sheet, and thickness of the FRP sheet. The relation between the flexural load cycles to fatigue failure and these nonlinear influencing parameters can be expressed by two equations, depending on the status of the beam (with or without FRP strengthening), as shown in Eqs. (4.25) and (4.26). For beams without strengthening,

$$\log N = C_1(A_{st}^{0.95}f_{yst}^{0.71}K_1^{0.94})/(p_r a) = C_1 D_1 \tag{4.25}$$

For FRP-strengthened beams,

$$\log N = C_1 A_{st}^{0.95}f_{yst}^{0.71}K_1^{0.94} + C_2 t_f^{0.97} w_f^{0.74} f_{frp} K_2^{1.07} = C_1 D_1 + C_2 D_2 \tag{4.26}$$

where the constants C_1 and C_2 are the slopes of the relationship between the increase in the fatigue life ($\log N$) determined from the FEA simulations, and D_1 and D_2 are integrated parameters, as shown in Fig. 4.55. For beams with and without strengthening with FRP sheets, the corresponding C_1 and C_2 values are 0.073 and 0.0036, respectively.

To examine the reliability and validity of the newly proposed model, an extensive verification was conducted using a series of experimental data available in the literature. The database under consideration is composed of the results of 181 experimental tests, including results for 65 beams without FRP strengthening and 116 beams strengthened with FRP sheets. The collected tests differ in terms of their geometry, concrete strength, shear span/depth ratio, and reinforcement ratios, with wide ranges of geometric and mechanical characteristics. The primary geometric

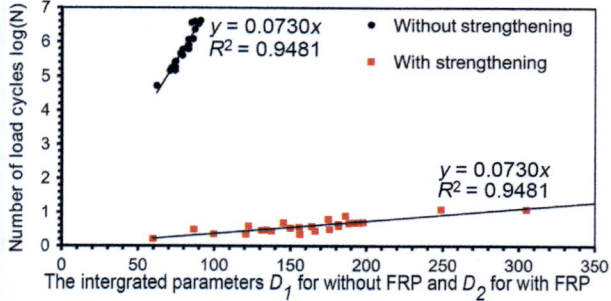

Figure 4.55 Relation between log numbers of cycle loads obtained from FEA simulations and integrated parameters for beams with or without FRP strengthening (Sayed, 2014).

and material properties and various possible failure modes of the experimental data can be found in Wang et al. (2015). A comparison of the predicted and actual failure modes indicates that the average limit proposed by Sayed et al. (2013) is the more suitable of the two limits for application. The experimental and predicted values are compared in Fig. 4.56. In this figure, the strengthening status of the beams is considered in the predicted values. For the beams with and without FRP strengthening, the mean values of $P_{r.Exp}/P_{r.Pred}$ are 1.05 and 1.02, respectively; the corresponding COVs are 17.12% and 16.06%, respectively. These values show that, from a statistical perspective, the proposed model is equally reliable for beams with and without FRP strengthening under the cyclic flexural fatigue loads considered in the analysis.

4.6   Performance enhancements for FRP flexural strengthening

4.6.1   Active strengthening with prestressed fiber sheets

Experimental work on RC beams strengthened with externally bonded nonprestressed FRP laminates showed that a limited increase can be achieved in stiffness, yielding load, and ultimate load-carrying capacity, owing to the limited ability of the FRP−concrete interface to transfer stresses from the concrete substrate to FRP laminates. This often greatly lowers the expected gains and results in a catastrophic, brittle failure such as delamination at the anchorage zone, peeling-off caused by shear cracks, or debonding initiated from intermediate flexural cracks. Generally, only a portion of the strength of FRP laminates is utilized in nonprestressed strengthening applications (Fig. 4.57). To take full advantage of FRP materials, FRP laminates should be prestressed before bonding them to concrete substrates. As in conventional prestressed structures, an initial prestress can be used to balance

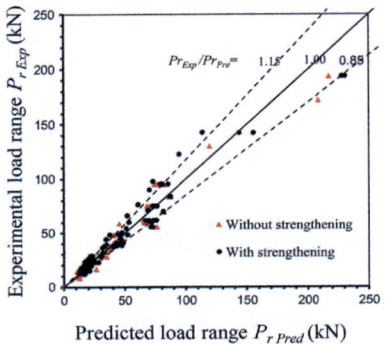

Figure 4.56 Predicted fatigue load by the new model in comparison with the experimental values (Sayed, 2014).

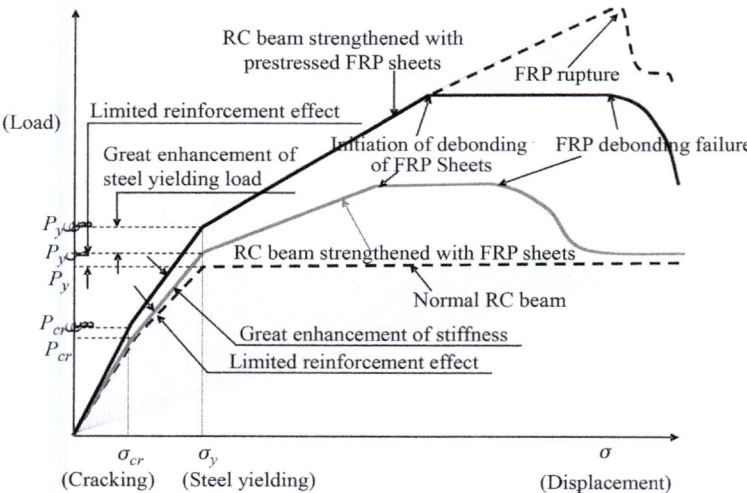

Figure 4.57 Comparison of strengthening methods.

the applied dead and live loads. In addition, an FRP prestressing method can improve the limited ability of the FRP-to-concrete interface without introducing additional interfacial shear stress, except at the anchorage zone, and thus achieve a better strengthening effect (Wu et al., 1999b). Furthermore, prestressing FRP laminates can delay the onset of cracking, reduce crack widths and deflections, relieve the strain in the internal steel reinforcement, and increase the yield load and ultimate load-carrying capacity (Triantafillou et al., 1992, Wu et al., 1999b, Meier et al., 2001, Wight et al., 2001). However, this may introduce another problem, that is, premature failure of the prestressed FRP system induced by a high shear stress concentration at the prestressed FRP end. This implies that the FRP prestressing technique should include some special treatments or additional anchors at the ends of the FRP laminates.

This section is mainly focused on establishing an effective prestressing method. Several anchorage treatments have been established to prevent anchorage bond failure due to high shear stress concentration after the release of prestress at the FRP ends (Wu et al., 1999b). In addition, the performance of RC beams externally strengthened with prestressed FRP sheets is experimentally studied under monotonic and fatigue loads (Wu et al., 2003a).

4.6.1.1 Anchorage treatment methods and prestressing system

The prestressing procedures include (1) pretension of FRP reinforcement, (2) bonding to the tension face of the concrete structures (including curing of adhesive and cutting the ends of FRP reinforcement), and (3) appropriate anchorage treatments, as shown in Fig. 4.58. A prestressing device is required for performing the work in the laboratory, as shown in Fig. 4.59. The two steel−FRP lapped joints are

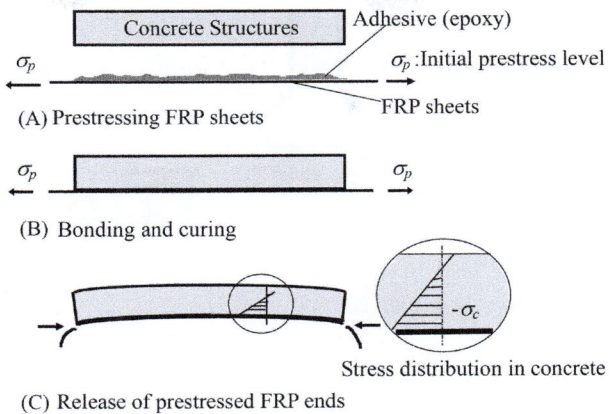

Figure 4.58 Schematic illustration of strengthening method with prestressing FRP sheets (Wu et al., 2003a).

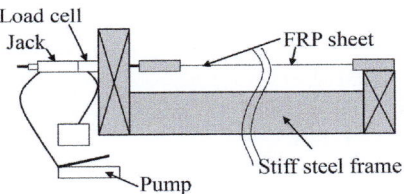

Figure 4.59 Prestressing system for lab experiments (Wu et al., 2003a).

connected to a fixed load cell at one end and a hydraulic jack at the other. Both the load cell and hydraulic jack are mounted to a stiff steel frame. To ensure a good bond condition in the case of an uneven or cambered concrete surface, an airbag system is developed to remove the air within the packaging film and produce a vacuum state between the FRP sheet and concrete surface to be bonded (Wu et al., 2003a), as shown in Fig. 4.60.

Stöcklin and Meier (2001) developed a prestressing device, in which to generate gradually anchored and prestressed CFRP strips. A stepwise approach is used as follows: the CFRP strip is clamped and prestressed and then covered with an adhesive. The device is temporarily mounted to the structure. The middle part of the strip is bonded. Then, the force is marginally reduced, and the following areas are bonded. The force is reduced further and the adjacent areas are bonded. This procedure is repeated until there is no remaining prestressing force at the ends of the strip. The prestressing device can then be removed immediately. The adhesive is completely cured using a heating device.

However, these external prestressing methods may result in a high interfacial shear stress concentration at the ends of the prestressed FRP sheets after the release

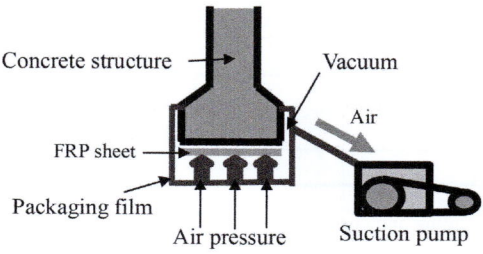

Figure 4.60 Air bag system (Wu et al., 2003a).

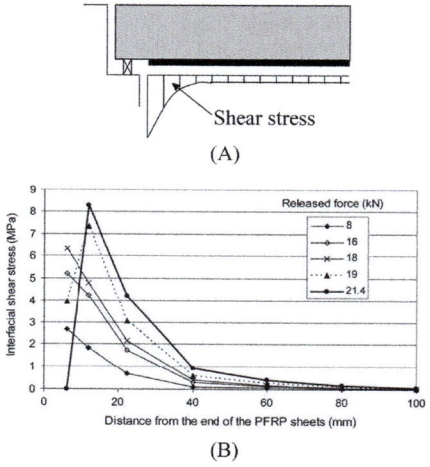

Figure 4.61 Stress concentration at FRP ends after the release of pretension. (A) Stress concentration. (B) Experimental interfacial shear stress (Diab and Wu et al., 2009).

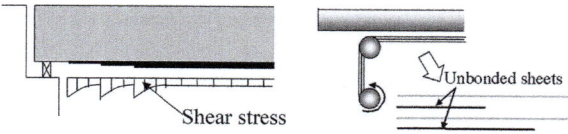

Figure 4.62 Taper method of releasing prestress towards the FRP ends (Wu et al., 2003a).

of pretension (Fig. 4.61), which may cause anchorage debonding at the ends of the FRP sheets. To avoid this failure, a certain balance should be considered to determine the prestress level to avoid the stress concentrations caused by either the release of the pretension or the external load. Two approaches can be used to relieve the stress concentration: one is to taper the FRP sheet toward the ends (Fig. 4.62) and the other is to gradually release the prestress until zero toward the

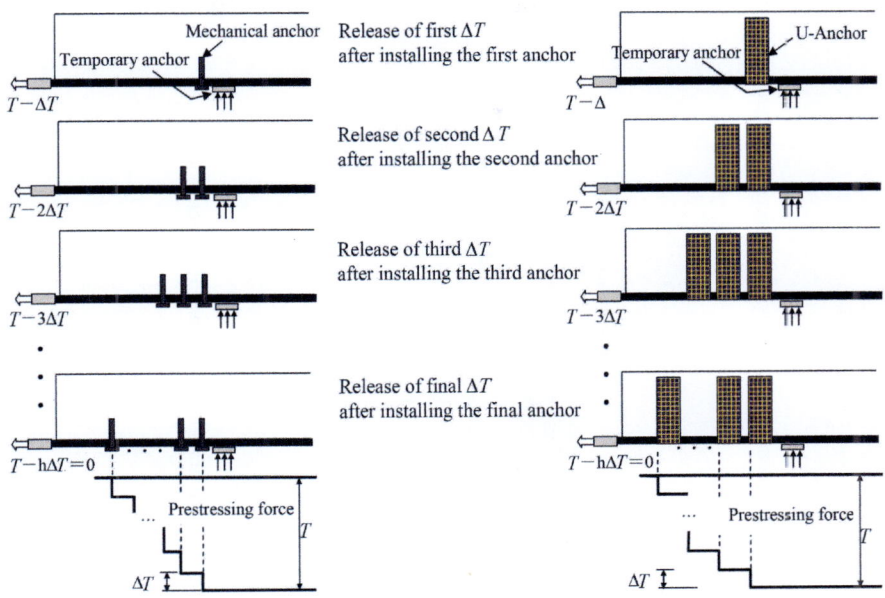

Figure 4.63 Stepwise method of releasing prestress towards the FRP ends.

ends to produce a stepwise distribution of prestress as shown in Fig. 4.63. In addition, some effective anchorage treatments can be combined to ensure the prestressing strengthening effect, as shown in Fig. 4.64.

Wight et al. (2001) developed a mechanical anchorage system, as shown in Fig. 4.65. The system consisted of steel roller anchors bonded to the sheets and steel anchor assemblies fixed to the beam. The roller anchors that gripped the sheet consisted of two stainless-steel rollers bonded to each end of the sheet. The rollers were 32 mm in diameter and 380 mm long, and the central 300 mm was knurled to improve the bond performance. At least 3 days before prestressing operations, the sheet was wrapped 2.5 times around the roller, and a two-part epoxy was applied to bond the sheet to the rollers.

Fig. 4.66 shows an example of a flat plate anchor for prestressing FRP sheets developed by El-Hacha et al. (2003). The designed metal anchorage system consisted of a movable flat steel plate anchor at the jacking end, a fixed steel angle anchor at the dead end, brackets, and steel rods.

4.6.1.2 PBO prestressing upgrading technique

Wu et al. (1999b) investigated the strengthening effect of prestressing carbon fiber sheets (CFSs) on RC beams and found that significant improvements in the flexural strength, ductility, stiffness, and crack resistance of the retrofitted beams can be achieved. However, the disadvantage of dry carbon fibers is that they cannot be

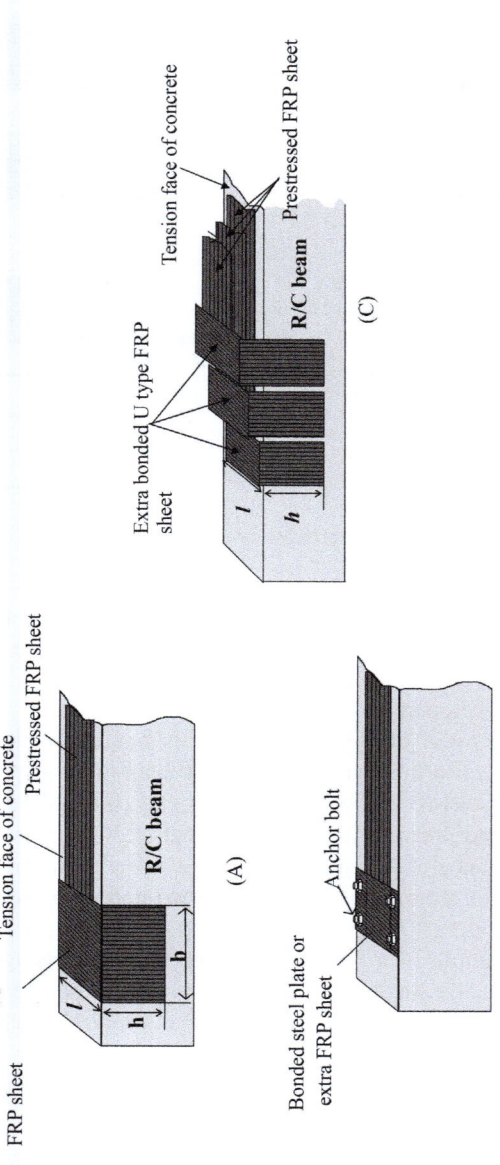

Figure 4.64 Anchorage treatments for the prestressed FRP laminates (Wu et al., 2003a). (A) Extra bonded U-type FRP sheet (normal). (B) Anchor bolts. (C) Extra bonded U-type FRP sheet (step by step).

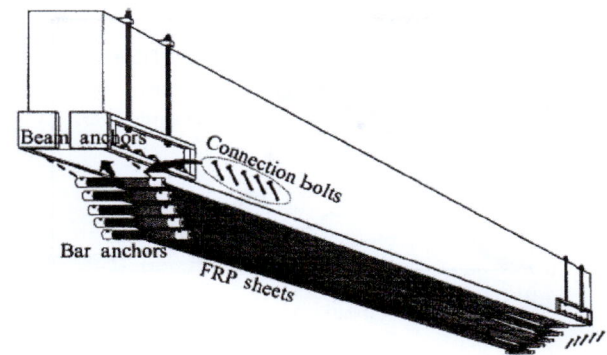

Figure 4.65 Mechanical anchorage system developed by Wight et al. (2001).

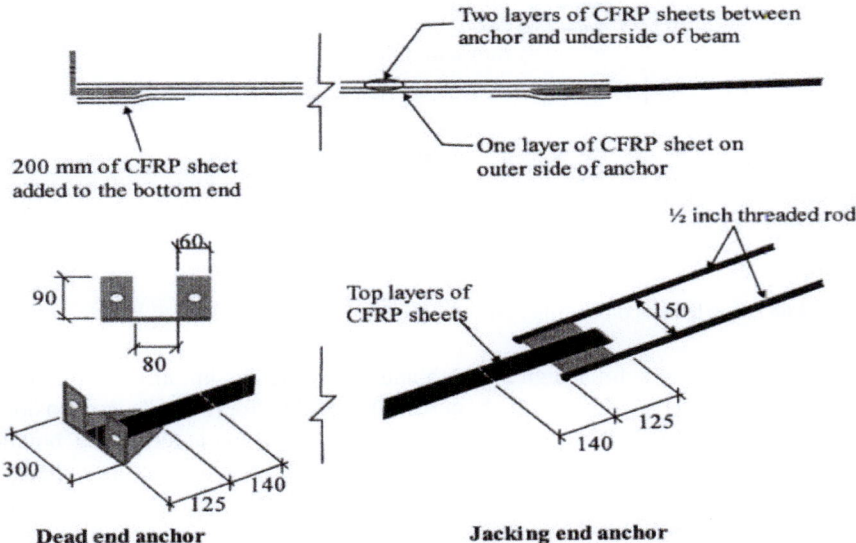

Two layers of CFRP sheets between anchor and underside of beam

One layer of CFRP sheet on outer side of anchor

200 mm of CFRP sheet added to the bottom end

½ inch threaded rod

Top layers of CFRP sheets

Dead end anchor

Jacking end anchor

Figure 4.66 Mechanical anchorage system developed by El-Hacha et al. (2003).

prestressed beyond a certain stress level (e.g., 30%) before the resin impregnation and curing procedure. This is mainly owing to the carbon fiber's poor energy absorption ability, which means that the fibers cannot maintain an even transfer of force without resin (Wu et al., 2007). This implies that we can only prestress FRP plates but not dry fiber sheets, which sometimes result in serious problems to achieve a complete bonding of the FRP-to-concrete interface.

To overcome this drawback and exploit the advantages of FRP sheets, a new type of fiber, namely PBO fiber sheet, has been applied for strengthening concrete beams (Wu et al., 2003a). PBO fiber sheets, in comparison with typical CFRP

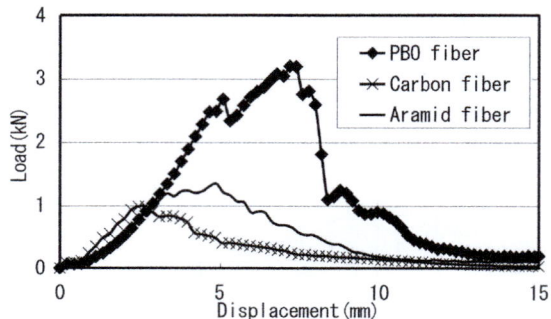

Figure 4.67 Energy absorption ability for different fibers (Wu et al., 2003a).

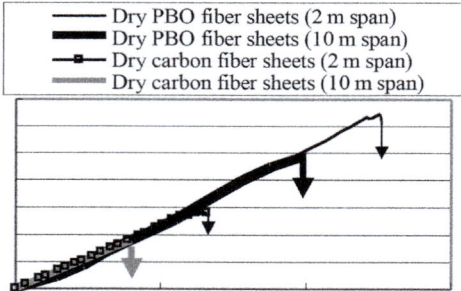

Figure 4.68 Tensile tests of dry fiber sheets (Wu, 2007).

sheets, have not only similar high strength and high modulus but also a high energy absorption ability (Fig. 4.67), which makes it possible for them to be tensioned before being impregnated with resin and, consequently, guarantees good bonding between the two materials. Fig. 4.68 compares tensile tests of dry fiber sheets in lengths of 2 and 10 m to simulate the prestressing process. Because of their relatively low energy absorption behavior, CFSs without resin impregnation break at 40% of their tensile strength when stretched in a length of 2 m, and can sustain only 27% when stretched in a length of 10 m. In contrast, 2-m-long dry PBO fibers can be tensioned to 90% of their capacity, and 10-m-long fibers hold up to 60% of the load.

A widely accepted procedure for upgrading concrete beams, slabs, and girders with bonded prestressed composites has been established as follows. PBO fiber sheets are first tensioned to a certain percentage of prestress, then coated with epoxy resin on-site and attached to the structural surface. After curing, the tensile load is released and the prestress is transferred to the concrete structure. The key to the success of this procedure lies in the bond at the interface. To achieve good bonding, an airbag system with vacuum capacity was developed for construction site applications. To facilitate this upgrading technique in structural retrofitting

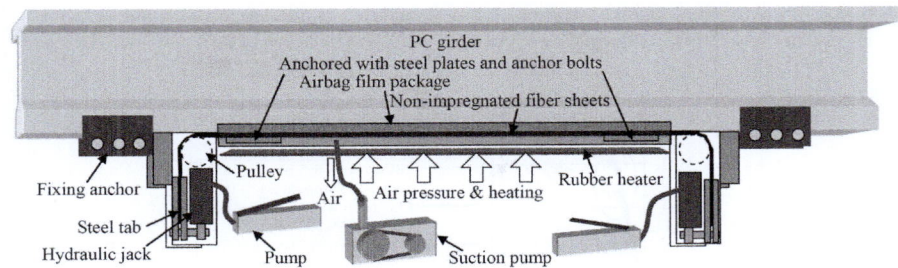

Figure 4.69 Prestressing system for real application.

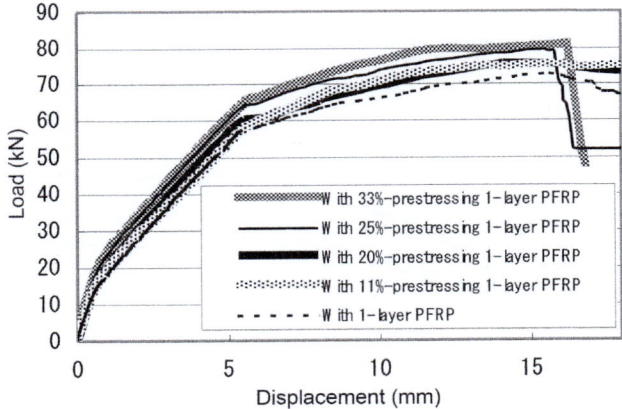

Figure 4.70 Load—deflection curves of 2-m-long T-girders.

using PBO fibers, a full-scale apparatus was developed with the capacity to apply the prestressing force, and simultaneously heat to a temperature of $6C-80°C$ for complete on-site curing in $3-6$ h (Fig. 4.69). In terms of the above concept and ideas, a new set of integrated operation systems were developed for this approach, including prestressing, curing, and auxiliary equipment. This technique creates high-quality interfacial bonding and achieves the goal of "completing construction in one day."

4.6.1.3 Strengthening effect

Typical load—deflection curves of 2-m-span beams and 10-m-span girders with PBO FRP (PFRP) sheets prestressed at different levels are shown in Figs. 4.70 and 4.71, respectively. The failure modes were bending for a specimen without reinforcement, peeling from the specimen center for a specimen without prestress and a specimen with one layer of 20%-prestressed sheet, and sheet rupture at the specimen center for a specimen with one layer of 25%-prestressed sheet and a specimen with

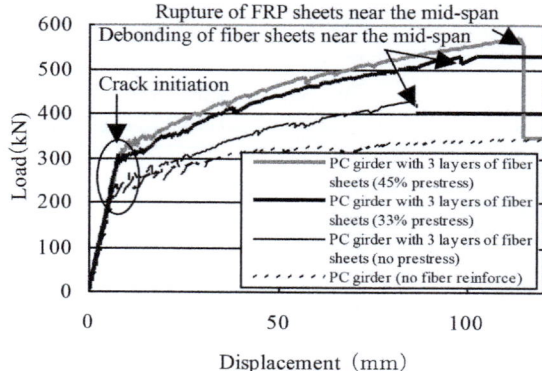

Figure 4.71 Load−deflection curves of 10-m-long T-girders.

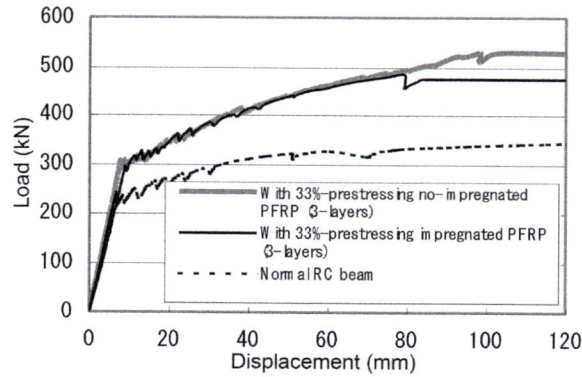

Figure 4.72 Load−Deflection Curves of 10-m-long T-girders.

one layer of 33%-prestressed sheet. Thus it can be confirmed that the introduction of prestress permits the PFRP sheet to display its higher strength. The control girder without external reinforcement failed owing to the yielding of tensile steel followed by concrete crushing. The girder strengthened by three-layer PFRP sheets without prestressing exhibited a similar bend-over-point as the control beam but obtained a 45% increase in yielding load and a 65% increase in ultimate load. A high percentage of prestressing could effectively change the failure mode from debonding to FRP rupture, leading to a significant increase in the load-carrying capacity. Fig. 4.72 shows the experimental results of the load versus deflection relationship for two 33%-prestressed FRP-strengthened beams with the resin impregnation procedure before and after the prestressing procedure of dry PBO fiber sheets (PFS). The initiation of FRP debonding was delayed for the case of resin impregnation after the prestressing procedure for FRP, compared to that of resin impregnation before the prestressing procedure for dry PFS. This result implies that excellent bonding performance can be

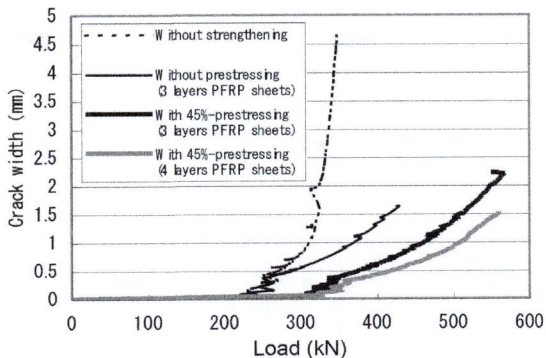

Figure 4.73 Crack width opening in T-girders.

achieved by prestressing dry PFS without resin impregnation. The crack opening width is also considerably reduced with prestressing, as shown in Fig. 4.73.

The fatigue performance of RC beams strengthened with prestressed PFS was experimentally investigated by Wu et al. (2003a). By considering the FRP strain at failure, it was found that the interfacial performances of beams strengthened with prestressed PFS have the same level as compared to those of beams strengthened with nonprestressed PFS. As shown in Fig. 4.74A, the FRP strain at failure decreases with the increase in the upper limit of the fatigue load for cases with both prestressed and nonprestressed PFRP. With the increase in loading number, no significant change can be observed in the displacement, FRP strain, and crack width in Fig. 4.74B–D, which shows that a satisfactory fatigue behavior can be provided by prestressing PFS. The experimental observations also indicate that the roughness of a debonded surface varies with the increase in the upper limit of the fatigue load. For cases of strengthening with both prestressed and nonprestressed PFS under fatigue loading, the larger the upper limit of fatigue load is, the shallower is the depth of debonded concrete. This shows that the interface performance becomes weaker with the increased upper limit of the fatigue load.

4.6.2 Strengthening with hybrid fibers

Unlike steel, FRP composites remain elastic until failure and fail in a noticeably brittle manner. Correspondingly, FRP-strengthened concrete structures can fail momentarily without any forewarning. Moreover, the gains in stiffness and yield load achieved by strengthening with aramid, glass, or basalt fiber composites are very limited. Even for the case of carbon fiber with a modulus similar to steel reinforcement, the gain in stiffness is insufficient because the reinforcing ratio is relatively low. Therefore a significant increase can hardly be achieved in the service load. To improve the performance of the strengthened structures and efficiently utilize the strengthening effect of FRPs, one idea to increase the ductility and stiffness is to use hybrid composites consisting of different types of fibers with different moduli and strengths, which fail at different strains during

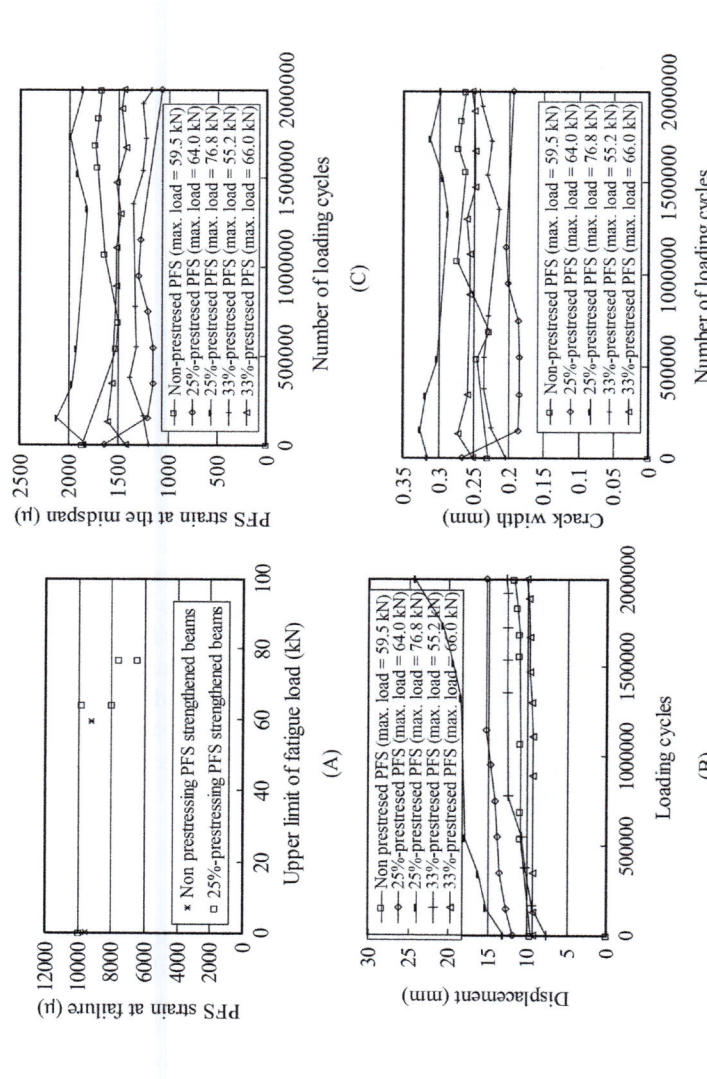

Figure 4.74 Fatigue performance of RC beams with prestressed PFRP sheets. (A) FRP strain at failure excluding initial prestress strain. (B) Displacement versus loading cycles. (C) FRP strain versus loading cycles. (D) Crack width versus loading cycles.

loading, thereby allowing a gradual failure of the composites. As shown in Fig. 1.14, high-modulus fibers can be used to ensure increases in the stiffness as well as the cracking and yield loads. To increase the structural strength, high-strength fibers can be used in combination with high-modulus fibers. The mixture of high-modulus and high-strength fibers presents a strain-hardening behavior until the rupture of the high-strength fibers, which may be used to control the deformation of the structure. In addition, the ductility can be increased by including some high-ductility fibers in the mixture. Theoretical studies have shown that the stiffness can be enhanced by higher-modulus fibers and ductile behavior may be obtained by hybrid fiber sheets in which carbon fibers are mixed in a certain proportion with aramid or glass fibers. However, a practical problem was often observed, whereby premature failure or significant load drops of the hybrid composites occur due to locally high stress concentrations at the location of fiber breakage causing damage to the surrounding fibers (Wu et al., 2006). Therefore the expected strengthening effect cannot be achieved by using the rule of mixtures. This requires rational mixture design and special efforts to control the stress drop. The following section introduces the experimental and numerical investigations on the flexural strengthening effects of RC beams strengthened with hybrid fiber sheets (Niu and Wu, 2003, Wu et al., 2007).

4.6.2.1. *Flexural test of RC beams strengthened with hybrid FRP sheets*
To investigate the hybrid effect on the FRP-strengthened RC beam, a series of RC beams shown in Fig. 4.75 were tested by Wu et al. (2007). Higher-modulus (C7) and higher-strength (C1) CFSs were adopted for the composition of hybrid fiber sheets. The mechanical properties of the CFSs are listed in Table 4.2. Fig. 4.76 shows that the stiffness, yield and failure load, and ductility can be greatly enhanced by hybrid fiber sheets with two layers of C1 sheet and one layer of C7 sheet. The progressive rupture of C7 sheet can lead to stress redistribution in the hybrid sheet and avoid early debonding, as shown in the cases with only C1 sheets.

4.6.2.2. *Numerical studies*

Numerical models were developed by Niu and Wu (2003) for simulating the mechanical behavior of hybrid fiber sheets in tension and bending tests, where a macromechanical damage constitutive model was proposed to model the stress transfer behavior caused by the progressive damage process of C7 fiber sheet in hybrid fiber sheets. FRP sheets of a single type generally show linear elastic

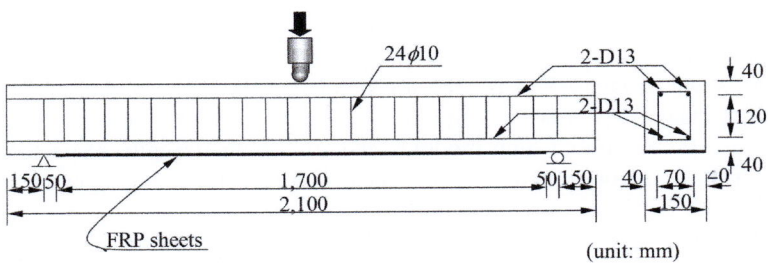

Figure 4.75 Details of FRP-strengthened RC beams (Wu et al., 2007).

Table 4.2 Material properties of carbon fiber sheet (Wu et al., 2007).

High-strength carbon fiber sheet (C1)	Modulus of elasticity	2.3×10^5 MPa
	Tensile strength	3400 MPa
	Thickness	0.111 mm
	Fiber areal weight	200 g/m^2
High-modulus carbon fiber sheet (C7)	Modulus of elasticity	5.4×10^5 MPa
	Tensile strength	1900 MPa
	Thickness	0.143 mm
	Fiber areal weight	300 g/m^2

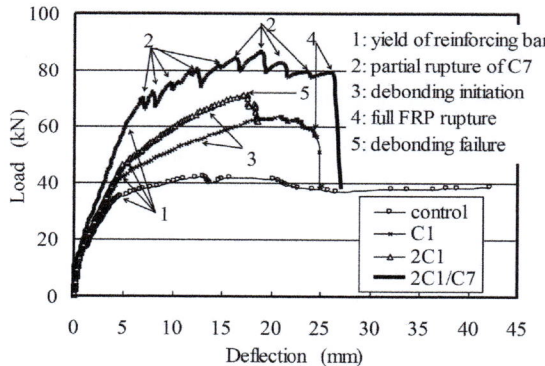

Figure 4.76 Load−deflection curves of tested beams (Wu et al., 2007).

behavior until rupture. However, when hybrid fiber sheets are subjected to loading, some progressive damage such as matrix cracking, fiber−matrix debonding, fiber rupture, and delamination may lead to a different mechanical behavior. Considering that the shocks from such damage may rupture a portion of unimpaired fibers, it is very important to evaluate the stress transfer capacity upon the partial rupture of some fibers. Herein, from a macroscopic point of view, the mechanical damage behavior shown in Fig. 4.77A is used for simulating the stress transfer from C7 to C1 sheet during progressive rupture of C7 sheet, where C7 and C1 sheets are expected to achieve initial high modulus and final high strength, respectively. As for C1 sheet, it is considered to show linear elastic behavior until brittle rupture (Fig. 4.77B).

The FE model for the FRP-strengthened RC beams is shown in Fig. 4.78. FRP sheets are modeled by truss elements whose behaviors are coded into user-defined material subroutines, and a perfect-bond condition is assumed between fiber sheets.

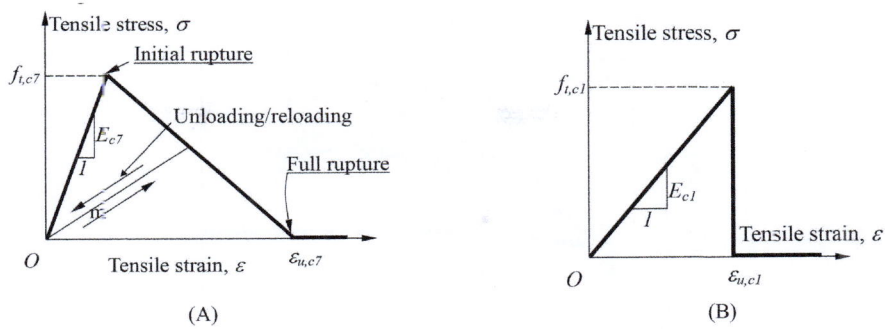

Figure 4.77 Stress−strain relations for C7 and C1 sheets (Niu and Wu, 2003). (A) High-modulus C7 sheet in hybrid sheets. (B) High-strength C1 sheet.

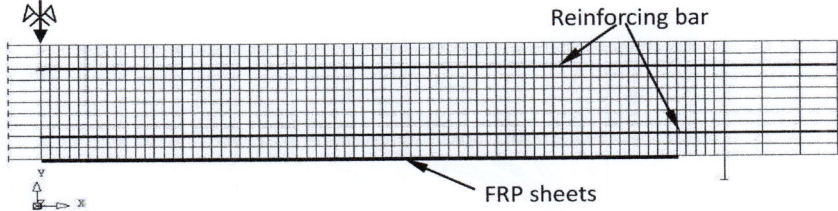

Figure 4.78 Finite-element meshes for FRP-strengthened RC beams (Niu and Wu, 2003).

Two series of beams with normal (N) and high-strength (H) concrete are used to investigate the hybrid effect of FRP-strengthened RC beams.

As shown in Fig. 4.79, numerical results with high-strength concrete are generally in good agreement with the experimental results. However, because the failure mode of 2C1/C7 is debonding within the concrete substrate shown in Fig. 4.80 and this is different from the experimental result (rupture of FRP sheets), the expected hybrid effect is not achieved. The numerical simulations also show that with the increased concrete properties, the final failure mode can be shifted from debonding within the concrete substrate to FRP rupture for the case of a C1 sheet-strengthened beam, which is very similar to the experimental result. The FRP stress distributions are illustrated in Fig. 4.81 for different concrete strengths to present the different failure modes. For the case of 2C1 sheets-strengthened beam, the debonding along the tension reinforcing bars remains unchanged, which was also observed in the experiment.

Based on the experimental observations, a macroscopic damage constitutive model is proposed to model the stress transfer behavior upon the progressive rupture of high-modulus fiber sheets in the hybrid sheets. This model can be used to interpret the experimental phenomena and investigate the hybrid behavior in the structural design. It is found that the hybrid ratio and stress transfer behavior can be used to control the load drop due to the rupture of high-modulus sheet.

2003]

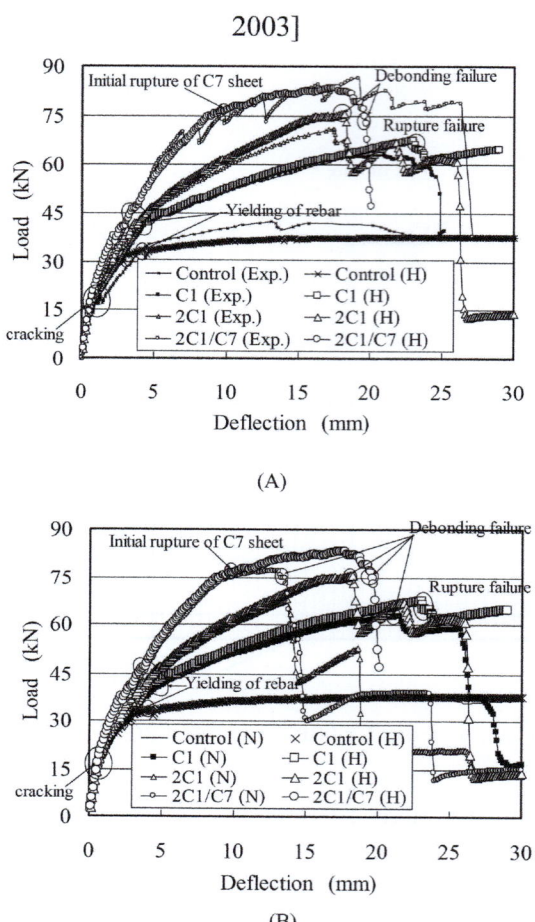

(A)

(B)

Figure 4.79 Comparison of numerical and experimental results for RC beams (Niu and Wu, 2003). (A) Numerical versus experimental results. (B) Effect of concrete properties.

4.7 Special field applications

4.7.1 Strengthening of concrete tunnel lining

Concrete tunnel linings often seriously suffer from progressive cracks at certain locations, caused by external forces, aging, and severe environmental conditions, as shown in Fig. 4.82A. Such progressive cracks are generally highly localized and eventually lead to the collapse of the lining. The application of bonding FRP sheets to the inner side of the damaged tunnel lining has become a solution to prevent further crack propagation in the concrete lining and also to enhance its structural load-carrying capacity, as shown in Fig. 4.82B. Because of their arch shape, concrete

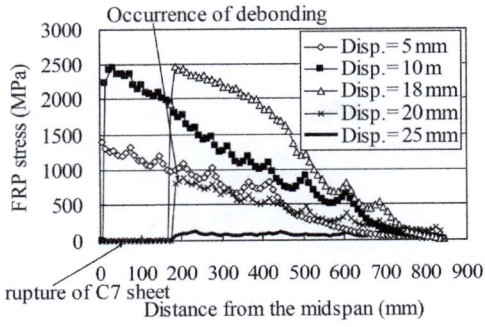

(A)

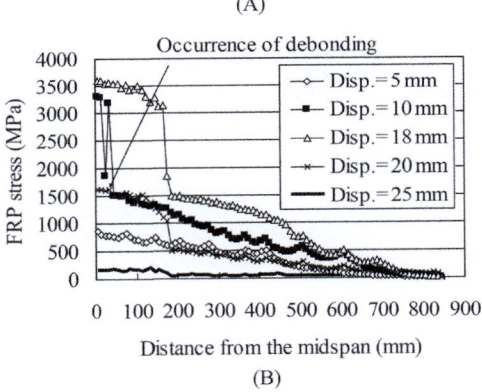

(B)

Figure 4.80 FRP stress distributions in 2C1/C7 hybrid sheet-strengthened RC beam with high-strength concrete (Niu and Wu, 2003). (A) FRP stress distributions in C7 Sheet. (B) FRP stress distributions in C1 sheet.

tunnel linings exhibit different fracturing behaviors from normal concrete members such as beams and slabs. To pursue an effective strengthening method for such damaged concrete tunnel linings, a series of experiments has been carried out (Wu et al., 2003b). It was found that externally bonded FRP sheets on the inner side could effectively enhance the structural load-carrying capacity and deformability. In this section, the structural responses of FRP-strengthened concrete tunnel linings are introduced.

4.7.1.1. *Experimental observations*
A 1/3-scale concrete tunnel model was constructed and tested with and without FRP reinforcement, as shown in Fig. 4.83. The vertical load was applied to the non-FRP-strengthened tunnel lining until the crack initiation at the inner crown, after which FRP sheets were bonded to the inner side of the tunnel lining while keeping the load unchanged. Then, the load application continued until the final failure.

During the loading process, the cracking behavior developed according to the following sequence. (1) The crack initiated at the inner crown where the vertical load

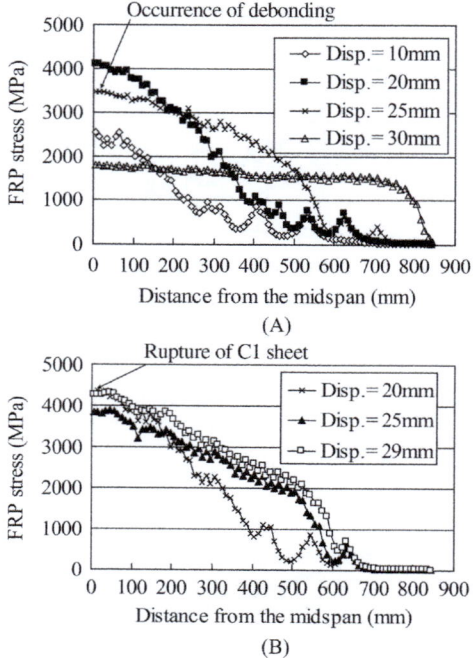

Figure 4.81 FRP stress distributions in C1 sheet-strengthened RC beam (Niu and Wu, 2003). (A) Normal-strength concrete. (B) High-strength concrete.

Figure 4.82 Retrofitting of concrete tunnel with externally bonded FRP.

was applied (Fig. 4.84A). (2) After bonding the FRP sheets on the inner side of the tunnel lining, the initial crack continued to propagate but the rate decreased (Fig. 4.84B). The bonded FRP sheets started to take strengthening effect through interfacial stress transfer. (3) With the formation of a crack at the loading point,

Figure 4.83 The concrete tunnel model.

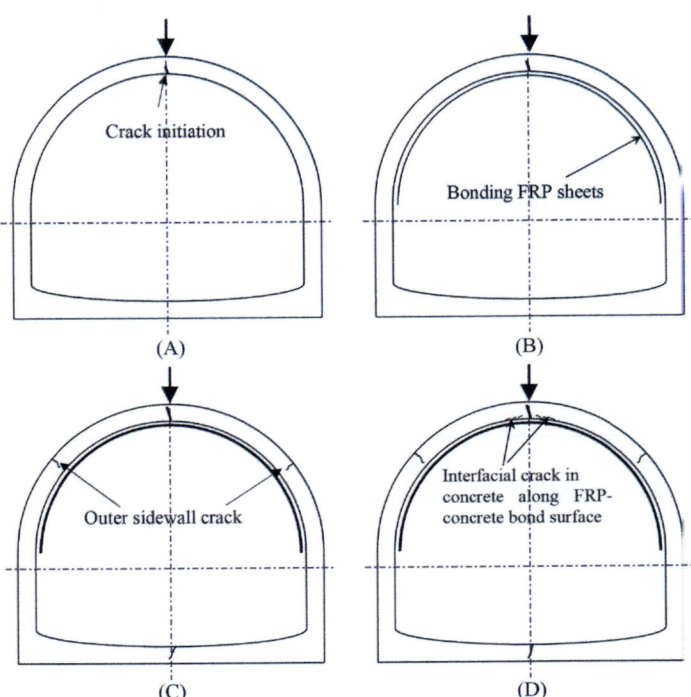

Figure 4.84 Crack evolution in experiment.

cracks at the outer sidewall occurred (Fig. 4.84C) and the load continued to increase. (4) Further loading made the outer sidewall cracks propagate further across the thickness of the tunnel lining. The load reached its maximum and began to decrease gradually with the occurrence and propagation of interfacial concrete cracks along the FRP—concrete interface at the inner crown (Fig. 4.84D). For all the cases with different types of FRP sheet reinforcement, the interfacial debonding that occurred in the interfacial concrete was the final failure mode that resulted in the collapse of the FRP-strengthened concrete tunnel lining.

Several cases with different reinforcing sheets were investigated. These include one-layer carbon FRP sheet, aramid sheet, two-layer carbon sheet, PBO sheet, and steel plate. The material properties and other experimental parameters are listed in Table 4.3.

The FRP strengthening effects may be seen by comparing the load—deflection curves between the cases with and without FRP strengthening, as shown in Fig. 4.85. It was found that the peak load increased with the increasing elastic modulus and the number of FRP sheets applied. Although the PBO strengthening enhanced the deformational behavior of the tunnel lining, the peak load was also dependent on its elastic modulus and the amount of FRP. The tunnel lining with steel plate strengthening can be seen as an extreme case. Because its thickness was much greater than that of general FRP sheet, the peak load was much higher than in other cases. In addition, distributed cracks at the inner crown were observed. This is also thought to have resulted from the increased strengthening degree of steel plate.

4.7.1.2. Numerical studies

An FE analysis of an FRP-strengthened concrete tunnel lining was performed by using interface elements and truss elements to simulate the FRP—concrete interface and the FRP sheets. The detailed material properties used in the numerical simulations can be found elsewhere (Wu et al., 2003b). The cases of FE 1 to FE 5 correspond to those of Exp. 1 to Exp. 5 in Table 4.3, respectively.

Fig. 4.86 shows a comparison of the load—deflection curves of each case, except for the case of PBO sheet, because its material properties are almost the same as one-layer carbon sheet and thus the load—deflection curves are also very similar. It can be seen that the peak load increases in the order of aramid sheet, one-layer carbon sheet, two-layer carbon sheet, and steel plate, which implies a relationship

Table 4.3 Experimental cases with various reinforcements.

Case No.	Type of reinforcement	Young's modulus E (kgf/cm^2)	Strength (kgf/cm^2)	Amount (g/m^2)
Exp. 1	Carbon sheet 1L	2.34×10^6	34,400	14.0
Exp. 2	Aramid sheet	1.20×10^6	21,000	200
Exp. 3	Carbon sheet 2L	$2.34. \times 10^6$	34,400	300
Exp. 4	Steel plate	2.10×10^6	784.0	3.2 mm
Exp. 5	PBO sheet	2.70×10^6	–	–

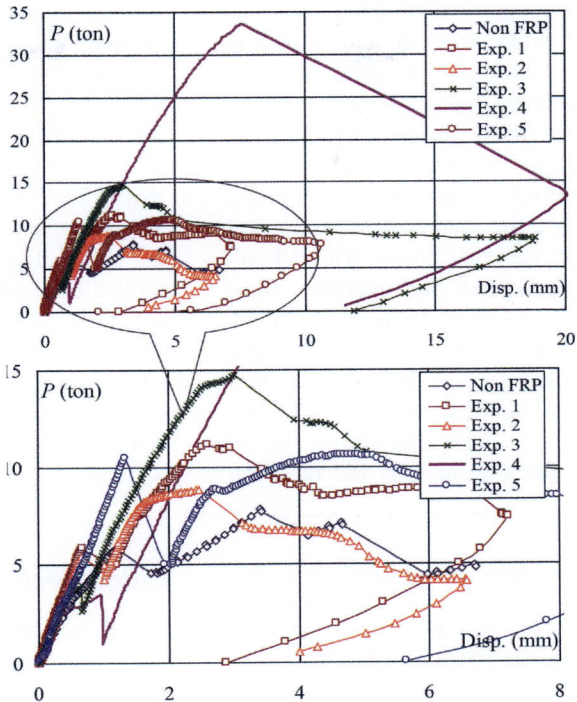

Figure 4.85 Comparison of load—deflection curves in different cases.

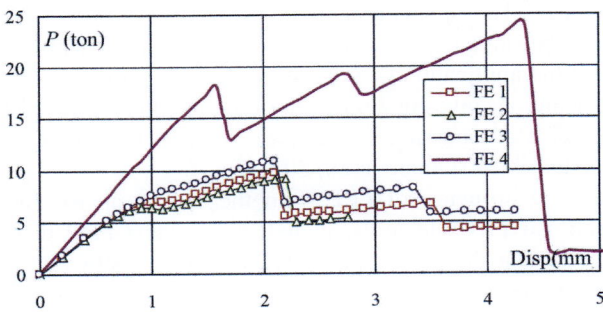

Figure 4.86 Load—deflection curve of simulation cases.

between the peak load and the product of the elastic modulus E_f and thickness t_f of FRP laminates.

As mentioned earlier, interfacial debonding between FRP sheets and concrete lining occurred in concrete adjacent to the bond interface in experiments. In the simulations, although the interface elements are used to model the debonding at the epoxy—concrete or epoxy—FRP surfaces, no debonding occurred in the interface elements. Similar to the experimental observations, the concrete elements near the FRP—concrete interface cracked, and these cracked concrete elements finally led to the interfacial debonding. The detailed crack patterns at failure are shown in Fig. 4.87. It was found that the crack at the inner crown tends to distribute along the FRP—concrete bond surface with the increase in the FRP sheet/steel plate Young's modulus and the strengthening amount.

From the numerical simulation, it can be deduced that in FRP-strengthened concrete tunnel linings, the final collapse follows the occurrence of interfacial debonding in concrete at the inner crown. The debonding does not likely propagate along the whole FRP—concrete bond interface. Once a complete debonding occurs, the strengthening takes no further effect. This may be due to the special arch shape of the tunnel lining. The higher the Young's modulus of the FRP sheet and the more strengthening amount is used, the longer is the obtainable effective bond length. In addition, the crack in concrete tends to distribute widely with the increased stiffness of the strengthening sheets and plates. Therefore high-modulus FRP sheets and using more strengthening amount are recommended to increase the strengthening performance.

4.7.2 Fatigue strengthening of concrete bridge deck

The deterioration of existing bridges is a critical issue all around the world. The effect of heavy trucks running on highways is particularly critical for the structural deficiency of bridges, and they result in serious fatigue damage of bridge decks (Yoshitake et al., 2010), as shown in Fig. 4.88. To maintain the performance of bridges, one of most important propositions is to retain the soundness of bridge decks. Many methods have been developed to repair and strengthen bridge decks with fatigue damage. The application of FRP laminates is one of the remarkable repairing methods because it can be applied to bridge decks without any influence on the traffic and the application of this method is easier than those of other methods. In the following, an investigation of the fatigue behavior of FRP-strengthened decks under wheel load running, in comparison with that of RC decks without strengthening, is discussed (Omote et al., 2011).

To investigate the influence of FRP strengthening on an RC bridge deck, two types of conventional FRP materials, namely CFS and CFRP plate (CFR), were bonded to the bottom of the RC deck, as shown in Fig. 4.89A and B, respectively. Detailed dimensions of the test specimens and properties of the materials can be found elsewhere (Omote et al., 2011).

Fig. 4.90A shows a photograph of the testing machine for simulating a wheel trucking machine acting directly on a deck slab. The testing machine is composed of a chassis running by applying a load on the test specimen and the railway on which the chassis is reciprocated by motor rotation. In the experiments, the loads were applied on the midspan of the deck specimens and increased stepwise, as shown in Fig. 4.90B.

Fig. 4.91 shows the main test results of the deck specimens. Under simulated wheel running cyclic loading, FRP-strengthened decks showed obviously lower

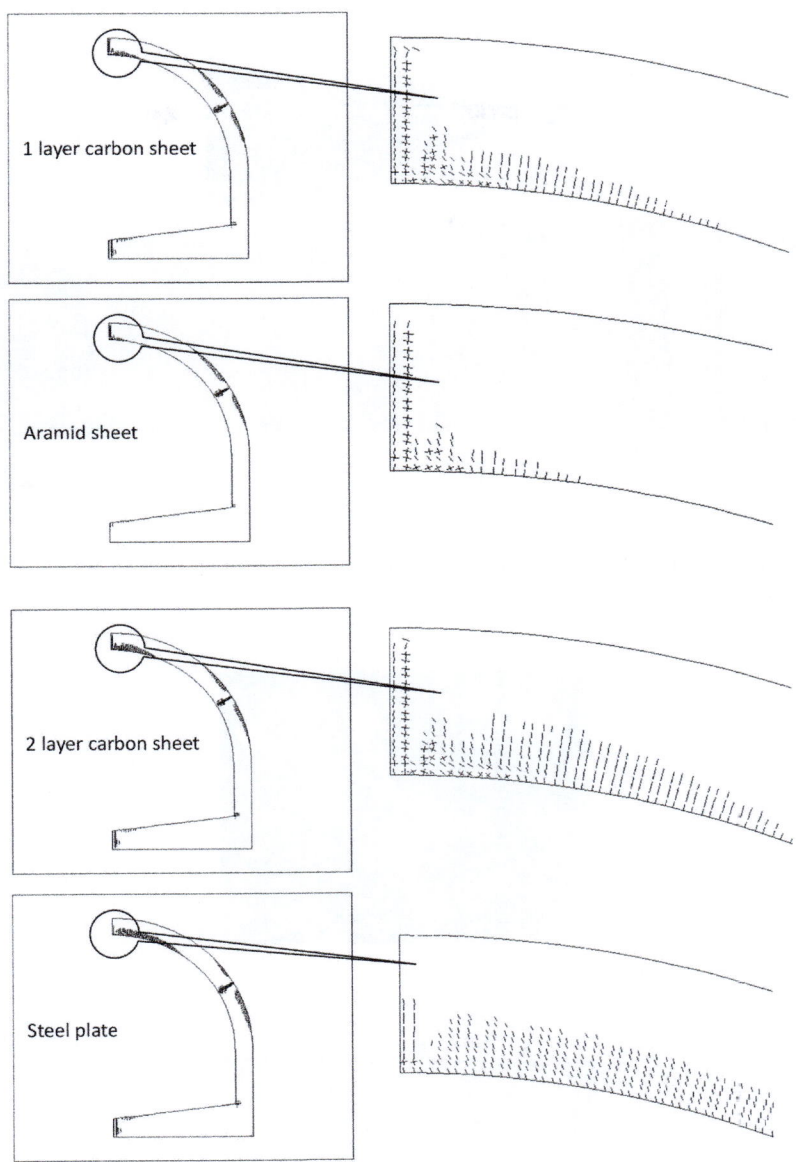

Figure 4.87 Crack patterns of FE simulation.

Figure 4.88 Deterioration of the bridge deck due to fatigue damage.

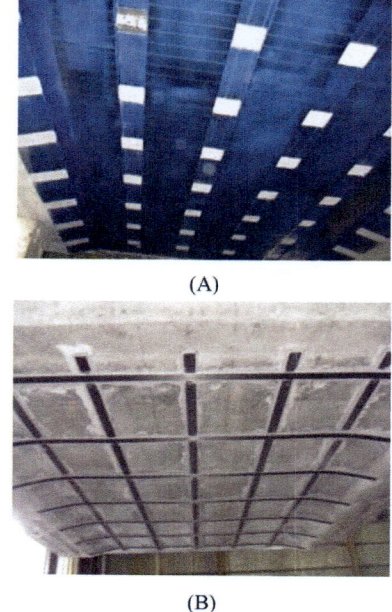

Figure 4.89 RC deck strengthened with FRP laminates (Omote et al., 2011). (A) Carbon fiber sheet. (B) CFRP plate.

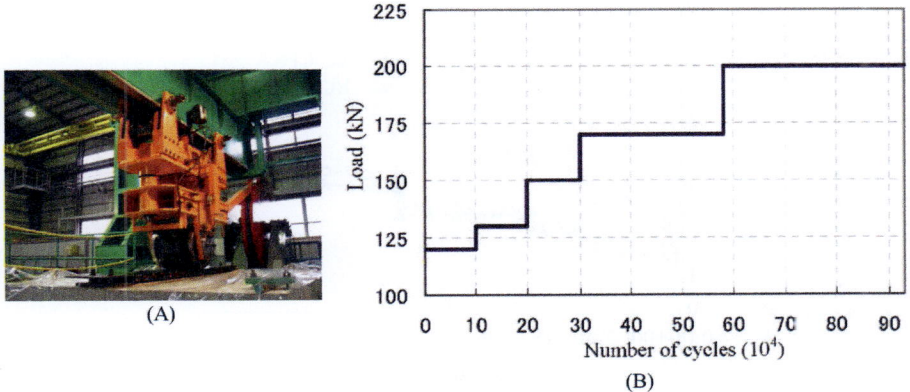

Figure 4.90 Cyclic loading of wheel load running (Omote et al., 2011). (A) Testing device. (B) Loading procedures.

deflection at the midspan and maintained a much longer fatigue life than the corresponding RC deck specimen (Fig. 4.91A). The recorded tensile strains of the midspan steel reinforcement were much lower for the two FRP-strengthened decks than that of the corresponding RC deck specimen (Fig. 4.91B), indicating that the application of FRP strengthening significantly decreased the steel strain in the concrete decks under simulated wheel running cyclic loading. The experiments verified that RC decks can be strengthened with externally bonded FRP laminates to have sufficient fatigue durability for practical use.

4.7.3 Retrofit of columns with longitudinal reinforcement cut-offs

A failure mode for columns with some of the longitudinal reinforcing bars cut off outside the potential plastic hinge regions at column ends has been identified under seismic attack (Priestley et al., 1996, Kobatake 1998, Teng et al., 2001). In these columns, longitudinal reinforcing bars were terminated at several levels based on the design moment envelope, without accounting for the effects of tension shift due to diagonal shear cracking. The flexural-shear failure of such columns may occur at a cut-off section rather than at the ends of the column as a consequence of premature termination of the longitudinal reinforcement. In Japan, a number of bridge columns developed such failures at column mid-height during the 1982 Urahawa-ohi and 1995 Kobe earthquakes. Fig. 4.92 shows the flexural-shear failure of bridge piers at mid-height due to the cut-off of the longitudinal reinforcement in the 1995 Kobe earthquake.

Longitudinal bonding of FRP is an effective retrofitting method to prevent such failure (Kobatake, 1998), as shown in Fig. 4.93. In this method, flexural reinforcement is enhanced by bonding fiber sheets in the direction of the cut-off rebars to concrete surfaces at the rebar cut-off sections. Shear reinforcement is also enhanced by

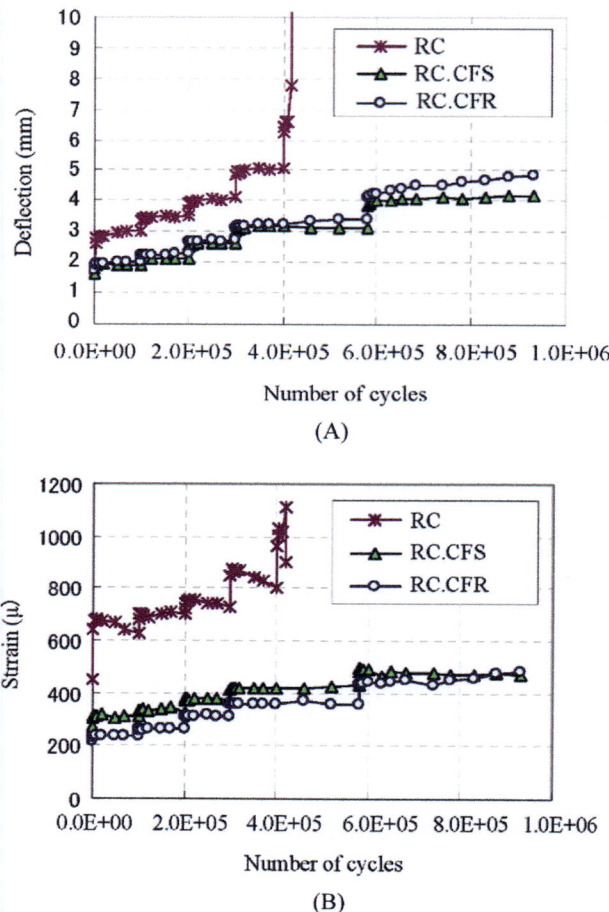

Figure 4.91 Test results (Omote et al., 2011). (A) Midspan deflection versus load cycles. (B) Midspan steel strain versus load cycles.

winding fiber sheets in the shear reinforcement direction of the rebar cut-off sections and column bases. This retrofitting prevents flexural failure by constructing flexural strengthening at the rebar cut-off sections. In addition, shear failure can also be prevented by shear strengthening at the rebar cut-off sections and column bases. Using these methods, the strength can be improved by causing flexural failure at column bases and earthquake forces can be resisted by the large deformation of columns.

4.7.3.1 Experimental verification

To verify the effectiveness of this strengthening method, the seismic behavior of the hollow structure of expressway bridge piers was investigated experimentally (Ogata and Osada, 2000). A specimen was created close to the actual structure as a

Figure 4.92 Flexural-shear failure at bridge pier mid-height due to the longitudinal reinforcement cut-offs (Priestley et al., 1996).

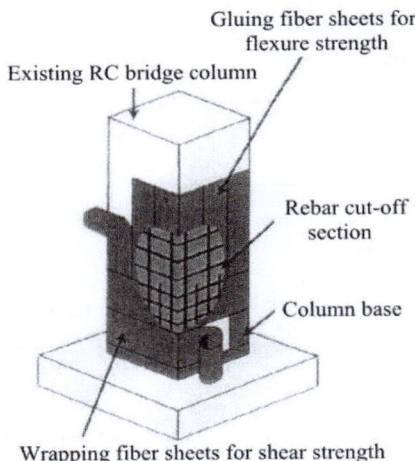

Figure 4.93 Seismic retrofitting method for bridge columns with longitudinal reinforcement cut-offs (Kobatake, 1998).

1/20-scale model. The shape and dimensions of the specimen are shown in Fig. 4.94. Details of the specimens and material properties can be found elsewhere (Ogata and Osada, 2000). A total of five specimens were fabricated. Two specimens had no additional reinforcement and three were reinforced with CFS. The area and weight of CFS for each reinforced specimen are shown in Fig. 4.95. A lateral force or lateral displacement was applied while exerting a certain axial force on the top of the column. The same loading method was adopted for all five specimens.

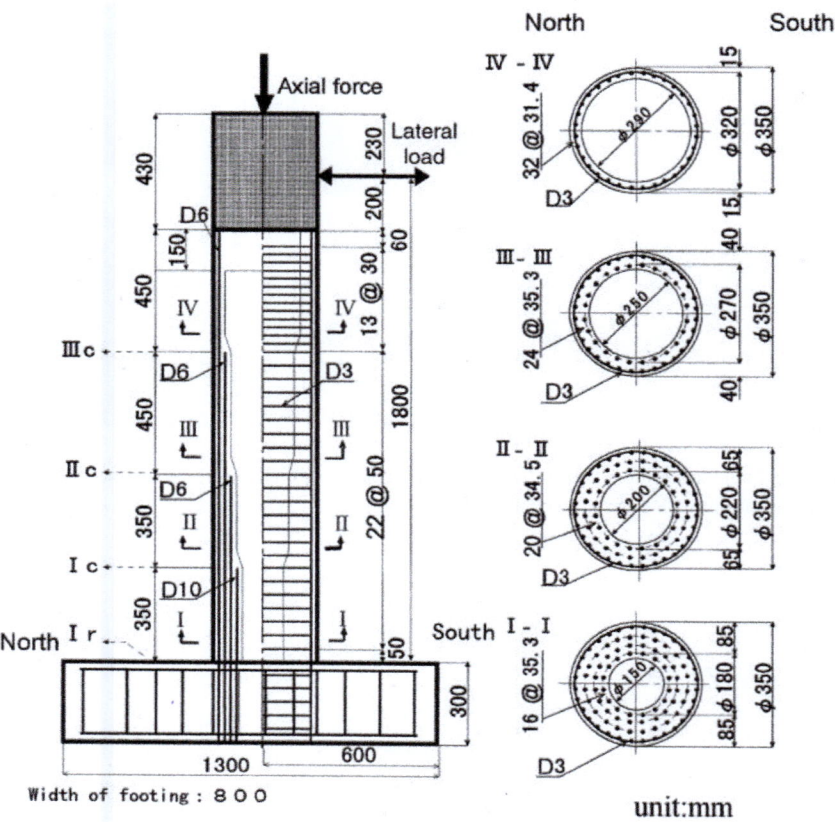

Figure 4.94 Shape of the test specimens with longitudinal reinforcement cut-offs (Ogata and Osada, 2000).

Fig. 4.96 shows the envelopes of the load—displacement curves of the specimens reinforced with CFS, together with that of a nonreinforced PS-NC-ST for comparison. As for PS-CF-ST, reinforced mainly longitudinally, flexural failure occurred near the bottom end of the CFS reinforcement in cross-section I when $-3\delta_y$ was applied, where δ_y is the yield displacement. The maximum strength was nearly identical to the calculated ultimate strength in cross-section Ir (Fig. 4.94). When $-4\delta_y$ was loaded, however, the CFS in the hoop direction ruptured at the overall height on the east and west sides, and the lateral load declined markedly. The final failure mode was shear failure, and the envelope showed a similar shape to that of PS-NC-ST, representing brittle behavior.

For specimen PS-CF2-ST with a longitudinal reinforcement content of $1/2-1/3$ of that for PS-CF-ST, and with reinforcement in the hoop direction five times as large, flexural failure occurred, in which the longitudinal reinforcement buckled at the bottom end of the CFS reinforcement in zone I, and the strength decreased. No

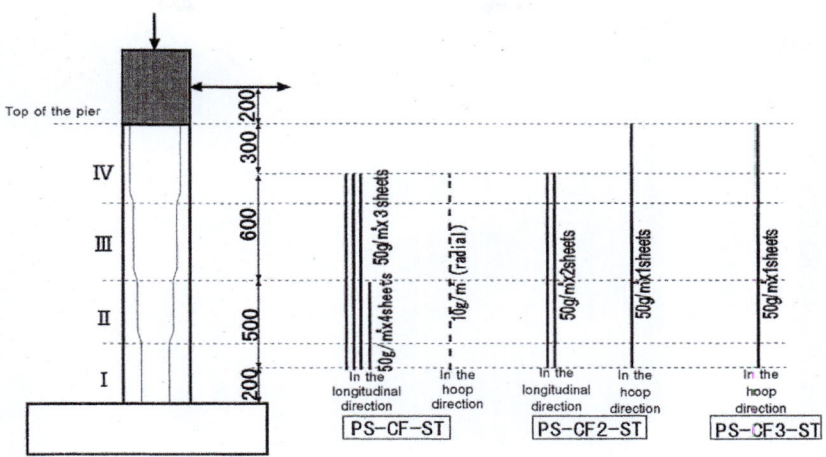

Figure 4.95 Weight and area of reinforcing carbon fiber sheet (Ogata and Osada, 2000).

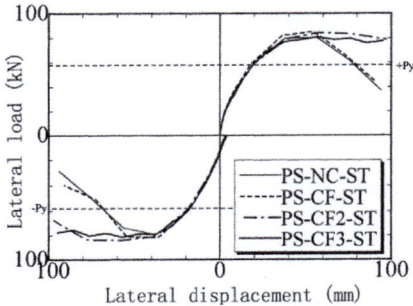

Figure 4.96 Envelope of load—displacement curve (Ogata and Osada, 2000).

shear failure was observed, and ductile behavior was found. The CFS reinforcement in the hoop direction was sufficient against shear but insufficient to increase the ductility, as it is obvious from existing studies (Osada et al., 1997). The CFS in the hoop direction finally fractured at the point where the longitudinal bars buckled. In the cases of specimens PS-CF-ST and PS-CF2-ST, which were reinforced in the longitudinal direction, the outermost longitudinal bars yielded at all of the cut-off points before the displacement of $1.5\delta_y$ as the strength exceeded the yield load by 20%−30%. No damage to the CFS, however, was confirmed at the cut-off points.

Specimen PS-CF3-ST, which was only reinforced in the hoop direction, suffered no serious damage such as shear failure or buckling of longitudinal bars, even under the ultimate loading of $-5\delta_y$, and displayed ductile behavior without any decrease in the lateral load. It was notable that the reinforcement at the cut-off points only in the hoop direction caused the cut-off points to be deformed more easily and resulted in better load-bearing behavior.

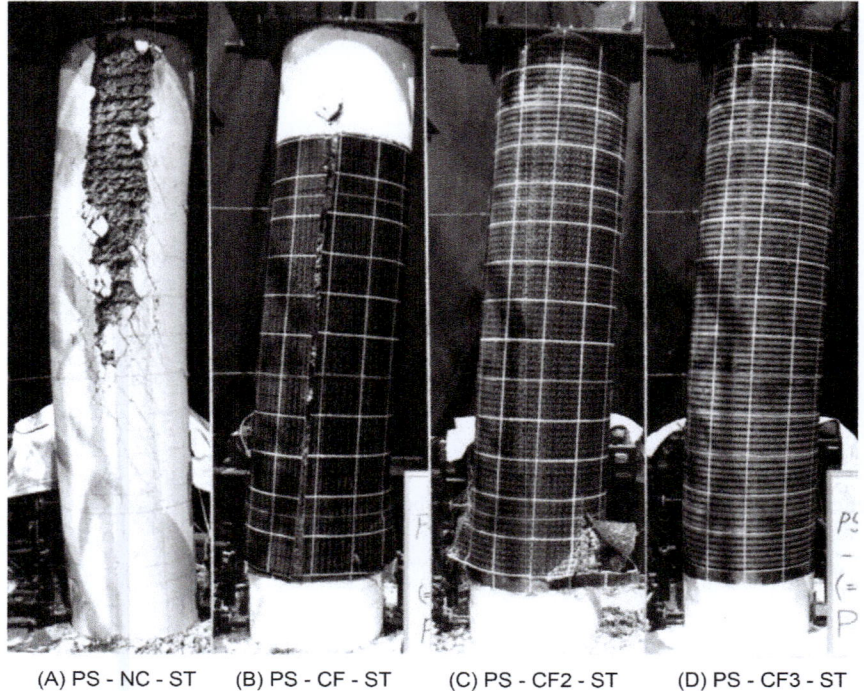

(A) PS - NC - ST   (B) PS - CF - ST   (C) PS - CF2 - ST   (D) PS - CF3 - ST

Figure 4.97 Cracking conditions at ultimate state (Ogata and Osada, 2000).

Fig. 4.97A−D show a comparison of the conditions among four specimens at
$-5\delta_y$ or under the ultimate load, except for specimen PS-N-ST. The ultimate
modes of failure were shear failure in zones III and IV for specimen PS-NC-ST,
overall shear failure following buckling of longitudinal bars at the bottom end
of CFS-reinforced portion for PS-CF-ST, flexural failure accompanied by buckling of longitudinal bars at the bottom end of CFS-reinforced portion for PS-
CF2-ST, and flexural failure at cut-off points for PS-CF3-ST. A review of the
effect of CFS on the reinforcement of the cut-off points, based on the ultimate
condition, shows that reinforcement in the longitudinal direction controlled the
growth of either flexural or diagonal cracks at the cut-off points, and thus the
response at the cut-off points was improved, and that the failure point moved to
the bottom of the column. In the case where the column was reinforced with
CFS in the hoop direction, however, it was shown that the growth of flexural
cracks from the cut-off points could not be controlled, but the growth of diagonal cracks could be prevented.

To summarize, the previous experimental results showed that longitudinally
bonded FRP over the cut-off region combined with laterally bonded FRP can effectively shift brittle shear failure back to flexural failure at the bottom of the columns.
In addition, even with a decrease in longitudinal CFS content, shear failure due to

the cut-off of reinforcement could be prevented if sufficient reinforcement against shear was applied.

References

American Concrete Institute (ACI), 2008. Guide for the design and construction of externally bonded FRP systems for strengthening concrete structures. ACI 440.2R-08, Farmington Hills, MI, USA.

American Society of Civil Engineers (ASCE), 2010, Minimum design loads for buildings and other structures. Author, ASCE/SEI 7-10. Reston, VA.

American Concrete Institute (ACI), 2011. Building code requirements for structural concrete. ACI 318M-11, Farmington Hills, MI, USA.

Atadero, A.A. and Karbhari, V.M. Calibration of resistance for reliability based design of externally bonded FRP composites, Composites B Eng. 39, 2008, 665− 679.

Bizindavyi, L., Neale, K.W., 1999. Transfer lengths and bond strengths for composites bonded to concrete. J. Compos. Constr. 3 (4), 153−160.

Breña, S.F., Bramblett, R.M., Wood, S.L. and Kreger, M.E., Increasing flexural capacity of reinforced concrete beams using carbon fiber − reinforced polymer composite, ACI Struct. J. 100, 2003, 36 − 46.

Diab, H., Wu, Z.S., Iwashita, K., 2009. Short and long-term bond performance of prestressed FRP sheet anchorages. Eng. Struct. 31 (5), 1241−1249.

El-Hacha, R., Green, M.F., Wight, R.G., 2003. Innovative system for prestressing FRP sheets. ACI Struct. J. 100, 239−252.

fib, 2001. Externally bonded FRP reinforcement for RC structures. Bulletin 14, 51−58.

Galambos, T.V., Ellingwood, B., MacGregor, J.G. and Cornell, C. A., Probability based load criteria: assessment of current design practice, J. Struct. Divi. 108, 1982, 959− 977.

GB, 2010. Technical Code for Infrastructure Application of FRP Composites. China Planning Press, Beijing, China, p. GB50608.

Harries, H., Aidoo, J., 2006. Debonding- and fatigue-related strain limits for externally bonded FRP. J. Compos. Constr. 10 (1), 87−90.

Heffernan, P. J. and Erki, M. A., Fatigue behavior of reinforced concrete beams strengthened with carbon fiber reinforced plastic laminates, J. Composit. Const 8(2), 2004, 132−140.

Japan Society of Civil Engineers (JSCE), 2001. Recommendations for upgrading of concrete structures with use of continuous fiber sheets. JSCE Concrete Engineering Series 41, Tokyo, Japan.

Kim, Y.J., Heffernan, J., 2008. Fatigue behavior of externally strengthened concrete beams with fiber-reinforced polymers: state of the art. J. Compos. Constr. 12 (3), 246−256.

Kobatake, Y., 1998. A seismic retrofitting method for existing reinforced concrete structures using CFRP. Adv. Compos. Mater. 7 (1), 1−22.

Larson, K.H., Peterman, R.J., Rashheed, H.A., 2005. Strength fatigue behavior of fiber reinforced polymer strengthened prestressed concrete T-beams. J. Compos. Constr. 9 (4), 313−326.

Meier, U., Winistorfer, A., 1995. Retrofitting of structures through external bonding of CFRP sheets. In: Proceedings of the Second International RILEM Symposium, Ghent, Belgium. E&FN Spon, London, UK, pp. 509−516.

Meier, U., Stocklin, I., Terrasi, G.P., 2001. Making better use of the strength of advanced materials in structural engineering. In: Proceedings of the International Conference on FRP Composites in Civil Engineering (CICE2001), Hong Kong, pp. 41-48.

Niu, H.D., 2002. Interfacial Fracture Mechanisms and Performance Evaluation of RC Structural System Strengthened with FRP Sheets. Doctoral thesis, Ibaraki University, Japan".

Niu, H.D., Wu, Z.S., 2002. Strengthening effects of RC flexural members with FRP sheets affected by adhesive layers. J. Appl. Mech., JSCE 5, 887−897.

Niu, H.D., Wu, Z.S., 2003. Numerical investigation on strengthened behavior of concrete structures strengthened with hybrid fiber sheets. JSCE J. Appl. Mech. 6, 1217−1226.

Niu, H.D., Wu, Z.S., 2005. Numerical analysis of debonding mechanisms in FRP-strengthened RC beams. Comput. Civ. Infrastruct. Eng. 20, 354−368.

Niu, H.D., Wu, Z.S., 2006. Effects of FRP-concrete interface bond properties on the performance of RC beams strengthened in flexure with externally bonded FRP sheets. J. Mater. Civ. Eng. 18 (5), 723−731.

Niu, H.D., Karbhari, V.M., Wu, Z.S., 2006. Diagonal macro-crack induced debonding mechanisms in FRP. Composites: Part B 37, 627−641.

Nowak, A.S. and Szerszen, M.M., 2003,Calibration of design code for buildings (ACI 318). Part 1 − Statistical models for resistance, ACI Structural Journal, 100, 377−382.

Nowak, A.S. and Collins, K.R., 2013, Reliability of structures (2nd ed.). CRC Press, Taylor & Francis Group, Boca Raton, FL.

Ogata, T., Osada, K., 2000. Seismic retrofitting of expressway bridges in Japan. Cem. Concr. Compos. 22, 17−27.

Omote, S., Mitamura, H., Watanabe, T., Matsui, S., 2011. A study about fatigue property for CFRP reinforcement of RC slab bottom. J. Struct. Eng. (A), JSCE 57A, 1273−1285.

Osada, K., Ohno, S., Yamaguchi, T., Ikeda, N., 1997. Seismic performance of reinforced concrete piers reinforced with carbon fiber sheets. Proc. Jpn. Concr. Inst. 8 (1), 189−203.

Oudah, F., El-Hacha, R., 2012. Research progress on the fatigue performance of RC beams strengthened in flexure using fiber reinforced polymers. Composites: Part B 47, 82−95.

Priestley, M.J.N., Seible, F., Calvi, G.M., 1996. Seismic Design and Retrofit of Bridges. John Wiley & Sons, New York.

Quattlebaum, J.B., Harries, K.A., Petrou, M.F., 2005. Comparison of three flexural retrofit systems under monotonic and fatigue loads. J. Bridge Eng. 10 (6), 731−740.

Said, H., Wu, Z.S., 2008. Evaluating and proposing models of predicting IC debonding failure. J. Compos. Constr. 12 (3), 284−289.

Sayed, A.M., 2014. Modeling Advancement for RC Beams Strengthened With FRP Composites. Doctoral thesis, Southeast University.

Sayed, A.M., Wang, X., Wu, Z.S., 2013. Modeling of fatigue life of FRP-concrete interface. In: Proc., 5th Int. Symp. on Innovation and Sustainability of Structures in Civil Engineering, China, pp. 245−254.

Shi, J.W., Wu, Z.S., Wang, X., Mohammad, N., 2015. Reliability analysis of intermediate crack-induced debonding failure in FRP-strengthened concrete members. Struct. Infrastruct. Eng. 11 (12), 1651−1671.

Stöcklin, I., Meier, U., 2001. Strengthening of concrete structures with pre-stressed gradually anchored CFRP strips. In: Proc., Fiber-Reinforced Polymer Reinforcement for Concrete Structures (FRPRCS-5), Cambridge, UK, 16−18 July, pp. 291−296.

Szerszen, M.M. and Nowak, A.S., Calibration of design code for buildings (ACI 318). Part 2 − Reliability analysis and resistance factors, ACI Struct. J. 100, 2003, 383− 391.

Teng, J.G., Chen, J.F., Smith, S.T., Lam, L., 2001. FRP Strengthened RC Structures. John Wiley & Sons, Chichester, UK.

Teng, J.G., Smith, S.T., Yao, J., Chen, J.F., 2003. Intermediate crack-induced debonding in RC beams and slabs. Constr. Build. Mater. 17 (6), 447−462.

Triantafillou, T.C., Deskovic, N., Deuring, M., 1992. Strengthening of concrete structures with prestressed fiber reinforced plastic sheets. ACI Struct. J. 89 (3), 235−244.

Ueda, T., Yamaguchi, R., Shoji, K., Sato, Y., 2002. Study on behavior in tension of reinforced concrete members strengthened by carbon fiber sheet. J. Compos. Constr. 6 (3), 168−174.

Wang, X., Sayed, A.M., Wu, Z., 2015. Modeling of the flexural fatigue capacity of RC beams strengthened with FRP sheets based on finite-element simulation. J. Struct. Eng. 141 (8), 04014189.

Wight, R.G., Green, M.F., Erki, M.A., 2001. Prestressed FRP sheets for post strengthening reinforced concrete beams. J. Compos. Constr. 5 (4), 214−220.

Wu, Z.S., 2007. Integrated strengthening of structures with bonded prestressed FRP reinforcement (keynote paper). In: Proceedings of Asia-Pacific Conference on FRP in Structures (APFIS 2007), Hong Kong, China, pp. 43−52.

Wu, Z.S., Niu, H.D., 2000. Study on debonding failure load of RC beams strengthened with FRP sheets. J. Struct. Eng. JSCE 46A, 1431−1441.

Wu, Z.S., Kurokawa, T., 2002. Strengthening effects and effective anchorage method for flexural members with externally bonded CFRP plates. J. Mater., Concr. Struct. Pavements, JSCE 56 (711), 1−14.

Wu, Z.S., Niu, H.D., 2007. Prediction of crack-induced debonding failure in R/C structures flexurally strengthened with externally bonded FRP composites. J. Mater., Concr. Struct. Pavements, JSCE 63 (4), 620−639.

Wu, Z.S., Matsuzaki, T., Tanabe, K., 1997. Interface crack propagation in FRP strengthened concrete structures. In: Proceedings of FRPRCS-3, Japan, pp. 319−326.

Wu, Z.S., Asakura, T., Yoshizawa, H., Hirahata, H., 1999a. Experimental study on the crack spacing of RC beam strengthened with FRP sheets. In: Proceedings of the 54th Annual Conference of the Japan Society of Civil Engineers, pp. 704−705.

Wu, Z.S., Matsuzaki, T., Yokoyama, K., Kanda, T., 1999b. Retrofitting method for reinforced concrete structures with externally prestressed carbon fiber sheet. In: Fourth International Symposium on Fiber Reinforced Polymer Reinforcement for Reinforced Concrete Structures (FRPRCS-4), ACI SP-188, pp. 751−765.

Wu, Z.S., Kurokawa, T., Yoshizawa, H., Hirahata, H., 2000. A study on debonding mechanism of bonded FRP laminates in beam members. J. Struct. Eng., JSCE 46A, 1469−1478.

Wu, Z.S., Iwashita, K., Hayashi, K., Higuchi, T., Murakami, S., Koseki, Y., 2003a. Strengthening prestressed-concrete girders with externally prestressed PBO fiber reinforced polymer sheets. J. Reinforced Plasics Compos. 22 (14), 1269−1285.

Wu, Z.S., He, W., Yin, J., Yokoyama, K., Asakura, T., 2003b. Strengthening performance of FRP sheets bonded to concrete tunnel linings. In: Proceedings of the Sixth International Symposium on FRP Reinforcement for Concrete Structures (FRPRCS-6), pp. 1157−1166.

Wu, Z.S., Sakamoto, K., Iwashita, K., Yue, Q., 2006. Hybridization of continuous fiber sheets as structural composites. J. Jpn. Soc. Compos. Mater. 32 (1), 12−21.

Wu, Z.S., Sakamoto, K., Iwashita, K., Kobayashi, A., 2007. Enhancement of flexural performances through FRP hybridization with high-modulus type carbon fibers. Struct. Eng./Earthq. Eng. JSCE 24 (2), 987−997.

Yoshitake, I., Kim, Y.J., Yumikura, K., Hamada, S., 2010. Moving-wheel fatigue for bridge decks strengthened with CFRP strips subject to negative bending. J. Compos. Constr. 14 (6), 784−790.

Further reading

Aidoo, J., Harries, K.A., Petrou, M.F., 2004. Fatigue behavior of carbon fiber reinforced polymer-strengthened reinforced concrete bridge girders. J. Compos. Constr. 8 (6), 501–509.

Kurokawa, T., Wu, Z.S., Yoshizawa, H., 1999. Experimental investigation on crack characteristics of RC beams strengthened with CFRP plate. In: Proceedings of the 54th Annual Conference of the JSCE, Part 5, pp. 702–703 (in Japanese).

Wu, Z.S., Yin, J., 2003. Fracturing behaviors of FRP-strengthened concrete structures. Eng. Fract. Mech. 70 (10), 1339–1355.

Wu, Z., Iwashita, K., Ishikawa, T., Hayashi, K., Hanamori, N., Higuchi, T., et al., 2003. Fatigue performance of RC beams strengthened with externally prestressed PBO fiber sheets. Fibre-Reinf. Polym. Reinf. Concr. Struct. (2 vols.) 885–894.

Wu, Z.S., Wang, X., Iwashita, K., Sasaki, T., Hamaguchi, Y., 2010. Tensile fatigue behaviour of FRP and hybrid FRP sheets. Compos. Part B 41, 396–402.

Yoshizawa, H., Wu, Z.S., 1999. Cracking behavior of plain concrete and reinforced concrete members strengthened with carbon fiber sheets. In: Fourth International Symposium on Fiber Reinforced Polymer Reinforcement for Reinforced Concrete Structures (FRPRCS-4), ACI SP-188, pp. 767–779.

Yoshizawa, H., Wu, Z.S., Yuan, H., Kanakubo, T., 2000. Study on FRP-Concrete Interface Bond Performance. J. Mater., Concr. Struct. Pavements, JSCE 49 (662), 105–119.

Shear and torsional strengthening of structures

5

5.1 Introduction

Additional web reinforcement (for enhancing the shear resistance of reinforced concrete (RC) beams) is a major strengthening application of fiber-reinforced polymer (FRP) composites. Shear strengthening is required when an RC element is shear deficient or when its shear capacity decreases below its flexural capacity after flexural strengthening. However, because the linear FRP differs from the nonlinear steel reinforcement, the design theory of FRP-strengthened RC members exhibits unique characteristics, particularly for shear capacity design. Substantial research has been conducted on the strengthening of RC beams with externally bonded FRP composites, in the past decade (e.g., Chajes et al., 1995; Chaallal et al., 1998; Khalifa et al., 1998; Takahashi et al.,1998; Malek et al., 1998; Triantafillou, 1998; Triantafillou and Antonopoulos, 2000; Chen and Teng, 2001a,b, 2003a,b; Täljsten, 2003; Adhikary et al., 2004; Bousselham and Chaallal, 2004; Cao et al., 2005; Ali-Ahmad et al., 2006, among others).

Research on FRP shear strengthening has been limited compared to that on FRP flexural strengthening of RC beams. However, the substantial available research has established a general understanding of structural behavior and has inspired several strength models, based on experimental observations and the corresponding theoretical assumptions. In particular, a few proposals or recommendations have been adopted in different countries for the design shear capacity of RC structures reinforced or strengthened with FRP sheets, such as those by Triantafillou and Antonopoulos (2000), Matthys and Triantafillou (2001), Deniaud and Cheng (2001a), Adhikary et al. (2004), Colotti et al. (2004), Carolin and Täljsten (2005a), Zhang and Hsu (2005), and ACI Committee 440 (2008). Recommendations on the design capacity of torsional strengthening of structures using FRP have also been proposed, such as those by Deifalla and Ghobarah (2005).

In this chapter, a state-of-the-art review of available shear and torsional strength models is presented. In addition, different failure mechanisms and factors influencing strengthening performance are presented briefly using finite element modeling. A prediction model of shear capacity developed based on numerical analyses is also proposed.

5.2 Why shear and torsion strengthening?

To utilize RC member ductility fully, it is necessary to ensure flexural rather than shear failure. Shear is a highly complex problem, because in shear failure the

Structures Strengthened with Bonded Composites. DOI: https://doi.org/10.1016/B978-0-12-821088-8.00005-9

structural element fails abruptly without sufficient warning. Shear deficiencies occur owing to several reasons, such as insufficient shear reinforcement.

Overloading of existing structures could be another reason that should motivate engineers to enhance the shear capacity of several elements. Furthermore, structures that were not designed based on provisions of the present seismic design code would be required to be strengthened against substantial impacts likely in the future (see Fig. 5.1).

5.3   What is shear and torsion strengthening with FRP?

The shear/torsion strengthening of beams with FRP can be carried out with three configurations: complete wrapping, U-wrapping, and side strips. Side strips (Fig. 5.2) can easily debond from the beam and are the least effective, whereas complete wrapping offers the highest effectiveness. However, complete wrapping is challenging in practice because the slab has to be slotted for the FRP to pass through. The U-wrapping configuration, which only requires access to the sides and bottom of a beam, is therefore more realistic for practical applications. It is noteworthy that unlike the significant advances in research on shear and flexural strengthening with FRP, the literature on torsional strengthening is limited and dates back only to 2001. Fig. 5.3 presents torsion strengthening schemes using externally bonded FRP jackets.

In the near-surface-mounting (NSM) method, FRP rods are embedded into precut grooves on the concrete cover of both sides of an RC beam. Research has

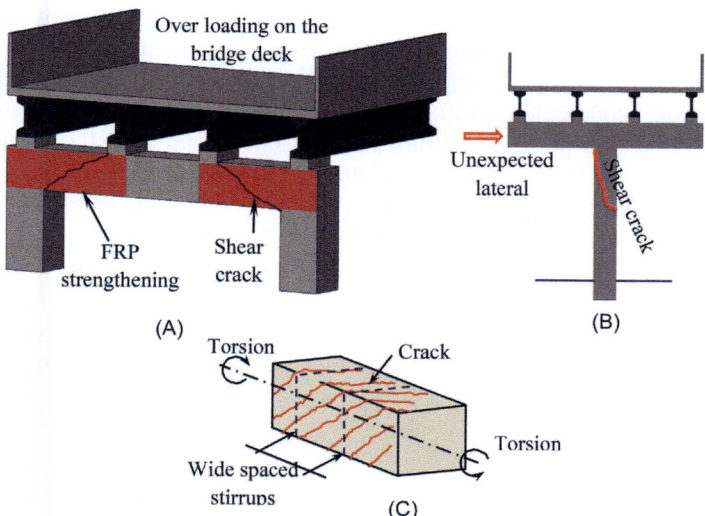

Figure 5.1 (A) Shear cracking of bridge piers due to overloading, (B) Shear cracking of bridge column due to unexpected lateral loadings, and (C) a girder with torsional deficiency.

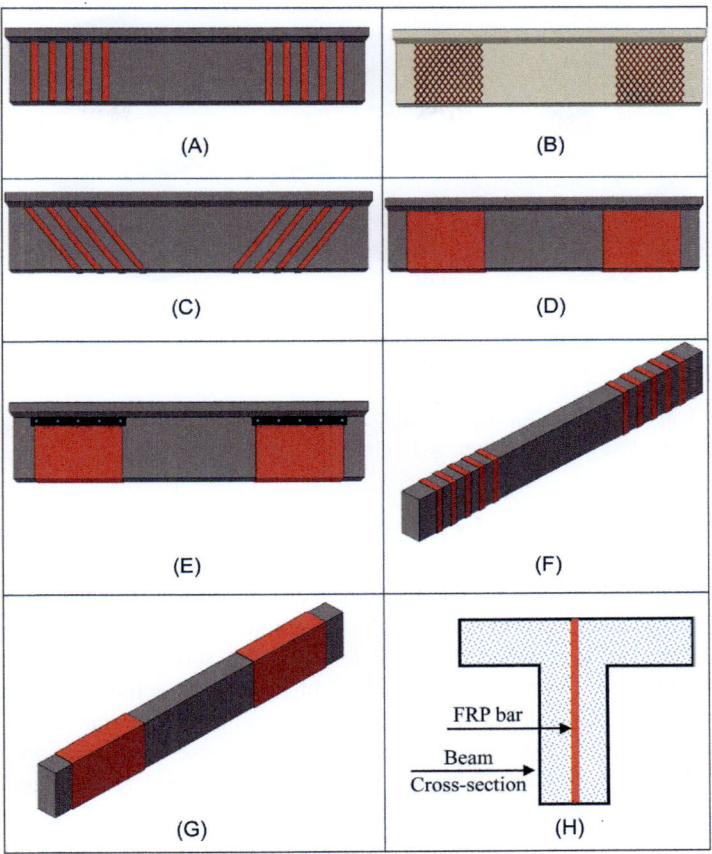

Figure 5.2 Schemes of shear strengthening using externally bonded FRP strips or sheets. (A) Side-bonded FRP strips. (B) Side-bonded FRP grid. (C) Inclined side-bonded FRP strips. (D) U-wrapped FRP sheets. (E) Anchored U-wrapped FRP sheets. (F) Fully wrapped beam with FRP strips. (G) Fully wrapped beam with FRP sheet. (H) Embedded-through-section.

established the effectiveness of this method in increasing the shear resistance of RC beams. However, the required surface and groove preparation are challenging. Moreover, debonding of FRP rods and delamination of concrete cover remain the main disadvantages of NSM methods. Another method developed recently to overcome this limitation is the embedded-through-section (ETS) method for strengthening RC beams in shear using FRP rods. In this method, FRP rods are epoxy-bonded to predrilled holes through the cross-section of the RC beam. Previous research has demonstrated the superior performance of the ETS method over the EB and NSM methods. This improvement is achieved because in the ETS method, the FRP relies on the concrete core of the RC beam, unlike the EB and NSM methods. This offers better confinement, and therefore improves bond performance.

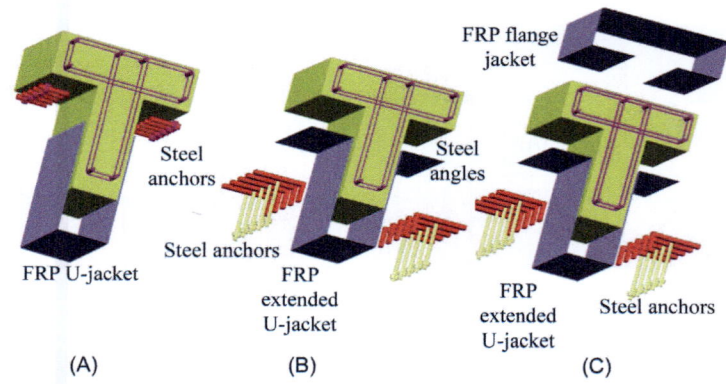

Figure 5.3 Schemes of torsion strengthening using externally bonded FRP jackets.
(A) U-jacket. (B) Extended U-jacket. (C) Full wrapping.

5.4 Failure modes of shear strengthened beams

Many research studies have established that the shear capacity of RC beams can be increased using externally bonded FRP strengthening techniques. FRP reinforcement configuration involves the (1) selection of the surfaces to be bonded (side bonding, U-wrap, or complete wrap), (2) selection of the orientation of the primary direction of fibers, (3) selection between continuous reinforcement and a series of discrete strips, and (4) selection between use and nonuse of an anchorage system (presented in Tables 5.1 and 5.2). In addition, the observed failure modes of RC beams strengthened with bonded FRP grid are listed in Table 5.3. The strengthening schemes have been tested experimentally under the effect of static loadings. Several failure modes have been observed: debonding or peeling of the FRP composites, shear cracking, and flexural failure. The observed failure modes were experienced by all the proposed strengthening schemes, as illustrated in Tables 5.1–5.3. The authors here cannot attribute any of the observed failure modes to a specific type of the proposed strengthening methods. It is generally concluded that all the adopted techniques could shift the shear failure mode to flexural failure and that FRP debonding can be delayed by installing mechanical anchors. However, the effectiveness of the strengthening method has been observed to depend on the failure mode. When strips or sheets are attached to the sides of a beam web, the debonding, which is the main cause of failure of such systems (Baggio et al., 2014; Khalifa and Nanni, 2000; Khalifa and Nanni, 2002; Pham and Hao, 2016; Ebead and Saeed, 2013; Islam et al., 2005; Panigrahi et al., 2014; Mofidi and Chaallal, 2011; Baghi and Barros, 2016; Mofidi and Chaallal, 2011; Norris et al., 1997; Foster et al., 2016), is observed to limit the efficiency of the corresponding strengthening method. The utilization of U-shaped layers of FRP composites improves the bond characteristics and further enhances the RC beam shear capacity. This is attributed to the fact that a better anchorage is achieved when the relatively ductile U-shaped

Table 5.1 Strips.

Side bonded	
Without anchor	**With anchor**
Shear cracking Chaallal et al. (1998) Al-Rousan et al. (2015) Debonding of CFRP strips  Khalifa and Nanni (2000), Islam et al. (2005), Ebead and Saeed (2013), Panigrahi et al. (2014) Flexural failure Sundarraja and Rajamohan (2009)	– Detachment combined with tension steel yielding 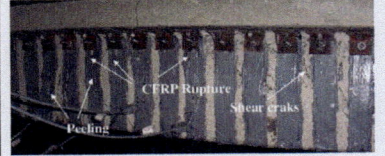 Ebead and Saeed (2013) –
U-Jacket	
Without anchor Debonding  Khalifa and Nanni (2002), Khalifa and Nanni (2000), Mofidi and Chaallal (2011), Mofidi et al. (2013), Panigrahi et al. (2014), Baggio et al. (2014), Pham and Hao (2017) Shear failure Anil (2008) Baggio et al. (2014) Flexure (concrete crushing) 	With anchor Peeling of CFRP straps and CFRP ruptured Anil (2008) Debonding (Mofidi et al., 2013) For partially embedded Shear compression Baggio et al. (2014) (GFRP reinforcement) Flexural

(Continued)

Table 5.1 (Continued)

Side bonded	
Without anchor	**With anchor**
Baggio et al. (2014)	Baggio et al. (2014) (CFRP reinforcement) Mofidi et al. (2013) (For fully embedded)
Full wrap	
Without anchor FRP rupture Cao et al. (2005), Teng et al. (2009), Li and Leung (2015)	With anchor Shear (PBO delamination) Trapko et al. (2015)

FRP composites are used. A few researchers adopted externally bonded systems with anchors to mitigate the debonding problem (Sundarraja and Rajamohan, 2009; Ebead and Saeed, 2013; Panigrahi et al., 2014; Mofidi and Chaallal, 2011; Anil, 2008). Consequently, the failure mode changed to either FRP rupture or flexural failure. Ultimately, other researchers (Cao et al., 2005; Teng et al., 2009; Li and Leung, 2015) examined the complete wrapping technique. It was observed that the failure mode was rupture of the FRP composites.

5.5 Fundamental mechanism for FRP retrofitting

5.5.1 Side bonded and U-jacket techniques

Most of the studies (Norris et al., 1997; Khalifa and Nanni, 2002; Bousselham and Chaallal, 2008; Teng et al., 2009; Mofidi and Chaallal, 2011; Mofidi et al., 2013; Baggio et al., 2014; Li and Leung, 2015; Pham and Hao, 2017; Baghi and Barros, 2016 among others) reported that before the initiation of the first shear crack, the contribution of the transverse steel reinforcement and the different FRP strengthening techniques are negligible. The formation of the first diagonal shear crack triggers the strain and the contribution of both the transverse steel reinforcement and FRP. In their experiments on RC beams reinforced with carbon FRP (CFRP) U-jackets, Bousselham and Chaallal (2008) reported that the maximum strain in the FRP jackets was larger in the middle of the failure zone, where the first diagonal cracks formed and propagated simultaneously toward the support and the compression flange. They also observed that the strain distribution in the CFRP over the failure zone was nonuniform. It tended to be high in the middle and low at the ends of the zone. The strains at the end of the critical zone near the load were highly marginal compared to those at the other end. This distribution was caused by the short anchorage length of the FRP jackets at the upper portion of the crack near the load, which caused this area to be vulnerable to debonding. As the load increased (after diagonal crack formation), the strains in the transverse steel reinforcement

Table 5.2 Sheets or plates.

Side bonded	
Without anchor Shear cracking, and detachment of the concrete cover  Pellegrino and Modena (2002) Separation of the fibers Norris et al. (1997) Debonding Baghi and Barros (2016) Panigrahi et al. (2014) Flexural failure Kachlakev and McCurry (2000)	**With anchor** Shear failure 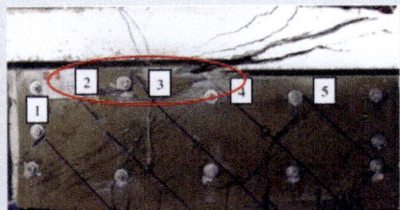 Baghi and Barros (2016) Peeling of the FRP plate Anil (2008) —
U-Jacket	
Without anchor Debonding Khalifa and Nanni (2002) Khalifa et al. (2000) Khalifa and Nanni (2000) Foster et al. (2016) Panigrahi et al. (2014) Mofidi et al. (2013) — Crushing of the concrete Islam et al. (2005) Bousselham and Chaallal. (2008)	**With anchor** Fabric separation/shear Foster et al. (2016) Shearing of CFRP plate Anil (2008)  —  FRP rupture Panigrahi et al. (2014)

Table 5.3 Grid.

Side bonded
Without anchor Separation of the concrete side covers preceded by multiple shear cracks Awani et al. (2016) Splitting of the bar fibers Islam et al. (2005)

and the FRP increased. However, the contribution of the transverse steel to the shear resistance was more significant than that of the FRP. Because the transverse steel was internal, a stronger bond developed at the stirrup—concrete interface compared to that at the surface-bonded FRP—concrete interface. Unlike the contributions of the FRP and transverse steel, that of concrete was effective from the first application of the loading. The contribution of concrete increased as the load increased, until it attained a value that matched the corresponding value at the occurrence of the first diagonal crack. Another difference between the contributions of the transverse steel reinforcement and FRP from one side and that of concrete from the other side was that the former increased as the load increased, whereas the latter remained unchanged.

Bousselham and Chaallal (2008) stated that the contribution of concrete at failure was equal to that at the first diagonal crack formation. At higher loading levels, the transverse steel reinforcement may yield if the debonding of the FRP is prevented. In this case, the transverse steel reinforcement yields gradually as the stirrup located in the critical section yields first, followed by the other adjacent stirrups. At this stage, the contribution of the transverse steel reinforcement becomes constant, and the FRP strain increases abruptly up to failure. Meanwhile, premature FRP debonding failure prevents the complete utilization of the transverse steel reinforcement.

Mofidi and Chaallal (2011) demonstrated that FRP U-jacket strips exhibit superior performance compared to a continuous FRP U-jacket. They explained that in a strengthened beam with FRP strips, two adjacent FRP strips can perform independently in the direction perpendicular to the major load-carrying direction of the FRP fibers. That is, local debonding in a FRP strip does not affect the performance of adjacent strips. However, in the continuous FRP U-jacket, the longitudinal tensile force in the U-jacket sheet facilitates a progressive debonding process in the U-jacket once local debonding has occurred in a particular FRP band.

5.5.2 Complete wrapping technique

Investigating the effect of the shear span—depth ratio on the behavior of RC beams fully wrapped with CFRP strips, Li and Leung (2015) reported that the FRP shear contribution increased as the shear span—depth ratios increased. The shear span—depth ratio affected the FRP shear contribution because the FRP strips played different roles in resisting shear under various failure modes. For a low shear span—depth ratio, the concrete arch resisted most of the applied load. Furthermore, the CFRP strips functioned as hoops to restrict the deformation of the beams in the vertical direction, which varied gradually until the shear capacity was attained. Hence, the FRP shear contribution was moderate. As the span—depth ratio increased, the effect of the arch action decreased and eventually became negligible. When shear cracks propagated, the concrete in the web started to lose its load-carrying capacity. Hence, the FRP strips could be utilized more effectively.

The angle of the inclined crack affected the effectiveness of the FRP strips in resisting shear. For small shear span—depth ratios, the angle of the inclined crack

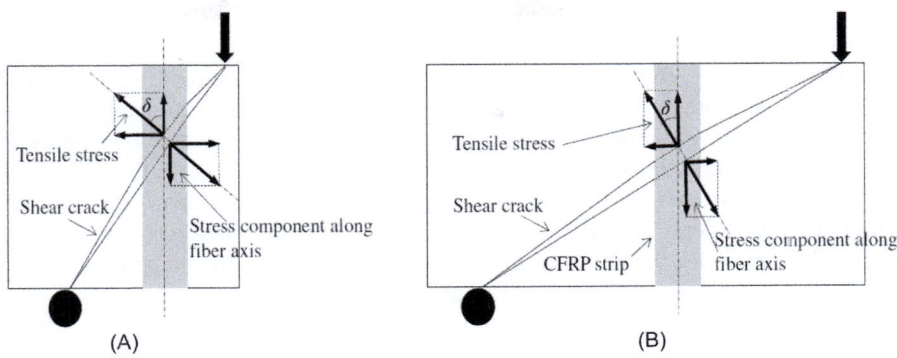

Figure 5.4 Effect of angle of critical shear crack on CFRP shear contribution: (A) low shear span—depth ratio; (B) high shear span—depth ratio (Li and Leung, 2015).

with respect to the fiber orientation was larger (Fig. 5.4A). Therefore the stress along the fiber direction was less effective in controlling the opening and propagation of the shear crack. As the shear span—depth ratios increased, the FRP shear contribution increased because the fibers were lying at a smaller angle to the opening direction of the crack (Fig. 5.4B).

Teng et al. (2009) explored the strengthening of RC beams with bonded and unbonded completely wrapped CFRP strips. For the beams strengthened with unbonded FRP strips, the strain in many of the FRP strips increased gradually after shear cracking. Meanwhile, for the beams with bonded FRP strips, the strains in only a few locations of the FRP strips, which were intersected by a crack, first displayed apparent increases. Once FRP debonding propagated to positions away from the crack, the FRP strains at these locations increased abruptly to a value nearly equal to that of the cracked position. The authors also state that certain adverse shear interaction between the FRP contribution and concrete contribution occurred in an RC beam strengthened with FRP wraps, near the ultimate state of the strengthened section. This was because the maximum crack width at failure could be large enough to compromise several of the concrete's shear-resisting mechanisms such as dowel action, aggregate interlock, or shear friction along the shear crack. Consequently, the concrete contribution at the ultimate state is less than that at the formation of the first shear crack.

5.6 Experimental investigation of parameters influencing shear capacities

The success of fibers in increasing the shear resistance of concrete beams through the application of many reinforcement methods has invoked several researchers to develop models based on a theoretical foundation or statistical evaluation of experimental results available from literature, to determine the shear resistance of RC beams after reinforcement. Consequently, design codes have furnished models considering the available observations to determine the shear resistance of FRP-strengthened beams. However, the available models cannot estimate the contribution of FRP reinforcements accurately owing to their deficiency in accounting for all the parameters influencing shear capacities. For example, available models of shear capacity omit the effect of the beam width, concrete strength, and shear span—depth ratio (*a/d*) on ultimate shear behavior. More importantly, they always regard the beam depth and FRP sheet height as identical. Therefore the ultimate shear behavior of RC beams strengthened with FRP should be studied further by considering the aforementioned factors and parameters.

A series of experimental data on strengthening effects of different FRP composites are available in the literature. The database considered in this section is composed of the results of 274 experimental tests. These include 87 beams strengthened by FRP side bonding, 115 beams strengthened by FRP U-jacketing, and 72 beams strengthened by FRP complete wrapping. The collected tests differ in terms of geometry, concrete strength, shear span—depth ratio, and reinforcement ratios, with a wide range of geometrical and mechanical characteristics. The main geometric and material properties of the experimental data and the failure modes described earlier are provided in Table 5.4.

5.7 Numerical investigation of parameters influencing shear capacities

In this part, a three-dimensional (3D) finite element (FE) simulation by Sayed (2014) to predict the ultimate shear behavior of FRP-strengthened RC beams is introduced. The parametric study was conducted to clarify their influence on shear capacity. Based on the study, a prediction model was proposed by considering all the common parameters that influence the ultimate shear capacity of a strengthened beam, including beam width, concrete strength, effective height of the beam, shear span—depth ratio, FRP thickness, height of the FRP sheet, strengthening configuration (completely wrapped, U-jacketing, and side bonding), and the elastic modulus of the FRP reinforcement. Furthermore, a database of 274 RC beams was retrieved from the available literature to examine the performance of the proposed model as well as the commonly used models and design guidelines.

5.7.1 Material modeling and properties

5.7.1.1 Concrete

A commercial FE program ANSYS is adopted for the numerical analysis. The Solid 65 element is used to model the concrete. This element has eight nodes with three degrees of freedom at each node in the nodal *x*-, *y*-, and *z*-directions. This element is capable of plastic deformation, cracking in three orthogonal directions, and being

Table 5.4 Properties of the tested beams and the results from the experimental studies (Sayed, 2014).

References	Beam specimen	Geometric data				External shear reinforcement						$V_{R,Exp}$ (kN)	Failure mode Exp.
		b (mm)	d (mm)	a (mm)	f'_c (MPa)	t_f (mm)	w_f (mm)	h_f (mm)	S_f (mm)	E_f (GPa)	Type		
Abdel-Jaber et al. (2003)	B6	150	165	400	47.2	0.270	20	200	50	230	C.2s	55.5	BF
	B9	150	165	400	52.4	0.270	20*	200	50	230	C.2s	58.0	BF
	B10	150	165	400	52.2	0.270	21*	200	60	230	C.2s	65.0	BF
	B12	150	165	400	47.5	0.270	1	200	1	230	C.2s	97.5	TR
	B13	150	165	400	43.0	0.270	1	200	1	230	C.2s	94.0	TR
Adhikary et al. (2004)	B-3	150	170	510	33.5	0.330	1	200	1	230	C.2s	63.6	TR
	B-4	150	170	510	31.5	0.170	1	150	1	230	C.2s	58.6	TR
Sharif et al. (1994)	SO	150	113	400	37.7	3.000	20	150	50	16	C.2s	41.5	TR
	WO	150	113	400	37.7	3.000	1	120	1	16	C.2s	42.0	TR
	SP	150	113	400	37.7	3.000	1	150	1	16	C.2s	41.2	TR
	WP	150	113	400	37.7	3.000	1	120	1	16	C.2s	45.2	TR
Allam and Ebeido (2003)	W2	120	175	450	24.0	0.260	1	200	1	230	C.2s	90.0	FL
	W3	120	175	450	24.0	0.260	1	200	1	230	C.2s	87.5	BF
	W4	120	175	450	24.0	0.260	1	200	1	230	C.2s	86.3	BF
	W-1	120	175	300	24.0	0.260	1	200	1	230	C.2s	115.0	NA
	S1	120	175	300	24.0	0.26	50	200	100	230	C.2s	100.0	BF
Beber and Campos Filho (2005)	V9-B	150	253	740	32.8	0.110	50	300	100	230	C.2s	104.3	BF
	V12-B	150	253	740	32.8	0.110	50*	300	100	230	C.2s	91.7	BF
	V14-B	150	253	740	32.8	0.110	50*	300	100	230	C.2s	91.7	BF
	V13-A	150	253	740	32.8	0.110	1	300	1	230	C.2s	122.0	BF
	V13-B	150	253	740	32.8	0.110	1	300	1	230	C.2s	125.8	BF
	V14-A	150	253	740	32.8	0.110	1	300	1	230	C.2s	128.4	BF
	V15-A	150	253	740	32.8	0.110	1	300	1	230	C.2s	120.6	BF
	V21-A	150	253	740	32.8	0.110	50	300	100	230	C.2s	115.2	BF
Bukhari et al. (2010)	C2	152	267	838	44.0	0.340	1	305	1	235	C.2s	115.4	BF
	C3	152	267	838	44.0	0.340	1	305	1	235	C.2s	127.0	BF
	C6	152	267	838	44.0	0.340	1	305	1	235	C.2s	144.3	BF
	D6	152	267	838	35.2	0.340	1	305	1	235	C.2s	138.5	BF

(Continued)

Table 5.4 (Continued)

References	Beam specimen	Geometric data				External shear reinforcement							Failure mode
		b (mm)	d (mm)	a (mm)	f_c (MPa)	t_f (mm)	w_f (mm)	h_f (mm)	S_f (mm)	E_f (GPa)	Type	Vn_{Exp} (kN)	Exp.
Carolin et al. (2005)	A145	180	430	1250	67.0	0.070	1	500	1	234	C.2s	247.0	TR
	A245a	180	430	1250	71.0	0.220	1	500	1	234	C.2s	257.0	BF
	A245b	180	430	1250	53.0	0.220	1	500	1	234	C.2s	305.0	BF
	A245Ra	180	430	1250	67.0	0.220	1	500	1	234	C.2s	306.0	BF
	A245Rb	180	430	1250	47.0	0.220	1	500	1	234	C.2s	251.0	BF
	A290a	180	430	1250	59.0	0.220	1	500	1	234	C.2s	256.0	BF
	A290b	180	430	1250	52.0	0.220	1	500	1	234	C.2s	298.0	BF
	A345	180	430	1250	71.0	0.510	1	500	1	234	C.2s	334.0	BF
	B290	180	330	1000	46.0	0.220	1	500	1	234	C.2s	298.0	BF
	B390	180	330	1000	46.0	0.510	1	400	1	234	C.2s	291.0	BF
Chaallal et al. (1998)	RS90a	150	215	600	35.0	1.000	50	250	100	150	C.2s	87.5	BF
	RS90b	150	215	600	35.0	1.000	50	250	100	150	C.2s	95.0	BF
	RS135a	150	215	600	35.0	1.000	50	250	150	150	C.2s	94.0	BF
	RS135b	150	215	600	35.0	1.000	50	250	150	150	C.2s	99.5	BF
Grande et al.(2007)	RS3s	250	405	1400	21.0	0.190	1	450	1	392	C.2s	270.0	BF
	RS2s	250	405	1400	21.0	0.190	1	450	1	392	C.2s	280.0	BF
Kamiharako et al. (1997)	4	250	400	1000	32.6	0.110	40	500	100	244	A.2s	183.9	BF
	5	250	400	1000	32.6	0.170	40	500	100	90	C.2s	180.9	BF
Khalifa and Nanni (2000)	C-BT5	150*	357	1070	35.0	0.330	50	305	125	228	C.2s	121.5	BF
Lim (2010)	CSF-12-N	200	272	835	40.0	1.200	50	300	125	165	C.2s	126.6	BF
	SCF-15-N	200	272	835	40.0	1.200	50	300	150	165	C.2s	124.2	BF
	SCF-25-N	200	272	835	40.0	1.200	50	300	250	165	C.2s	111.8	BF
Pellegrino and Modena (2002)	TR30D10	150	250	750	31.4	0.33	1	300	1	234	C.2s	193.0	BF
	TR30D2	150	250	750	31.4	0.495	1	300	1	234	C.2s	213.3	BF
	TR30D20	150	250	750	31.4	0.495	1	300	1	234	C.2s	247.5	BF
	TR30D3	150	250	750	31.4	0.165	1	300	1	234	C.2s	161.4	BF
	TR30D4	150	250	750	31.4	0.330	1	300	1	234	C.2s	208.8	BF
	TR30D40	150	250	750	31.4	0.33	30	300	1	234	C.2s	212.0	BF
Sato et al. (1996)	s2	200	260	700	45.2	0.220	30	300	60	230	C.2s	160.6	BF
	s4	200	260	700	37.5	0.110	1	300	1	230	C.2s	156.4	BF
Täljsten (2003)	RC1	180	430	1250	67.4	0.070	1	500	1	234	C.2s	306.1	TR
	C1	180	430	1250	67.4	0.110	1	500	1	234	C.2s	246.7	TR
	C2	180	430	1250	71.4	0.110	1	500	1	234	C.2s	257.2	TR
	C3	180	430	1250	58.7	0.110	1	500	1	234	C.2s	260.6	TR
	C5	180	430	1250	71.4	0.165	1	500	1	234	C.2s	334.3	TR

Reference	ID												
Triantafillou (1998)	S1A	70	100	320	30.0	0.160	30	110	60	235	C.2s	21.8	BF
	S1B	70	100	320	30.0	0.160	30	110	60	235	C.2s	19.5	BF
	S145	70	100	320	30.0	0.160	30*	110	60	235	C.2s	22.3	BF
	S2A	70	100	320	30.0	0.160	45	110	60	235	C.2s	24.1	BF
	S2B	70	100	320	30.0	0.160	45	110	60	235	C.2s	21.1	BF
	S245	70	100	320	30.0	0.160	45*	110	60	235	C.2s	23.7	BF
Uji (1992)	6	100	170	425	27.0	0.100	1	200	1	230	C.2s	45.5	BF
	7	100	170	425	27.0	0.140	1	200	1	230	C.2s	58.0	BF
Zhang et al. (2004)	Z11-S90	102	203	381	42.5	1.200	40	229	102	165	C.2s	84.7	BF
	Z11-S45	102	203	381	42.5	1.200	40*	229	102	165	C.2s	97.0	BF
	Z22-S90	102	203	254	42.5	1.200	40	229	102	165	C.2s	104.5	BF
	Z22-S45	102	203	254	42.5	1.200	40	229	102	165	C.2s	121.4	BF
	Z31-F90	102	203	381	42.5	0.330	1	229	1	73	C.2s	77.5	BF
	Z31-FD	102	203	381	42.5	0.660	1	229	1	73	C.2s	87.7	BF
	Z42-F90	102	203	254	42.5	0.330	1	229	1	73	C.2s	128.8	BF
	Z42-FD	102	203	254	42.5	0.660	1	229	1	73	C.2s	133.4	BF
	ZC4-R45	102	203	533	43.9	1.200	40	229	102	238	C.2s	59.8	BF
	Z11-SCR45	203	203	254	42.5	1.200	40*	229	102	238	C.2s	133.4	BF
	Z22-SCR45	203	203	381	42.5	1.200	40*	229	102	238	C.2s	82.7	BF
	Z31-FCRU	203	203	381	42.5	0.330	1	229	1	73	C.2s	107.4	BF
Zhang and Hsu (2005)	Z4-90	152	195	533	43.9	1.500	40	229	127	165	G.2s	73.7	BF
	Z4-45	152	195	533	43.9	1.500	40*	229	127	165	G.2s	82.8	BF
	Z6-90	152	195	533	43.9	1.500	40	229	127	165	G.2s	63.9	BF
	Z6-Mid	152	195	533	43.9	0.330	1	229	1	73	C.2s	60.3	TR
Adhikary et al. (2004)	C-1	300	245	1000	37.2	0.170	1	300	1	230	C.3s	165.0	TR
	C-2	300	245	1000	41.0	0.170	1	300	1	230	C.3s	228.5	TR
	C-3	300	245	1000	41.1	0.170	1	300	1	230	C.3s	237.5	TR
	A-1	300	245	1000	39.6	0.290	1	300	1	120	A.3s	155.0	TR
	A-2	300	245	1000	41.8	0.290	1	300	1	120	A.3s	200.0	TR
	A-3	300	245	1000	43.9	0.290	1	300	1	120	A.3s	245.0	TR
Adhikary et al. (2004)	B-7	150	170	510	34.4	0.165	1	150	1	230	C.3s	68.5	NA
	B-8	150	170	510	35.4	0.165	1	200	1	230	C.3s	85.8	NA
Allam and Ebeido (2003)	US-2	120	175	450	24.0	0.260	50	200	100	230	C.3s	87.5	BF
	US-3	120	175	450	24.0	0.200	50	200	100	230	C.3s	86.3	BF
	US-4	120	175	450	24.0	0.260	50	200	100	230	C.3s	85.0	BF
	UW-2	120	175	450	24.0	0.260	1	200	1	230	C.3s	90.0	FL
	UW-3	120	175	450	24.0	0.260	1	200	1	230	C.3s	88.8	FL
	UW-4	120	175	450	24.0	0.260	1	200	1	230	C.3s	88.8	FL
	US-1	120	175	300	24.0	0.260	50	200	100	230	C.3s	115.0	NA
	UW-1	120	175	300	24.0	0.260	1	200	1	230	C.3s	115.0	NA

(Continued)

Table 5.4 (Continued)

References	Beam specimen	Geometric data				External shear reinforcement						Vn_{Exp} (kN)	Failure mode Exp.
		b (mm)	d (mm)	a (mm)	f_c (MPa)	t_f (mm)	w_f (mm)	h_f (mm)	S_f (mm)	E_f (GPa)	Type		
Altin (2009)	2	120*	330	1650	25.2	0.120	50	285	125	231	C,3s	101.4	BF
	3	120*	330	1650	24.9	0.120	50	285	150	231	C,3s	91.6	BF
	4	120*	330	1650	24.8	0.120	50	285	200	231	C,3s	84.6	BF
	5	120*	330	1650	24.8	0.120	50	285	125	231	C,3s	87.7	FL
	6	120*	330	1650	25.0	0.120	50	285	150	231	C,3s	86.3	FL
	7	120*	330	1650	24.9	0.120	50	285	200	231	C,3s	85.6	FL
Beber and Campos Filho (2005)	V10-A	150	253	740	32.8	0.110	50	300	100	230	C,3s	107.5	BF
	V10-B	150	253	740	32.8	0.110	50	300	100	230	C,3s	106.0	BF
	V17-A	150	253	740	32.8	0.110	50	300	100	230	C,3s	102.8	BF
	V11-A	150	253	740	32.8	0.110	50	300	100	230	C,3s	98.4	BF
	V11-B	150	253	740	32.8	0.110	50	300	100	230	C,3s	124.8	BF
	V17-B	150	253	740	32.8	0.110	50	300	100	230	C,3s	92.9	BF
	V19-A	150	253	740	32.8	0.110	50*	300	100	230	C,3s	118.4	BF
	V19-B	150	253	740	32.8	0.110	50*	300	100	230	C,3s	115.1	BF
	V15-B	150	253	740	32.8	0.110	1	300	1	230	C,3s	138.4	BF
	V16-B	150	253	740	32.8	0.110	1	300	1	230	C,3s	112.4	BF
Bousselham and Chaalal (2006)	DB-S0-05L	152*	350	525	25.0	0.050	1	284	1	231	C,3s	134.1	NA
	DB-S0-1L	152*	350	525	25.0	0.110	1	284	1	231	C,3s	142.8	NA
	DB-S0-2L	152*	350	525	24.0	0.210	1	284	1	231	C,3s	144.5	NA
	DB-S1-2L	152*	350	525	25.0	0.210	1	284	1	231	C,3s	357.7	NA
	DB-S2-2L	152*	350	525	24.0	0.110	1	284	1	231	C,3s	389.7	NA
	DB-S2-1L	152*	350	525	24.0	0.210	1	284	1	231	C,3s	404.8	NA
	SB-S0-05L	152*	350	525	24.0	0.050	1	284	1	231	C,3s	102.4	NA
	SB-S0-1L	152*	350	525	24.0	0.110	1	284	1	231	C,3s	120.0	NA
	SB-S0-2L	152*	350	525	24.0	0.210	1	284	1	231	C,3s	121.7	NA
	SB-S1-05L	152*	350	525	25.0	0.050	1	284	1	231	C,3s	282.0	NA
	SB-S1-1L	152*	350	525	24.0	0.110	1	284	1	231	C,3s	255.0	NA
	SB-S1-2L	152*	350	525	24.0	0.210	1	284	1	231	C,3s	267.0	NA
	SB-S2-1L	152*	350	525	24.0	0.110	1	284	1	231	C,3s	309.4	NA
Deniaud and Cheng (2001b)	T6NS-C45	140*	540	1480	44.1	0.110	50*	450	100	230	C,3s	213.6	BF
	T6S4-C90	140*	540	1480	44.1	0.110	50	450	100	230	C,3s	272.8	BF
	T6S2-C90	140*	540	1480	44.1	0.110	50	450	100	230	C,3s	309.8	BF

References	Beam specimen	Geometric data				External shear reinforcement						Vn_{Exp} (kN)	Proposed new model		Failure mode Exp.
		b (mm)	d (mm)	a (mm)	f'_c (MPa)	t_f (mm)	w_f (mm)	h_f (mm)	S_f (mm)	E_f (GPa)	Type		Vn (kN)	$\frac{V_{Exp}}{V_{Pred}}$	
Diagana et al. (2003)	PU1	130	425	900	38.0	0.430	40	450	200	105	C,3s	142.5	156.1	0.91	BF
	PU2	130	425	900	38.0	0.430	40	450	250	105	C,3s	130.0	149.4	0.87	BF
	PU3	130	425	900	38.0	0.430	40*	450	300	105	C,3s	154.5	144.3	1.07	BF
	PU4	130	425	900	38.0	0.430	40*	450	300	105	C,3s	150.0	144.3	1.04	BF
Gamino et al. (2010)	VTC1	120*	265	700	60.0	0.110	50	220	150	221	C,3s	110.0	114.3	0.96	TR
	VTC2	120*	265	700	59.0	0.110	60	220	150	219	C,3s	152.5	126.1	1.21	BF
	VTC3	120*	265	700	60.0	0.170	50	220	175	219	C,3s	121.5	118.6	1.02	TR
	VTC4	120*	265	700	57.0	0.110	50	220	175	221	C,3s	137.5	108.7	1.26	TR
	VTC5	120*	265	700	55.0	0.110	50	220	150	221	C,3s	103.5	110.2	0.94	TR
Grande et al. (2007)	RS3U	250	405	1400	21.0	0.190	1	450	1	392	C,3s	270.0	317.3	0.85	BF
	RS2U	250	405	1400	21.0	0.190	1	450	1	392	C,3s	280.0	351.4	0.80	BF
Islam et al. (2005)	B1	120	720	600	37.8	0.330	1	800	1	230	C,3s	701.0	648.8	1.08	BF
Khalifa and Nanni (2000)	A-SW3-2	150	253	760	19.3	0.330	1	305	1	228	C,3s	177.0	123.3	1.44	NA
	A-SW4-2	150	253	1020	19.3	0.330	1	305	1	228	C,3s	180.5	229.0	0.79	NA
	A-SO3-2	150	253	760	27.5	0.165	50	305	125	228	C,3s	131.0	100.9	1.30	NA
	A-SO3-3	150	253	760	27.5	0.165	75	305	125	228	C,3s	133.5	111.9	1.19	BF
	A-SO3-4	150	253	760	27.5	0.165	1	305	1	228	C,3s	144.5	128.4	1.13	BF
	A-SO3-5	150	253	760	27.5	0.165	1	305	1	228	C,3s	169.5	128.4	1.32	BF
	A-SO4-2	150	253	1020	27.5	0.165	50	305	125	228	C,3s	127.5	94.32	1.35	NA
	A-SO4-3	150	253	1020	27.5	0.165	1	305	1	228	C,3s	155.0	119.9	1.29	BF
	B-CW2	150	253	915	27.5	0.330	1	305	1	228	C,3s	214.0	255.9	0.84	NA
	B-CO2	150	253	915	19.5	0.165	50	305	125	228	C,3s	88.0	81.77	1.08	NA
	B-CO3	150	253	915	19.5	0.165	1	305	1	228	C,3s	113.0	103.5	1.09	BF
	B-CF3	150	253	915	52.0	0.330	1	305	1	228	C,3s	131.0	195.1	0.67	NA
	B-CF2	150	253	915	52.0	0.165	1	305	1	228	C,3s	119.0	165.3	0.72	NA
	C-BT2	150*	357	1070	35.0	0.165	1	305	1	228	C,3s	155.0	157.4	0.98	BF
	C-BT3	150*	357	1070	35.0	0.330	1	305	1	228	C,3s	157.5	183.5	0.86	BF
	C-BT4	150*	357	1070	35.0	0.165	50	305	125	228	C,3s	162.0	126.0	1.29	BF
	C-BT6	150*	357	1070	35.0	0.165	1	305	1	228	C,3s	221.0	159.1	1.39	FL

(Continued)

Table 5.4 (Continued)

References	Beam specimen	Geometric data b (mm)	d (mm)	a (mm)	f_c (MPa)	External shear reinforcement t_f (mm)	w_f (mm)	h_f (mm)	S_f (mm)	E_f (GPa)	Type	Vin_{Exp} (kN)	Proposed new model Vn (kN)	V_{Exp}/V_{Pred}	Failure mode Exp.
Leung et al. (2007)	MB-U1	150	305	900	27.4	0.220	40	360	120	235	C,3s	154.6	174.5	0.89	BF
	MB-U2	150	305	900	27.4	0.220	40	360	120	235	C,3s	159.8	174.5	0.92	BF
	LB-U1	300	660	1800	27.4	0.440	80	720	240	235	C,3s	563.4	652.4	0.86	BF
	LB-U2	300	660	1800	27.4	0.440	80	720	240	235	C,3s	559.8	652.4	0.86	BF
Monti and Liotta (2007)	US45+	250	402	1400	11.2	0.220	150	300	300	390	C,3s	126.0	172.6	0.73	BF
	US90(2)*	250	402	1400	11.2	0.220	150	300	300	390	C,3s	90.0	172.6	0.52	BF
	UF90	250	402	1400	11.2	0.220	1	300	1	390	C,3s	125.0	193.2	0.65	BF
	US45++	250	402	1400	11.2	0.220	50	300	100	390	C,3s	133.5	172.6	0.77	BF
	US45++B	250	402	1400	11.2	0.220	1	300	1	390	C,3s	172.0	193.2	0.89	NA
	US45++C	250	402	1400	11.2	0.220	1	300	1	390	C,3s	183.0	193.2	0.95	NA
	US45+D	250	402	1400	11.2	0.220	150	300	225	390	C,3s	164.0	180.5	0.91	NA
	US45++E	250	402	1400	11.2	0.220	150	300	225	390	C,3s	163.5	180.5	0.91	NA
Norris et al. (1997)	IE	127	167	458	36.5	1.000	1	203	1	34	C,3s	68.0	90.48	0.75	BF
	IIE	127	167	458	36.5	1.000	1	203	1	33	C,3s	68.0	89.98	0.76	BF
Pellegrino and Modena (2006)	A-U1-C17	150	250	750	41.4	0.170	1	300	1	230	C,3s	238.1	243.4	0.98	BF
	A-U1-C20	150	250	750	41.4	0.170	1	300	1	230	C,3s	225.0	231.5	0.97	BF
	A-U1-S17	150	250	750	41.4	0.170	1	300	1	230	C,3s	247.3	243.5	1.02	BF
	A-U1-S20	150	250	750	41.4	0.170	1	300	1	230	C,3s	235.1	231.6	1.02	BF
	A-U2-C17	150	250	750	41.4	0.330	1	300	1	230	C,3s	243.0	270.7	0.90	BF
	A-U2-C20	150	250	750	41.4	0.330	1	300	1	230	C,3s	229.7	258.8	0.89	BF
	A-U2-S17	150	250	750	41.4	0.330	1	300	1	230	C,3s	218.9	270.7	0.81	BF
	A-U2-S20	150	250	750	41.4	0.330	1	300	1	230	C,3s	207.5	258.8	0.80	BF
Sato et al. (1996)	s3	200	260	700	41.3	0.220	30	300	60	230	C,3s	202.0	163.0	1.24	BF
	s5	200	260	700	39.7	0.111	1	300	1	230	C,3s	198.2	165.4	1.20	BF
	s6	200	260	700	39.7	0.111	1	300	1	230	C,3s	211.2	209.2	1.01	BF
Taerwe et al. (1997)	BS2	200	420	1250	36.2	0.110	50	450	400	230	C,3s	247.0	217.2	1.14	BF
	BS4	200	420	1250	36.2	0.110	1	450	1	230	C,3s	252.0	263.8	0.96	BF
	BS5	200	420	1250	36.2	0.110	50	450	400	230	C,3s	170.0	183.7	0.93	BF
	BS6	200	420	1250	35.8	0.110	50	450	400	230	C,3s	167.0	182.9	0.91	BF

References	Beam specimen	Geometric data				External shear reinforcement						Vn_{Exp} (kN)			Failure mode Exp.
		b (mm)	d (mm)	a (mm)	f_c (MPa)	t_f (mm)	w_f (mm)	h_f (mm)	S_f (mm)	E_f (GPa)	Type				
Xue song and Zhong fan (2004)	SB1-3	150	300	864	40.8	0.220	1	360	1	235	C,3s	240.0	251.6	0.95	BF
	SB1-4	150	300	864	40.8	0.220	1	360	1	235	C,3s	253.0	251.6	1.01	BF
	SB1-5	150	300	864	40.8	0.220	40	360	120	235	C,3s	246.0	199.4	1.23	BF
	SB1-6	150	300	864	40.8	0.220	40	360	120	235	C,3s	230.0	199.4	1.15	BF
	SB1-7	150	300	864	40.8	0.220	40	360	120	235	C,3s	240.0	199.4	1.20	BF
	SB1-8	150	300	864	40.8	0.220	40	360	120	235	C,3s	239.0	199.4	1.20	BF
	SB1-9	150	300	864	40.8	0.440	40	360	120	235	C,3s	240.0	223.3	1.07	BF
	SB1-10	150	300	864	40.8	0.440	40	360	120	235	C,3s	243.0	223.3	1.09	BF
Zhang et al. (2004)	Z31-FU	102	203	381	42.5	0.330	1	229	1	73	C,3s	96.1	87.93	1.09	BF
Adhikary et al. (2004)	C-4	300	245	1000	42.4	0.170	1	300	1	250.0	C,4s	251.9		0.99	FL
	A-4	300	245	1000	43.5	0.290	1	300	1	244.0	A,4s	249.1		0.98	FL
Alagusundaramoorthy et al. (2003)	SB9-45	230	330	915	31.0	0.180	1	380	1	228	C,4s	264.5			TR
	SB10-45	230	330	915	31.0	0.180	1	380	1	228	C,4s	306.0			TR
	SB11-45	230	330	915	31.0	0.180	1	380	1	228	C,4s	254.5			TR
	SB12-45	230	330	915	31.0	0.180	1	380	1	228	C,4s	265.5			TR
Araki et al. (1997)	CF045	200	340	524	24.8	0.110	40	400	170	235	C,4s	236.0			TR
	CF064	200	340	524	24.9	0.110	70	400	170	235	C,4s	262.0			TR
	CF097	200	340	524	25.2	0.110	120	400	170	235	C,4s	306.0			TR
	CF131	200	340	524	25.4	0.110	1	400	1	235	C,4s	358.0			TR
	CF243	200	340	524	25.6	0.220	1	400	1	235	C,4s	406.0			TR
	AF060	200	340	524	25.8	0.144	70	400	170	87	A,4s	236.0			TR
	AF090	200	340	524	25.9	0.144	120	400	170	87	A,4s	258.0			TR
	AF120	200	340	524	26.1	0.144	1	400	1	87	A,4s	312.0			TR
Beber and Campos Filho (2005)	V12-A	150	253	740	32.8	0.110	50	300	100	230	C,4s	116.4			TR
	V18-A	150	253	740	32.8	0.110	50	300	100	230	C,4s	127.3			TR
	V20-A	150	253	740	32.8	0.110	50	300	100	230	C,4s	140.1			TR
Cao et al. (2005)	A6	150	223	400	31.0	0.200	30	250	50	249	C,4s	217.0			TR
	Bb	150	223	550	30.5	1.300	20	250	40	21	G,4s	136.0			TR
	BC	150	223	550	30.5	1.300	20	250	80	21	G,4s	121.0			TR
	Be	150	223	300	30.5	1.300	20	250	40	21	G,4s	178.0			TR

(Continued)

Table 5.4 (Continued)

References	Beam specimen	Geometric data				External shear reinforcement						Vn_{Exp} (kN)	Failure mode Exp.
		b (mm)	d (mm)	a (mm)	f'_c (MPa)	t_f (mm)	w_f (mm)	h_f (mm)	S_f (mm)	E_f (GPa)	Type		
Diagana et al. (2003)	PC1	130	425	900	38.0	0.430	40	450	200	105	C,4s	177.5	TR
	PC2	130	425	900	38.0	0.430	40	450	250	105	C,4s	155.0	TR
	PC3	130	425	900	38.0	0.430	40*	450	424	105	C,4s	145.5	TR
Funakawa (1997)	S3	600	510	1275	27.0	0.330	1	600	1	240	C,4s	795.0	TR
	S4	600	510	1275	27.0	0.500	1	600	1	240	C,4s	942.0	TR
Carolin et al. (2005)	A245W	180	430	1250	46.0	0.220	1	500	1	234	C,4s	338.0	TR
	A290W	180	430	1250	52.0	0.220	1	500	1	234	C,4s	367.0	NA
	A290WR	180	430	1250	46.0	0.220	1	500	1	234	C,4s	388.0	TR
Grande et al. (2007)	RS4W	250	405	1400	21.0	0.190	1	450	1	392	C,4s	250.0	TR
	RS3W	250	405	1400	21.0	0.190	1	450	1	392	C,4s	330.0	TR
	RS2W	250	405	1400	21.0	0.190	1	450	1	392	C,4s	300.0	TR
Iamniruberto and Imbimbo (2004)	ST1a	150	300	900	41.0	0.120	1	350	1	76	G,4s	242.0	TR
	ST1b	150	300	900	41.0	0.120	1	350	1	76	G,4s	242.0	TR
	ST2a	150	300	900	41.0	0.240	1	350	1	76	G,4s	278.0	TR
	ST2b	150	300	900	41.0	0.240	1	350	1	76	G,4s	270.0	TR
	ST3a	150	300	900	41.0	0.360	1	350	1	76	G,4s	318.0	TR
	ST3b	150	300	900	41.0	0.360	1	350	1	76	G,4s	279.0	TR
Kamiharako et al. (1997)	2	250	400	1000	32.6	0.110	40	500	100	244	C,4s	285.2	TR
	3	250	400	1000	32.6	0.170	40	500	100	90	A,4s	236.0	TR
	7	400	600	1000	34.6	0.110	40	700	100	244	C,4s	568.8	TR
	8	400	600	1000	34.6	0.170	40	700	100	90	A,4s	529.6	TR
Leung et al. (2007)	MB-f1	150	305	900	30.9	0.220	40	360	120	235	C,4s	236.4	TR
	MB-f2	150	305	900	30.9	0.220	40	360	120	235	C,4s	250.3	TR
	LB-F1	300	660	1800	30.9	0.440	80	720	240	235	C,4s	871.6	TR
	LB-F2	300	660	1800	30.9	0.440	80	720	240	235	C,4s	881.2	TR
Miyajima et al. (2005)	Case3	340	375	1100	29.9	0.110	75	440	150	253	C,4s	253.0	TR
	Case4	340	375	1100	29.9	0.110	88	440	150	253	C,4s	263.0	TR
	Case5	340	375	1100	29.9	0.110	100	440	150	253	C,4s	293.0	TR

Reference	Specimen												Failure
Monti and Liotta (2007)	WS45++	250	402	1400	11.2	0.220	50*	300	141	390	C,4s	158.5	NA
Sato et al. (1997)	SB2	300	260	400	32.3	0.110	1	150	1	248	C,4s	267.0	NA
	SB3	300	260	400	32.3	0.110	1	300	1	248	C,4s	358.0	NA
	SB4	300	260	400	32.3	0.220	1	300		248	C,4s	446.0	NA
	SB5	300	260	400	33.2	0.220	1	150	1	230	C,4s	441.0	NA
Taerwe et al. (1997)	BS7	200	420	1250	34.7	0.110	50	450	200	230	C,4s	235.5	NA
Umezu et al. (1997)	CS1	300	257	760	40.5	0.110	1	300	1	244	C,4s	214.0	TR
	CS2	300	257	760	40.5	0.110	100	300	200	244	C,4s	159.0	TR
	CS3	300	272	800	44.8	0.110	100	300	200	244	C,4s	116.0	TR
	AS1	150	257	800	43.0	0.040	1	300	1	73	A,4s	91.2	TR
	AS2	150	272	800	43.0	0.040	100	300	200	73	A,4s	89.7	TR
	AS3	150	272	800	44.8	0.090	100	300	200	73	A,4s	114.0	TR
	AB1	150	253	760	41.9	0.040	1	300	1	73	A,4s	110.0	TR
	AB2	300	258	760	45.6	0.040	1	300	1	73	A,4s	173.0	TR
	AB3	300	253	760	41.9	0.090	1	300	1	73	A,4s	209.0	TR
	AB4	300	269	760	41.9	0.090	1	300	1	73	A,4s	224.0	TR
	AB5	300	253	760	42.7	0.140	1	300	1	73	A,4s	254.0	TR
	AB6	300	253	760	43.1	0.220	1	300	1	73	A,4s	247.0	FL
	AB7	300	253	760	43.5	0.290	1	300	1	73	A,4s	240.0	FL
	AB8	600	253	760	43.5	0.140	1	300	1	73	A,4s	424.0	TR
	AB9	450	399	1200	39.9	0.140	1	450	1	73	A,4s	379.0	TR
	AB10	550	499	1500	39.9	0.140	1	550	1	73	A,4s	569.0	TR
	AB11	550	499	1500	40.6	0.290	1	550	1	73	A,4s	662.0	TR

Note: BF, debonding failure; TR, tensile rupture failure; FL, flexural failure; NA, not reported.

crushed. To simulate real concrete behavior, ANSYS requires linear isotropic and multilinear isotropic material as well as certain additional concrete material properties. The shear transfer coefficient β_t represents the condition of the cracked face. β_t ranges from 0.0 to 1.0, with 0.0 representing a smooth crack and 1.0 representing a rough crack (Kwan et al., 1999; Terec et al., 2010). In this study, β_t is considered to be 0.2 for an open crack (Kachlakev et al., 2001) and 0.8 for a closed crack (Raongjant and Jing 2008). The modulus of elasticity of the concrete can be calculated by $(E_c = 4700\sqrt{f_c'})$ and the tensile strength by $(f_r = 0.62\sqrt{f_c'})$. Poisson's ratio of 0.2 was used for concrete. The compressive uniaxial stress—strain relationship was obtained by the following equations (Kachlakev et al., 2001; Raongjant and Jing, 2008):

$$E_c = f/\varepsilon, \varepsilon_0 = \frac{2f_{co}'}{E_c}, \text{ and } f = \frac{E_c\varepsilon}{1 + (\varepsilon/\varepsilon_o)^2} \tag{5.1}$$

where f = stress at any strain ε, ε_o = strain at the ultimate compressive strength, and f_{co}' = cylinder compressive strength of concrete.

The maximum dimensions for concrete elements were set as follows: 50 mm for length, 20 mm for height, and 20 mm for width.

5.7.1.2 Steel reinforcement and steel plates

A Link 8 element was used to model steel reinforcement. This element is a 3D spar element. It has two nodes with three degrees of freedom in each node. This element is also capable of plastic deformation. A perfect bond between the concrete and steel reinforcement is considered. The finite element model for the rebar was assumed to be a bilinear isotropic, elastic—perfectly plastic material, and identical in tension and compression with an elastic modulus of 210,000 MPa and Poisson's ratio of 0.3. The yield stress was considered equal to 360 MPa for the longitudinal reinforcement and 240 MPa for the vertical reinforcement.

Solid 45 element was used for the steel plates at the supports for the beam. This element has eight nodes with three degrees of freedom at each node. The steel plates were assumed to be linear elastic with an elastic modulus of 210,000 MPa and Poisson's ratio of 0.3.

5.7.1.3 FRP composites

Solid 46, a layered solid element, was used to model the FRP composites. The element allows for up to 100 material layers, with different orientations and orthotropic material properties in each layer (Kaw, 1997). The element has three degrees of freedom at each node.

Because FRP composites are orthotropic materials, their properties are different in each direction. Eqs. (5.2) and (5.3) express the relationship between ν_{xy} and ν_{yx} (Kaw, 1997; Ansys User's Manual, 2009).

$$1 - \nu_{xy}^2(E_y/E_x) - \nu_{yz}^2(E_z/E_y) - \nu_{xz}^2(E_z/E_x) - 2\nu_{xy}\nu_{yz}\nu_{xz}(E_z/E_x) = \text{positive}$$
$$(5.2)$$

$$G_{xy} = G_{xz} = \frac{E_x E_y}{E_x + E_y + 2\nu_{xy}E_x}, \quad G_{yz} = \frac{E_z \text{ or } E_y}{2(1 + \nu_{yz})}, \quad \text{and} \quad \nu_{yx} = \frac{E_y}{E_x}\nu_{xy}$$
$$(5.3)$$

In this study, the elastic properties of FRP composites are assumed linear. Poisson's ratios of $\nu_{xy} = \nu_{xz} = 0.22$ and $\nu_{yz} = 0.30$ (Kachlakev and McCurry, 2000), which are widely used in previously published literature on this subject, are adopted.

The contact between FRP and concrete is modeled by the contact elements TARGE170 and CONTA174. These enable a Coulomb friction model to be considered while producing a gap between interface elements when tensile stresses are detected. In the basic Coulomb friction model, two contacting surfaces can carry shear stresses. The Coulomb friction model is defined as ($\tau_{lim} = a + \mu p$). Here, τ_{lim} is the limit shear stress, a is the contact adhesion, p is the normal contact pressure, and μ is the coefficient of friction. For the case without normal pressure ($p = 0$), the contact cohesion, $a = \tau_{max}$ is considered.

The behavior of the FRP—concrete interface is simulated by a relationship between the bond stress τ and relative slip S. The maximum bond stress ($\tau_{max} = 1.5\beta_w f_t$) and the corresponding slip ($S_o = 0.0195\beta_w f_t$) developed in Lu et al. (2005) has received wide acceptance. It is considered an accurate bond stress—slip model that can be incorporated into an FE analysis. In the equation, the width ratio factor is $\beta_w = ((2.25 - W_f /S_f)/(1.25 + W_f /S_f))^{0.5}$, and f_t is the concrete tensile strength ($f_t = 0.62\sqrt{f_c'}$). The contact slip at the completion of debonding is $S_{max} = 2G_f/\tau_{max}$, where G_f is the interfacial fracture energy $= 0.308\beta_w^2\sqrt{f_t}$.

5.7.2   Parametric study of shear capacity

A parametric study was carried out to evaluate the effects of those parameters on the shear capacity. The parameters included the beam width, concrete strength, effective height of the beam, shear span—depth ratio, FRP thickness, strengthening configuration, height of the FRP sheet, and elastic modulus of the FRP sheet. Meanwhile, four typical models of shear strength (ACI Committee 440, 2008; Triantafillou and Antonopoulos, 2000; Matthys and Triantafillou, 2001; Carolin and Täljsten, 2005a) were utilized for comparison.

5.7.2.1   Modeling RC beam

Fifty five reinforced concrete beams of length 2400 mm and with variable cross-sections (Fig. 5.5) were considered in this study. The beam was designed to be simply supported over a span of 2300 mm and under two-point loading. The longitudinal reinforcement of the beam consisted of 2.05% tension bars and 0.41%

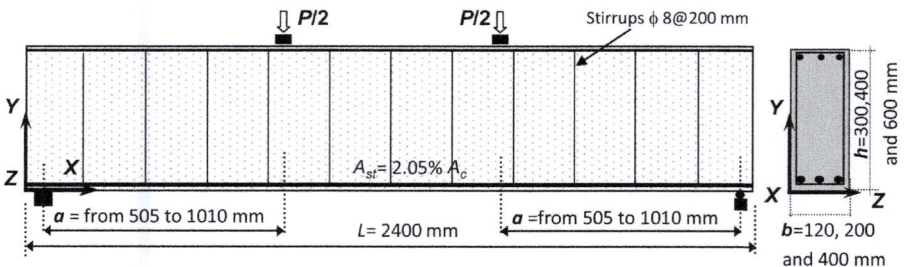

Figure 5.5 Geometrical details of the RC beams (Sayed, 2014).

compression bars with a yield strength of 360 MPa. The spacing between the 8 mm diameter stirrups (with a yield strength of 240 MPa) was 200 mm. Table 5.5 presents all the beams evaluated in this study, the numerically determined values of ultimate shear loads of the tested beams, and the likely failure modes that have been considered.

The parametric studies in the following sections are based on the numerical cases presented in Table 5.5.

5.7.2.2   Influence of the beam width

The beam width is an important factor that affects the strength and stiffness of a strengthened beam directly. The beam width significantly influences the lateral strain of beams, as shown in Fig. 5.6. For the case of side bonding, the debonding failure mode occurs at the top or bottom surfaces of the beam. When the beam width increases, the lateral strain also increases. This results in faster debonding failure. However, in cases of U-jacketing and completely wrapped confinement, because the lateral strain is controlled by the horizontal part of the FRP sheet, no debonding occurs. For side bonding confinement (Fig. 5.6), when the beam width increases, the total amount of shear load increases. However, when ($V_c + V_s$) is subtracted from V_n, the shear capacity carried by FRP sheets V_f decreases.

Fig. 5.7 shows the influence of beam width on shear capacity by FE simulations and comparisons with previous design guidelines. As illustrated by the horizontal lines, ACI Committee 440 (2008) and Carolin and Täljsten (2005) do not consider the effect of beam width. According to the FE simulation, the beam width affects the strength by a factor of $b^{-0.21}$, $b^{0.25}$, and $b^{0.30}$ for side bonding, U-jacketing, and completely wrapped FRP, respectively. These values were obtained using a power regression line, to determine the influence of beam width. The power regression analysis resulted in values of R^2 higher than those from the other regression types.

To verify the influence of beam width in FE simulations, results of 10 beams from previous studies in the literature were considered for comparison: for side bonding, beam specimens Z22-S45 (Zhang et al., 2004), C3 (Bukhari et al., 2010), and SCF-12 (Lim, 2010); for U-jacketing, Z31-FU (Zhang et al., 2004), A-U1-S20 (Pellegrino and Modena, 2006), S5 (Sato et al., 1996), and C-3 (Adhikary et al.,

Table 5.5 Summary of beams evaluated in Sayed (2014) and results.

Beam specimen	Geometric data						External shear reinforcement						V_n Num. (kN)	Failure mode
	b (mm)	h (mm)	d (mm)	a (mm)	a/d	Concrete f'_c (MPa)	t_f (mm)	w_f (mm)	h_f (mm)	s_f (mm)	f_{fu} (MPa)	E_f (GPa)		
2S-1	200	400	360	650	1.80	30	0.1	100	400	200	3000	200	163.1	S(TR)
2S-2	200	400	360	650	1.80	30	0.2	100	400	200	3000	200	174.3	S(BF)
2S-3	200	400	360	650	1.80	30	0.3	100	400	200	3000	200	180.8	S(BF)
2S-4	200	400	360	650	1.80	30	0.2	50	400	150	3000	200	163.0	S(BF)
2S-5	200	400	360	650	1.80	30	0.2	50	400	200	3000	200	152.3	S(BF)
2S-6	200	400	360	650	1.80	30	0.2	1	400	1	3000	200	218.8	S(BF)
2S-7	200	400	360	650	1.80	30	0.2	100	400	200	2000	120	162.9	S(TR)
2S-8	200	400	360	650	1.80	30	0.2	100	400	200	3500	230	178.8	S(BF)
2S-9	200	400	360	540	1.50	30	0.2	100	400	200	3000	200	178.7	S(BF)
2S-10	200	400	360	810	2.25	30	0.2	100	400	200	3000	200	172.9	S(BF)
2S-11	120	400	360	650	1.80	30	0.2	1	400	1	3000	200	198.3	S(BF)
2S-12	400	400	360	650	1.80	30	0.2	1	400	1	3000	200	280.2	S(TR)
2S-13	200	400	360	650	1.80	20	0.1	100	400	200	3000	200	141.4	S(BF)
2S-14	200	400	360	650	1.80	50	0.1	100	400	200	3000	200	190.4	S(TR)
2S-15	200	400	360	650	1.80	30	0.2	1	240	1	3000	200	182.1	S(TR)
2S-16	200	400	360	650	1.80	30	0.2	1	320	1	3000	200	197.8	S(BF)
2S-17	200	300	280	505	1.80	30	0.1	1	300	1	3000	200	143.2	S(TR)
2S-18	200	400	360	650	1.80	30	0.1	1	300	1	3000	200	172.9	S(TR)
2S-19	200	600	560	1010	1.80	30	0.1	1	300	1	3000	200	247.9	S(TR)
U-jac.-1	200	400	360	650	1.80	30	0.1	100	400	200	3000	200	206.5	S(TR)
U-jac.-2	200	400	360	650	1.80	30	0.2	100	400	200	3000	200	229.1	S(BF)
U-jac.-3	200	400	360	650	1.80	30	0.3	100	400	200	3000	200	247.0	S(BF)
U-jac.-4	200	400	360	650	1.80	30	0.2	50	400	150	3000	200	211.7	S(BF)
U-jac.-5	200	400	360	650	1.80	30	0.2	50	400	200	3000	200	205.9	S(BF)
U-jac.-6	200	400	360	650	1.80	30	0.2	1	400	1	3000	200	268.8	S(BF)
U-jac.-7	200	400	360	650	1.80	30	0.2	100	400	200	2000	120	201.9	S(TR)
U-jac.-8	200	400	360	650	1.80	30	0.2	100	400	200	3500	230	231.8	S(BF)

(Continued)

Table 5.5 (Continued)

Beam specimen	Geometric data						External shear reinforcement							
	b (mm)	h (mm)	d (mm)	a (mm)	a/d	Concrete f'_c (MPa)	t_f (mm)	w_f (mm)	h_f (mm)	s_f (mm)	f_{fu} (MPa)	E_f (GPa)	Vn Num. (kN)	Failure mode
U-jac.-9	200	400	360	540	1.50	30	0.2	100	400	200	3000	200	242.7	S(BF)
U-jac.-10	200	400	360	810	2.25	30	0.2	100	400	200	3000	200	225.0	S(BF)
U-jac.-11	120	400	360	650	1.80	30	0.2	1	400	1	3000	200	209.1	S(BF)
U-jac.-12	400	400	360	650	1.80	30	0.2	1	400	1	3000	200	360.2	S(TR)
U-jac.-13	200	400	360	650	1.80	20	0.1	100	400	200	3000	200	172.1	S(BF)
U-jac.-14	200	400	360	650	1.80	50	0.1	100	400	200	3000	200	243.9	S(TR)
U-jac.-15	200	400	360	650	1.80	30	0.2	1	400	1	3000	200	211.9	S(TR)
U-jac.-16	200	400	360	650	1.80	30	0.2	1	240	1	3000	200	246.7	S(BF)
U-jac.-17	200	300	280	505	1.80	30	0.1	1	320	1	3000	200	185.2	S(TR)
U-jac.-18	200	400	360	650	1.80	30	0.1	1	300	1	3000	200	210.5	S(TR)
U-jac.-19	200	600	560	1010	1.80	30	0.1	1	300	1	3000	200	278.7	S(TR)
Wrapped-1	200	400	360	650	1.80	30	0.1	100	400	200	3000	200	255.4	S(TR)
Wrapped-2	200	400	360	650	1.80	30	0.2	100	400	200	3000	200	301.8	S(TR)
Wrapped-3	200	400	360	650	1.80	30	0.3	100	400	200	3000	200	322.8	F(CC)
Wrapped-4	200	400	360	650	1.80	30	0.2	50	400	150	3000	200	272.2	S(TR)
Wrapped-5	200	400	360	650	1.80	30	0.2	50	400	200	3000	200	252.8	S(TR)
Wrapped-6	200	400	360	650	1.80	30	0.2	1	400	1	3000	200	355.8	F(CC)
Wrapped-7	200	400	360	650	1.80	30	0.2	100	400	200	3000	200	270.5	S(TR)
Wrapped-8	200	400	360	650	1.80	30	0.2	100	400	200	3500	230	305.7	F(CC)
Wrapped-9	200	400	360	540	1.50	30	0.2	100	400	200	3000	200	307.6	S(TR)
Wrapped-10	200	400	360	810	2.25	30	0.2	100	400	200	3000	200	272.1	F(CC)
Wrapped-11	120	400	360	650	1.80	30	0.2	1	400	1	3000	200	288.3	F(CC)
Wrapped-12	400	400	360	650	1.80	30	0.2	1	400	1	3000	200	482.0	S(TR)
Wrapped-13	200	400	360	650	1.80	20	0.1	100	400	200	3000	200	209.1	S(TR)
Wrapped-14	200	400	360	650	1.80	50	0.1	100	400	200	3000	200	313.1	S(TR)
Wrapped-15	200	300	280	505	1.80	30	0.1	1	300	1	3000	200	217.8	S(TR)
Wrapped-16	200	400	360	650	1.80	30	0.1	1	400	1	3000	200	282.8	S(TR)
Wrapped-17	200	600	560	1010	1.80	30	0.1	1	600	1	3000	200	435.0	S(TR)

Note: BF, debonding failure; *CC*, concrete crushing; *TR*, tensile rupture failure; *F*, flexural failure; *S*, shear failure.

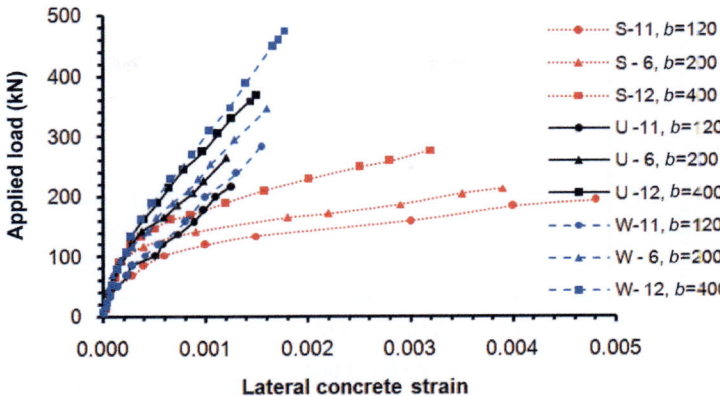

Figure 5.6 Load—concrete strain at point from bottom surface for beams show the influence of beam width (Sayed, 2014).

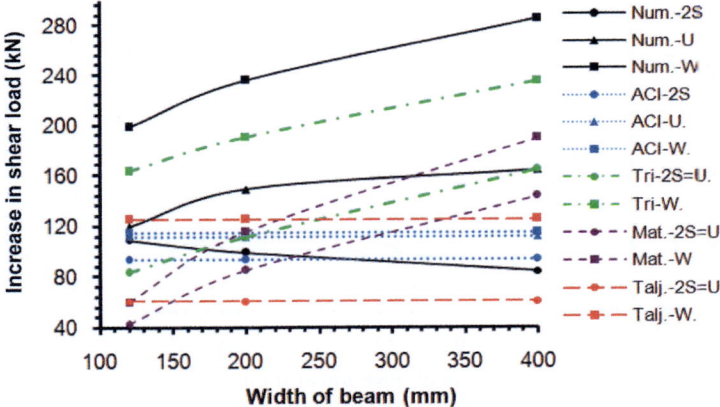

Figure 5.7 Influence of the beam width by the existing models and the FE simulations (Sayed, 2014).

2004); and for complete wrapping, AB5 and AB8 (Umezu et al., 1997) and AF090 (Araki et al., 1997). The wrapping scheme was identical for all the cases. As shown in Fig. 5.8, for side bonding, the shear capacity of FRP decreases with increase in the beam width. Meanwhile, for U-jacketing and complete wrapping, the shear capacity increases when the beam width increases.

5.7.2.3 Influence of FRP thickness

The thickness of FRP is the main factor that affects the strength and stiffness of the strengthening material directly. The variations in the predicted FRP contribution versus FRP thickness from design guidelines are shown in Fig. 5.9. As is evident

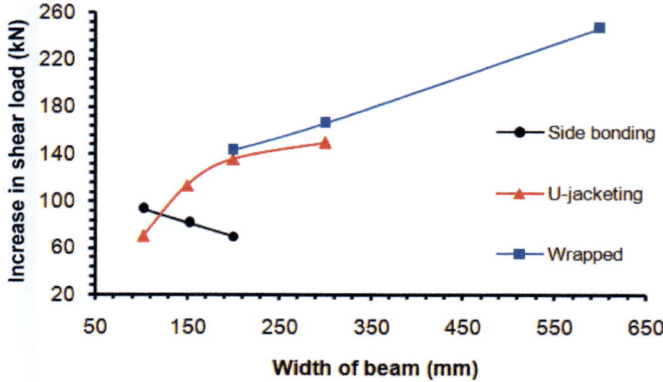

Figure 5.8 Influence of the beam width for the beams previously studied in the literature (Sayed, 2014).

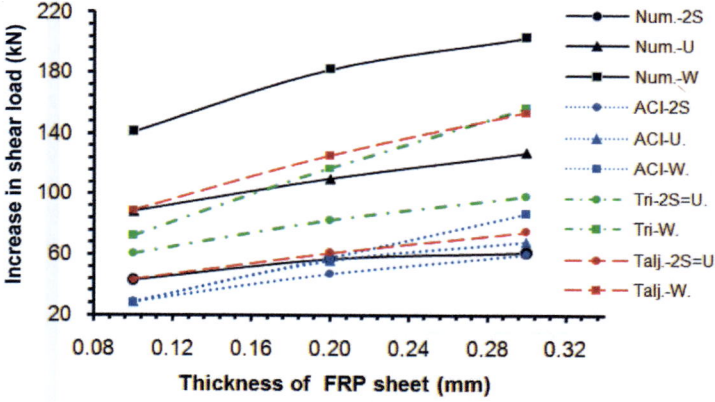

Figure 5.9 Influence of the FRP thickness by the existing models and the FE simulations (Sayed, 2014).

from the similarity in the trends of all the curves, the predictions of the models of ACI Committee 440 (2008) and of Triantafillou and Antonopoulos (2000) are similar. Based on the FE simulations, the thickness of FRP also affects strengthened beams by $(t_f^{0.334})$ for side bonding, complete wrapping, and U-jacketing.

5.7.2.4 Influence of concrete strength

The concrete strength is also an important factor that directly affects the strength and the stiffness of the strengthened beam. The variations of the predicted FRP contributions versus concrete strength from design guidelines are shown in Fig. 5.10. As is evident from this figure, ACI Committee 440 (2008) for a completely wrapped case does not take into account the effect of the concrete strength.

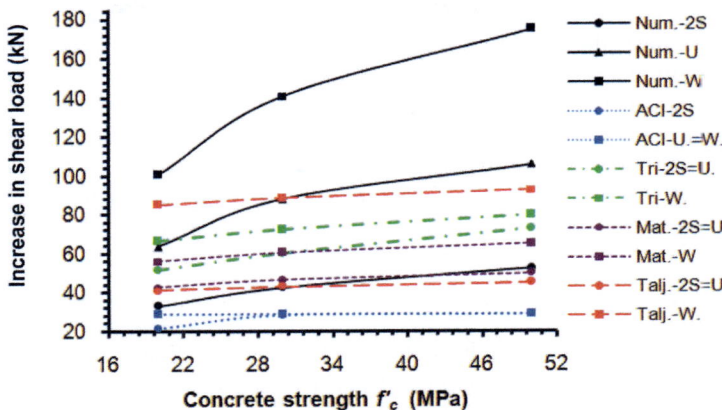

Figure 5.10 Influence of the concrete strength as predicted by the existing models and the FE simulations (Sayed, 2014).

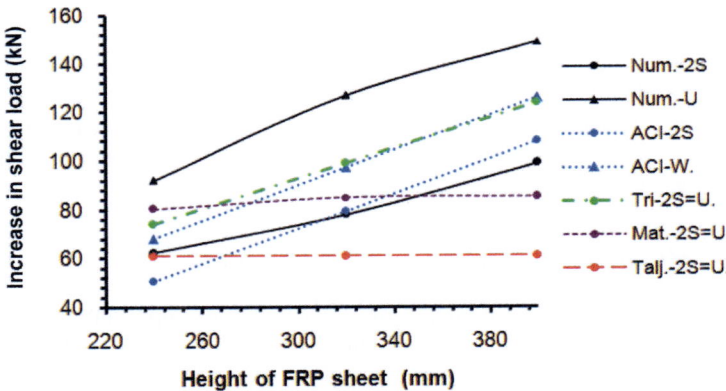

Figure 5.11 Influence of the height of FRP as studied by the existing models and the FE simulations (Sayed, 2014).

According to the FE simulations, the concrete strength results in an increase in the shear strength of the strengthened beams by a factor of $f_c^{0.5}$ for side bonding, $f_c^{0.55}$ for U-jacketing, and $f_c^{0.6}$ for complete wrapping.

5.7.2.5 Influence of height of FRP sheet

This is arguably the main parameter that directly affects FRP contribution. Predictions revealed an almost linear increase in FRP contributions when effective height (which is the distance between the centroid of the tensile reinforcement and the FRP curtailment point at the beam compression side) is increased, as shown in Fig. 5.11. The model of Carolin and Täljsten (2005) does not consider the effect of the height of FRP sheet. However, the model of Matthys and Triantafillou (2001) demonstrated that the effective height adversely affects side bonding and U-jacketing configurations. The FE simulations reveal that the effective height of FRP sheet causes an increase in the strength of the strengthened beams by a factor of $h_f^{0.90}$ and $h_f^{0.95}$ for side bonding and U-jacketing, respectively, as shown in Fig. 5.11. However, for completely wrapped beams, the height of the FRP sheet depends on the depth of the beam; hence, both beam depth and FRP height are considered as one parameter. This is discussed in the next section.

5.7.2.6 Influence of beam depth

Previous studies on this case have revealed an almost linear increase in shear strength because of FRP contributions (Carolin and Täljsten, 2005), see Fig. 5.12. The models by ACI Committee 440 (2008), Triantafillou and Antonopoulos (2000), and Matthys and Triantafillou (2001) do not consider the effect of beam depth, as shown in Fig. 5.12. However, as the FE simulations demonstrated, variations in beam depth may result in an increase in the FRP contribution of shear strength by a factor of $d^{0.30}$ for side bonding. However, it does not affect beams strengthened with U-jacketing. For the completely wrapped case, the beam depth and FRP height are considered as one parameter. Because the configuration of completely wrapped beams is approximately similar to that of U-jacketing, the beam depth is assumed to have no effect on the shear capacity. Therefore it is considered that the shear capacity of a completely wrapped beam is affected by an increase in the height of the FRP sheet, by a factor of $h_f^{1.0}$.

5.7.2.7 Influence of shear span−depth ratio

The influence of the shear span−depth ratio a/d is highly important and could govern the ultimate shear capacity. The influence of this ratio has not been considered

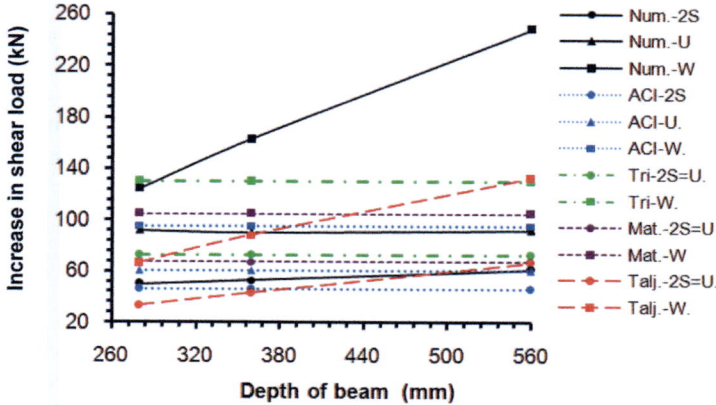

Figure 5.12 Influence of the beam depth by the existing models and the FE simulations (Sayed, 2014).

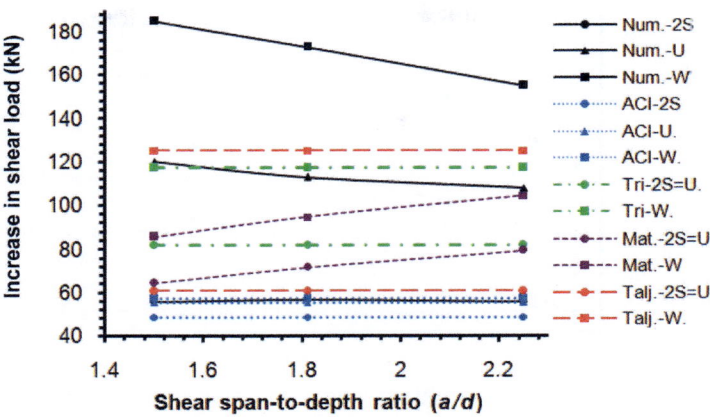

Figure 5.13 Influence of the shear span-to-depth ratio by the existing models and the FE simulations (Sayed, 2014).

previously by ACI Committee 440 (2008), Triantafillou and Antonopoulos (2000), or Carolin and Täljsten (2005), as shown in Fig. 5.13. From the FE simulations, the influence of the shear span—depth ratio does not affect beams strengthened by side bonding. However, it affects the result by a factor of $(a/d)^{-0.25}$ and $(a/d)^{-0.45}$ for U-jacketing and complete wrapping, respectively, as shown in Fig. 5.13. Accordingly, as a/d increases, FRP contributions to the shear strength of a strengthened beams decrease.

5.7.2.8 Influence of type of wrapping scheme

The influence of W_f/S_f plays a key role in design guidelines on FRP contribution. Here, W_f is the width of an FRP strip, and S_f is the spacing between the centerlines of adjacent FRP strips, respectively. According to the slope of the curves in Fig. 5.14, there is an almost linear increase in FRP contribution when W_f/S_f increases, based on the models of ACI Committee 440 (2008), Triantafillou and Antonopoulos (2000), and Carolin and Täljsten (2005). However, this change is not linear in the case of Matthys and Triantafillou (2001), as shown in Fig. 5.14. As this figure shows, the highest contribution from the strengthening material is obtained when the shear span is covered with FRP sheets. This effect is identical for all guideline predictions. From the FE simulations, the increase in W_f/S_f results in an increase in V_f by a factor of $(W_f/S_f)^{0.79}$ for side bonding, and $(W_f/S_f)^{0.41}$ for U-jacketing and complete wrapping, as shown in Fig. 5.14.

5.7.2.9 Influence of elastic modulus of FRP configuration

The influence of the elastic modulus of FRP (E_f) significantly influences the FRP contribution of shear strength based on design guidelines, as shown in Fig. 5.15. According to the slope of the curves, there is an almost linear increase in the FRP

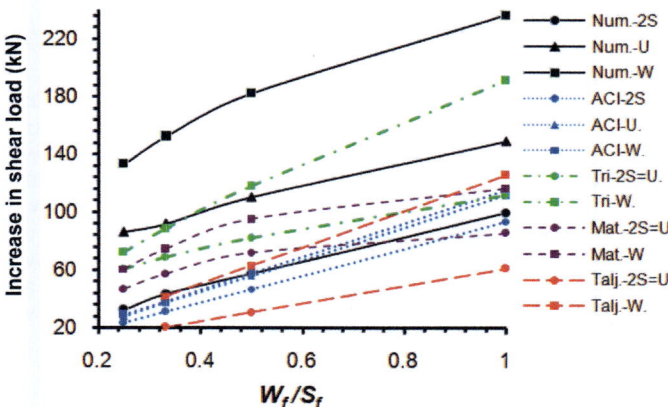

Figure 5.14 Influence the type of wrapping scheme by the existing models and the FE simulations (Sayed, 2014).

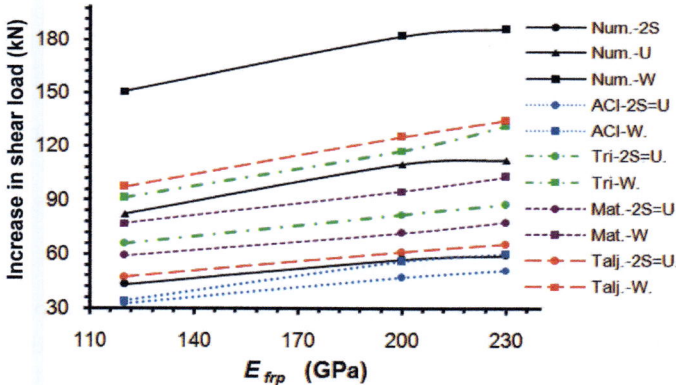

Figure 5.15 Influence of the elastic modulus of FRP by the existing models and the FE simulations (Sayed, 2014).

contribution. However, this increase is nonlinear according to the models by Matthys and Triantafillou (2001) and Triantafillou and Antonopoulos (2000), particularly for complete wrapping. From the FE simulations, the elastic modulus of the FRP shows a linear increase in the strengthened beams by a factor of $E_f^{0.5}$ for side bonding and U-jacketing, and $E_f^{0.334}$ for complete wrapping, as shown in Fig. 5.15.

5.8   Shear strength models

Ideally, the shear capacity of a beam could be predicted by a detailed examination of the shear transfer mechanisms, crack propagation, and failure of the beam

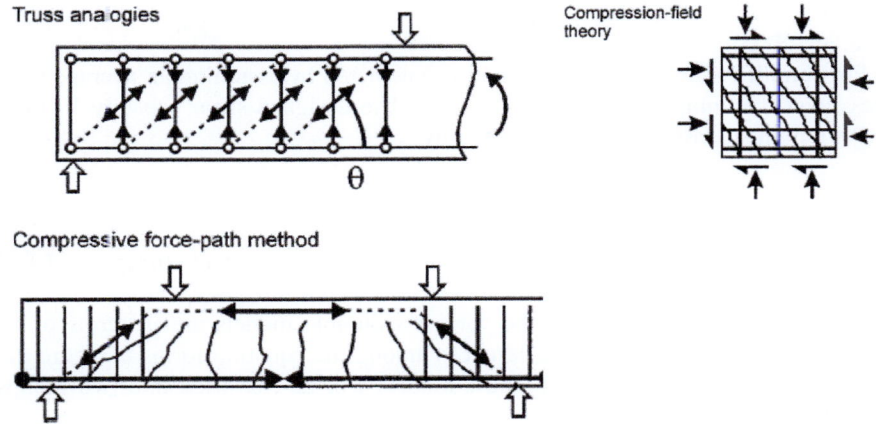

Figure 5.16 Influence of the elastic modulus of FRP by the existing models and the FE simulations (Stratford and Burgoyne, 2003).

components. Instead, shear design is based on simplistic models of equilibrium conditions within the beam. For example, Fig. 5.16 shows the truss analogy with a fixed strut angle (Mörsch, 1909) or variable strut angle (Nielsen et al., 1978), compression-field theory (Collins et al., 1996), and compressive force-path method (Kotsovos and Pavlović, 1999).

A different equilibrium state within the beam is assumed to exist in each shear model. None of these is based on the actual stress distribution. Nonetheless, all the theories have been used effectively to design steel-reinforced concrete. They rely on the lower-bound (or safe load) theorem of plasticity: "If any stress distribution throughout the structure can be found, which is everywhere in equilibrium internally and balances certain external loads and at the same time does not violate the yield condition, those loads will be carried safely by the structure" (Calladine, 1969). Attempts have been made to extend a few of these models to FRP-strengthened RC beams. These include the modified shear friction method, the compression field theory, various truss models, and the design code approach.

5.8.1   Models based on experimental observations

Because the linear FRP differs from the nonlinear steel reinforcement, the design theory of FRP-strengthened RC member exhibits unique characteristics, particularly for shear capacity design.

The shear capacity of RC beams strengthened externally with FRP sheets or strips has been proposed on the basis of the shear design method of conventional RC beams, and based on experimental observations and the corresponding theoretical assumptions (Triantafillou and Antonopoulos, 2000; Matthys and Triantafillou, 2001; Carolin and Täljsten, 2005; Mofidi and Chaallal, 2011; Colotti et al., 2004;

Adhikary et al., 2004; Zhang and Hsu, 2005; Deniaud and Cheng, 2001a, 2001b, Ye et al., 2005).

For simplification, the shear strength is calculated by assuming an average shear stress over the entire effective cross-section. The designed shear capacity V_n of an FRP-strengthened RC beam is calculated by

$$V_n = V_c + V_s + V_f \qquad (5.4)$$

where V_c, V_s, and V_f are the contributions of the concrete, steel stirrups, and FRP composites, respectively.

The main differences among the various available models are in terms of the evaluation of V_f. Therefore the differences among the equation forms of V_f are discussed in the next section. There, these models will be classified into two groups. The first comprises models from the previous researches, and the second comprises those that have been used in the different codes.

A review of the relevant literature up to early 1997 is available in Triantafillou (1998). It is the first systematic attempt to characterize the contribution of externally bonded FRPs to the shear capacity of RC members, based on both analysis and all available experimental results. It was reported in that study that the effectiveness of external FRP shear reinforcement and its contribution to the shear capacity of RC members depends on the mode of failure, which may occur either by peeling off through the concrete near the concrete−FRP interface or by FRP tensile fracture at a stress that may be lower than the FRP tensile strength (e.g., because of stress concentrations at debonded areas or at rounded corners).

Whether peeling off or fracture occurs first depends on the bond conditions, available anchorage length, type of attachment at the FRP curtailment, FRP thickness, FRP elastic modulus, concrete strength, etc. It was also stated that in many cases, the actual mechanism is a combination of peeling off at certain areas and fracture at others. Based on mostly qualitative arguments, Triantafillou (1998) derived a polynomial function that relates the strain in the FRP at shear failure of the member, defined as effective strain ($\varepsilon_{f,e}$), to the axial rigidity of externally bonded strips or sheets.

1. Triantafillou and Antonopoulos (2000) proposed a model. It was calibrated using 75 published experimental test results. This resulted in an expression for predicting the effective strain in FRP laminates. The model proposed by Triantafillou and Antonopoulos (2000) of the contribution of FRP (V_f) has the following form:

$$V_f = 0.9\varepsilon_{f,e}E_f\rho_f(\sin\beta + \cos\beta)bd_f \qquad (5.5)$$

where β = angle between the principal fiber orientation and longitudinal axis of the member.

For members completely wrapped with CFRP (shear failure accompanied or followed by CFRP fracture):

$$\varepsilon_{f,e} = 0.17\left(\frac{f'_{co}{}^{2/3}}{E_f\rho_f}\right)^{0.3}\varepsilon_{f,u} \qquad (5.6)$$

For two or three side laminated with CFRP (CFRP debonding failure mode):

$$\varepsilon_{f,e} = 0.00065 \left(\frac{f'^{2/3}_{co}}{E_f \rho_f} \right)^{0.56} \tag{5.7}$$

For beams completely wrapped with aramid fiber-reinforced polymer (AFRP) (shear failure combined with or followed by AFRP fracture):

$$\varepsilon_{f,e} = 0.048 \left(\frac{f'^{2/3}_{co}}{E_f \rho_f} \right)^{0.47} \varepsilon_{f,u} \tag{5.8}$$

$$\rho_f = \frac{2t_f W_f}{bS_f} \quad \text{and} \quad (E_f \text{ in GPa}) \tag{5.9}$$

The key parameter of the analytical expression ($\varepsilon_{f,e}$) was obtained through regression analysis of experimental data from beam tests. This may indicate a narrow coverage of the shear problem. This effective strain has been observed to be dependent on both the axial rigidity of the composite and is used as the minimum of strain caused by debonding, and strain corresponding to shear failure accompanied or followed by FRP rupture.

2. Khalifa and Nanni (2000) recommended a modified effective strain both for fiber rupture and debonding failure, based on the model of Triantafillou and Antonopoulos (2000). Eq. (5.10) was proposed to define the FRP contribution to the shear strength:

$$V_f = \frac{2nt_f W_f f_{f,e}(\sin\beta + \cos\beta)d_f}{S_f} \tag{5.10}$$

$$f_{f,e} = Rf_{f,u} \tag{5.11}$$

The effective stress $f_{f,e}$ in fibers was established to be a function of FRP stiffness. The ultimate strain was obtained by regression analysis of experimental data. The equation is valid only for CFRP continuous sheets and depends on the failure mechanism.

For FRP sheet rupture:

$$R = 0.5622(\rho_f E_f)^2 - 1.2188\rho_f E_f + 0.788 \le 0.5 \tag{5.12}$$

For debonding failure:

$$R = \frac{f_c^{2/3} w_{f,e}}{\varepsilon_{f,u} d_f} \left[738.93 - 4.06(t_f E_f) \right] \times 10^{-6} \le 0.5 \tag{5.13}$$

$$w_{f,e} = \begin{cases} d - 2L_e & \text{for side bonding} \\ d - L_e & \text{for U-jacketing} \end{cases} \tag{5.14}$$

$$L_e = e^{6.134 - 0.58\ln(t_f E_f)} \le 75 \text{mm} \tag{5.15}$$

where L_e and $w_{f,e}$ are the effective bond length and effective width of the FRP sheet.

3. Matthys and Triantafillou (2001) modified the Triantafillou and Antonopoulos (2000) model further and incorporated the effect of shear span—beam depth ratio in the calculation of the effective strain of FRP. According to their modification, the effective strain could be expressed as

$$\varepsilon_{f,e} = \begin{cases} 0.72\varepsilon_{fu}e^{-0.0431\Gamma_f} \text{ for completely wrapped} \\ 0.56\varepsilon_{fu}e^{-0.0455\Gamma_f} \text{ for two or three sides laminated} \end{cases} \tag{5.16}$$

$$\Gamma_f = \frac{E_f\rho_f}{f_{co}'^{2/3}(a/d)} \tag{5.17}$$

4. Chen and Teng (2003a and 2003b) proposed another model, which is widely used for FRP shear strengthened elements. They evaluated shear capacity according to two failure modes: FRP rupture and FRP debonding. In the case of side-bonding configuration, debonding failure mode occurs. In the case of wrapped configuration, rupture failure mode should be considered. In the case of U-jacketing, the shear capacity should be evaluated according to both the modes, and the smallest value has to be used. The model was based on the truss model, with the remark that discrete FRP strips were modeled as equivalent continuous FRP sheets/plates. Moreover, a reduction factor for the stress (rather than strain) was used, as in the previous models. The FRP contribution to shear strength could be expressed as

$$V_f = 2D_{frp}\sigma_{f.max}t_fW_f\frac{h_{f.e}(\cot\alpha + \cot\beta)\sin\beta}{S_f} \tag{5.18}$$

where α is the critical shear crack inclined to the beam longitudinal axis.
   FRP rupture failure mode, Chen and Teng (2003a):

$$D_{frp} = \frac{\int_{z_t}^{z_b} \varepsilon_z dz}{h_{f.e}\varepsilon_{z.max}} \tag{5.19}$$

$$h_{f.e} = z_b - z_t \tag{5.20}$$

where $h_{f.e}$ is the effective height of the FRP. z_t and z_b are the co-ordinates of the top and bottom ends of the effective FRP, respectively.

$$z_t = d_{f.t} \text{ and } z_b = 0.9d - (h - d_f) \tag{5.21}$$

in which $d_{f.t}$ is the distance from the compression face to the top edge of the FRP.

$$\sigma_{f.max} = \begin{cases} 0.8f_f & \text{if } \dfrac{f_f}{E_f} \le \varepsilon_{max} \\[3mm] 0.8\varepsilon_{max}E_f & \text{if } \dfrac{f_f}{E_f} > \varepsilon_{max} \end{cases} \tag{5.22}$$

The maximum stress in the FRP (when fiber rupture occurred) was considered to be the ultimate tensile strength. Owing to the loss of aggregate interlocking, the ultimate tensile failure of the fiber may be attained before shear failure of the beam.

FRP debonding failure mode, Chen and Teng (2003b):

The debonding model considers "an effective bond length beyond which an extension of the bond length does not increase the bond strength." The maximum stress in the FRP at debonding was considered to be

$$
\sigma_{f.\max} = \min \begin{cases} f_f \\ 0.315\beta_w\beta_L\sqrt{\dfrac{E_f\sqrt{f'_{co}}}{t_f}} \end{cases}
\tag{5.23}
$$

The normalized maximum bond length parameter, maximum bond length L_{max}, and effective bond length L_e are expressed as

$$
\lambda = \frac{L_{\max}}{L_e}
\tag{5.24}
$$

$$
L_e = \sqrt{\frac{E_f t_f}{\sqrt{f'_{co}}}}
\tag{5.25}
$$

$$
L_{\max} = \begin{cases} h_{f,e}/\sin\beta & \text{for U} - \text{jacketing} \\ h_{f,e}/(2\sin\beta) & \text{for side bonding} \end{cases}
\tag{5.26}
$$

$$
\beta_w = \sqrt{\frac{2 - (W_f/(S_f\sin\beta))}{1 + (W_f/(S_f\sin\beta))}}
\tag{5.27}
$$

$$
\beta_L = \begin{cases} 1 & \text{if } \lambda \geq 1 \\ \sin\dfrac{\pi\lambda}{2} & \text{if } \lambda < 1 \end{cases}
\tag{5.28}
$$

5. Adhikary et al. (2004) developed the model based on experiments performed on rectangular beams, after a series of tests on RC beams strengthened by U-jacketing with different anchorage lengths on top of the beam. The proposed equations are valid only for cases where the axial rigidity is in the interval $0 < \rho_f E_f \leq 1.0$. Eq. (5.29) was adopted to calculate the shear contribution of FRP:

$$
V_f = \rho_f E_f \varepsilon_{f,e} d_f b_w (\sin\beta + \cos\beta)
\tag{5.29}
$$

Adhikary et al. (2004) proposed two equations to predict the contributions of CFRP and AFRP systems to the shear resistance of RC beams.

The effective strain when debonding occurs is

$$
\frac{\varepsilon_{f,e1}}{\varepsilon_{f,u}} = \begin{cases} 0.038 f'^{1/3}_{co}/\sqrt{\rho_f E_f} & \text{for CFRP} \\ 0.034 f'^{1/3}_{co}/\sqrt{\rho_f E_f} & \text{for AFRP} \end{cases}
\tag{5.30}
$$

The model considers the bonded anchorage provided to the top of the surface of the beam by using Eq. (5.31):

$$
\frac{\varepsilon_{f,e2}}{\varepsilon_{f,u}} = \begin{cases} 0.0043 f'^{2/3}_{co} & \text{for CFRP} \\ 0.0046 f'^{2/3}_{co} & \text{for AFRP} \end{cases}
\tag{5.31}
$$

The effective strain in the FRP was assumed to increase owing to this anchorage. Therefore in this case, the effective strain at failure is the sum of the effective strain in the FRP in the debonding mode $\varepsilon_{f,e1}$ and the increase in effective strain in FRP owing to bond anchorage $\varepsilon_{f,e2}$.

6. Colotti et al. (2004) developed a theoretical model based on the truss analogy method in conjunction with the theory of plasticity. They proposed that the total contribution of the shear strength of a beam specimen V_n could be expressed as

$$V_n = (\tau/f_c'')bd_v f_c'' \tag{5.32}$$

where b = thickness of the web, d_v = depth of the web, $f_c'' = \nu f_{co}'$, and $\nu = 0.8 - f_{co}'/200$ (f_{co}' in MPa). By assuming that the effectiveness of an external shear reinforcement can be evaluated similarly as for internal steel stirrups, the total degree of shear reinforcement ψ could be considered as the sum of two terms:

$$\psi = \psi_i + \psi_e \tag{5.33}$$

where the subscripts i and e refer to the internal and external shear reinforcement, respectively. The internal shear strength contribution by an internal steel reinforcement is expressed by Eq. (5.34), and the external contribution of FRP is expressed by Eq. (5.35):

$$\psi_i = \frac{A_{st}f_{ys,t}}{bsf_c''} \tag{5.34}$$

$$\psi_e = \begin{cases} \dfrac{2W_f t_f}{bS_f f_c''}f_{ft} & \text{for completely wrapped} \\[3mm] \min \dfrac{2W_f t_f}{bS_f f_c''}f_{bt} \; ; \dfrac{2W_f t_f}{bS_f f_c''}f_{ft} & \text{for side bonding or U-jacketing} \end{cases} \tag{5.35}$$

where the effective stress, $f_{ft} = 0.5 \approx 0.75 f_{f,u}$. (τ/f_{co}') is the minimum of the values obtained from the following failure cases:

Failure mode of shear web-crushing:

$$\tau/f_c'' = \begin{cases} \dfrac{1}{2}\left[\sqrt{1+\alpha^2} - (1-2\psi)\alpha\right] & \text{for } 0 \le \psi \le \psi_o = \dfrac{\sqrt{1+\alpha^2}-\alpha}{2\sqrt{1+\alpha^2}} \\[3mm] \sqrt{\psi(1+\psi)} & \text{for } \psi_o \le \psi \le 0.5 \\[3mm] \dfrac{1}{2} & \text{for } \psi > 0.5 \end{cases} \tag{5.36}$$

Failure mode of flexure–shear interaction:

$$\tau/f_c'' = \begin{cases} \psi\left[\sqrt{\dfrac{2\eta}{\psi} + \alpha^2} - \alpha\right] & \text{for } \psi > \psi_o = \dfrac{2\eta}{2\alpha^2} \\[3mm] \dfrac{\psi\alpha}{4} + \dfrac{\eta}{2\alpha} & \text{for } \psi < \psi_o \end{cases} \tag{5.37}$$

$$\eta = \frac{A_{si} f_{ys.l}}{b d_v f_c''}; \quad \alpha = \frac{a}{d_v} \tag{5.38}$$

Finally, the actual load-carrying capacity of the strengthened RC beam is given by the minimum of the two failure load values obtained from Eqs. (5.36) and (5.37), which correspond to the failure mechanisms considered in the analysis.

7. Carolin and Täljsten (2005) proposed a model. Herein, the superposition principal has been applied to derive the governing equations of the model. Here, V_f represents the contribution of the externally bonded composite as illustrated by Eq. (5.39):

$$V_f = \eta_f.\varepsilon_{cr}.E_f.t_f.r_f.z.\frac{\sin(\alpha + \beta)}{\sin\alpha} \tag{5.39}$$

The reduction factor η_f considers the nonuniform distribution of stresses over the cross-section. The appropriate values for η_f were indicated to be 0.6 for side bonding and U-jacketing, and 1.0 for completely wrapped confinement. α is the angle of crack inclination, and β is the angle made by the fiber direction on the beam's longitudinal axis. The effective strain used in this model is

$$\eta_f = \frac{\int_{-h/2}^{h/2} \varepsilon_f(y)dy}{\varepsilon_{max}h} \tag{5.40}$$

$$\varepsilon_{cr} = \min \begin{cases} \varepsilon_{f,u} \\ \varepsilon_{bond}.\sin^2(\alpha + \beta) \end{cases} \tag{5.41}$$

$$\varepsilon_{bond} = \frac{\sqrt{2E_f.t_f.G_f}}{E_f.t_f} \begin{cases} \sin(\omega L_{cr}) \text{ for } L_{cr} \le \dfrac{\pi}{2\omega} \\ 1 \qquad\quad \text{ for } L_{cr} > \dfrac{\pi}{2\omega} \end{cases} \tag{5.42}$$

As in Eq. (5.41), the critical strain, ε_{cr}, is limited by the ultimate permissible fiber capacity, ε_{fu}, maximum permissible strain without achieving anchor failure, ε_{bond}, and maximum permissible strain to achieve concrete contribution, $\varepsilon_{c\ max}$.

$$\omega = \sqrt{\frac{\tau_{max}^2}{2E_f.t_f.G_f}} \tag{5.43}$$

$$r_f = \begin{cases} \sin\beta & \text{continuous wrapping} \\ \dfrac{W_f}{S_f} & \text{for discrete strips} \end{cases} \tag{5.44}$$

where $G_f \approx 0.644 f_{co}'^{\,0.19}$, $\tau_{max} = 3.5 f_{co}'^{\,0.19}$, and z is the length of a vertical tension tie in the truss equal to either $0.9d$ or the beam height for complete wrapping. The reduction of $\sin^2(\alpha + \beta)$ to the anchorage and concrete contribution arises from the anisotropic behavior of the composite.

8. Cao et al. (2005) proposed a simple model to predict the contribution of FRP to the shear capacity of RC beams and the effective strain of beams where "complete debonding of

the critical strips occurs." Furthermore, the maximum strain in the FRP at debonding was analyzed. The average and maximum strains along the critical shear crack were also determined by regression analysis, considering the discrete strain observations. To estimate the contribution of FRP to the shear resistance at debonding, the interaction between the shear span—effective depth ratio and the critical shear crack angle was analyzed. This yielded the following equations:

$$V_f = 2D_{f\theta}t_f W_f \varepsilon_{f.max} E_f \frac{h_{f,e}}{S_f(\tan\alpha)} \tag{5.45}$$

$$D_{f\theta} = \left(1 - \frac{\pi-2}{\lambda_{frp}\pi}\right) \times \begin{cases} 1.00 & \text{for } \lambda \le 1.4 \\ \dfrac{1}{1-0.2(\lambda'-1.4)^2} & \text{for } 1.4 < \lambda < 3 \\ 2.05 & \text{for } \lambda \ge 3 \end{cases} \tag{5.46}$$

where the normalized FRP bond length $\lambda_{frp} = L_{max}/L_e$, the effective bond length $L_e = (E_f t_f/(f_{co})^{0.5})^{0.5}$, the maximum bond length $L_{max} = h_{f,e}$ for U-jacketing, $L_{max} = h_{f,e}/2$ for side bonding, $h_{f,e}$ = effective height of FRP on the sides of the beam, and λ = the shear span—effective depth ratio.

$$\varepsilon_{fmax} = 0.427\beta_w \sqrt[4]{f'_{co}} \frac{1}{\sqrt{E_f.t_f}} \tag{5.47}$$

$$\beta_w = \sqrt{\frac{2.25 - (W_f/S_f)}{1.25 + (W_f/S_f)}} \tag{5.48}$$

9. Ye et al. (2005) developed a theoretical model of shear debonding strength. The proposed average $\varepsilon_{f,e}$ depends on the FRP strain for an infinite bond length when debonding is a dominant failure mode. Therefore Eq. (5.49) was adopted:

$$\varepsilon_{f.e} = k_v \varepsilon_{f.inf} \tag{5.49}$$

$$k_v = \begin{cases} 0.77(1-e^{-\lambda/0.79}) & \text{for side bonding} \\ 0.96(1-e^{-\lambda/0.62}) & \text{for U-jacketing} \end{cases} \tag{5.50}$$

$$\varepsilon_{f.inf} = \beta_w \sqrt{(0.616\sqrt{f_t})/(E_f t_f)} \tag{5.51}$$

The bond length ratio (Eq. 5.52) is expressed as the ratio of the FRP effective bond height $h_{f,e}$ to the effective bond length of FRP L_e:

$$\lambda = \frac{h_{f,e}}{2L_e \sin\beta} \tag{5.52}$$

$$L_e = 1.33 \frac{\sqrt{E_f t_f}}{f_t} \tag{5.53}$$

$$\beta_w = \sqrt{\frac{2.25 - (W_f/S_f)}{1.25 + (W_f/S_f)}}$$ (5.54)

10. Monti and Liotta (2007) developed a complete design method by considering different strengthening schemes and failure modes. The following three main aspects were addressed in developing the model: (1) a generalized FRP–concrete bond constitutive law was defined, (2) boundary limitations were considered, and (3) the stress field in the FRP crossing a shear crack was determined analytically. In addition, the following assumptions were considered:

- Shear cracks are uniformly spaced along the beam axis and inclined at an angle α.
- At the ultimate limit state, the cracks depth is equal to the internal lever arm $= 0.9d$.
- For U-jacketing (U) and wrapping (W), the resisting shear mechanism is based on the Mörsch truss. Meanwhile, for side bonding (S), a different resisting mechanism ("crack-bridging") is considered for development. This is because the Mörsch truss cannot be formed as the tensile diagonal tie is absent.

The effective debonding strength $f_{f,ed}$ was specified for side bonding, U-jacketing, and wrapping.

Side bonding:

$$V_f = \frac{1}{\gamma_{Rd}} 2t_f f_{fed} \frac{\sin\beta}{\sin\alpha} \cdot \frac{W_f}{S_f} \cdot \min(0.9d, h_w)$$ (5.55)

$$f_{fed} = f_{fdd} \frac{z_{rid.eq}}{\min(0.9d, h_w)} \cdot \left(1 - 0.6\sqrt{\frac{l_{eq}}{z_{rid.eq}}}\right)^2$$ (5.56)

$$z_{rid.eq} = \min(0.9d, h_w) - \left(l_e - \frac{S_f}{f_{fdd}/E_f}\right)\sin\beta$$ (5.57)

U-jacketing or complete wrapping:

$$V_f = 2\frac{1}{\gamma_{Rd}} 0.9 d t_f f_{fed} (\cot\alpha + \cot\beta) \cdot \frac{W_f}{S_f}$$ (5.58)

Design effective stress for U-jacketing:

$$f_{fed} = f_{fdd} \cdot \left(1 - \frac{l_e \sin\beta}{3\min(0.9d, h_w)}\right)$$ (5.59)

Design effective stress for complete wrapping:

$$f_{fed} = f_{fdd} \cdot \left(1 - \frac{l_e \sin\beta}{6\min(0.9d, h_w)}\right) + \frac{1}{2}(\varphi_R f_{fd} - f_{fdd}) \cdot \left(1 - \frac{l_e \sin\beta}{\min(0.9d, h_w)}\right)$$

(5.60)

$$L_e = \sqrt{\frac{E_f t_f}{2f_{ctm}}} \text{ where } f_{ctm} = 0.27 R_{ck}^{2/3}$$

(5.61)

$$f_{fdd} = \frac{0.80}{\gamma_{f.d}} \sqrt{\frac{2E_f \Gamma_{Fk}}{t_{ctm}}}$$

(5.62)

$$\Gamma_{Fk} = 0.03 k_b \sqrt{f_{ck} f_{ctm}} \ , k_b = \sqrt{\frac{2 - (W_f/S_f)}{1 + (W_f/400)}} \ge 1$$

(5.63)

$$W_f \le \min(0.9d, h_w) \sin(\alpha + \beta)/\sin\alpha$$

(5.64)

$$f_{fdd}(l_b) = f_{fdd} \frac{l_b}{l_e}\left(2 - \frac{l_b}{l_e}\right)$$

(5.65)

A reduction coefficient considering the radius of the corner (r_c) of the beam was applied when U-jacketing or wrapping were used:

$$\varphi_R = 0.2 + 1.6\frac{r_c}{b_w} \text{ where } 0 \le \frac{r_c}{b_w} \le 0.5$$

(5.66)

11. Pellegrino and Modena (2008) proposed a model based on the observed experimental shear failure caused by the peeling of a triangular portion of concrete cover in U-jacketing and side bonding FRP-sheet-strengthened RC beams. It considers the interaction between the external FRP and internal transverse steel reinforcement. The FRP strains were assumed to be equal to those of internal stirrups. The FRP shear contribution V_f was obtained from the rotational equilibrium of the forces F_f and F_c operating in the FRP and concrete surface, respectively, at failure:

$$V_f = \frac{2nt_f W_f L_f \varepsilon_{f,e} E_f h_f}{S_f}$$

(5.67)

$$\varepsilon_{f,e} = \frac{2f_{ct} A_c b_{c,v} \cos^2\varphi}{nt_f L_f E_f b_f \frac{h_f - l_e}{h_f}}$$

(5.68)

where f_{ct} is the tensile strength of concrete, A_c is the area of the beam cross-section, L_f is the length of the failure surface, and l_e is the effective bond length. ϕ is the angle characterizing the conventional roughness of the interface, which was assumed to be equal to 79 degrees according to the experimental results obtained by the authors. All the parameters are presented in Figs. 5.17 and 5.18.

12. Mofidi and Chaallal (2011) model reveals that the effect of transverse steel on the shear contribution of FRP is important. Separate design equations were proposed for the U-jacketing and side-bonding FRP configurations. The effective strain in this model is

$$V_f = 2t_f \varepsilon_{f,e} E_f W_f d_f \frac{(\cot\alpha + \cot\beta)\sin\alpha}{S_f} \tag{5.69}$$

$$\varepsilon_{f,e} = 0.31 \beta_c \beta_L \beta_w \sqrt{\frac{\sqrt{f'_c}}{t_f E_f}} \leq \varepsilon_{f,u} \tag{5.70}$$

$$\beta_c = \begin{cases} \dfrac{0.60}{\sqrt{\rho_f E_f + \rho_s E_s}} & \text{for U-jacketing} \\[2ex] \dfrac{0.43}{\sqrt{\rho_f E_f + \rho_s E_s}} & \text{for side-bonding} \end{cases} \tag{5.71}$$

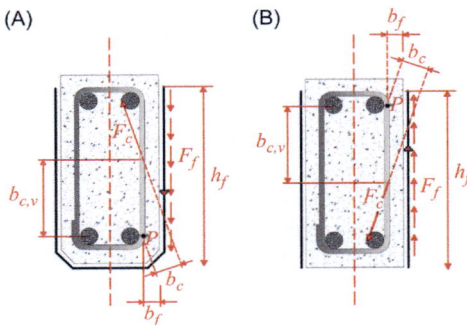

Figure 5.17 Forces acting in the cross section of U-jacketed and side-bonded beams (Pellegrino and Modena, 2008).

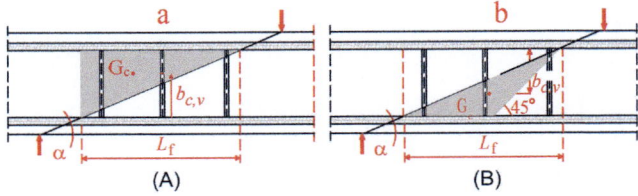

Figure 5.18 Shape of the fracture surface of (A) U-jacketed (B) side-bonding beams (Pellegrino and Modena, 2008).

$$L_e = \sqrt{\frac{t_f \, E_f}{2 \, f_t}} \tag{5.72}$$

$$\beta_w = \sqrt{\frac{2 - (W_f / S_f)}{1 + (W_f / S_f)}} \tag{5.73}$$

$$\beta_L = \begin{cases} 1 & \text{if } \lambda = \dfrac{L_{\max}}{L_e} \geq 1 \\[2ex] \sin\dfrac{\pi\lambda}{2} & \text{if } \lambda = \dfrac{L_{\max}}{L_e} < 1 \end{cases} \tag{5.74}$$

where β_L is a decreasing coefficient (FRP effective anchorage length ratio), β_c is the cracking modification factor, and E_f and E_s are the elastic moduli (in GPa) at βc.

13. Belarbi et al. (2012) conducted an experimental study on this topic using full-scale RC T-beams to understand the behavior of full-scale bridge beams strengthened in shear with FRP. The effective strain was calculated from Eq. (5.75) based on the experimental results of V_f. It was concluded that the effective strain decreases as the rigidity of FRP increases. The effective strain values of the full-scale beams are similar to those drawn from a database on FRP rigidity collected by the authors. This indicates that the effectiveness of FRP shear strengthening of full-scale RC beams matches with the range of FRP rigidity observed in this study.

$$V_f = 2t_f \varepsilon_{f,e} E_f W_f d_f \frac{(\cos\beta + \sin\beta)}{S_f} \tag{5.75}$$

14. Chen et al. (2013) proposed a shear strength model for evaluating FRP shear contribution by introducing a suitable shear interaction factor to consider the interaction between internal and external reinforcements. It is implicitly assumed that the contribution of concrete (V_c) to shear capacity is identical to that for ordinary RC beams given the reference code. The model is valid only for strengthening beams with side bonding and U-jacketing schemes. The shear resistance of an externally strengthened RC beam can be expressed as

$$V_{frp} = K \left(2t_f f_{f,e} E_f W_f h_f \frac{(\cot\alpha + \cot\beta)\sin\beta}{S_f} \right) \tag{5.76}$$

$$f_{f,e} = \sigma_{f,\max} D_{frp} \tag{5.77}$$

$$\sigma_{f,\max} = \min \begin{cases} f_f \\ \sigma_{db,\max} \end{cases} \tag{5.78}$$

$$\sigma_{db,\max} = \begin{cases} \sqrt{\dfrac{2E_f G_f}{t_f}} & L_{\max} \geq L_e \\[3ex] Sin(\dfrac{\pi}{2} \cdot \dfrac{L}{L_e}) \sqrt{\dfrac{2E_f G_f}{t_f}} & L_{\max} < L_e \end{cases} \tag{5.79}$$

$$L_{max} = \begin{cases} \dfrac{h_f + h_t + h_b}{2\sin\beta} & \text{for side strips} \\[2em] \dfrac{h_f + h_t + h_b}{\sin\beta} & \text{for U strips} \end{cases} \qquad (5.80)$$

where h_b = thickness of the concrete cover (from the beam bottom to the crack end) and h_t = vertical distance from the top of the FRP strips to the crack tip. K is the shear interaction factor that reflects the reduction in the efficiency of FRP strengthening owing to the adverse interaction effect between steel stirrups and FRP strips, as illustrated by Eq. (5.95).

L_{max} = maximum bond length of FRP strips intersected by the critical shear crack, L_e = effective bond length of FRP strips as defined by Eq. (5.89), h_b = thickness of the concrete cover (from the beam bottom to the crack end), and h_t = vertical distance from the top of the FRP strips to the crack tip. The expression of the stress distribution factor, D_{frp}, for FRP side strips is

$$D_{frp} = 1 - \left(1 - \frac{\pi}{4}\right).\left(\frac{h_{df}}{d_f}\right) - \left(\frac{h_{db}}{d_f}\right) \qquad (5.81)$$

$$h_{df} = \frac{h_f - h_{db}}{1 + \frac{\pi}{2}(K_h - 1)} \qquad (5.82)$$

$$h_{db} = \sin\beta \,.L_m - h_b \qquad (5.83)$$

$$L_m = K_h.L_e \qquad (5.84)$$

$$K_h = \frac{2}{\pi}\sqrt{\left(1 - \frac{\pi}{4}\right).\left(1 - \frac{\pi}{2}\right) + \frac{\pi}{2}.\left(1 - \frac{\pi}{4}\right).\left(\frac{h_f + h_b}{L_e\,.\sin\beta}\right) + 1} - \frac{2}{\pi} \qquad (5.85)$$

The expression of D_{frp} for FRP U strips is

$$D_{frp} = 1 - \left(1 - \frac{\pi}{4}\right).\left(\frac{h_{df}}{h_f}\right) \qquad (5.86)$$

$$h_{df} = \frac{2\delta_f.h_f}{w_{e,p}\sin(\alpha + \beta)} \qquad (5.87)$$

$$w_{e,p} = \delta_f \frac{1 + \frac{\pi}{2}\left(\frac{h_f}{L_e + \sin\alpha} - 1\right)}{\sin(\alpha + \beta)} \qquad (5.88)$$

$$L_e = \sqrt{\frac{\tau_f}{\delta_f.E_t.t_f}} \tag{5.89}$$

$$\delta_f = 2Gf/\tau_f \tag{5.90}$$

$$Gf = 0.308\beta_w\sqrt{f_t} \tag{5.91}$$

$$\tau_f = 1.5\beta_w f_t \tag{5.92}$$

$$\beta_w = \sqrt{\frac{2 - w_f/(s_f\sin\beta)}{1 + w_f/(s_f\sin\beta)}} \tag{5.93}$$

$$f_t = 0.395f_{cu}^{0.35} \tag{5.94}$$

$$K = \begin{cases} \dfrac{B}{B + \mu} & \text{for side strips} \\ 1 - \mu(1 - k_s) & \text{for U strips} \end{cases} \tag{5.95}$$

$$\mu = \frac{f_y A_s}{f_{f,e} A_f} \tag{5.96}$$

$$k_s = \frac{w_{ep}^{1.4}}{\alpha_0 + w_{ep}^{1.4}} \tag{5.97}$$

$$\alpha_0 = \frac{A}{(\cos\alpha)^{1.4}} \tag{5.98}$$

$$A = \begin{cases} 1.4\left[Ln(d_f) - 4.52\right](f_y - 173)(\varphi_s - 0.935)/10000 & \text{for plain bar stirrups} \\ 4.94\left[Ln(d_f) - 3.34\right](f_y - 245)(\varphi_s - 0.767)/10000 & \text{for deformed bar stirrups} \end{cases} \tag{5.99}$$

$$B = \begin{cases} \dfrac{1.01 \times 10^5}{\varphi_s^{0.834} f_y^{1.88}}\left[(d_f/L_e) + 2.11\right] & \text{for plain bar stirrups} \\ \dfrac{2.05 \times 10^5}{\varphi_s^{1.13} f_y^{1.71}}\left[(d_f/L_e) + 1.58\right] & \text{for deformed bar stirrups} \end{cases} \tag{5.100}$$

In the above equations, $\sigma_{db,max}$ is the maximum stress in the FRP strips crossing the critical shear crack as governed by debonding failure, L_{max} is the maximum bond length of the FRP strips intersected by the critical shear crack, and L_m is the maximum mobilized bond length.

The effectiveness of a hybrid FRP-fabric-reinforced cementitious matrix (FRCM) in shear strengthening was investigated through an experimental study by Jung et al. (2015). The FRP materials of FRCM are generally fabricated in the form of a fabric to enhance the bond strength between the FRP material and cementitious matrix. The hybrid FRP fabric used in this study consisted of CFRP and glass FRP (GFRP) in the warp and weft directions, respectively. A total of 11 beams were fabricated, with 8 of them being strengthened in shear with externally bonded hybrid FRP-FRCM. The number of plies, bond types, and spacing of the hybrid FRP fabric were considered as experimental variables. In addition, a shear capacity model of a FRCM shear strengthened beam was proposed. Therefore the contribution of the hybrid FRP-FRCM to the nominal shear strength can be defined as follows:

$$V_{frp} = 2\tau_b d_f W_f d \frac{\cot\alpha}{w} \tag{5.101}$$

where τ_b is the bond strength of the FRCM, d is the distance from the extreme compression fiber to the centroid of the longitudinal tension reinforcement, α is the inclination of the shear crack, w is the length of the shear span, and W_f is the width of the hybrid FRP fabric bonded to the shear span.

5.8.2 The models for design codes

In this section, different design codes for the FRP contribution V_f are explained.

1. ACI-440.2 (2008)

$$V_f = \frac{2W_f t_f E_f \varepsilon_{fe} d_f}{S_f}(\sin\beta + \cos\beta) \tag{5.102}$$

$$\varepsilon_{f,e} = \begin{cases} 0.004 \leq 0.75\varepsilon_{f,u} \text{ for completely wrapped} \\ K_v\varepsilon_{f,u} \leq 0.004 \text{ for two or three sides laminated} \end{cases} \tag{5.103}$$

$$K_v = \frac{k_1 k_2 L_e}{11,900\varepsilon_{f,u}} \leq 0.75 \tag{5.104}$$

$$L_e = \frac{23,300}{(t_f E_f)^{0.58}}, \quad k_1 = \left(\frac{f'_{co}}{27}\right)^{2/3} \tag{5.105}$$

$$
k_2 = \begin{cases} \dfrac{d_f - L_e}{d_f} & \text{for U-jacketing} \\[4mm] \dfrac{d_f - 2L_e}{d_f} & \text{for two sides l aminated} \end{cases}
\tag{5.106}
$$

2. Chinese code(Chines 2010)
a. Wrapping

$$
V_{frp} = \psi_v \sigma_{f,vd}(2t_f W_f d_f) \left(\frac{(\sin\theta + \cos\theta)}{(S_f + W_f)} \right)
\tag{5.107}
$$

$$
\sigma_{f,vd} = \min \begin{cases} f_f \\ E_f \varepsilon_{fev} \end{cases}
\tag{5.108}
$$

$$
f_{fd} = f_{fk}/\gamma_e \gamma_f
\tag{5.109}
$$

$$
\varepsilon_{fev} = \frac{8}{\sqrt{\lambda_{Ef}} + 10} \frac{f_{fd}}{E_f}
\tag{5.110}
$$

$$
\lambda_{Ef} = \frac{2t_f W_f d_f}{b(S_f + W_f)} \frac{E_f}{f_t}
\tag{5.111}
$$

b. U-jacketed and side bonded

$$
V_{frp} = k_f \tau_b W_f \frac{h_{fe}^2}{(S_f + W_f)} (\sin\theta + \cos\theta)
\tag{5.112}
$$

$$
k_f = \varphi \frac{\sin\alpha \sqrt{E_f t_f}}{\sin\alpha \sqrt{E_f t_f} + 0.3 h_{fe} f_t}
\tag{5.113}
$$

$$
\tau_b = 1.5 \beta_w f_t
\tag{5.114}
$$

$$
\beta_w = \sqrt{\frac{2 - w_f/(s_f \sin\beta)}{1 + w_f/(s_f \sin\beta)}}
\tag{5.115}
$$

$\phi = 1$ for side bonded, and 1.3 for U-jacketing.

f_{fd} is the design value of the tensile strength of FRP materials, f_{fk} is the standard value of the tensile strength of FRP materials, γ_e is the environmental reduction factor, and γ_e is the breakdown coefficient of the material ($=1.4$ for fiber cloth and FRP reinforcement, and 1.25 for other FRP products).

3. FIB Bulletin 35 (2006) and Italian Code CNR-DT200/2004 (2004)

$$
V_{frp} = \begin{cases} \dfrac{1}{\gamma_{rd}} \sigma_{f,vd} \min\{h_w,\ 0.9*d_f\} \dfrac{(\sin\beta)}{\sin\alpha} \dfrac{2t_f W_f}{S_f} & \text{for side bonding} \\[2em] \dfrac{1}{\gamma_{rd}} \sigma_{f,vd}(2t_f W_f 0.9*d_f) \dfrac{(\cot\beta + \cot\alpha)}{S_f} & \text{for U-jacketing and wrapping} \end{cases}
\tag{5.116}
$$

$$
\gamma_{rd} = 1.2
\tag{5.117}
$$

$$
\sigma_{f,vd} = f_{fdd} \cdot \frac{Z_{rid,req}}{\min\{h_w,\ 0.9*d_f\}} \cdot \left(1 - 0.6\sqrt{\frac{l_{eq}}{Z_{rid,req}}}\right)^2 \quad \text{for side bonding}
\tag{5.118}
$$

where

$$
Z_{rid,req} = Z_{rid} + l_{eq} \quad Z_{rid} = \min\{h_w,\ 0.9*d_f\} - l_{eq}.\sin\alpha \quad l_{eq} = \frac{s'_f}{f_{fdd}/E_f}.\sin\alpha
\tag{5.119}
$$

l_{eq} is the optimal anchorage length, and s'_f is the ultimate debonding slip (assumed to be 0.2 mm).

$$
\sigma_{f,vd} = f_{fdd} \cdot \left[1 - \frac{1}{3} \frac{l_{eq}\sin\beta}{\min\{h_w,\ 0.9*d_f\}}\right] \qquad \text{For U-jacketing}
\tag{5.120}
$$

$$
\sigma_{f,vd} = f_{fdd} \cdot \left[1 - \frac{1}{6} \frac{l_{eq}\sin\beta}{\min\{h_w,\ 0.9*d_f\}}\right] + \frac{1}{2}(\varphi_R.f_{fd} - f_{fdd})
$$

$$
\left[1 - \frac{l_{eq}\sin\beta}{\min\{h_w,\ 0.9*d_f\}}\right] \quad \text{for wrapping}
\tag{5.121}
$$

$$
\varphi_R = 0.2 + 1.6\frac{r_c}{b_w}, \qquad 0 \le \frac{r_c}{b_w} \le 0.5
\tag{5.122}
$$

where r_c is the corner rounding radius and b_w is the web width.

4. JSCE Code (Japan Society of Civil Engineers (Japan Society of Civil Engineers JSCE, 2001))

In Method (1), the coefficient expressing the shear reinforcing efficiency of the continuous fiber sheet is used to evaluate the ultimate mean stress of the sheet and to determine the shear contribution of the sheet:

$$
V_f = K \cdot \left[A_f \cdot f_{fud} \frac{(\sin\beta + \cos\beta)}{(S_f)}\right] \cdot Z/\gamma_b
\tag{5.123}
$$

$$
K = 1.68 - 0.67R(0.4 \le K \le 0.8)
\tag{5.124}
$$

where K is the shear reinforcing efficiency of the continuous fiber sheets.

$$R = (\rho_f \cdot E_f)^{0.25} \left(\frac{f_{fud}}{E_f}\right)^{2/3} \left(\frac{1}{f'_{co}}\right)^{1/3} \quad (0.5 \le R \le 2.0) \tag{5.125}$$

$$\rho_f = A_f / (b_w \cdot S_f) \tag{5.126}$$

A_f: Total cross-sectional area of continuous fiber sheets in space s_f
s_f: Spacing of continuous fiber sheet
f_{fud}: Design tensile strength of continuous fiber sheet (N/mm^2)
E_f: Modulus of elasticity of continuous fiber sheet (kN/mm^2)
β: Angle formed by continuous fiber sheet on the member axis
γ_b: Member factor (generally set to 1.25)

In Method (2), the stress distribution of the continuous fiber sheets is evaluated based on the bond constitutive law to determine the shear contribution of the sheet. This method uses numerical calculation based on the following hypothesis to evaluate the stress distribution of the continuous fiber sheet in upgraded members for determining the shear contribution of the sheet.

i. The shear crack forms an angle of 35 degrees on the member axis.
ii. The member deformation after the occurrence of shear crack is expressed by a rigid body rotation model with the end of a shear crack as the center of rotation.
iii. The progress of sheet peeling that traverses the shear crack is evaluated through stress analysis. The assumptions are that the concrete is a rigid body, the sheet is an elastic body, and there is a linear relationship between the relative displacement and bond stress between the sheet and e concrete (the bond constitutive law).
iv. The strain of concrete in compression sections is expressed as a function of the angle of rotation of the members, for which rigid body rotation is assumed.

The member factor generally used for this method is 1.25.

5.9 Numerical-based model for prediction of shear capacity

5.9.1 Prediction model

It was previously observed based on FE simulation of 55 beams that the increase in shear capacity varied with the variations in b, t_f, f'_c, h_f, d, a/d, W_f /S_f, and E_f (Sayed et al., 2013). Therefore the relationship between the increase in shear capacity and those parameters is formulated as an equation for each type of strengthening configuration (side bonding, U-jacketing, and completely wrapped), as expressed by Eqs. (5.19)−(5.21):

For side bonding:

$$V_f = C_S \left(\frac{W_f^{0.79} t_f^{0.334} E_f^{0.50} h_f^{0.90} d^{0.30} f'_c^{0.50}}{S_f^{0.79} b^{0.21}} \right) \tag{5.127}$$

For U-jacketing:

$$V_f = C_U \left(\frac{W_f^{0.41} t_f^{0.334} E_f^{0.50} h_f^{0.95} d^{0.25} b^{0.25} f_c'^{0.55}}{S_f^{0.41} a^{0.25}} \right) \tag{5.128}$$

For completely wrapped:

$$V_f = C_W \left(\frac{W_f^{0.41} t_f^{0.334} E_f^{0.334} h_f d^{0.45} b^{0.30} f_c'^{0.60}}{S_f^{0.41} a^{0.45}} \right) \tag{5.129}$$

where the constants C_S, C_U, and C_W are the slopes of the relationship between the increase in the shear capacity V_f obtained from the FE simulation and the corresponding parameters, for side bonding, U-jacketing, and completely wrapped, respectively (Fig. 5.19). C_S, C_U, and C_W are 0.156, 0.089, and 0.57, respectively.

5.9.2 Comparison with the existing experimental results

To examine the reliability and validity of the proposed model, an extensive verification is carried out using a series of experimental data available in the literature. The database considered is composed of the results of 274 experimental tests: it includes 87 beams strengthened by FRP side bonding, 115 beams strengthened by FRP U-jacketing, and 72 beams strengthened by FRP complete wrapping. The tests differ in terms of geometry, concrete strength, shear span—depth ratio, and reinforcement ratios, with a wide range of geometrical and mechanical characteristics. The main geometric and material properties covered by the experimental data and the likely failure modes described earlier have been considered in Table 5.4.

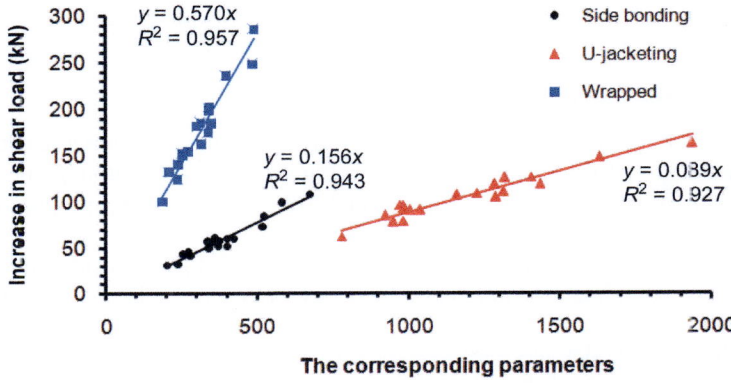

Figure 5.19 The relation between the increases in the shear capacity obtained from the FE simulation and the corresponding parameters (Sayed, 2014).

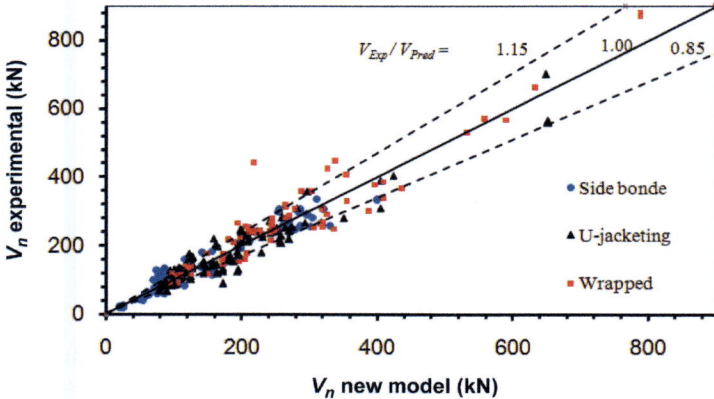

Figure 5.20 Predicted shear strength by the new model in comparison with the experimental values (Sayed, 2014).

A graphical representation of the comparison between the experimental and numerical values is shown in Fig. 5.20. In this figure, the proposed model is shown to consider the different types of shear strengthening configurations. For the beams strengthened with side bonding, U-jacketing, and complete wrapping, the mean values of V_{Exp}/V_{Pred} are 1.03, 0.98, and 1.03, respectively. The corresponding coefficients of variation (COV_S) are 18.9%, 17.0%, and 18.3%, and coefficients of correlation (R) are 0.96, 0.95, and 0.95, respectively. These values reveal that from a statistical perspective, the proposed model is equally reliable for all the shear strengthening configurations considered in the analysis.

In addition, in this figure, two lines limited to $\pm 15\%$ deviation from the new model predictions and from the experimental values are also reported. As is observed, almost all the results fall within these bounds. The average of the new model predictions is observed to be very close to the experimental results.

5.9.3   Comparison with available models

To evaluate the trustworthiness of the predicted results, the interactive results of the design models (ACI Committee 440, 2008; Triantafillou and Antonopoulos, 2000; Matthys and Triantafillou, 2001; Carolin and Täljsten, 2005) and proposed model were compared with the experimental results.

Table 5.6 presents the average, R, and $COVs$ of V_{Exp}/V_{Pred} for side bonding, U-jacketing, and complete wrapping configurations, both separately and in a total validation set.

The proposed model has the best average for V_{Exp}/V_{Pred}. The corresponding COV is less than those of the other models, and R is higher than those of the other models for beams strengthened with side bonding, U-jacketing, and complete wrapping.

Table 5.6 Average, r, and COV of V_{Exp}/V_{Pred} for different models (Sayed, 2014).

		ACI440	Tri. & Ant.	Matthys & Tri.	Täljsten	Sayed et al. (2013)
Side bonding	Average	1.16	0.98	1.18	1.18	1.03
	R	0.89	0.94	0.88	0.87	0.96
	COV %	28.4	24.7	54.7	32.0	18.9
U-jacketing	Average	1.22	1.11	1.16	1.33	0.98
	R	0.90	0.88	0.87	0.78	0.95
	COV %	25.5	22.5	24.3	25.6	17.0
Complete wrapped	Average	1.67	1.06	1.19	1.08	1.03
	R	0.83	0.88	0.85	0.86	0.95
	COV %	23.8	28.1	32.6	29.1	18.3
Total set of validation	Average	1.32	1.06	1.17	1.22	1.01
	R	0.87	0.91	0.89	0.86	0.96
	COV %	30.3	25.1	38.7	29.5	18.1

5.10 Failure modes of torsional strengthened beams

5.10.1 Complete wrapping strips and sheets

Ghobarah et al. (2002) observed that for beams strengthened with strips placed at 45 degrees and spiraling around each beam, after the ultimate load is attained, the carbon fibers failed in tension with highly rapid deterioration of the load-carrying capacity of the beam (Fig. 5.21A). Hii and Al-Mahaidi (2006, 2007) and Al-Mahaidi and Hii (2007) tested a solid and hollow box section beam strengthened with two layers of CFRP strips. They observed that at higher load levels, cracking propagated along the strips. It resulted in partial debonding of the CFRP strips (Fig. 5.21B). High-speed video recordings revealed that failure rupture of the CFRP strips generally occurred first at the corners. This was followed by peeling away, with a thin layer of concrete underneath. For the solid section, partial rupture in the middle of the CFRP strip first occurred at the top face. Following this, an explosive succession of ruptures occurred at the corners.

Ameli et al. (2007) observed that all the beams wrapped with CFRP failed highly abruptly after attaining their peak torque; Fig. 5.21C−D. However, the GFRP wrapped beams displayed a longer post-peak unloading response and developed larger angles of twist. The failure of both CFRP- and GFRP-wrapped beams was initiated similarly by the FRP rupture at the peak. Then, GFRP strengthened beams deformed significantly as they unloaded.

The failure mode for beams that were completely wrapped with FRP sheets or strips was FRP rupture. The major crack angles with respect to the longitudinal axis of the beam were approximately 45 degrees, whereas the minor crack angles varied between 40 and 49 degrees. The cracks were more uniformly distributed in beams that were wrapped completely around the whole cross-section. In contrast, a local helical pattern of cracks was generally observed in the unwrapped beams.

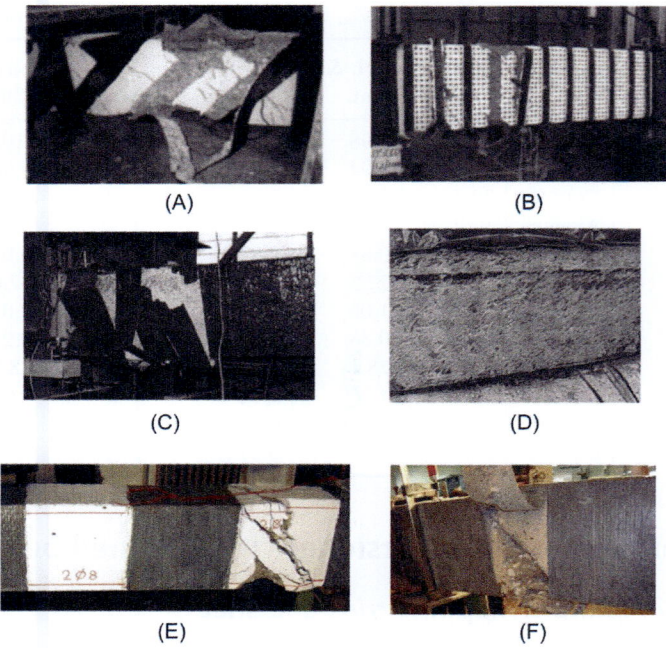

Figure 5.21 Failure modes of torsional strengthened beams. (A) Ghobarah et al. (2002). (B) Hii and Al-Mahaidi (2006). (C) Ameli et al. (2007), FRP rupture. (D) Ameli et al. (2007), cracks on the beam. (E) Chalioris (2008), diagonal cracks. (F) Chalioris (2008), FRP rupture.

Chalioris (2008) observed that the failure of the wrapped beams with FRP strips was partially delayed with respect to the failure of the control specimens. However, eventually, diagonal torsional cracks occurred and widened in the unwrapped concrete part of the beams between the strips. No fiber rupture was detected (Fig. 5.21E). Beams wrapped with continuous FRP sheets exhibited a completely different failure mode because the fibers inhibited the propagation of cracks. Thereby, the completely wrapped beams presented higher values of torque moment at cracking. Subsequently, torsional failure occurred in conjunction with tensile rupture of FRP sheet, at a high level of loading (Fig. 5.21F). The failure started in the most highly stressed fibers. This was followed rapidly by the rupture of the part of the FRP sheet intersected by the main torsional crack.

5.10.2   U-Jacket sheets and strips

Salom et al. (2004) tested a spandrel beam with [0/90] composite laminate and without anchorage, where the 90 degrees lamina was placed perpendicular to the longitudinal axis of the RC spandrel beam. The 0 degrees lamina was applied to delay diagonal cracking by resisting a part of the torsional forces, and functioned similarly to the longitudinal steel rebars used in conventional torsional reinforcing.

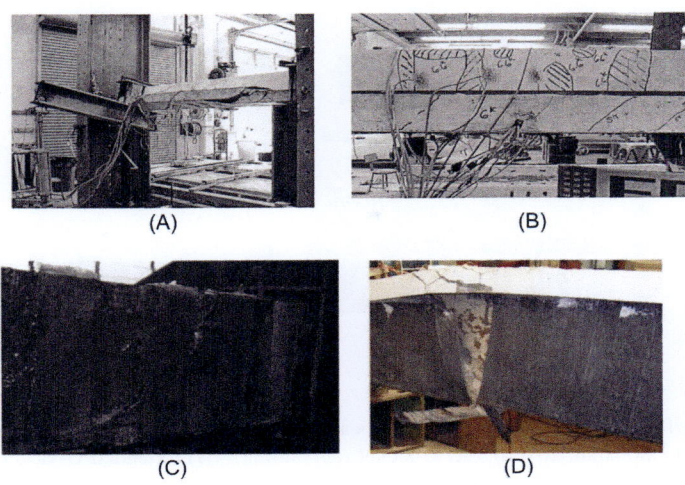

(A) (B)

(C) (D)

Figure 5.22 Failure modes of torsional strengthened beams. (A) Salom et al. (2004) [0/90] composite laminate. (B) Salom et al. (2004) [90] composite laminate with anchorage. (C) Ameli et al. (2007). (D) Chalioris (2008) debonding.

The 90 degrees lamina provided the primary torsional strength to the specimen and performed similarly to the stirrups in an RC spandrel beam. It was observed that the failure occurred close to the fixed end of the sample due to an extensive torsional crack, which was followed by composite delamination (Fig. 5.22A). When the authors used [90] composite laminate with anchorage, hairline cracks developed in the beam, then a general composite delamination occurred, followed by extensive concrete cracking, which ultimately resulted in specimen failure (Fig. 5.22B).

Ameli et al. (2007) found that for beams with U-jacket configuration, debonding governed the failure (Fig. 5.22C). Chalioris (2008) observed also that the U-jacketed flanged beams exhibited premature debonding failure at the concrete and the FRP adhesive interface (Fig. 5.22D). The failure process is characterized by a sequential debonding that has been initiated from the most vulnerable and highly stressed region of the concrete—FRP adhesive interface. This kind of failure is the reason for the limited increase on the torsional capacity of the flanged beams.

5.10.3 Combined torsion and shear

Deifalla and Ghobarah (2010b) tested RC T-beams subjected to combined torsion and shear. When the beam was strengthened using U-Jacket, the beam failed by flange cracks and FRP debonding, or rupture as the flange was not covered with FRP (Fig. 5.23A). The FRP debonding occurred first near the west end of the test region and propagated in a spiral form around the perimeter of the beam. The pattern of the debonding propagation was similar to the torsion crack propagation. FRP debonding occurred following parallel spirals at approximately equal distances, which was also similar to the torsion crack pattern. As the load increased,

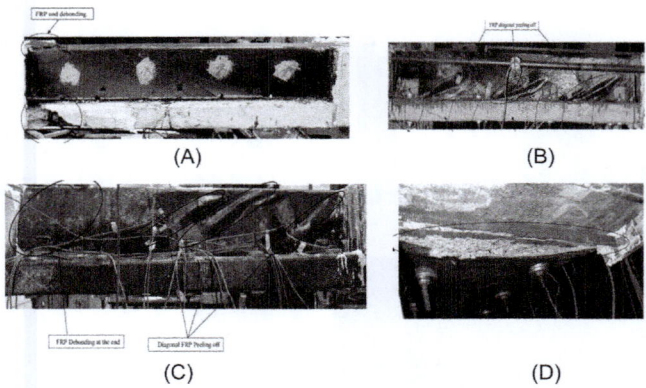

Figure 5.23 Failure modes of strengthened beams subjected to combined torsion and shear (Deifalla and Ghobarah, 2010b). (A) U-jacket (debonding). (B) Extended U-jacket (debonding). (C) Full wrapping (debonding). (D) Combined full wrapping and extended U-jacket (debonding).

diagonal cracks continued to widen and new ones were initiated in the concrete flange and debonding continued in the FRP sheets. Spalling occurred in the flange, where the concrete cover was falling out in chunks.

For beam strengthened with extended U-Jacket, sounds indicating debonding in the FRP along with cracks in the flange were heard (Fig. 5.23B). The flange cracks continued to propagate and widen. One of the concrete cracks crossed the location of the anchor bolt. Spalling occurred in the flange and around the bolt where small pieces of concrete started falling down on the floor. Anchor bolts and transverse steel reached yield.

For beam strengthened with full wrapping, debonding sounds from the FRP were heard. The FRP debonded in the east side of the test region of the beam. The debonding propagated in a spiral form around the beam similar to the torsion crack pattern. Parallel spiral debonding with an almost equal spacing continued (Fig. 5.23C). For beam strengthened with combined full wrapping and extended U-jacket, debonding noises from the FRP were heard (Fig. 5.23D).

5.10.4   Combined torsion and bending

Jariwala et al. (2013) tested rectangular beams subjected to combined torsion and bending strengthened with different GFRP configurations. Failure of fibers in specimen with full wrapping was observed from the edge of central portion of beam specimen (Fig. 5.24A). Inclined diagonal cracks were observed inside the wrapping. Sudden failure of RC beam was observed after generation of first crack in fiber. In specimen with full longitudinal wrapping (Fig. 5.24B), debonding of fiber was observed from many places. Failure of fibers started in longitudinal direction and propagated throughout the length of beam.

Initially vertical crack was observed at bottom of the side face in specimen with strip wrapping (Fig. 5.24C). Diagonal torsional cracks occurred and widened in the

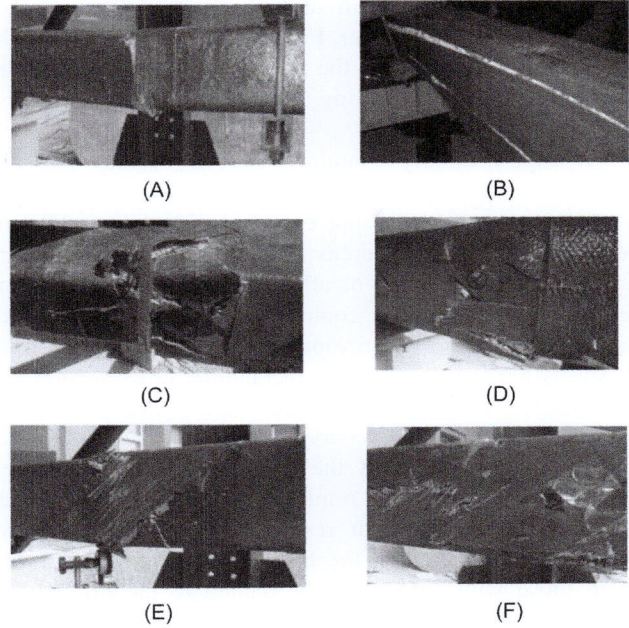

Figure 5.24 Failure modes of strengthened beams subjected to combined torsion and bending (Jariwala et al., 2013). (A) Full wrapping. (B) Full longitudinal wrapping. (C) Strip wrapping. (D) Corner and strip wrapping. (E) Diagonal strip wrapping. (F) Corner and diagonal strip wrapping.

unwrapped concrete part of the beams. Failure of beam was exhibited by extended diagonal cracks started from the vertical face. Maximum twisting and central displacement was observed in specimen with corner and strip wrapping (Fig. 5.24D). The failure was partially delayed compared to control specimens. Diagonal torsional cracks occurred and widened in the unwrapped concrete part between the strips on all four sides of the beam. Failure of GFRP started by tearing of bottom corner longitudinal strip. In specimen with diagonal strip wrapping (Fig. 5.24E), first crack started from the bottom of vertical face of the beam on concrete.

Cracks were developed perpendicular to the fiber orientation. Failure of GFRP started from the edge as well as from the center of vertical side by tearing of diagonal fibers and corner longitudinal fibers in specimen with corner and diagonal strip wrapping (Fig. 5.24F).

5.11 Resistance mechanism of FRP torsional strengthening

Shearing stresses and principal stresses develop when a concrete member is loaded in pure torsion. One or more inclined cracks develop when the maximum principal tensile stress attains the tensile strength of the concrete. The onset of cracking causes failure of an unreinforced member. Furthermore, the addition of longitudinal steel without stirrups negligibly affects the strength of a beam loaded in pure torsion because it is effective only in resisting the longitudinal component of the diagonal tension forces.

Chalioris (2007) experimentally examined 12 RC beams with a rectangular cross-section. These were tested under pure torsion and sorted into two groups of six. Eight beams were strengthened using epoxy-bonded carbon FRP materials as external transverse reinforcement, whereas the other four were the control specimens. The longitudinal reinforcement of all the tested beams was identical. Each group comprised two unstrengthened control specimens: one without transverse reinforcement and one with closed stirrups. Three beams were retrofitted using continuous FRP sheets wrapped around the rectangular cross-section of the beams along their entire length, whereas five beams were strengthened using discrete FRP strips wrapped around the cross-section. The fiber direction was oriented perpendicular to the longitudinal axis of the beam. As anticipated, the retrofitted beams with FRP as external transverse reinforcement displayed increased torsional strength and higher performance with respect to the control specimens. This improvement was more significant in the FRP beams strengthened with continuous sheets than in those with strips. This is related to the different failure processes of these beams. The strips prevented crack propagation at the retrofitted parts of the beam, and failure was observed mainly at the unwrapped concrete. In contrast, wrapping with continuous sheets shifted the failure to fiber rupture. Moreover, the advantages of the high strength composite material were utilized further.

With regard to the alignment of FRP, Salom et al. (2004) stated that when the fibers were aligned with the principal tensile stress (45 degrees), the strain in the FRP attained the highest level, and the strengthening efficiency was the maximum.

In a study of the torsional behavior of solid and box-sectioned beams strengthened with CFRP strips, Hii and Al-Mahaidi (2006) demonstrated that crack growth was impeded by the presence of the FRP strips. It resulted in higher aggregate interlock and higher postcracking stiffness. According to these authors, the increased torsional capacity was owing to the direct contribution of the CFRP strips and the indirect higher aggregate interlock effect.

Al-Mahaidi and Hii (2007) used photogrammetry to study the strain distribution of FRP strips used to strengthen solid and box RC cross-sections. After the formation of torsional cracks, the strain at the crack locations increased significantly. With further increase in the load, the crack extended and the area of active bonding shifted until debonding of the whole FRP strip occurred. This altered the strain distribution of the FRP strips and resulted in a flat profile where FRP strips functioned as an unbonded belt.

Ameli et al. (2007) explained a noteworthy aspect in their study on the use of externally bonded CFRP and GFRP sheets for torsional strengthening of RC beams. Although the CFRP strengthened beams exhibited higher peak torsional moment, the GFRP strengthened beams offered higher peak twist angles. The authors explained this by stating that the GFRP sheets are softer, attract less of the cross-sectional torque, and deform more straightforwardly in harmony with the RC beam.

5.12 Evaluation of torsional strength of FRP-strengthened elements

The internal steel reinforcements in most cases requiring strengthening are yielded or deteriorated. Therefore the assumption of no interaction between the steel and FRP for torsional strength may be adopted for simplification. The total contribution of reinforcement (T_r) to the torsion capacity of RC beams can be computed by

$$T_r = T_s + T_f \tag{5.130}$$

where T_s and T_f are the steel and FRP contributions, respectively, to the torsion capacity of the cross-section.

5.12.1 Available models

The models for predicting the behavior of FRP-strengthened RC beams subjected to torsion are limited. The following models are available from different resources.

I. Deifalla and Ghobarah (2005) and Deifalla and Ghobarah (2010a):

Applying the Mohr circle equilibrium for space truss in conjunction with the hollow tube analogy, Deifalla and Ghobarah (2005) proposed that the torsion contribution of inclined steel reinforcements (T_s) to the total torsion capacity of an RC beam be computed as

$$T_S = \frac{2A_o f_y A_t \left[\cot(\beta_s) + \cot(\alpha) \right] \sin(\beta_s)}{S_s} \tag{5.131}$$

where A_o is the area enclosed inside the center line of the shear flow path, f_y is the yield strength of the reinforcement, A_t is the area of the reinforcement resisting torsion, α is the angle of inclination of the principal cracks, and β_s is the angle of inclination of the steel reinforcement. Similarly, the FRP contribution (T_f) to the torsion capacity can be calculated by

$$T_f = \frac{2A_{of} f_f A_f \left[\cot(\beta_f) + \cot(\alpha) \right] \sin(\beta_f)}{S_f} \tag{5.132}$$

where A_{of} is the area enclosed inside the critical shear flow path owing to the strengthening, f_f is the stress in the FRP sheets at failure, β_f is the angle of orientation of the fiber direction to the longitudinal axis of the beam, and s_f is the spacing between the center line of the FRP strips. A_f is the effective area of the FRP resisting torsion and is calculated using

$$A_f = n_f t_f w_f \tag{5.133}$$

where n_f is the number of FRP layers and w_f is the width of the FRP strips. Eq. (5.133) implies that the use of a number of FRP layers is equivalent to that of an FRP layer of equal total area. Experimental observations have revealed that this is not necessarily the case (Zhang et al., 2001; Panchacharam and Belarbi, 2002). This assumption will be used to simplify the design. Eqs. (5.130), (5.132), and (5.133) accounts for the effects of parameters such as the strengthening techniques, number of FRP layers, thickness of each layer, spacing between FRP strips, fiber orientation of FRP sheets, average stress level of FRP sheets, and angle of principal crack.

5.12.1.1   FRP effective stress f_f

To use the previously mentioned equations, the effective stress of FRP, f_f, has to be computed. Assuming that FRP sheets are elastic up to failure, the constitutive model would be a straight line bounded by the effective FRP strain level. The effective strain level of the FRP depends on several parameters such as the dominating failure mode of the beam, FRP properties, beam dimensions, concrete properties, and strengthening scheme configuration. The FIB (2001) task force report indicated that the development of a rigorous analytical model to calculate the effective FRP strain level is exceptionally challenging. In a study conducted by Deifalla and Ghobarah (2005), the FRP constitutive model was developed. The constitutive model is elastic up to failure. This is determined by limiting the effective FRP strain. Three limits that were derived will be used in this proposed simplified model.

Beams strengthened in torsion fail either by debonding or rupture of the FRP. In both cases, the effective strain level is lower than the ultimate strain level. For the case of debonding failure, the effective strain is lower than the ultimate strain due to the existence of an effective bond length, which governs the maximum interfacial force between the FRP and concrete. Teng et al. (2002) proposed a formula for determining the effective bond length (L_e) of FRP-strengthened RC beams:

$$L_e = \sqrt{\frac{E_f t_f}{\sqrt{f'_{co}}}} \tag{5.134}$$

This formula was used in this study to calculate the development length. Debonding failure could occur due to excessive concrete cracking or bond slip. Deifalla and Ghobarah (2005) limited the effective FRP strain to account for the failure due to FRP peeling off as a result of concrete cracking, as shown in

Eq. (5.135). This limit was derived based on the maximum shear stress that the RC beam can sustain under torsion loading.

$$\varepsilon_f = \frac{0.33}{L_e} \frac{w_f}{S_f} \tag{5.135}$$

Another limit for the effective FRP strain was introduced to account bond-slip failure as shown in Eq. (5.136). The limit was developed based on the nonlinear fracture mechanics bond model introduced by Teng et al. (2002). The original model was developed to predict the strength of RC beams strengthened in shear. It was modified and calibrated to fit the proposed model to the case of torsion strengthening of RC beams.

$$\varepsilon_f = \frac{0.02\alpha_f}{L_e} \tag{5.136}$$

where α_f is a constant that considers the difference in the stress distribution between the continuous FRP sheets and the strips. This can be calculated as

$$\alpha_f = \sqrt{\frac{\left(2 - \frac{w_f}{S_f.\sin(\beta_f)}\right)}{\left(1 + \frac{w_f}{S_f.\sin(\beta_f)}\right)}} \tag{5.137}$$

For the case of rupture failure, strengthened beams fail at an effective strain level lower than the ultimate strain level. This is due to the nonuniform distribution of the strain across the concrete cracks. Another limit was derived by Deifalla and Ghobarah (2005) by considering the axial rigidity of the FRP with respect to the concrete section. The model is being used to predict the strain limit for beams failing in rupture as shown in Eq. (5.138). Here, the effective FRP strain can be computed by

$$\varepsilon_f = 0.1(E_{fu}\rho_{ft})^{-0.86}\varepsilon_{ft} \tag{5.138}$$

where ρ_{ft} is the FRP reinforcement ratio and is calculated as

$$\rho_{ft} = \frac{A_f}{t_c S_f} \tag{5.139}$$

where t_c is the thickness of the equivalent hollow tube section, considered as equal to A_o/p_f. Eqs. (5.135), (5.136), and (5.138) represent the three proposed limits for the effective FRP strain.

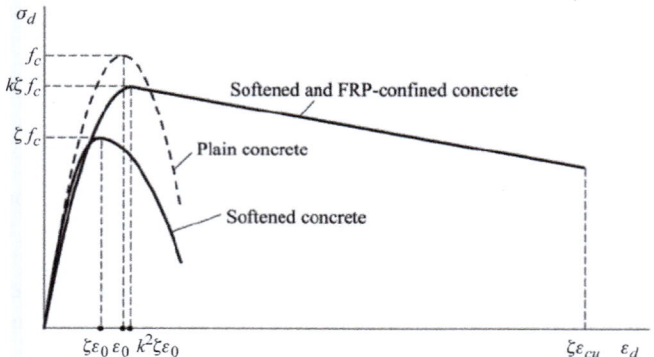

Figure 5.25 Constitutive stress−strain laws for concrete (Chalioris, 2007).

II. Chalioris (2007):

The first analytical model was proposed by Chalioris (2007), where the classical softened truss model forms the basis. This model utilizes softened and FRP confined compression curve for concrete, as shown in Fig. 5.25. In addition, the author concluded that the proposed model is effective in predicting postcrack stiffness. The proposed model is discussed in detail in the next section:

The strength of concrete struts is reduced substantially by diagonal cracking caused by tension in the perpendicular direction (concrete softening). The appropriate constitutive law with softening effect was developed by Belarbi and Hsu (1995):

$$\sigma_d = \zeta f'_{co}\left[2\frac{\varepsilon_d}{\zeta\varepsilon_o} - \left(\frac{\varepsilon_d}{\zeta\varepsilon_o}\right)^2\right] \text{ for } \varepsilon_d \leq \zeta\varepsilon_o \tag{5.140}$$

$$\sigma_d = \zeta f'_{co}\left[1 - \left(\frac{\varepsilon_d/(\zeta\varepsilon_o)-1}{2/\zeta-1}\right)^2\right] \text{ for } \varepsilon_d > \zeta\varepsilon_o \tag{5.141}$$

$$\zeta = \frac{0.9}{\sqrt{1+400\varepsilon_r}} \tag{5.142}$$

where f_{co} is the concrete cylinder compressive strength, $\varepsilon_o = -0.002$ (concrete strain corresponding to peak strength), ζ is the softening coefficient, and ε_r = principal tensile strain.

III. Chalioris (2007) model for confined concrete

$$\sigma_d = k\zeta f'_{co}\left[2\frac{\varepsilon_d}{k^2\zeta\varepsilon_o} - \left(\frac{\varepsilon_d}{k^2\zeta\varepsilon_o}\right)^2\right] \text{ for } \varepsilon_d \leq k^2\zeta\varepsilon_o \tag{5.143}$$

$$\sigma_d = \varepsilon_d\left(\frac{k\zeta f'_{co}-f_{cu}}{\zeta\varepsilon_{cu}-k^2\zeta\varepsilon_o}\right) \text{ for } \varepsilon_d > \zeta\varepsilon_o \tag{5.144}$$

$$\varepsilon_o = 0\,003k^2 \tag{5.145}$$

$$k = 1 + 1.4\alpha_n\omega_\omega \tag{5.146}$$

$$\alpha_n = 1 - \frac{b^2 + h^2}{3A} \tag{5.147}$$

$$\omega_\omega = \frac{\text{volume of FRP material}}{\text{volume of the confined concrete core}}\frac{f_{fu}}{f'_{co}} \tag{5.148}$$

$$f_{fe} = \varepsilon_{fe}E_f \tag{5.149}$$

where ε_{cu} is the point of ultimate strain, f_{cu} is the point of ultimate stress, α_n is the in-section coefficient confinement calculated using the cross-sectional dimensions (b and h), f_{fu} is the ultimate tensile strength of the FRP, E_f is the elastic modulus of the FRP, A is the area of the cross-section, and ω_w is the volumetric mechanical ratio for external FRP-confinement. The effective strain in the principal material direction ε_{fe} is calculated as follows:

a. For the case of CFRP

for wrapping and rupture failure:

$$\varepsilon_{fe} = 0\,17\left(\frac{f'_{co}2/3}{E_f\rho_f}\right)^{0.3}\varepsilon_{fu} \tag{5.150}$$

for U-jacketing and rupture or peeling-off failure:

$$\varepsilon_{fe} = \min\begin{cases} 0.17\left(\frac{f'_{co}2/3}{E_f\rho_f}\right)^{0.3}\varepsilon_{fu} \\ 0.65\left(\frac{f'_{co}2/3}{E_f\rho_f}\right)^{0.56} \times 10^{-3} \end{cases} \tag{5.151}$$

b. For the case of GFRP

$$\varepsilon_{fe} = D_f\sigma_{f.max}/E_f \tag{5.152}$$

where ρ_f is the FRP ratio for torsion. The maximum FRP stress $\sigma_{f,max}$ and stress distribution factor D_f are expressed as

1. For rupture failure (wrapped beams):

$$D_f = 0.5 \tag{5.153}$$

$$\sigma_{f.max} = \begin{cases} 0.8 f_{fu} & \text{for} \quad \dfrac{f_{fu}}{E_f} \le 1.5\% \\[3mm] 0.8 \cdot (0.015 f_{fu}) & \text{for} \quad \dfrac{f_{fu}}{E_f} > 1.5\% \end{cases} \tag{5.154}$$

2. For debonding failure (U-jacketed beams):

$$\sigma_{f.max} = \min \begin{cases} f_{fu} & \text{for} \quad \dfrac{f_{fu}}{E_f} \le 1.5\% \\[3mm] 0.427 \beta_L \beta_w \sqrt{\dfrac{E_f \sqrt{f_{co}'}}{t_f}} & \text{for} \quad \dfrac{f_{fu}}{E_f} > 1.5\% \end{cases} \tag{5.155}$$

$$D_f = \begin{cases} \dfrac{2}{\pi\lambda} \dfrac{1 - \cos\left(\dfrac{\pi\lambda}{2}\right)}{\sin\left(\dfrac{\pi\lambda}{2}\right)} & \lambda \le 1 \\[5mm] 1 - \dfrac{\pi - 2}{\pi\lambda} & \lambda > 1 \end{cases} \tag{5.156}$$

$$\lambda = \frac{L_{max}}{L_e} = \frac{0.9\min(b, h)}{\sqrt{\dfrac{E_f t_f}{\sqrt{f_{co}'}}}} \tag{5.157}$$

$$\beta_L = \begin{cases} 1 & \text{if} \quad \lambda \ge 1 \\[3mm] \sin\left(\dfrac{\pi\lambda}{2}\right) & \text{if} \quad \lambda < 1 \end{cases} \tag{5.158}$$

$$\beta_w = \sqrt{\frac{\left(2 - \dfrac{w_f}{S_f}\right)}{\left(1 + \dfrac{w_f}{S_f}\right)}} \tag{5.159}$$

$$T = -\sigma_d 2 A_o t_d \sin\alpha \cos\alpha \tag{5.160}$$

where T is the torsional moment, A_o is the area enclosed by the center line of the shear flow, t_d is the effective thickness of the compression zone, and α is the inclination of the diagonal crack.

IV. Ganganagoudar et al. (2016):

In the recent study by Ganganagoudar et al. (2016), the authors proposed analytical and FE studies on the behavior of FRP-composite-strengthened RC beams under torsional loading. They proposed a modified concrete constitutive law under compression by incorporating the confinement effect of the outer FRP wrapping. In addition, they recommended a new tension stiffening model of concrete for improved predictions of the torsional response of FRP-strengthened beams. A

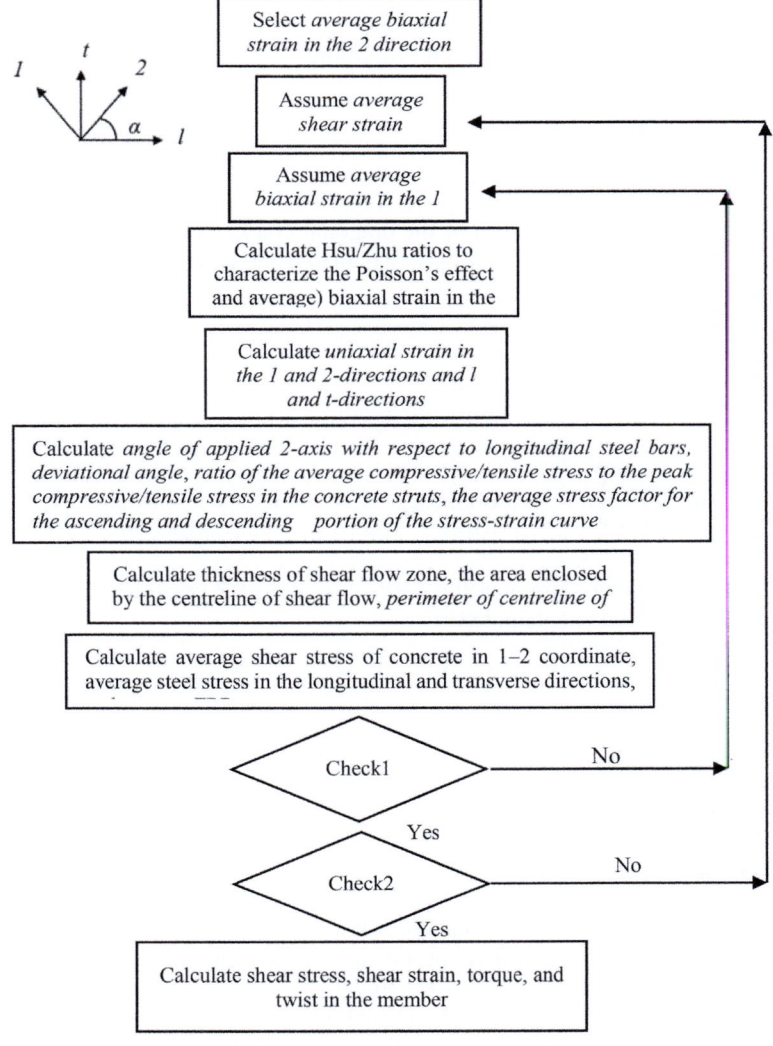

Figure 5.26 Solution algorithm of SMMT-FRP (Ganganagoudar et al., 2016).

strain-controlled algorithm for the softened membrane model for torsion (SMMT)-FRP, which includes the modified constitutive laws for concrete in compression and tension, was implemented in MATLAB, as shown in Fig. 5.26.

5.12.2 Design codes

FIB (2001) assumed an absence of interaction between the steel and FRP contributions to the torsion resistance of a section. The contribution of FRP to the torsion capacity of a beam (T_f) is computed as

$$T_f = \frac{2E_f \varepsilon_f t_f w_f bh\cot(\alpha)}{S_f} \text{(Complete wrapping)} \tag{5.161}$$

$$T_f = \frac{E_f \varepsilon_f t_f w_f bh\cot(\alpha)}{S_f} \text{(Uwrapping)} \tag{5.162}$$

where α is the angle of inclination of the diagonal cracks to the longitudinal axis of the beam. The effective FRP strain ε_f is calculated for CFRP and GFRP using the following formulas:

$$\varepsilon_f = 0.17 \left(\frac{f_{cm}^{2/3}}{E_f \rho_f} \right)^{0.3} \varepsilon_{fu} \text{ for CFRP} \tag{5.163}$$

$$\varepsilon_f = 0.048 \left(\frac{f_{cm}^{2/3}}{E_f \rho_f} \right)^{0.47} \varepsilon_{fu} \text{ for GFRP} \tag{5.164}$$

where ε_{fu} is the ultimate strain in the FRP. ρ_f is the FRP reinforcement ratio with respect to concrete and is calculated as

$$\rho_f = \frac{2t_f w_f}{bs_f} \tag{5.165}$$

References

Abdel-Jaber, M.S., Walker, P.R., Hutchinson, A.R., 2003. Shear strengthening of reinforced concrete beams using different configurations of externally bonded carbon fiber reinforced plates. Mater. Struct. 36 (5), 291–301.

ACI Committee 440, 2008. Guide for the Design and Construction of Externally Bonded FRP Systems for Strengthening Concrete Structures (ACI 440.2R-08) [S]. American Concrete Institute, Farmington Hills, MI.

Adhikary, B.B., Mutsuyoshi, H., Ashraf, M., 2004. Shear strengthening of reinforced concrete beams using fiber-reinforced polymer sheets with bonded anchorage. Struct. J. 101 (5), 660−668.

Alagusundaramoorthy, P., Harik, I.E., Choo, C.C., 2003. Flexural behavior of R/C beams strengthened with carbon fiber reinforced polymer sheets or fabric. J. Compos. Constr. 7 (4), 292−301.

Ali-Ahmad, M., Subramaniam, K., Ghosn, M., 2006. Experimental investigation and fracture analysis of debonding between concrete and FRP sheets. J. Eng. Mech. 132 (9), 914−923.

Allam, S.M., Ebeido, T.I., 2003. Retrofitting of RC beams predamaged in shear using CFRP sheets. Alex. Eng. J. 42 (1), 87−101.

Al-Mahaidi, R., Hii, A.K., 2007. Bond behaviour of CFRP reinforcement for torsional strengthening of solid and box-section RC beams. Compos. Part B: Eng. 38 (5), 720−731.

Al-Rousan, R.Z., Haddad, R.H., Swesi, A.O., 2015. Repair of shear-deficient normal weight concrete beams damaged by thermal shock using advanced composite materials. Compos. Part B: Eng. 70, 20−34.

Altin, M., 2009. The comparison of the two same type school buildings before and after strengthening. Eng. Sci. 4 (4), 504−517.

Ameli, M., Ronagh, H.R., Dux, P.F., 2007. Behavior of FRP strengthened reinforced concrete beams under torsion. J. Compos. Constr. 11 (2), 192−200.

Anil, Ö., 2008. Strengthening of RC T-section beams with low strength concrete using CFRP composites subjected to cyclic load. Constr. Build. Mater. 22 (12), 2355−2368.

Ansys User's Manual, 2009. [Computer Software] Version (12) [CP]. Swanson Analysis Systems, Canonsburg, PA.

Araki, N., Matsuzaki, Y., Nakano, K., Kataoka, T., Fukuyama, H., October 1997. Shear capacity of retrofitted RC members with continuous fiber sheets. In: Non-Metallic (FRP) Reinforcement for Concrete Structures, Proceedings of the Third Symposium, vol. 1, pp. 515−522.

Awani, O., El-Maaddawy, T., El Refai, A., 2016. Numerical simulation and experimental testing of concrete beams strengthened in shear with fabric-reinforced cementitious matrix. J. Compos. Constr. 20 (6), 04016056.

Baggio, D., Soudki, K., Noel, M., 2014. Strengthening of shear critical RC beams with various FRP systems. Constr. Build. Mater. 66, 634−644.

Baghi, H., Barros, J.A., 2016. Shear strengthening of reinforced concrete T-beams with hybrid composite plate. J. Compos. Constr. 20 (6), 04016036.

Beber, A.J., Campos Filho, A., 2005. CFRP composites on the shear strengthening of reinforced concrete beams. Rev. IBRACON Estrut. 1 (2).

Belarbi, A., Hsu, T.T., 1995. Constitutive laws of softened concrete in biaxial tension compression. Struct. J. 92 (5), 562−573.

Belarbi, A., Bae, S.-W., Brancaccio, A., 2012. Behavior of full-scale RC T-beams strengthened in shear with externally bonded FRP sheets. Constr. Build. Mater. 32, 27−40.

Bousselham, A., Chaallal, O., 2004. Shear strengthening reinforced concrete beams with fiber-reinforced polymer: assessment of influencing parameters and required research. Struct. J. 101 (2), 219−227.

Bousselham, A., Chaallal, O., 2006. Effect of transverse steel and shear span on the performance of RC beams strengthened in shear with CFRP. Compos. Part B: Eng. 37 (1), 37−46.

Bousselham, A., Chaallal, O., 2008. Mechanisms of shear resistance of concrete beams strengthened in shear with externally bonded FRP. J. Compos. Constr. 12 (5), 499–512.

Bukhari, I.A., Vollum, R.L., Ahmad, S., Sagaseta, J., 2010. Shear strengthening of reinforced concrete beams with CFRP. Mag. Concrete Res. 62 (1), 65–77.

Calladine, C.R., 1969. Engineering Plasticity. Pergamon, Oxford, UK.

Cao, S.Y., Chen, J.F., Teng, J.G., Hao, Z., Chen, J., 2005. Debonding in RC beams shear strengthened with complete FRP wraps. J. Compos. Constr. 9 (5), 417–428.

Carolin, A., Täljsten, B., 2005. Theoretical study of strengthening for increased shear bearing capacity. J. Compos. Constr. 9 (6), 497–506.

Carolin, A., Täljsten, B., Hejll, A., 2005. Concrete beams exposed to live loading during carbon fiber reinforced polymer strengthening. J. Compos. Constr. 9 (2), 178–186.

Chaallal, O., Nollet, M.J., Perraton, D., 1998. Shear strengthening of RC beams by externally bonded side CFRP strips. J. Compos. Constr. 2 (2), 111–113.

Chajes, M.J., Thomson, T.A., Farschman, C.A., 1995. Durability of concrete beams externally reinforced with composite fabrics. Constr. Build. Mater. 9 (3), 141–148.

Chalioris, C.E., 2007. Analytical model for the torsional behaviour of reinforced concrete beams retrofitted with FRP materials. Eng. Struct. 29 (12), 3263–3276.

Chalioris, C.E., 2008. Torsional strengthening of rectangular and flanged beams using carbon fiber-reinforced-polymers – experimental study. Constr. Build. Mater. 22 (1), 21–29.

Chen, J.F., Teng, J.G., 2001a. A shear strength model for FRP-strengthened RC beams. FRPRCS-5: Fiber-Reinforced Plastics for Reinforced Concrete Structures, Volume 1: Proceedings of the Fifth International Conference on Fiber-Reinforced Plastics for Reinforced Concrete Structures. Thomas Telford Publishing, Cambridge, UK, pp. 205–214, 16–18 July 2001.

Chen, J.F., Teng, J.G., 2001b. Anchorage strength models for FRP and steel plates bonded to concrete. J. Struct. Eng. 127 (7), 784–791.

Chen, J.F., Teng, J.G., 2003a. Shear capacity of fiber-reinforced polymer-strengthened reinforced concrete beams: fiber reinforced polymer rupture. J. Struct. Eng. 129 (5), 615–625.

Chen, J.F., Teng, J.G., 2003b. Shear capacity of FRP-strengthened RC beams: FRP debonding. Constr. Build. Mater. 17 (1), 27–41.

Chen, G.M., Teng, J.G., Chen, J.F., 2013. Shear strength model for FRP-strengthened RC beams with adverse FRP-steel interaction. J. Compos. Constr. 17 (1), 50–66.

CNR-DT(200/2004), 2004. Guide for the design and construction of externally bonded FRP systems for strengthening concrete structures. Reported by ACI Committee, 440(2002).

Collins, M.P., Mitchell, D., Adebar, P., Vecchio, F.J., 1996. A general shear design method. ACI Struct. J. 93 (1), 36–45.

Colotti, V., Spadea, G., Swamy, R.N., 2004. Analytical model to evaluate failure behavior of plated reinforced concrete beams strengthened for shear [J]. ACI Struct. J 101 (6), 755–764.

Deifalla, A., Ghobarah, A., 2005. Simplified analysis for torsionaly strengthened RC beams using FRP. In: Chen, Teng (Eds.), Proceedings of the International Symposium on Bond Behavior of FRP in Structures (BBFS 2005). International Institute for FRP in Construction, pp. 377–378.

Deifalla, A., Ghobarah, A., 2010a. Full torsional behavior of RC beams wrapped with FRP: analytical model. J. Compos. Constr. 14 (3), 289–300.

Deifalla, A., Ghobarah, A., 2010b. Strengthening RC T-beams subjected to combined torsion and shear using FRP fabrics: experimental study. J. Compos. Constr. 14 (3), 301–311.

Deniaud, C., Cheng, J.R., 2001a. Review of shear design methods for reinforced concrete beams strengthened with fiber reinforced polymer sheets. Can. J. Civ. Eng. 28 (2), 271−281.

Deniaud, C., Cheng, J.R., 2001b. Shear behavior of reinforced concrete T-beams with externally bonded fiber-reinforced polymer sheets. Struct. J. 98 (3), 386−394.

Diagana, C., Li, A., Gedalia, B., Delmas, Y., 2003. Shear strengthening effectiveness with CFF strips. Eng. Struct. 25 (4), 507−516.

Ebead, U., Saeed, H., 2013. Hybrid shear strengthening system for reinforced concrete beams: an experimental study. Eng. Struct. 49, 421−433.

FIB Bulletin 35, 2006. Retrofitting of concrete structures by externally bonded FRPs. Bulletin 14, 138.

Foster, R.M., et al., 2016. Experimental investigation of reinforced concrete T-beams strengthened in shear with externally bonded CFRP sheets. J. Compos. Constr. 04016086.

Funakawa, I., 1997. Experimental Study on Shear strenghthening with Continuous Fiber Reinforcement Sheet and Methacrylate Resin. In: Proceeding of Third International Symposium of Non-Metallic (FRP) Reinforcement for Concrete Strutures, vol. 1, pp. 475−482.

Gamino, A.L., Sousa, J.L.A.O., Manzoli, O.L., Bittencourt, T.N., 2010. R/C structures strengthened with CFRP Part II: Analysis of shear models. Rev. IBRACON Estrut. Mater. 3 (1), 24−49.

Ganganagoudar, A., Mondal, T.G., Prakash, S.S., 2016. Analytical and finite element studies on behavior of FRP strengthened RC beams under torsion. Compos. Struct. 153, 876−885.

Ghobarah, A., Ghorbel, M.N., Chidiac, S.E., 2002. Upgrading torsional resistance of reinforced concrete beams using fiber-reinforced polymer. J. Compos. Constr. 6 (4), 257−263.

Grande, E., Imbimbo, M., Rasulo, A., 2007. Experimental behaviour of RC beams strengthened in shear by FRP sheets. In: Proc 8th Int Symp FRP Reinforcement for Concrete Structures, Patras, Greece.

Hii, A.K., Al-Mahaidi, R., 2006. Experimental investigation on torsional behavior of solid and box-section RC beams strengthened with CFRP using photogrammetry. J. Compos. Constr. 10 (4), 321−329.

Hii, A.K., Al-Mahaidi, R., 2007. Torsional capacity of CFRP strengthened reinforced concrete beams. J. Compos. Constr. 11 (1), 71−80.

Ianniruberto, U., Imbimbo, M., 2004. Role of fiber reinforced plastic sheets in shear response of reinforced concrete beams: experimental and analytical results. J. Compos. Constr. 8 (5), 415−424.

Islam, M.R., Mansur, M.A., Maalej, M., 2005. Shear strengthening of RC deep beams using externally bonded FRP systems. Cem. Concr. Compos. 27 (3), 413−420.

Japan Society of Civil Engineers (JSCE), 2001. Recommendations for upgrading of concrete structures with use of continuous fiber sheets. Research Committee on Upgrading of Concrete Structures with Use of Continuous Fiber Sheets. Japan Society of Civil Engineers, Japan.

Jariwala, V.H., Patel, P.V., Purohit, S.P., 2013. Strengthening of RC beams subjected to combined torsion and bending with GFRP composites. Procedia Eng. 51, 282−289.

Jung, K., Hong, K., Han, S., Park, J., Kim, J., 2015. Shear Strengthening performance of hybrid FRP-FRCM. Adv. Mater. Sci. Eng. 2015, 1−11.

Kachlakev, D., McCurry, D.D., 2000. Behavior of full-scale reinforced concrete beams retrofitted for shear and flexural with FRP laminates. Compos. Part B: Eng. 31 (6), 445–452.

Kachlakev, D., Miller, T., Yim, S., Chansawat, K., Potisuk, T., 2001. Finite element modeling of concrete structures strengthened with FRP laminates. Final Report, SPR, 316.

Kamiharako, A., Maruyama, K., Takada, K., Shimomura, T., 1997. Evaluation of shear contribution of FRP sheets attached to concrete beams. Non-Metallic (FRP) Reinf. Concr. Struct. 1, 467–474.

Kaw, A.K., 1997. Mechanics of Composite Materials [D]. CRC Press LLC, Boca Raton, Florida.

Khalifa, A., Nanni, A., 2000. Improving shear capacity of existing RC T-Section beams using CFRP composites. Cem. Concr. Compos. 22 (3), 165–174.

Khalifa, A., Nanni, A., 2002. Rehabilitation of rectangular simply supported RC beams with shear deficiencies using CFRP composites. Constr. Build. Mater. 16 (3), 135–146.

Khalifa, A., Gold, W.J., Nanni, A., MI, A.A., 1998. Contribution of externally bonded FRP to shear capacity of RC flexural members. J. Compos. Constr. 2 (4), 195–202.

Khalifa, A., Belarbi, A., Nanni, A., January 2000. Shear performance of RC members strengthened with externally bonded FRP wraps. In: Proceedings (CD-ROM) of Twelfth World Conference on Earthquake, Auckland, New Zealand.

Kotsovos, M.D., Pavlović, M.N., 1999. Ultimate Limit-State Design of Concrete Structures —A New Approach. Thomas Telford Ltd., London.

Kwan, A.K.H., Dai, H., Cheung, Y.K., 1999. Non-linear seismic responce of reinforced concrete slit shear walls. J Sound Vib. 226, 701–718.

Leung, C.K., Chen, Z., Lee, S., Ng, M., Xu, M., Tang, J., 2007. Effect of size on the failure of geometrically similar concrete beams strengthened in shear with FRP strips. J. Compos. Constr. 11 (5), 487–496.

Li, W., Leung, C.K., 2015. Shear span–depth ratio effect on behavior of RC beam shear strengthened with full-wrapping FRP strip. J. Compos. Constr. 20 (3), 04015067.

Li, W., Leung, C.K.Y., 2015. Shear span–depth ratio effect on behavior of RC beam shear strengthened with full-wrapping FRP strip. J. Compos. Constr. 20 (3), 04015067.

Lim, D.H., 2010. Shear behaviour of RC beams strengthened with NSM and EB CFRP strips. Mag. Concr. Res. 62 (3), 211–220.

Lu, X.Z., Teng, J.G., Ye, L.P., Jiang, J.J., 2005. Bond–slip models for FRP sheets/plates bonded to concrete. Eng. Struct. 27 (6), 920–937.

Malek, A.M., Saadatmanesh, H., Ehsani, M.R., 1998. Prediction of failure load of R/C beams strengthened with FRP plate due to stress concentration at the plate end. ACI Struct. J. 95, 142–152.

Matthys, S., Triantafillou, T.C., 2001. Shear and torsion strengthening with externally bonded FRP reinforcement [C]. In: Reality, A., Cosenza, E., Manfredi, G., Nanni, N. (Eds.), Proc. Int. Workshop on Compos. in Constr., Capri, Italy, July 20–21, pp. 203–210.

Miyajima, H., Kosa, K., Tasaki, K., Matsumoto, S., 2005. Shear strengthening of RC beams using carbon fiber sheets & its resistance mechanism. In: Proceedings, pp. 114–125.

Mofidi, A., Chaallal, O., 2011. Shear strengthening of RC beams with EB FRP: influencing factors and conceptual debonding model [J]. J. Compos. Constr. 15 (1), 62–74.

Mofidi, A., Chaallal, O., 2011. Shear strengthening of RC beams with externally bonded FRP composites: effect of strip-width-to-strip-spacing ratio. J. Compos. Constr. 15 (5), 732–742.

Mofidi, A., et al., 2013. Behavior of reinforced concrete beams strengthened in shear using L-shaped CFRP plates: experimental investigation. J. Compos. Constr. 18 (2), 04013033.

Monti, G., Liotta, M.A., 2007. Tests and design equations for FRP-strengthening in shear [J]. Constr. Build. Mater. 21, 799−809.

Mörsch, E., 1909. Concrete steel construction (Der Eisenbetonbau). English translation of the 3rd German edition.

Nielsen, M.P., Braestrup, M.W., Bach, F., 1978. Rational analysis of shear in reinforced concrete beams. In: IABSE Proceedings, P-15/78, Zurich, Switzerland.

Norris, T., Saadatmanesh, H., Ehsani, M.R., 1997. Shear and flexural strengthening of R/C beams with carbon fiber sheets. J. Struct. Eng. 123 (7), 903−911.

Panchacharam, S., Belarbi, A., October 2002. Torsional behavior of reinforced concrete beams strengthened with FRP composites. In: First FIB Congress, Osaka, Japan, vol. 1, pp. 1−110.

Panigrahi, A.K., Biswal, K.C., Barik, M.R., 2014. Strengthening of shear deficient RC T-beams with externally bonded GFRP sheets. Constr. Build. Mater. 57, 81−91.

Pellegrino, C., Modena, C., 2002. Fiber reinforced polymer shear strengthening of reinforced concrete beams with transverse steel reinforcement. J. Compos. Constr. 6 (2), 104−111.

Pellegrino, C., Modena, C., 2006. Fiber-reinforced polymer shear strengthening of reinforced concrete beams: experimental study and analytical modeling. ACI Struct. J. 103 (5), 720.

Pellegrino, C., Modena, C., 2008. An experimentally based analytical model for shear capacity of FRP strengthened reinforced concrete beams. Mech. Compos. Mater. 44 (3), 231−244.

Pham, T.M., Hao, H., 2016. Impact behavior of FRP-strengthened RC beams without stirrups. J. Compos. Constr. 04016011.

Pham, T.M., Hao, H., 2017. Behavior of fiber-reinforced polymer-strengthened reinforced concrete beams under static and impact loads. Int. J. Protect. Struct. 8 (1), 3−24.

Raongjant, W., Jing, M., 2008. Nonlinear model on torsional behaviors of CFRP strengthened reinforced concrete beams, Advanced Materials Research, vol. 47. Trans Tech Publications, pp. 881−885.

Salom, P.R., Gergely, J., Young, D.T., 2004. Torsional strengthening of spandrel beams with fiber-reinforced polymer laminates. J. Compos. Constr. 8 (2), 157−162.

Sato, Y., Ueda, T., Kakuta, Y., Tanaka, T., 1996. Shear reinforcing effect of carbon fiber sheet attached to side of reinforced concrete beams. In: Proceedings of the 2nd International Conference on Advanced Composite Materials In Bridges And Structures, ACMBS-II, Montreal.

Sato, Y., Ueda, T., Kakuta, Y., Ono, S., 1997. Ultimate shear capacity of reinforced concrete beams with carbon fiber sheets. In: Non-Metallic (FRP) Reinforcement for Concrete Structures, Proceedings of the Third Symposium, vol. 1, pp. 499−506.

Sayed, A.M., 2014. Modeling advancement for RC beams strengthened with FRP composites. Doctoral dissertation, Southeast University, Nanjing, China.

Sayed, A.M., Wang, X., Wu, Z., 2013. Modeling of shear capacity of RC beams strengthened with FRP sheets based on FE simulation. J. Composit. Construct 17 (5), 687−701.

Sharif, A., Al-Sulaimani, G.J., Basunbul, I.A., Baluch, M.H., Ghaleb, B.N., 1994. Strengthening of initially loaded reinforced concrete beams using FRP plates. Struct. J. 91 (2), 160−168.

Stratford, T., Burgoyne, C., 2003. Shear analysis of concrete with brittle reinforcement. J. Compos. Constr. 7 (4), 323−330. Available from: https://doi.org/10.1061/(ASCE)1090-0268(2003)7:4(323).

Sundarraja, M.C., Rajamohan, S., 2009. Strengthening of RC beams in shear using GFRP inclined strips − an experimental study. Constr. Build. Mater. 23 (2), 856−864.

Taerwe, L., Khalil, H., Matthys, S., 1997. Behaviour of RC beams strengthened in shear by external CFRP sheets. In: Proceedings of the Third International Symposium on Non-Metallic (FRP) Reinforcement for Concrete Structures (FRPRCS-3), Sapporo (Japan), 14−16 October, vol. 1.

Takahashi, Y., Hata, C., Maeda, T., Sato, Y., 1998. Experimental study of flexural behavior of aramid FRP rods reinforced concrete beam with externally bonded carbon fiber sheet. Proc. Jpn. Concr. Inst. 20 (1), 509−514.

Täljsten, B., 2003. Strengthening concrete beams for shear with CFRP sheets. Constr. Build. Mater. 17 (1), 15−26.

Teng, J.G., Chen, J.F., Smith, S.T., Lam, L., 2002. FRP: strengthened RC structures. Front. Phys. 266.

Teng, J.G., et al., 2009. Behavior of RC beams shear strengthened with bonded or unbonded FRP wraps. J. Compos. Constr. 13 (5), 394−404.

Terec, L., Bugnariu, T., Păstrav, M., 2010. Non-linear analysis of reinforced concrete frames strengthened with infilled walls. Rev. Rom. Mater./Rom. J. Mater. 40 (3), 214−221.

Trapko, T., Urbańska, D., Kamiński, M., 2015. Shear strengthening of reinforced concrete beams with PBO-FRCM composites. Compos. Part B: Eng. 80, 63−72.

Triantafillou, T.C., 1998. Shear strengthening of reinforced concrete beams using epoxy-bonded FRP composites. Struct. J. 95 (2), 107−115.

Triantafillou, T.C., Antonopoulos, C.P., 2000. Design of concrete flexural members strengthened in shear with FRP [J]. J Compos. Constr. 4 (4), 198−205. ASCE.

Uji, K., 1992. Improving shear capacity of existing reinforced concrete members by applying carbon fiber sheets. Trans. Jpn. Concr. Inst. 14 (253), 66.

Umezu, K., Fujita, M., Nakai, H., Tamaki, K., October 1997. Shear behavior of RC beams with aramid fiber sheet. In: Non-Metallic (FRP) Reinforcement for Concrete Structures, Proceedings of the Third Symposium, vol. 1, pp. 491−498.

Ye, L.P., Lu, X.Z., Chen, J.F., 2005. Design proposals for debonding strengths of FRP strengthened RC beams in the Chinese design code [C]December Proc., Int. Symp. on Bond Behaviour of FRP in Structures. International Institute for FRP in Construction (IIFC), Hong Kong, China, pp. 45−54.

Zhang, Z., Hsu, C.T.T., 2005. Shear strengthening of reinforced concrete beams using carbon-fiber-reinforced polymer laminates. J. Compos. Constr. 9 (2), 158−169.

Zhang, J.W., Lu, Z.T., & Zhu, H. (2001). Experimental study on the behaviour of RC torsional members externally bonded with CFRP. In: FRP Composites in Civil Engineering. Proceedings of the International Conference on FRP Composites in Civil Engineering, vol. 1.

Zhang, Z., Hsu, C.T.T., Moren, J., 2004. Shear strengthening of reinforced concrete deep beams using carbon fiber reinforced polymer laminates. J. Compos. Constr. 8 (5), 403−414.

Further reading

ACI Committee 318, 2011. Building Code Requirements for Structural Concrete (ACI 318M-11) and Commentary. American Concrete Institute.

Anil, Ö., 2008. Strengthening of RC T-section beams with low strength concrete using CFRP composites subjected to cyclic load. Constr. Build. Mater. 22 (12), 2355−2368.

Baghi, H., Barros, J.A., Menkulasi, F., 2016. Shear strengthening of reinforced concrete beams with hybrid composite plates (HCP) technique: experimental research and analytical model. Eng. Struct. 125, 504−520.

Bakis, C.E., Ganjehlou, A., Kachlakev, D.I., Schupack, M., Balaguru, P.N., Gee, D.J. et al., 2002. Guide for the design and construction of externally bonded FRP systems for strengthening concrete structures. Reported by ACI Committee, 440 (2002).

Breveglieri, M., Aprile, A., Barros, J.A., 2015. Embedded through-section shear strengthening technique using steel and CFRP bars in RC beams of different percentage of existing stirrups. Compos. Struct. 126, 101−113.

Busel, J., Lindsay, K., 1997. On the road with John Busel: a look at the world's bridges. CDA/Compos. Des. Appl. 14−23.

Canadian Standards Association, 1996. CHBDC—Canadian Highway Bridge Design Code, Section 16: Fiber Reinforced Structures. Canadian Standards Association, Mississauga, Ont..

Canadian Standards Association (CSA), 2012. Design and construction of building structures with fiber-reinforced polymers. S806-12, Mississauga, ON, Canada. See more at: <http://ascelibrary.org/doi/abs/10.1061/(ASCE)CC.1943-5614.0000491#sthash. XgUq1kRv.dpuf/>.

Chaallal, O., Mofidi, A., Benmokrane, B., Neale, K., 2010. Embedded through-section FRP rod method for shear strengthening of RC beams: Performance and comparison with existing techniques. J. Compos. Construct. 15 (3), 374−383.

CNR, D., 2004. 200/2004. Guide for the Design and Construction of Externally Bonded FRP Systems for Strengthening Existing Structures.

Crasto, A.S., et al., 1996. Rehabilitation of concrete bridge beams with externally-bonded composite plates. Part I. In: First International Conference on Composites in Infrastructure.

Dong, J., Wang, Q., Guan, Z., 2013. Structural behaviour of RC beams with external flexural and flexural−shear strengthening by FRP sheets. Compos. Part B: Eng. 44 (1), 604−612.

Hawileh, R.A., et al., 2015. Effect of flexural CFRP sheets on shear resistance of reinforced concrete beams. Compos. Struct. 122, 468−476.

International Federal for Prestressing (FIB), 2001. Externally bonded FRP reinforcement for RC structures, International Federation for Structural Concrete, Bull., 14. Sprint-Digital-Druck, Stuttgart, Germany, p. 138.

ISIS-M04, 2009. FRP Rehabilitation of Reinforced Concrete Structures Design Manual. ISIS, Winnipeg, Canada.

Karaca, E., 2002. Diss FRP Strengthening of RC Beams in Flexure and Shear: Failure Modes and Design. Massachusetts Institute of Technology.

Mofidi, Amir, et al., 2012. Experimental tests and design model for RC beams strengthened in shear using the embedded through-section FRP method. J. Compos. Constr. 16 (5), 540−550.

Moslehy, Y., Labib, M., Mullapudi, T.R.S., Ayoub, A., 2015. Development of a new constitutive model for analysis of RC elements retrofitted with FRP. ACI Spec. Publ. 301, 1−18.

Popov, E.G., 1998. Engineering Mechanics of Solids [R], second ed. Prentice-Hall, Upper Saddle River, NJ.

Press, C.P., 2010. Technical Code for Infrastructure Application of FRP Composites. GB-50608. Beijing.

Replacement, Highway Bridge, 1993. Rehabilitation Program. Eleventh Report of the Secretary of Transportation to the United States Congress.

Sayed, A.M., Wang, X., Wu, Z., 2013. Modeling of shear capacity of RC beams strengthened with FRP sheets based on FE simulation. J. Compos. Constr. ASCE 17 (5), 687−701.

Teng, J.G., Lam, L., Chen, J.F., 2004. Shear strengthening of RC beams with FRP composites. Prog. Struct. Eng. Mater. 6 (3), 173−184.

Xue Song, F., Jie, L., Zhong Fan, C., 2004. Experimental research on shear strengthening of reinforced concrete beams with externally bonded CFRP sheets. Ind. Constr. 34, 89−93.

FRP strengthening of concrete columns

6

6.1 Performance of under-designed existing structures

In many countries, most of the existing buildings, bridges, and other types of structures are substandard and deficient, and novel materials and techniques are necessary for retrofitting/repairing (see Fig. 6.1). Figs. 6.2 and 6.3 show that bridges and buildings are susceptible to severe damage and potential collapse due to large lateral displacements.

Concrete columns have an important function in many structures. Often, these columns are vulnerable to exceptional loads (such as impact, explosion, or seismic loads), load increase (increasing use or a change in the function of structures), and degradation (corrosion of steel reinforcement and alkali—silica reaction) (Matthys et al., 2005). Under a seismic load/deformation input, several failure modes, such as lap-splice failure at the connection between the footing and the column, shear failure, and confinement failure in the flexural plastic hinge region, were observed in existing reinforced concrete (RC) bridge columns (Seible et al., 1997). These three failure modes are potentially related to poor detail in the longitudinal lap splices, improper transverse confinement, and insufficient shear strength, as shown in Fig. 6.4 (Xiao and Ma, 1997; Chang et al., 2004; Harries et al., 2006).

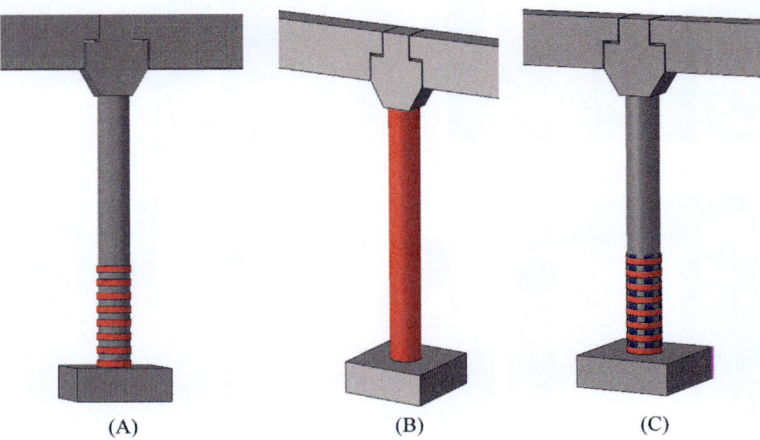

(A)   (B)   (C)

Figure 6.1 Different configurations for strengthening of existing column structures using external FRP composites. (A) Partial FRP wrapping. (B) Full FRP wrapping. (C) Attached FRP sheet/plate + FRP wrapping.

Structures Strengthened with Bonded Composites. DOI: https://doi.org/10.1016/B978-0-12-821088-8.00006-0

Figure 6.2 Damage levels for building after seismic actions.

Figure 6.3 Damage levels for bridges after seismic actions.

Figure 6.4 Failure pattern of RC bridge pier with (A) lap-splice deficiency, (B) shear deficiency, and (C) flexural deficiency.

6.1.1 Columns with shear deficiency

Shear failure is identified as the most dangerous failure mode because it is most likely to cause a total collapse in existing bridges (Xiao et al., 1999). Shear deficiency may occur due to many factors, such as insufficient shear reinforcement or reduction in steel area due to corrosion, increased service load, and construction errors. In addition, there is an urgent need to upgrade the shear resistance of older RC bridges to meet the current seismic design standards in regions with high seismicity. For circular columns with shear deficiency, Xiao et al. (1999) reported that the inclined shear cracks became dominant during the early loading cycles just after yielding column longitudinal reinforcement (Fig. 6.5A). Xiao et al. (1999) and Li and Sung (2004) demonstrated that shear failure occurred at a very low drift ratio (1%−1.5%). Haroun and Elsanadedy (2005a) found that these columns underwent shear failure without showing ductile behavior; for example, in Fig. 6.5B, the column failed at a lateral drift of approximately 0.5%.

6.1.2 Columns with lap-splice deficiency

Some existing bridge columns constructed before the 1970s feature lap splices in the column reinforcement located at the column base to form the connection between the column and footing. Starter bars are placed during the footing construction for column reinforcement and lapped with the longitudinal column reinforcement with a typical lap length of 20 bar diameters. In addition, the transverse column reinforcement is widely spaced, independent of column size, strength, or deformation demands. Lap-splice debonding occurs once vertical microcracks

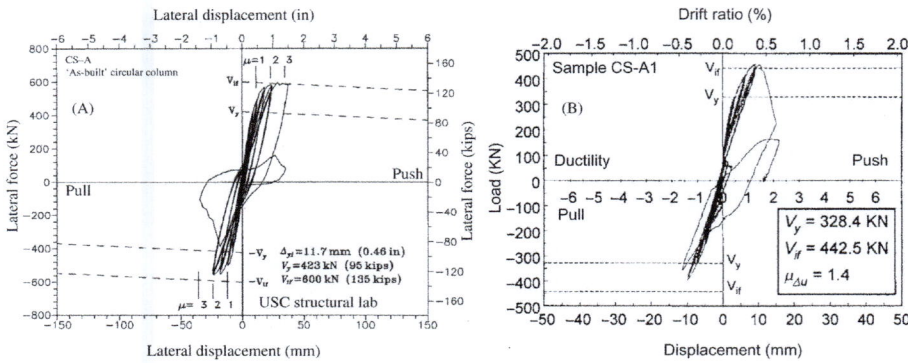

Figure 6.5 Load−deformation response of RC columns with shear deficiencies: (A) sample CS-A (Xiao et al., 1999) and (B) sample CS-A1 (Haroun and Elsanadedy, 2005).

develop in the cover concrete and it gets progressively worse with increased vertical cracking and cover concrete spalling in regions where short lap splices are present and insufficient clamping action by transverse reinforcement is provided. Consequently, the flexural strength of such columns degrades rapidly under lateral cyclic loading at low flexural ductility. Fig. 6.4A displays the damages to the base of a bridge column in the 1989 Loma Prieta earthquake, attributed to lap-splice bond failure. The lateral load−deformation responses of two RC columns with lap-splice deficiencies are illustrated in Fig. 6.6. The as-built columns tested by Xiao and Ma (1997), Haroun and Elsanadedy (2005a,b), Elsanadedy (2002), Ghosh and Sheikh (2007), and Breña and Schlick (2007) developed an unstable response due to the bond deterioration in the lap-spliced longitudinal reinforcement. At low load levels, flexural cracks were observed on the column surface above the footing. As the lateral load increased, the flexural cracks were extended in the entire lap-splice region.

6.1.3 Columns with flexural deficiency

Plastic hinge failures typically occur with some displacement ductility and are limited to shorter regions in the column. Thus these failures are less destructive and, owing to their inelastic flexural deformations (Fig. 6.7), are more desirable than the brittle column shear failure in the entire column (Seible et al., 1997). Typically, flexural yield can be observed at the base of the column and after yielding the column main reinforcement, the cover concrete spalls and confinement is lost, thereby resulting in buckling in the longitudinal bars between the exposed ties, leading to a loss in the remaining confinement. Loss of cover concrete, some core concrete, and buckling in the longitudinal bars result in a loss of both axial and flexural resistance in the column at a low lateral drift capacity (Chai et al., 1991; Seible et al., 1997; Gallardo-Zafra and Kawashima, 2009).

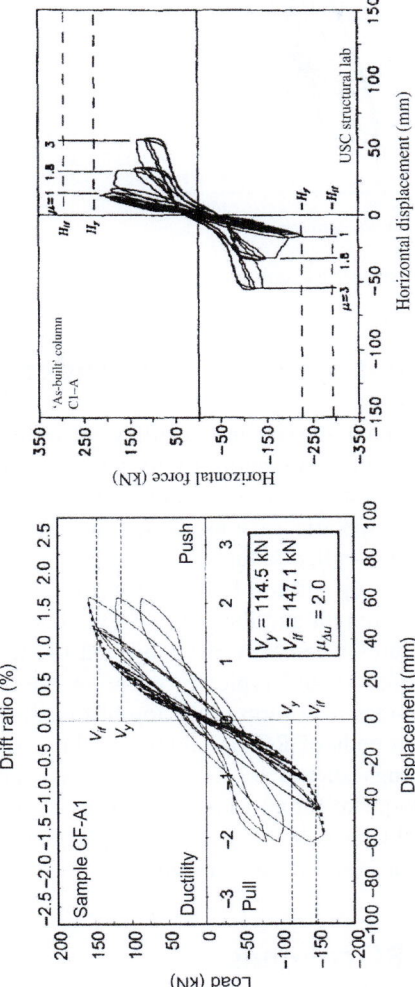

Figure 6.6 Load–deformation response of RC columns with lap-splice deficiencies: (A) sample CF-A1 (Haroun and Elsanadedy, 2005b) and (B) sample C1-A (Xiao and Ma, 1997).

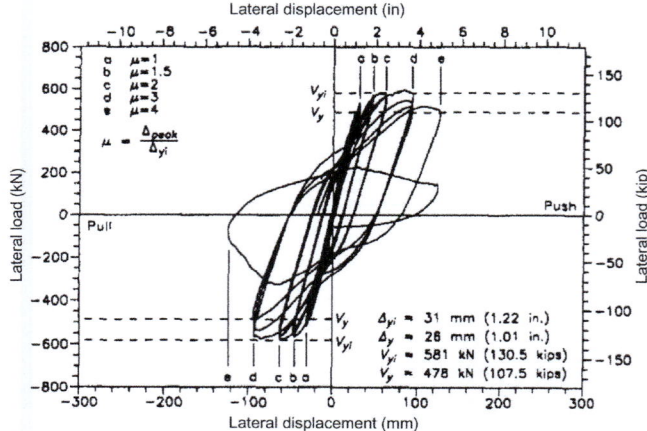

Figure 6.7 Load—deformation response of RC column with flexural deficiency (Seible et al., 1997).

6.2 Provisions of current seismic design codes

Modern seismic design codes provide design and detailing provisions for both RC bridges and buildings to minimize their susceptibility to damage from earthquakes. These provisions necessitate the limitation of damage under the effect of the moderate-to-strong earthquake to specific regions, which are called plastic hinge zones. In other words, shear and lap-splice failures are not allowed in current codes and plastic hinge failure is the only allowable failure mode. The plastic hinge should be designed for a certain strength and ductility. These zones are responsible for ductility demand from structures when they are subjected to lateral shaking. It should be noted that plastic hinge should typically occur at the column end of bridges, and secondary plastic hinges may form at the top of foundation piles; however, it should occur at the beam ends of RC buildings (see Fig. 6.8). In addition, conventional bridges and buildings should be designed to have adequate lateral strength (capacity) to resist earthquake forces (demands). This implies that modern conventional constructions should have both adequate displacement capacity and strength resistance to accommodate earthquake demands.

6.3 Strengthening of RC columns

Over the years, design engineers have implemented several methods and techniques to retrofit/repair existing structures. One popular solution to the problem is to place jackets around the structural elements (see Fig. 6.9). Jackets can be constructed using concrete, steel elements, and fiber-reinforced polymer (FRP) composites (Vandoros and Dritsos, 2008). In previous decades, the concept of external confining (jacketing) was investigated to define the appropriate conditions for a design

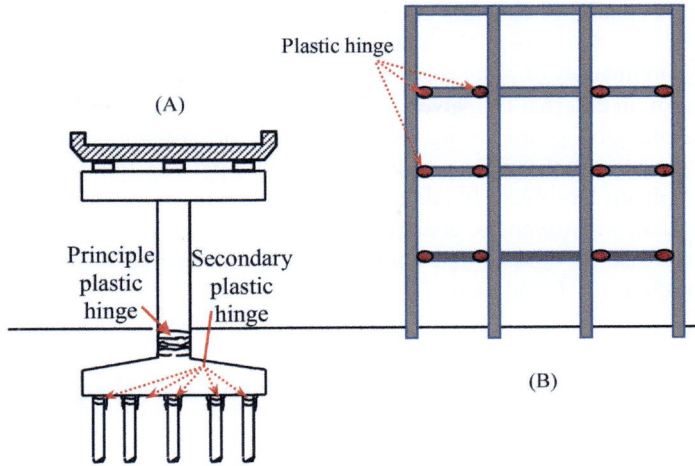

Figure 6.8 Location of primary plastic hinges in conventional constructions: (A) bridges and (B) buildings.

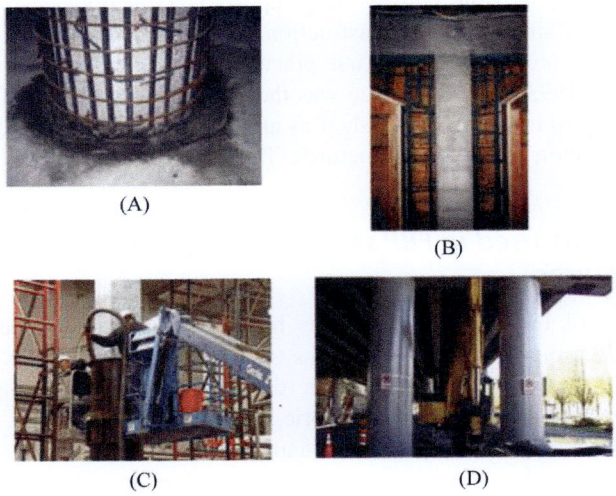

Figure 6.9 Strengthening of existing deficient columns using concrete and steel jackets. (A) Insertion of steel cage around the existing concrete column. (B) Outer wooden framework for casting of new concrete. (C) Filling the gap between steel jacket and existing concrete column with concrete grout. (D) Finalized steel-jacketed column.

according to the materials adopted in the retrofitting work. Traditional materials, such as concrete and steel, have many disadvantageous for different reasons. (1) Reinforced concrete jacketing means enlargement of the original column section through the placement of concrete and steel to achieve the desired column mechanical characteristics that satisfy the strength demand of whether the applied load is axial or a combination of axial and lateral. Moreover, this technique requires the construction of an external framework. (2) Although steel jacketing proved to be an effective technique, guaranteeing enhancements in column axial and lateral strengths, corrosion is the main problem of widely applying this technique, which remains completely unsolved. In addition, the disadvantages of RC and steel jacketing include labor intensiveness, construction inconvenience and longer construction time, and increase in rigidity that could attract more seismic loading to adjacent structures, such as the foundation.

6.4 FRP as external reinforcement for existing RC columns

In recent years, the use of FRP composites in concrete columns has significantly increased owing to their several advantages against the traditional reinforcement materials. First, their resistance to electrochemical deterioration: fiber composites do not corrode and are not affected by salt spray and other aggressive environmental factors. Second, the low density of composites (typically one-fifth of that of steel) significantly simplifies the construction procedure and reduces cost. The use of fiber-wrapping technology was first practiced on concrete chimneys in Japan (Ballinger et al., 1993). The technique was then extended to the retrofit of concrete columns. FRP wrap has been established as an effective method for the strengthening and rehabilitation of concrete structures (Toutanji, 1999).

6.4.1 Methods used in FRP strengthening/upgrading of existing structures

Different types of FRP-wrapping systems have been investigated and developed based on material types, form, and process of application (Fig. 6.10). They can be classified into four categories based on the method of processing/installation, namely (A) wet lay-up process using fabric, tape, or individual tow; (B) prefabricated shells; (C) resin infusion processes; and (D) automated winding.

In technique (A), the column can be wrapped with either one or multilayer FRP sheets (Fig. 6.10A) or even with FRP strips placed in spirals or rings (Fig. 6.10B). Applications of this technique are widely reported in both buildings columns and bridge piers (e.g., ACI440R-96 1996, Neale and Labossiere, 1997; Tan, 1997). Laying-up the prepreg (a reinforcing fabric preimpregnated with a resin system) tape is a straightforward and very fast construction approach but it is more difficult to control because it is carried out by hand completely. Additionally, there are concerns related to the quality control of resin mix, attainment of good wet-out of fibers with uniform resin impregnation without entrapment of excessive voids, good compaction of fibers without excessive wrinkling of the predominantly hoop-

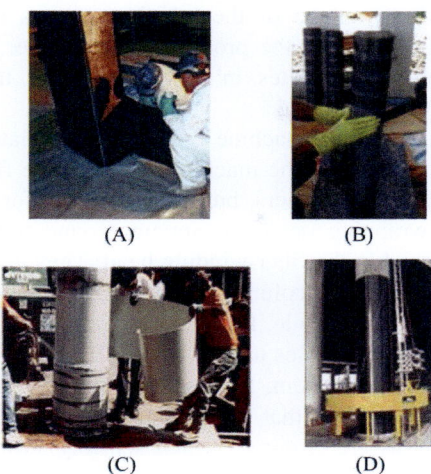

(A)   (B)

(C)   (D)

Figure 6.10 Methods of FRP strengthening for deficient RC columns. (A)Wet lay-up technique (full wrapping). (B) Wet lay-up technique (partial wrapping). (C) Prefabricated jackets. (D) Automated winding.

directed fibers, control of cure kinetics and achieving a full cure, and aspects related to environmental durability during and after curing.

In technique (B), a column is wrapped with two prefabricated half-shells (Fig. 6.10C) that can be, depending on its cross-sectional shape, either circular (e.g., Nanni and Norris, 1995) or rectangular. Alternatively, full circular shells with a vertical slit can be opened and placed around the column (e.g., Xiao and Ma, 1997). This is a very simple system for in situ applications, ensuring a high level of material quality control owing to the controlled factory-based fabrication of shells. However, the shells must be fabricated with strict tolerance with respect to column dimensions. In the case of multiple layers, they must be properly positioned to ensure the desired jacketing effect. Prefabricated shells can be used as formworks as well as transverse reinforcement. In the case of rectangular sections, the confining action of shells is less effective and it is generally preferred to modify the column cross-sectional shape.

In the resin-infusion process (Karbhari et al., 1993), the resin is injected under a vacuum film using a resin distribution medium to achieve full and rapid infusion of the fabric. Dry fibers are preformed in a mold in a fabrication shop and are sent to the site. The preform is then attached to the structure and a resin supply is channeled into it (Hollaway and Head, 2001). As the resin flows through the dry fiber preform, it forms the composite material and the adhesive bond between the fibers and the structure. The entire assembly is evacuated before the resin is introduced and a vacuum is maintained until the resin is cured. With the use of multiple injection ports, very large parts can be processed efficiently at a minimal cost. Because this is a closed process, there are practically no emissions, thereby reducing worker exposure to fumes (such as styrene) from the resin system to an almost negligible level. However, one disadvantage of the infusion process is that a considerable amount of resin is wasted during the production procedure. The process provides high fiber volume fraction composites, in the order of 55%, that have high strength and stiffness values.

In technique (D), a wrapping machine is used to automatically wind the fibers around the column (Fig. 6.10D). The machine, built for the first time in Japan, has been designed to upgrade bridge piers, but it can be used for buildings columns as well. The automated wrapping device is set up around the column. The fibers, wound on reels, are placed in the fiber winding head. They are preimpregnated with the resin and wound around the column moving upward. After winding, a curing blanket is placed. The winding angle, fiber volume fraction, and thickness are computer controlled. This device enables us to wind the fibers while pretensioning them to obtain an active jacketing system, independent of concrete lateral dilation. The disadvantage of such a device is that in the presence of nonleveled soils, preliminary calibration operations are required, which sensibly slow down its usage. The same pretensioning effect can be obtained with other systems by injecting either expanding mortar or epoxy in pressure between the jacket and the column surface.

6.4.2 Concrete confinement mechanism and objectives

It is well known that by confining a concrete element in two out of three mutually perpendicular directions, the ultimate compressive strength of the element in the third direction considerably increases. Lateral confinement to concrete, generally, can be categorized into two types: active confinement, such as by hydraulic pressure, and passive confinement. Steel stirrups before yielding provide passive confinement to concrete; owing to the elongation imposed on steel stirrups by the expansion of concrete, they induce compressive stresses in the element. However, after the steel stirrups are yielded, the provided confinement is nearly constant and could be considered active. FRP behaves linearly elastic until failure; hence, for FRP-confined concrete, the inward radial pressure increases with the lateral expansion of concrete such that the assumption of a constant confining pressure is no longer valid. So, FRP jackets are known to offer passive confinement to columns.

In terms of shear resistance, the outer FRP jacket, when forming an additional closed hoop, is generally effective in enhancing the shear strength of existing substandard columns by providing additional shear reinforcement within the jackets. In terms of flexural resistance, when concrete columns are subjected to axial compressive loading (concentric or eccentric) or both axial compressive and lateral loading, the confining concrete columns with FRP composites provide support to the longitudinal reinforcement and restrain the lateral expansions on the core concrete, thereby enabling higher strains sustained by the compression concrete before failure, as shown in Fig. 6.11. Along with this, confining the core concrete increases the concrete compression strength, which further improves the concrete postcracking behavior (enhancing the flexural strength). Another primary effect of applying the external FRP jacket is improving the behavior of short lap-splice reinforcements by increasing the bond stress along the splice (see Fig. 6.11) to postpone the onset

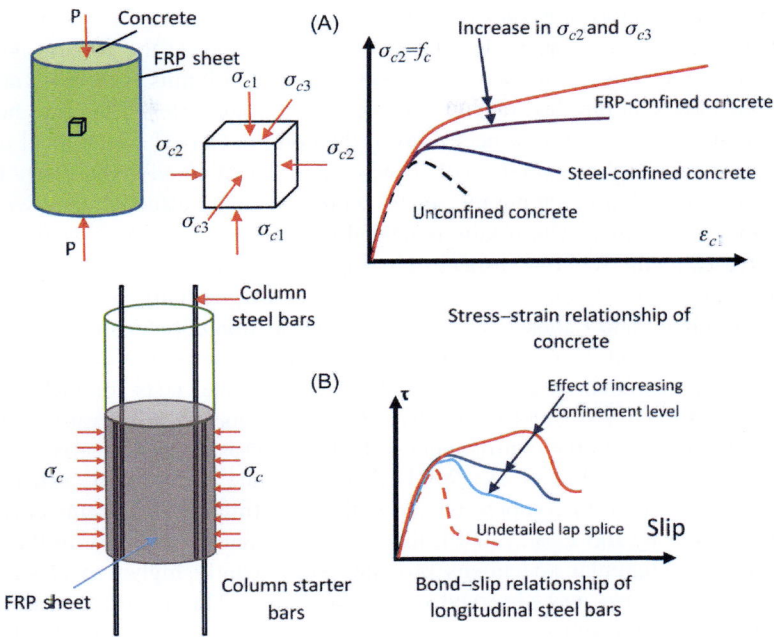

Figure 6.11 Effect of FRP confinement on (A) concrete stress—strain relationship and (B) bond—slip relationship of the longitudinal steel bars.

of splitting and reduce the severity of subsequent deterioration (enhance the performance of lap-splice deficient columns).

For concrete confined by active hydrostatic fluid pressure, early studies by Richart et al. (1928) resulted in the formulation of a simple equation to define the strength of the confined concrete as follows.

$$f'_{cc} = f'_{co} + k_1 f_{lu} \tag{6.1}$$

where f'_{cc} is the ultimate strength of confined concrete under a maximum lateral pressure of f_{lu} and f'_{co} is the unconfined concrete compressive strength.

6.4.3 Parameters affecting behavior of FRP-confined concrete

6.4.3.1 Concrete cross-section

Rectangular sections are generally less effective than their circular counterparts in confining concrete. This may be explained by the distribution of confining pressure in circular and rectangular sections (Mirmiran et al., 1998; Rochette and Labossière, 2000; Pessiki et al., 2001; Shehata et al., 2002; Campione and Miraglia, 2003; Harajli, 2006; Wu et al., 2007). In circular sections, the confining pressure is uniform and a function of hoop strength of the jacket. However, in rectangular sections, the confining pressure varies. The confining pressure at the corners is due to the membrane action in the transverse sides of the columns; whereas, at other points, it depends on the flexural rigidity of the FRP composites. Therefore both the corner radius and the section aspect ratio of the column can affect the level of confinement exerted on the concrete core. It was shown by Wu and Wei (2010) and De Luca et al. (2010) that wrapping the square sections is more effective than wrapping the rectangular sections. The results indicated that the confinement effectiveness of FRP increases as the sectional aspect ratio approaches unity (Ozbakkaloglu, 2013a, b). Moreover, the confinement effectiveness increases gradually as the corners are rounded (Rochette and Labossière, 2000); thus the section corners should always be rounded off sufficiently to prevent premature failure due to stress concentration in the FRP close to the sharp corners. Based on experimental tests for FRP-confined columns, Wang and Wu (2008) addressed that as the corner radius ratio increases, the strength gain due to the confinement of FRP increases almost linearly.

It was reported by Ilki et al. (2004) that while the increase in corner radius resulted in an increase in compressive strength, the ultimate axial strains decreased. In another study by Ilki et al. (2008), they indicated that the increase in the corner radius of the rectangular specimens provided significantly higher axial strengths, but no clear influence was observed on deformability. This result is considerably consistent with the results recently presented by Pham and Hadi (2014a,b,c) for FRP-confined rectangular columns, as they showed that the corner radius significantly affects the compressive strength but marginally affects the compressive strain. Wang and Wu (2008) stated the beneficial effect of providing a large corner radius, that is, larger values of corner radius result in higher postpeak stiffness. Wu et al. (2007) showed that a corner radius of 25.4 mm (1 in.) and higher ensures suitable confinement for rectangular concrete columns. In addition, Sharma et al. (2013) recommended that a corner radius equal to the concrete cover yields better results in terms of ultimate load-carrying capacity, than the corner radius less than the cover of confined RC columns.

Modifying the shape of the columns from square to circular and rectangular to elliptical is another solution proposed by several researchers to eliminate the corner stress concentration in prisms and improve confinement effectiveness. Subsequent FRP wrapping of shape-modified columns will substantially improve the axial load and pseudo ductility. For instance, Lam and Teng (2002) examined the behavior of FRP-confined concrete in elliptical columns. They reported that the axial compressive strength of FRP-confined concrete in elliptical specimens is controlled by the amount of confining FRP and the major-to-minor axis length ratio, a/b, of the column section. The confining FRP becomes increasingly less effective as the section becomes more elliptical, but substantial strength gains from FRP confinement can still be achieved even for strongly elliptical sections. Hadi and Widiarsa (2012) proposed the circularization of the square columns with four segmental circular covers. In comparison to square columns with rounded corners, it was evident that the efficiency of carbon fiber-reinforced polymer (CFRP) confinement of the circularized columns was higher.

6.4.3.2   Scale size of test specimens

Laboratory investigations of the compressive behavior of FRP-confined concrete columns have generally been performed using relatively small-scale specimens, and the majority of theoretical models that have been developed so far are based on test data from such specimens. Thériault et al. (2004) experimentally examined the influence of slenderness ratio and specimen size on axially loaded FRP-confined concrete columns. According to the statistical analysis of the results, it was shown that conventional FRP-confined concrete cylinders can effectively be used to model the axial behavior of short columns. Size effects, however, were evident in very small (diameter = 50 mm) specimens. Moreover, no slenderness effects were observed between the medium- and large-scale specimens; this finding was also confirmed by Park et al. (2008). In contrast, Issa et al. (2009) concluded that the confinement effectiveness decreased with the increase in column diameter and the size effect was more pronounced for columns with low concrete compressive strength and FRP-confinement ratio.

Several works have been conducted to determine the effect of column aspect ratio, section aspect ratio, and column size on the stress—strain behavior of FRP-confined rectangular columns. For the effect of length-to-diameter ratios within the range of 2:1 and 5:1, Mirmiran et al. (1998) found that this range of variation had no significant effect on either the strength or ductility of the section; thus both eccentricities and strength reductions are within the limits prescribed by ACI 318-95 for tied columns. Maalej et al. (2003) also showed that when the column's aspect ratio increases, the reduction in the strengthening ratio (defined as the ultimate strength of a strengthened RC column divided by the ultimate strength of a control column) becomes more insignificant. For the aspect ratio of the cross-section, there is a general consensus among several researchers that the increase in its value results in a lower ultimate strength for an FRP-jacketed RC rectangular column (Cole and Belarbi, 2001; Lam and Teng, 2003a,b; Hadi, 2006; Harajli et al., 2006; Kumutha et al., 2007; Tao et al., 2008; De Luca et al., 2010; Ozbakkaloglu, 2013a,b; Teng et al., 2013; Mostofinejad et al., 2015). It was also shown by Wu and Wei (2010) that as the aspect ratio increases from one to two, the strength gain in confined concrete columns (f'_{cc}/f'_{co}) decreases until it becomes insignificant at an aspect ratio of greater than two. Moreover, the stress—strain response changes from a monotonically increasing and strain-hardening type to a strain-softening type when the aspect ratio increases. This effect could be attributed to the rectangular sections that develop higher hoop rupture strains along their short spans than their long spans at ultimate, and the ratio of the short-to-long span strains increase with an increase in the aspect ratio (Ozbakkaloglu, 2013a,b).

6.4.3.3   Grade of unconfined concrete

Several published works addressed the behavior of concrete with different grades (low, moderate, and high) confined with FRP and concluded that FRP confinement on low-strength concrete specimens produced higher gains in terms of strength and strains than that on high-strength concrete specimens. Therefore the confinement effectiveness reduces with an increase in the unconfined concrete strength of both circular and square FRP-confined concrete specimens (Saadatmanesh et al., 1996; Shahawy et al., 2000; Pessiki et al., 2001; Chaallal et al., 2003; Ilki et al., 2006; Almusallam, 2007; Wu et al., 2007; Sadeghian et al., 2008; Wang and Wu, 2008; Ozbakkaloglu, 2013a,b; Parvin and Brighton, 2014). For instance, Shahawy et al. (2000) showed that for concrete specimens with a compressive strength of 20.7 MPa, one layer of carbon could effectively curtail the dilation tendency of the concrete because no volume expansion was observed; however, for the 41.4 MPa concrete specimens, three and even four layers were required to provide adequate confinement pressure to resist the dilation tendency of concrete. In comparison to normal strength concrete externally confined with FRP, the higher the concrete compressive strength, the higher the reduction in the efficiency of FRP confinement, for example, the decrease in the confinement efficiency reached a value of 15% and 25% for a concrete grade of 100 and 200 MPa, respectively (Berthet et al., 2005). In the study by Moran and Pantelides (2012), it was concluded that the amount of FRP (i.e., the volumetric jacket reinforcement ratio (ρ_j) or the total thickness of FRP jacket (t_j)) required to achieve strain-hardening behavior increases as the unconfined concrete compressive strength f'_{co} increases.

6.4.3.4   Fiber direction

For circular concrete columns, the study by Karbhari and Gao (1997) showed that a comparison of the sets of architectures with the same number of layers always indicates that the 0 degrees (fibers oriented along the hoop direction) architecture results in the highest increase. Li (2006) concluded that the FRP jacket with 0 degrees fibers shows slightly higher confinement effectiveness than other angleply jackets (fibers oriented at 30, 45, and 60 degrees with respect to the circumferential direction) owing to its smaller radial expansion. However, the efficiency of longitudinal CFRP layers is dependent on the arrangement of the added longitudinal fibers with respect to the transversely wrapped fibers. This implies that the efficiency of longitudinal fibers should be neglected if they are wrapped outside the transverse CFRP layers, and considered to be approximately 50% of that of transverse CFRP layers if they are wrapped inside the transverse CFRP layers (Issa et al., 2009). Sadeghian et al. (2008) observed that the hoop orientation of fibers (0/ 0 degrees) results in the largest gain in ultimate stress, while the angle orientation (± 45 degrees) leads to the largest ultimate strain (i.e., ductility). Kabir and Shafei (2012) recommended taking fiber angles between 0 (circumferential) and 30 degrees to improve the ultimate strength and ductility of confined short concrete columns. However, for slender concrete columns, the optimum fiber orientation can be set between 15 and 30 degrees. The efficiency of angled ply is also affected by the configuration of FRP plies. Parvin and Brighton (2014) demonstrated that the cylinders with "hoop−angle−hoop" ply configuration generally exhibited higher axial stress and strain capacities as compared to those with the "angle−hoop−angle" ply configuration.

In contrast, for rectangular concrete specimens, Rochette and Labossière (2000) reported that angle-ply wrap configurations should be investigated as a potential way to achieve more strength and ductility when confining sections with sharp corners. Similar to circular columns, Tan (2002) showed that longitudinal fiber sheets are effective only when confined by transverse fiber sheets. In addition, the study by Tan (2002) proved that increasing the number of longitudinal fiber sheets leads to higher strength and ductility. However, increasing the transverse fiber sheets leads to an increase in ductility only.

6.4.3.5 Lateral rigidity (stiffness) and confinement ratio

The behavior of a deficient column strengthened with FRP is not a function of the type of fiber used. However, column behavior is mainly dependent on the amount of rigidity, $E_f * t_f$, of the jacket. When different types of fibers are available for application as external reinforcement, the column behavior is dependent on the design of lateral stiffness provided by the external reinforcement instead of the type of FRP used (Fahmy and Wu, 2010); see Fig. 6.12. A certain curve of the stress—-strain relationship of FRP-confined concrete is expected at a specified confinement rigidity, and only the endpoint of this relationship depends on the ultimate rupture strain of the fiber used. E_2 would change with different values of confinement rigidity. This indicates that with the available range of FRP types, the preferred FRP material would be selected from the various options based on the required ultimate compressive strength and strain of the confined concrete, which is mainly dependent on the maximum lateral pressure that could be provided by the FRP type used. The stress—strain behavior of FRP-confined concrete can be related to the effective confinement ratio, which is the ratio of the effective confining pressure to the strength of unconfined concrete (Lam and Teng, 2002). Consequently, the

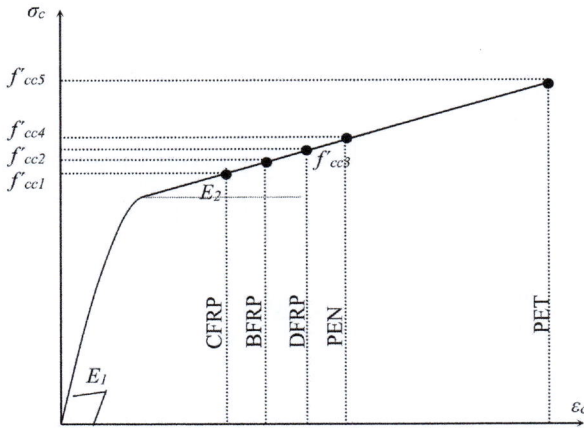

Figure 6.12 Relationships among E_2, FRP-confinement rigidity, ultimate FRP-confinement pressure, and concrete axial strength and strain (Fahmy and Wu, 2010).

maximum increase in the axial strength from FRP wrapping is expected with the fiber providing the highest confinement pressure, as shown in Fig. 6.12.

In contrast, when different types of FRP were designed to provide the same ultimate confining force to concrete, Rousakis et al. (2007) showed that carbon confinement was more effective than glass in terms of strength enhancement. Furthermore, the increase in concrete ductility was similar for either of the two confining materials. Karbhari and Gao (1997) reported that the maximum effect of confinement can be observed through an increase in the number of layers with fibers oriented in the hoop direction, with the addition of each layer in the hoop (0 degrees) direction, thereby increasing the ultimate strength of confined concrete by approximately 30%−33%.

It is worth mentioning that the stress−strain behavior of confined concrete changes from strain softening to strain hardening at certain confinement rigidity. The following table (Table 6.1) summarizes several definitions for the limit of FRP-confined concrete with strain-hardening behavior. There is also a general consensus among several studies that the rate of increase in confinement effectiveness decreases nonlinearly as confinement ratio increases (Li and Wang, 2006; Wu et al., 2006; Park et al., 2008).

Table 6.1 Summary of limits for FRP-confined concrete with strain-hardening behavior.

Reference	Limit
ACI(440.2-08) (2008)	$f_{l,eff}/f'_{co} > 0.08$ $f_{l,f}/2f'_{co} > 0.073$
Chines (2010)	$E_{l,f}/f'_{co} > 6.5$
Mirmiran et al. (1998)	$f_{l,f}/f'_{co} > 0.15$
Spoelstra and Monti (1999)	$f_{l,f}/f'_{co} > 0.07$
Xiao and Wu (2000)	$E_{l,f}/(f'_{co})^2 > 0.2$
Lam and Teng (2003a,b)	$f_{l,eff}/f'_{co} > 0.07$
Binici (2008)	$f_{l,f}/f'_{co} > 0.14$
Jiang and Teng (2006)	$E_{l,f}/(f'_{co}/\varepsilon_{co}) > 0.01$
Wu et al. (2006)	$f_{l,f}/f'_{co} > \begin{cases} 0.13\sqrt{\dfrac{250}{E_f}}E_f > 250\text{GPa} \\ 0.13E_f \leq 250\text{GPa} \end{cases}$
Pantelides and Yan (2007)	$f_{l,f}/f'_{co} > 0.2$
Ozbakkaloglu and Lim (2013)	$E_{l,f} > (f'_{co})^{1.65}$
Kawashima et al. (2000)	$E_2 = -0.658\dfrac{f'_{co}2}{2E_{l,f}\varepsilon_{cf}+0.098(2f_{l,s})}+0.078\sqrt{\dfrac{2E_{l,f}}{E_f}}E_f > 0$
Hu (2013)	$f_{l,f}/f'_{co} > 0.11$
Lim and Ozbakkaloglu (2014a,b)	$\dfrac{K_{s1}*E_{l,f}}{K_{Lo}} > 1$ $E_{l,f}/f'_{co} > 6.3$

For rectangular columns, the geometry of the corner of rectangular columns could play an important role in the efficiency of the confinement ratio provided. Rochette and Labossière (2000) showed that when the rounding of corners for square sections cannot be increased, additional confinement can be achieved by adding more wrapping layers. In general, it can be concluded from the results of several studies that the ultimate strength and ductility of FRP-confined concrete increase with an increase in the number of confining layers (Chaallal et al., 2003; Maalej et al., 2003; Ilki et al., 2004; Harajli, 2006; Ilki et al., 2006; Kumutha et al., 2007; Benzaid et al., 2008; Wang and Wu, 2008; Wang and Wu, 2010; Wu and Wei, 2010; Sundarraja and Prabhu, 2012; Tan et al., 2013; Vasumathi et al., 2014; Feng et al., 2015). However, it was reported by Rochette and Labossière (2000) that excessive confinement will lead to sudden and destructive compressive failures, which must be avoided.

6.4.3.6 Bonding condition between FRP and concrete

Because the adhesive bond between concrete and the wrap did not appear to significantly affect the confinement behavior of fiber-wrapped concrete cylinders, Shahawy et al. (2000) recommended that the same confinement model can be applied for both fiber-wrapped concrete cylinders and concrete-filled FRP tubes. This would be an indication that the adhesive bond between concrete and the wrap would not significantly affect the confinement behavior (Issa et al., 2009). This result was also confirmed through a numerical study conducted by Sadeghian et al. (2008) where they proved that interfacial bonding has minimal effect on the stress—strain response.

Examining the effect of this parameter on the behavior of FRP-confined rectangular concrete columns showed that adhesive bonds did not affect the load-carrying capacity of FRP-confined concrete. However, mechanical bonds (shear connectors) could significantly improve the performance of the section by distributing the confinement pressure more effectively around the circumference of the column (Mirmiran et al., 1998). However, it would not be possible to generalize this conclusion for rectangular columns with a high aspect ratio of the section because it was shown by Doran et al. (2009) that the deterioration of the bond between concrete and FRP could lead to the loss of confinement and an unstable postpeak behavior. For RC columns with a section aspect ratio of 3.65, the fiber sheets delaminated from the wider faces of the columns, at between 60% and 90% of the ultimate load, depend on the installation skill (Tan, 2002). However, anchoring the transverse fiber sheets along the wider faces of the column could lead to better confinement of the concrete and restrain the longitudinal fiber sheets, thereby leading to a higher axial load capacity. It was interesting to find in the same study that plastering the surface of confined columns affected the stress—strain behavior. The plastered columns possessed higher stiffness than the nonstrengthened columns, and eventually carried higher ultimate loads. However, they generally exhibited lower ductility at failure than nonplastered columns for the same fiber configuration, probably owing to the spalling off of the plaster finishes at high loads. In a recent study by Beddiar et al. (2015), the concrete columns confined by glass fiber-reinforced polymer (GFRP) prefabricated shells and shrinkage-compensating cement mortar were examined. One of the main findings was that the unsticking phenomenon between the shells and the mortar helps in redistributing the stresses over the height and prevents premature localized failure.

6.4.3.7   Existence of internal steel stirrups

Demers and Neale (1999) considered a range of steel stirrup spacing for concrete cylinders confined with external FRP. Test results showed that the stirrup spacing had no effect on the compressive behavior of the confined columns tested by them because the stirrups could not provide effective confinement. However, Hoshikuma et al. (1997) displayed that as the volumetric ratio of hoop reinforcement increases, both the peak stress and the strain at the peak stress increase, thereby preventing the deterioration of the concrete after the peak stress. Moreover, the effect of cross ties is significant in increasing the ductility of concrete piers with wall-type cross-sections. In addition, comparing the behavior of plain concrete columns confined by FRP sheet to the behavior of steel-reinforced columns confined by FRP, Karabinis et al. (2008) noted that the contribution of bars is significant, thereby revealing an increase in strength, whereas the deformability levels were quite similar. In a study by Pellegrino and Modena (2010), the overall increment in strength, with respect to concrete, for the columns with internal steel reinforcement and external FRP was found to be more than that for those without steel reinforcement owing to the additional contribution of confining steel stirrups. Ghorbi et al. (2013) reported that with the same volumetric stirrup ratio (ρ_s), the increase in stirrup spacing ratio (s/D) leads to the decrease in mean axial stress—strain curve of confined concrete. If this dimensionless ratio exceeds a certain value, the behavior of the concrete core will be similar to that of unconfined concrete. Triantafyllou et al. (2015) reported that the columns with internal steel stirrups presented higher FRP-wrapping efficiency than their plain concrete counterparts. Even in the cases of partially wrapped columns, the dual confining action of steel and FRP altered the behavior from softening to hardening despite the inadequate steel reinforcement detailing.

All codes and design guidelines focus on the compressive strength of FRP-wrapped concrete but neglect the effect of transverse steel reinforcement. In reality, however, most research studies proved that the actual concrete columns are under two actions of confinement, that is, the action of the FRP and that of steel reinforcement. Therefore existing codes and guidelines cannot provide accurate estimations. The minimum amount of steel reinforcement must also be added to FRP-confined concrete to avoid generating a brittle failure mode. Thus researchers note the need for a concrete confinement model that considers the interaction between internal steel reinforcement and external FRP (Shirmohammadi et al., 2015).

6.4.3.8   Loading condition [(concentric, eccentric), (monotonic and cyclic), (static, dynamic)]

Various experimental studies have been conducted on confined concrete under the effect of axial loading; however, there is a lack of studies that focus on confined concrete under the effect of eccentric loadings, especially for rectangular columns (Fahmy and Farghal, 2016). In contrast, several experimental studies regarding FRP-confined concrete columns were conducted under the effect of constant axial load and reversed cyclic lateral loadings. The reported test results prove that FRP confinement ensures a high ductility level in the retrofitted columns before the capturing rupture of FRP. However, stress—strain models developed for axially loaded FRP-confined concrete cannot ensure the accurate evaluation of the failure mode of confined RC under the effect of both axial and lateral loadings (Fahmy et al., 2017). Parvin and Brighton (2014) stated that for eccentrically loaded columns, a smaller enhancement factor of FRP confinement, than that of axially loaded columns, should be considered in the design.

Rousakis et al. (2007) demonstrated that repeated load—unload cycles of axial compressive load in each of the three distinct regions of the response curve of rectangular FRP-confined columns caused no appreciable decrease in strength, ductility, or energy absorption of concrete. FRP jackets partially restore the mechanical properties of the disintegrated concrete core during unloading and provide a behavior similar to that of the specimens subjected to monotonic loads. In addition, although cyclic loading did not influence the compressive strength, the ultimate axial deformations were higher in the case of cyclic loading in comparison to monotonic loading (Ilki et al., 2008). Rousakis (2014) also showed that the stress—strain response of circular columns under monotonic or cyclic compression is almost identical. Rousakis and Karabinis (2012) noticed that the behavior of columns with modified confinement ratio (MCR = 0.153), as a measure of confinement effectiveness under monotonic loading, displays an ideally plastic or softening behavior and is considered marginally acceptable. However, the same column under cyclic loading presents a softening inelastic branch, with higher strain at failure. Ultimately, there are very limited studies available on FRP-confined concrete under the effect of dynamic loadings.

6.4.3.9   Predamaged level of the unconfined concrete

FRP-confinement can be applied to structural elements that are already damaged after they were loaded beyond their original design capacity. In this case, it is referred to as the repairing technique for predamaged elements. The effect of damage-level of concrete on the performance of FRP-repaired concrete columns, to some extent, has been examined in some studies. Recently, Wu et al. (2014) studied the mechanical properties of FRP-repaired concrete columns through an extensive experimental and analytical investigation. The test program included 102 concrete cylinders involving variations in damage degree, concrete grade, and confinement pressure. Repairing the damaged cylinders with FRP jackets was highly effective for all damaged levels in terms of strength recovery and general enhancement in the stress—strain behavior. However, the initial stiffness of repaired cylinders was significantly lowered than that of undamaged cylinders. Similarly, Demers and Neale (1999) showed that concrete damage, produced by loading the column specimens to failure before FRP confinement, had an observed effect on the elastic modulus of cracked concrete. In contrast, when concrete samples were loaded until the axial strain level of 0.003—0.004 and the specimens had lost a small portion of their strengths, no practical consistent difference in terms of strength, axial and transverse deformations, and axial stiffness was observed between the performances of predamaged and undamaged specimens (Ilki et al., 2008).

6.5   Modeling of stress—strain behavior of FRP-confined concrete

To accurately model the behavior of reinforced concrete structures, it is obligatory to understand the material behavior of the constituent concrete. Although the uniaxial unconfined stress versus strain relationship for plain concrete is well established, most in situ concrete is confined in some manner. Confinement may take the form of conventional internal reinforcing steel, external steel, or FRP composite jackets. Accurate structural modeling, therefore, requires a sound understanding of the stress—strain behavior of confined concrete.

The axial stress—strain behaviors of unconfined and confined concrete differ significantly; see Fig. 6.13. In conventionally reinforced and externally jacketed concrete columns, the confining pressure is passive in nature. This indicates that the confining pressure is engaged by the transverse dilation of concrete resulting from principal axial strains—the Poisson effect. There are cases where an initial active confining pressure is present, such as when an expansive grout is injected between

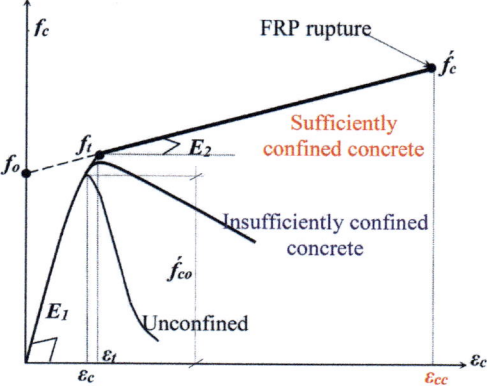

Figure 6.13 Typical stress—strain responses for unconfined and FRP-confined concrete (Fahmy and Wu, 2010).

a column and an external jacket. In these cases, however, the active pressure is generally quite small in comparison to the additional passive pressure generated by concrete dilation (Harries and Kharel, 2003).

Passive confinement may be constant or variable depending on the axial load history. Constant confining pressure is generated in cases where the confining material behaves in a plastic manner. This is typically assumed to be the case where confinement is provided by conventional transverse reinforcing steel. Variable confining pressure is generated when the confining material is predominantly elastic. FRP jackets and steel, before yielding, generate variable confining pressures. Variable passive confinement is dependent on the axial and transverse deformation behavior of the concrete, as well as the amount and stiffness of confinement provided (Harries and Kharel, 2003).

It has been shown that concrete behaves very differently when confined by linear elastic materials, such as fiber composites, in comparison to elastoplastic materials, such as steel (Mirmiran and Shahawy, 1997a,b). The implementation of steel confined concrete models to fiber-wrapped concrete results in an unsafe overestimation of strength. There are three distinct modeling techniques available for FRP-confined concrete. A brief description of each technique is given below.

Direct application of steel-based confinement models

A typical example of models for steel-confined concrete is the one proposed by Mander et al. (1988), in which a single equation defines the entire stress—strain response. This model was used for fiber-wrapped columns by Saadatmanesh et al. (1996), who generated moment-thrust interaction charts based on its results. However, other studies (Nanni and Bradford, 1995; Mirmiran and Shahawy, 1997a, b) have shown Mander's model to grossly overestimate the strength of FRP-confined concrete.

Design-oriented models for FRP-confined concrete

Many investigations have been conducted into the behavior of FRP-confined concrete and consequently, several design-oriented stress—strain models have been proposed. In these models, the compressive strength, ultimate axial strain (hereafter, often referred to as the ultimate strain for brevity), and stress—strain behavior of FRP-confined concrete are predicted using closed-form equations based directly on the interpretation of experimental results. The first hyperbolic stress—strain model for FRP-encased concrete was developed by Fardis and Khalili (1982). Later, Nanni and Bradford (1995) demonstrated that the model grossly underestimates the ductility of fiber-wrapped columns whereas predicts the strength reasonably well.

Finite element models with plasticity approach

The stress—strain curves of FRP-confined concrete are also generated using an incremental numerical procedure. In this approach, an active confinement model for concrete is used to evaluate the axial stress and strain of passively confined concrete at a given confining pressure. The interaction between the concrete and the confining material is explicitly accounted for by equilibrium and radial displacement compatibility considerations (Rochette and Labossiere, 1996, Karabinis and Rousakis, 2002, Rousakis et al., 2007, Eid and Paultre, 2007, Jiang et al., 2011, Jiang and Wu, 2012, Kabir and Shafei, 2012). This type of stress—strain model is suitable for computer-based numerical analysis, such as nonlinear finite element analysis.

6.5.1   Stress—strain models of FRP-confined circular columns

6.5.1.1   Design-oriented models

In design-oriented models, experimental data have been used to investigate the stress—strain models and ultimate axial stress/strain by closed-form equations. Regression analyses or artificial neural network methods are the main methods used to develop most of these models that consist of different forms of equations. Therefore the accuracy of these models depends on the size of test samples as well as the parametric range of the test data used for model development.

By scanning the available design-oriented models, it has been recognized that they could be categorized into four groups, that is, Group "A," Group "B," Group "C," and Group "D." All FRP-confined concrete models describing the strain-hardening behavior are designated to group "A" and those describing the strain-softening behavior are labeled into group "B." Some studies focused in building stress—strain models that consider both the strain-hardening and strain-softening behaviors of FRP-confined concrete; therefore in this study, they are designated to group "C." Other works focused on predicting the ultimate strength and/or strain of FRP-confined concrete, and those models are labeled into group "D." A brief discussion for each group is given in the next sections.

Models of group A (ascending slope of second branch)

In this group, the design-oriented models with strain-hardening behavior (ascending stress—strain relationship) are classified into five broad categories based on the geometric form of the curves, namely, parabolic for all trends (Type I), two parabolic parts (Type II), parabolic curve of three parts (Type III), parabolic in the first branch and linear part in the second branch (Type IV), and bilinear relationship (Type V), as exhibited in Fig. 6.14.

Models of group B (descending slope of second branch)

The stress—strain response of FRP-confined concrete sections may have a strain-hardening component (strong confinement) or a strain-softening component (week confinement), as shown in Fig. 6.15. Therefore some researchers developed models providing a limit for the change from strain-softening to strain-hardening behaviors. Most of the available studies proposed a limit based on a definition of FRP confinement ratio (f_l/f'_{co}), wherein its values were 0.08, 0.13, 0.14, and 0.2, as reported by Wu et al. (2006), Binici (2008), Pantelides and Yan (2007), and Xiao and Wu (2000), respectively. Another way to determine the FRP confinement ratio is by using

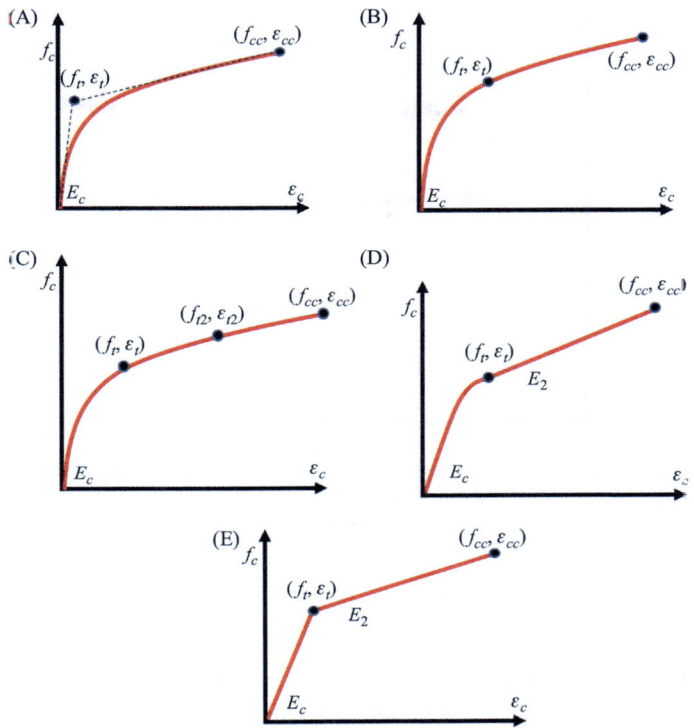

Figure 6.14 Designations of different stress–strain relationships of Group "A" of FRP-confined concrete with strain-hardening behavior (Fahmy et al., 2017).

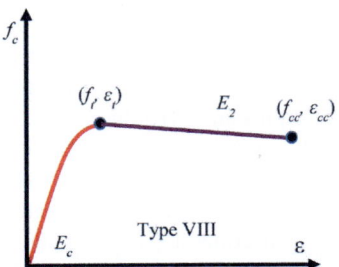

Figure 6.15 Stress–strain relationship of FRP-confined concrete with strain-softening behavior (descending second branch).

confined stiffness (E_l) to unconfined stiffness (f'_{co}/ε_{co}) (Teng et al., 2009), with which this value equals 0.01. Jiang and Wu (2016) suggested another criterion based on the Drucker–Prager plasticity. They recognized that the confinement stiffness ratio determines the characteristic curve of the plastic work modulus, which

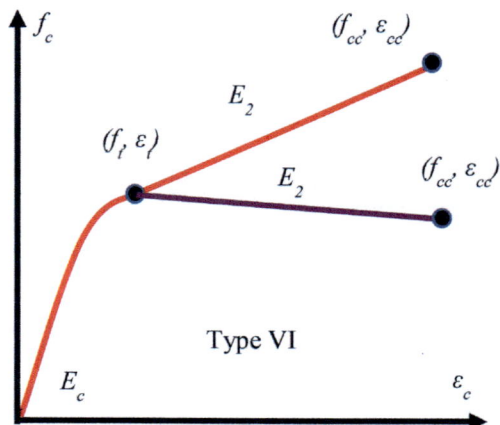

Figure 6.16 Models describing both strain-hardening and strain-softening behaviors of FRP-confined concrete.

determines the energy dissipation behavior, and hence, the strain-hardening or strain-softening behaviors. Table 6.1 summarizes all the proposed forms to differentiate between FRP-confined concrete with strain-hardening or strain-softening behavior.

Models of group C (descending or ascending slope of second branch)

Some stress—strain models of FRP-confined concrete were investigated to predict the behavior of FRP-confined concrete with strain-hardening or strain-softening behavior (Fig. 6.16), (Li et al., 2003, and Wu and Wei, 2014).

Models of group D (ultimate stress/strain)

Numerous studies proposed models evaluating the ultimate condition of FRP-confined concrete (peak point of stress—strain curve) without any recommendation for mathematical or empirical forms describing its stress—strain behavior. Most of the existing strength models adopted the concept of "strength at failure for concrete confined by a hydrostatic fluid pressure" (Richart et al., 1929). Based on the studies by Richart et al. (1928, 1929), several forms for modifications were considered to increase the accuracy in predicting the ultimate strength of FRP-confined concrete, as shown in Table. 6.2. The second important parameter of the stress—strain curve is the ultimate strain of confined concrete, beyond which the confined concrete is assumed to fail. The ultimate strain is assumed by many researchers to be a function of the ultimate confining pressure exerted by FRP. This could be acceptable as the FRP jacket and its mechanical properties significantly affect the ultimate strain of confined concrete.

6.5.1.2   Analysis-oriented models

Analysis-oriented models explicitly consider the responses of the concrete and FRP jacket, as well as their interaction. Analysis-oriented models provide a unified

Table 6.2 Summary of state of arts on modeling stress–strain behavior of FRP-confined concrete.

State of art	Samples/models (cross-section)	Models	Best models		Authors proposed new model
			Ultimate stress models	Ultimate strain models	
Rousakis et al. (2012a,b)	471/20 (circular columns)	Fardis and Khalili (1982), Saadatmanesh et al. (1996), Mirmiran and Shahawy (1997a,b), Karbhari and Gao (1997), Kono et al. (1998), Spoelstra and Monti (1999), Toutanji (1999), Xiao and Wu (2000), Xiao and Wu (2003), Lam and Teng (2003a,b), Campione and Miraglia (2003), CNR-DT(200/2004) (2004), Rousakis (2005), Matthys et al. (2006), Kumutha et al. (2007), Vintzileou and Panagiotidou (2008), Wu and Wang (2009), Teng et al. (2009), Girgin (2009), Wu and Zhou (2010), Fahmy and Wu (2010)	Best model (Rousakis et al., 2012a,b) Fair models (Lam and Teng, 2003a,b; Campione and Miraglia, 2003; Girgin, 2009; Wu and Wang, 2009; Wu and Zhou, 2010)	No	Yes
Ozbakkaloglu et al. (2013)	705/68 (circular columns)	Fardis and Khalili (1982), Saadatmanesh et al. (1996), Mirmiran (1996), Karbhari and Gao (1997), Miyauchi (1997), Jolly and Lilistone (1998), Kono	a. Best model (Lam and Teng, 2003a,b). b. Faire models (Bishby et al., 2005; Teng et al., 2007),	(a) Best model (Tamuzs et al., 2006a,b), (b) Faire models (Jiang and	No

(Continued)

Table 6.2 (Continued)

State of art	Samples/ models (cross-section)	Models	Best models		Authors proposed new model
			Ultimate stress models	Ultimate strain models	
		et al. (1998), Samaan et al. (1998), Miyauchi et al. (1999), Saafi et al. (1999), Spoelstra and Monti (1999), Toutanji (1999), Jolly and Lillistone (2000), Thériault and Neale (2000), Xiao and Wu (2000), Lin and Chen (2001), Ilki et al. (2002), Lam and Teng (2002), Moran and Pantelides (2002), Shehata et al. (2002), De Lorenzis and Tepfers (2003), Lam and Teng (2003a,b), Li et al. (2003), Xiao and Wu (2003), Ilki et al. (2004), Bisby et al. (2005), Saiid Saiidi et al. (2005), Berthet et al. (2005), Guralnick and Gunawan (2006), Jiang and Teng (2006), Matthys et al. (2006), Tamuzs et al. (2006a,b), Tamuzs et al. (2006a, b), Wu et al. (2006), Al-Tersawy et al. (2007), Cupala et al.		Teng, 2006; Teng et al., 2009),	

| Lim and Ozbakkaloglu (2014a,b) | 231/10 (circular columns) | (2007), Shehata et al. (2007), Tabbara and Karam (2007), Vintzileou and Panagiotidou (2007), Pantelides and Yan (2007), Youssef et al. (2007), Binici (2008), Al-Salloum and Siddiqui (2009), Girgin (2009), Teng et al. (2009), Wu and Wang (2009), Wu et al. (2009), Benzaid et al. (2010), Fahmy and Wu (2010), Mohamed and Masmoudi (2010), Wu and Wang (2010), Wu and Zhou (2010), Wang and Wu (2010), Yu and Teng (2011), Cevik (2011), Park et al. (2011), Realfonzo and Napoli (2011), Mirmiran and Shahawy (1997a,b), Spoelstra and Monti (1999), Fam and Rizkalla (2001), Park (2002), Harries and Kharel (2003), Marques et al. (2004), Binici (2005), Albanesi et al. (2007), Jiang and Teng (2007), Teng et al. (2007), Aire et al. (2010), Xiao et al. (2010) | Miyauchi et al. (1999), Mandal et al. (2005), Tamuzs et al. (2006a,b), Berthet et al. (2006), Youssef et al. (2007), Jiang and Teng | Lim and Ozbakkaloglu (2014a,b) | Yes |

(Continued)

Table 6.2 (Continued)

State of art	Samples/models (cross-section)	Models	Best models		Authors proposed new model
			Ultimate stress models	Ultimate strain models	
Ozbakkaloglu and Lim (2013)	a. 753/9 (ultimate stress) b. 511/9 (ultimate strain) (circular columns)	(2007), Teng et al. (2007), Xiao et al. (2010), Wei and Wu (2012) a. For ultimate stress (Lam and Teng, 2003a,b; Bisby et al., 2005; Jiang and Teng, 2007; Teng et al., 2007; Wu and Wang, 2009; Al-Salloum and Siddiqui, 2009; Wu and Zhou, 2010; Realfonzo and Napoli, 2011; Wei and Wu, 2012) b. For ultimate strain (De Lorenzis and Tepfers, 2003; Binici, 2005; Tamuzs et al., 2006a,b; Jiang and Teng, 2006; Youssef et al., 2007; Teng et al., 2009; Fahmy and Wu, 2010; Wei and Wu, 2012)	Ozbakkaloglu and Lim (2013)		Yes
Nisticò et al. (2014)	655/25 (circular and	Fardis and Khalili (1982), Saadatmanesh et al. (1996), Mirmiran and Shahawy (1997a,b), Karbhari and Gao (1997), Kono	a. For circular section (Rousakis et al., 2012a,b).	For circular section (De Lorenzis and Tepfers, 2003;	No

	rectangular columns)	et al. (1998), Spoelstra and Monti (1999), Saafi et al. (1999), Toutanji (1999), Miyauchi et al. (1999), Thériault and Neale (1999), Xiao and Wu (2000), De Lorenzis and Tepfers (2003), Matthys et al. (2006), Girgin (2009), Teng et al. (2009), Fahmy and Wu (2010), Realfonzo and Napoli (2011), Rousakis et al. (2012a,b), ACI(440.2-02) (2002), Campione and Miraglia (2003), Wu and Wang (2009), Nisticò and Monti (2013), Lam and Teng (2003a,b), CNR-DT(200/2004) (2004), Wei and Wu (2011)	b. For circular and square (Wu and Wang, 2009; Nisticò and Monti, 2013)	Rousakis et al., 2012a,b) For circular and square section (CNR-DT (200/2004) 2004; Wei and Wu, 2011)	Yes
Sadeghian and Fam (2015)	518/7 (circular cols.)	Fardis and Khalili (1982), Mander et al. (1988), Karbhari and Gao (1997), Samaan et al. (1998), Toutanji (1999), Lam and Teng (2003a,b), Teng et al. (2009)	Sadeghian and Fam (2015)		
Shirmohammadi et al. (2015)	22/4 (circular columns)	Samaan et al. (1998), Lee et al. (2010), Youssef et al. (2007), Kawashima et al. (2000)	Proposed model (Shirmohammadi et al., 2015)		Yes

treatment of well-confined concrete featuring a bilinear stress—strain curve and weakly confined concrete with a descending branch, and can potentially be used to predict the behavior of concrete confined with different materials. These features make analysis-oriented models more versatile and powerful than design-oriented models (Teng et al., 2007). In analysis-oriented models, an incremental numerical procedure is used to generate the stress—strain curves of FRP-confined concrete. With this approach, an active confinement model for concrete is used to evaluate the axial stress and strain of passively confined concrete at a given confining pressure, and the interaction between the concrete and the confining material is explicitly accounted for by equilibrium and radial displacement compatibility (Lam and Teng, 2003a,b). Analysis-oriented models assume that at a given lateral strain, the axial compressive stress and strain of FRP-confined concrete are the same as those of the same concrete when it is actively confined under a confining pressure equal to that supplied by the FRP jacket (see Fig. 6.17). Examples of these models include those developed by Ferracuti and Savoia (2005); Megalooikonomou et al. (2012); Hu and Seracino (2014); Teng et al. (2014); Rousakis and Tourtouras (2015), Fakharifar et al. (2015). Analysis-oriented models better account for the interaction between concrete and confining materials, including steel and FRP composites. In contrast, based on the assessment of 68 FRP-confined concrete

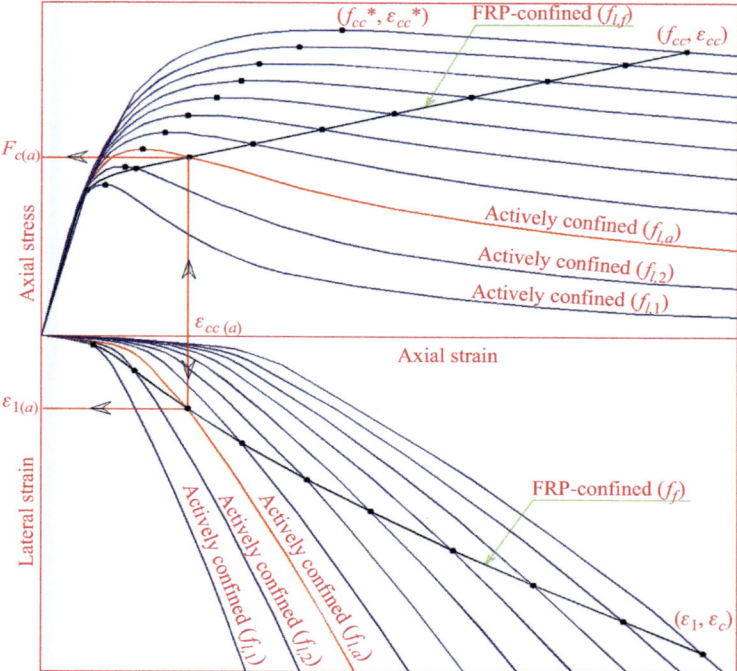

Figure 6.17 Determination of stress—strain curves of FRP-confined concrete by analysis-oriented models (Ozbakkaloglu et al., 2013).

stress—strain models, Ozbakkaloglu et al. (2013) concluded that design-oriented models predict the ultimate conditions of FRP-confined concrete (i.e., the ultimate strength and strain) more accurately.

6.5.2   Stress—strain models of FRP-confined rectangular columns

6.5.2.1   Design-oriented models

In this group, the design-oriented models with strain-hardening behavior (ascending stress—strain relationship) are classified into three broad categories based on the geometric form of the curves, that is, parabolic for all trends (Type I), two parabolic parts (Type II), and parabolic in the first branch as well as a linear part in the second branch (Type III); as exhibited in Fig. 6.18. The first type has been used in many studies, such as Pantelides and Yan (2007); Pellegrino and Modena (2010); and Faustino et al. (2014). The second type was proposed by Harajli (2006) and the third type was applied in several studies (Kawashima et al., 2000; ACI (440.2-08), 2008; Wei and Wu, 2012).

Many studies attempted to compare the available design-oriented models by evaluating the ultimate condition (ultimate stress and strain) only. These studies are summarized in Table 6.3. These studies suggested models that perform better than other examined models.

6.5.2.2   Analysis-oriented models

Marques et al. (2004) presented a theoretical incremental-iterative model for evaluating the behavior of both rectangular and circular short concrete columns confined

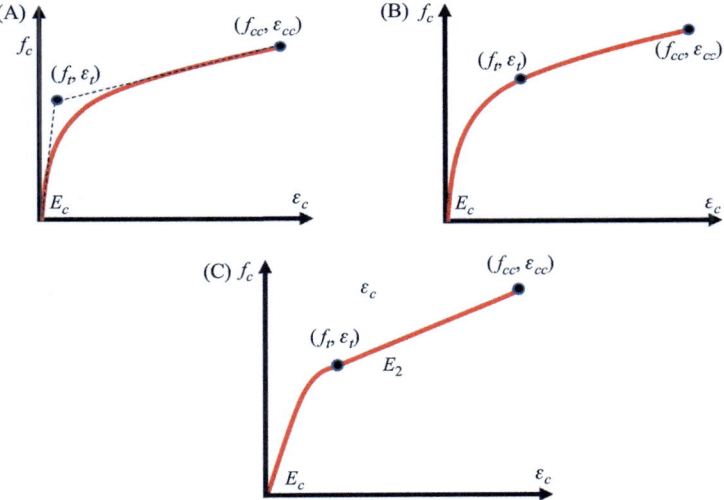

Figure 6.18 Different stress—strain relationships of Group "A" of FRP-confined concrete with strain-hardening behavior (Ismail et al., 2019).

Table 6.3 Comparison of stress−strain models for FRP-confined rectangular concrete columns.

Reviewed papers	Samples/ models	models	Best models		Proposed new model
			Ultimate Stress models	Ultimate strain models	
Lim and Ozbakkaloglu (2014a,b)	484/15	Restrepol and DeVino (1996), ACI(440.2-02) (2002), Shehata et al. (2002), Chaallal et al. (2003), Lam and Teng (2003a,b), Ilki et al. (2004), Harajli (2005) Harajli (2006), Harajli et al. (2006), Pantelides and Yan (2007), Wu et al. (2007), Youssef et al. (2007), ACI(440.2-08) (2008), Toutanji et al. (2010), Wei and Wu (2012)	Lim and Ozbakkaloglu (2014a,b)		Yes
Nisticò et al. (2014)	655/25	Fardis and Khalili (1982), Saadatmanesh et al. (1996), Mirmiran and Shahawy (1997a,b), Karbhari and Gao (1997), Kono et al. (1998), Spoelstra and Monti (1999), Saafi et al. (1999), Toutanji (1999), Miyauchi et al. (1999), Thériault and Neale (2000), Xiao and Wu (2000), De Lorenzis and Tepfers (2003), Matthys et al. (2006), Girgin (2009), Teng et al. (2009), Fahmy and Wu (2010), Realfonzo and Napoli (2011), Rousakis et al. (2012a,b)	For circular and square (Wu and Wang, 2009; Nisticò and Monti 2013)	For circular and square section (CNR-DT (200/2004), 2004; Wei and Wu, 2011)	No

Pham and Hadi (2014a,b,c)	Stress 104/5 Strain 69/5	ACI(440.2-02) (2002), Campione and Miraglia (2003), Wu and Wang (2009), Nisticò and Monti (2013), Lam and Teng (2003a,b), CNR-DT(200/2004) (2004), Wei and Wu (2011) a. For strength (Lam and Teng, 2003a,b; Wu and Wang, 2009; Wu and Wei, 2010; Toutanji et al., 2010; Pham and Hadi, 2014a,b,c) b. For strain (Shehata et al., 2002; Lam and Teng, 2003a,b; ACI(440.2-08), 2008; Ilki et al., 2008; Pham and Hadi, 2013)	ANN Model (Pham and Hadi, 2014a,b,c)	Yes
Faustino and Chastre (2015)	7/6	Manfredi (2001), Campione and Miraglia (2003), Lam and Teng (2003a,b), Wu et al. (2007), Pellegrino and Modena (2010), Faustino et al. (2014)	Faustino et al. (2014) and Manfredi (2001)	No

by FRP composites. The proposed formulation considers a uniaxial constitutive relation that utilizes the area strain as a parameter for measuring the material secant axial stiffness. The stress−strain curves were compared to experimental and analytical results available in the literature. They concluded that the proposed model produced acceptable results for both rectangular and circular columns.

6.5.3 Plasticity approach for modeling circular/rectangular FRP-confined concrete

The plasticity theory is widely used for modeling the stress−strain relationship of concrete. Because the use of fiber composites for the confinement of concrete is relatively new, theoretical work in this area is, so far, limited. Few studies focused on developing a nonlinear finite element model using the Drucker−Prager plasticity for FRP-confined concrete. Rochette and Labossiere (1996) used an incremental finite element technique to evaluate the response of fiber-wrapped concrete columns. They modeled concrete as an elastically perfect plastic material and adopted an associative Drucker−Prager failure criterion. The model compared favorably with the results of their uniaxial compression tests (Picher et al., 1996). However, a more complex formulation of concrete behavior is required to enhance the model for various cross-sections, fiber orientations, and load combinations. Mirmiran et al. (2000) used a nonassociative Drucker−Prager plasticity model for FRP-confined concrete and examined its applicability to cyclic loading. The sensitivity analysis showed that the response of FRP-confined concrete can best be modeled by a zero dilatancy angle. It was shown that while the Drucker−Prager plasticity can be calibrated fairly well for predicting the axial stress−strain response, it does not properly establish the true dilation tendencies of FRP-confined concrete because the Drucker−Prager model corresponds to an elastically perfect plastic material. Karabinis and Rousakis (2002) applied a constitutive model based on the plasticity theory. The model was originally proposed for steel-confined concrete but they modified and calibrated it to incorporate the dilation characteristics of FRP-confined concrete. For the analytical model description, 14 material parameters were used to reproduce the concrete behavior. The study by Doran et al. (2009) dealt with the nonlinear finite element modeling of rectangular/square concrete columns wrapped with FRP to simulate the compressive behavior under concentric loading. Adopting the cohesion and internal friction of the Drucker−Prager criterion from a previous study conducted by the authors, the emphasis was placed on determining the confining stress and the lateral-to-axial strain relation.

Drucker−Prager yield criterion was also deployed by Becque et al. (2003); Eid et al. (2007); Eid and Paultre (2007); Karabinis et al. (2008); Jiang et al. (2011); Jiang and Wu (2012); and Jiang and Wu (2014). Becque et al. (2003) modified the octahedral stress−strain model by Gerstle (1981) and adopted the expression of confinement effectiveness coefficient (k) proposed by Samaan et al. (1998) to define the failure surface of FRP-confined concrete. Some research works used the FE software to predict the behavior of FRP-confined concrete under axial or lateral monotonic loading. Youssf et al. (2014) used LS-DYNA and Ramkrishna Dandapat and Bhattacharyya (2012) used ABAQUS to predict the stress—strain behavior of concrete encased in FRP under axial load. Gambarelli et al. (2014) used the 3D FE software, MASA, which is based on the microplane approach.

6.6 Ductility enhancement

6.6.1 Rei nforced concrete columns confined with FRP

The results of a research program that evaluated the confinement effectiveness of the type and amount of FRP used to retrofit circular concrete columns were reported by Gu et al. (2010). A total of 17 circular concrete columns were tested under combined lateral cyclic displacement excursions and constant axial load. It was demonstrated that a high axial load level has a detrimental effect and a large aspect ratio has a positive effect on drift capacity. All specimens were tested under lateral cyclic load reversals while simultaneously being subjected to constant axial load throughout the test. The specimen represented the part of a bridge column or a building column from the section of the maximum moment to the point of contra flexure. The test specimens were divided into two groups according to their diameters. The diameter, D, of the first group was 300 mm and that of the second group was 360 mm.

The first group included nine specimens. One specimen was tested under the "as-built" condition, while the others were tested after being retrofitted with FRP jacketing. The height of the specimens, H, measured from the top of the footing to the application point of the horizontal force, was 850 mm (Fig. 6.19 and Table 6.4). The aspect ratio H/D was 2.8. The specimens were reinforced with 12 deformed 19 mm diameter bars, and the yield stress was 400 MPa. The longitudinal steel bars were evenly distributed in a circle with a constant clear cover of 20 mm. Hoops

Figure 6.19 Test setup (Gu et al., 2010).

Table 6.4 Details of test specimens (Gu et al., 2010).

	D (mm)	L (mm)	f_c' (MPa)	n	FRP treatment	Test mode	E_l (GPa)	λ_f
J0	300	850	28	0.05	—	Test 1	—	—
J1	300	850	28	0.05	1 layer DFRP	Test 1	0.103	0.113
J2	300	850	28	0.05	1 layer CFRP-a	Test 1	0.169	0.111
J3	300	850	28	0.05	1 layer CFRP-a	Test 2	0.169	0.111
J4	300	850	28	0.05	2 layers DFRP	Test 2	0.206	0.225
J5	300	850	28	0.05	2 layers CFRP-a	Test 2	0.337	0.222
J6	300	850	28	0.05	1 layer CFRP-a + 1 layer DFRP	Test 2	0.272	0.223
J7	300	850	28	0.05	0.5 layer CFRP-a + 1 layer DFRP	Test 2	0.188	0.168
J8	300	850	28	0.05	1 layer CFRP-a + 2 layer DFRP	Test 2	0.375	0.336
CH0	360	1100	34.9	0.36	—	Test 1	—	—
CH1	360	1100	34.9	0.36	0.5 layer CFRP-a + 1 layer DFRP	Test 1	0.207	0.127
CH2	360	1100	34.9	0.36	2.5 layers DFRP	Test 1	0.215	0.188
CH3	360	1100	34.9	0.36	1.5 layers CFRP-b	Test 1	0.362	0.157
CL0	360	800	34.9	0.36	—	Test 1	—	—
CL1	360	800	34.9	0.36	4 layers DFRP	Test 1	0.344	0.3
CL2	360	800	34.9	0.36	2.5 layers CFRP-b	Test 1	0.603	0.261
CL3	360	800	34.9	0.36	3.5 layers CFRP-b	Test 1	0.844	0.366

Note: Two cycles of the displacement corresponding to the drift ratios of 1/800, 1/400, 1/200, 1/100, 1/50, 1/25, and 1/20 were conducted. Then, the specimens were monotonically pushed to failure. The first load sequence is termed "Test 1," and the second load sequence is termed "Test 2."

(diameter = 6.5 mm and yield stress = 350 MPa) spaced at 160 mm were used as transverse reinforcement. The equivalent cylinder strength, f_c, was 28.0 MPa, which was calculated by $f_c = 0.8f_{cu}$, where f_{cu} was the 28-day mean compressive strength of 10 cubes (150 × 150 × 150 mm) cast together with the specimens. The axial load was 100 kN and the axial load ratio, $n = P/(A_g f_c)$, was 0.05, where A_g represented the gross area of the column.

The second group included eight specimens. The height, H, of four of the specimens was 1100 mm (Fig. 6.19 and Table 6.4) and that of other four specimens was 800 mm, resulting in aspect ratios of 3.1 and 2.2, respectively. Two specimens were tested under the as-built condition. The columns were reinforced with 12 deformed 25 mm diameter bars, and the yield stress was 382 MPa. The clear cover had a thickness of 25 mm. Hoops (diameter = 6.5 mm and yield stress = 320 MPa) spaced at 150 mm were used as transverse reinforcement. The region of 100 mm from the stub face was strengthened with additional deformed 25 mm diameter bars, and the 10 mm ties were placed at a spacing of 30 mm within this region to minimize the chances of failure at the section adjacent to the stub face. The equivalent concrete cylinder strength was 34.9 MPa. All material mechanical properties were tested using the same standard as the first group. The axial load was 1200 kN and the axial load ratio was 0.36. In this test program, the axial load was applied via steel strands running through the center of the column.

Two types of fabrics were used in this test series, that is, CFRP and Dyneema FRP (DFRP). DFRP has a large fracture strain and low elastic modulus compared to those of CFRP. A flat coupon test was conducted to determine the mechanical properties of FRP. At least three flat coupons of each type of FRP were tested, and the average results are presented in Table 6.5.

The type of fabric and the number of layers used for each column specimen were designed to study a range of parameters and their effects on column behavior. When 0.5 layer of fabric was used, 20 mm wide CFRP bands were wrapped at a clear spacing of 20 mm. For DFRP, one-half of the fiber was drawn from the sheet and the remnant fiber was wrapped around the column. The effect of CFRP and DFRP hybrid fibers was also studied in this program. When hybrid FRP was used, CFRP was wrapped first, followed by wrapping the DFRP outside. It was reported that the most extensively damaged region always shifted away by approximately 150 mm from the column footing interface owing to the additional confining effect of the footing (Sheikh and Yau, 2002; Ozbakkaloglu and Saatcioglu, 2006). Thus to ensure that the section adjacent to the footing is not damaged for all columns, three layers of additional 100 mm wide CFRP bands were wrapped at the end of every column. It was assumed that this does not affect the lateral displacement capacity of the columns.

The shear failure that occurred in the as-built columns was entirely prevented by composite wrapping in the retrofitted columns. Composite wrapping provided additional shear capacity and wrapping at the potential plastic hinge region allowed the specimens to undergo large inelastic displacements. The failure modes for all specimens were dominated by the flexural effects. Fig. 6.20 shows the different failure modes of the specimens. Each retrofitted column exhibited similar behavior under revised cyclic loading. Specimen J2 was selected as a representative sample. There were no visual signs of damage to the column until the lateral drift reached approximately 2.0% when an obvious horizontal crack was found. The position of the first crack was approximately 100 mm above the column—footing interface. As the applied lateral displacement increased, horizontal flexural cracks developed evenly above the first crack, spaced at approximately 50 mm. The opening of the first

Table 6.5 Material properties of fibers (Gu et al., 2010).

FRP	Thickness (mm)	Ultimate strength (MPa)	Ultimate strain (%)	Young's modulus (GPa)
CFRP-a	0.111	4232	1.84	230
CFRP-b	0.167	3945	1.52	260
DFRP	0.258	1832	3.05	60

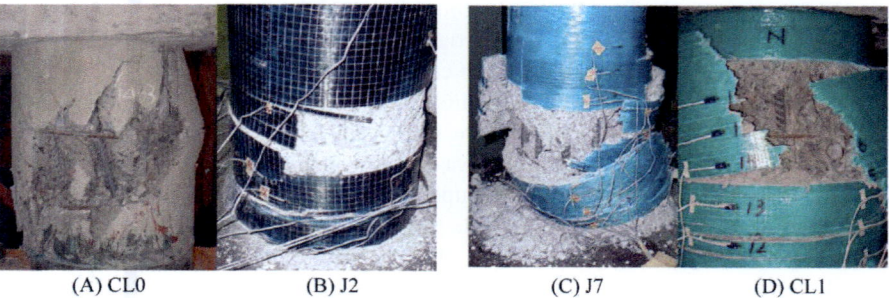

(A) CL0	(B) J2	(C) J7	(D) CL1

Figure 6.20 Failure modes of specimen (Gu et al., 2010).

crack was significant whereas those of other horizontal flexural cracks did not appear to increase with the increase in latera; displacement excursion. Strips of fibers of approximately 10 mm first ruptured where the first horizontal crack developed when the lateral displacement drift reached around 6.8%. This local fiber rupture did not significantly influence the column lateral load-carrying capacity. When the third cycle of 9.2% drift was imposed on the column, more fiber ruptured, while the concrete was crushed and the longitude reinforcement buckled. The column could no longer support the axial load and lateral load, and the test was terminated. The most extensive damage was concentrated within a segment of 100−400 mm from the column−footing interface, which coincides with the location of the initial fiber rupture. The plastic hinge shift from the column−footing interface was attributed to the confining effect of the footing and the additional three CFRP bands. Similar observations were previously reported (Sheikh and Yau, 2002; Ozbakkaloglu and Saatcioglu, 2006).

The J3-to-J8 columns, which were monotonically pushed to failure after the lateral displacement reached 5% drift, exhibited similar behavior. Thus Specimen J3 was chosen as a representative sample. Specimen J3 and Specimen J2 had identical test parameters and were both retrofitted with one layer of CFRP. Specimen J3 initially behaved very similarly to Specimen J2 (up to 5% drift). Then, Specimen J3 was monotonically loaded in the push direction. The lateral resistance continuously increased as the lateral displacement increased. Local fiber rupture occurred at 7.2% drift. At this time, the column lateral resistance dropped to approximately 14 kN, which could potentially increase as the lateral displacement increased. When the lateral drift reached approximately 10.8%, an explosive rupture occurred in the CFRP, which resulted in a significant reduction of lateral load; subsequently, the test was stopped (see Fig. 6.21).

It was found that the confinement of FRP does not affect the yield state of the columns (Binici, 2008). Thus the yield displacement of the columns with the same axial load ratio and aspect ratio can be averaged to eliminate the difference in test results due to material property scatter. The determined yield displacement, ultimate drift ratio, and displacement ductility of each column are presented in Table 6.6.

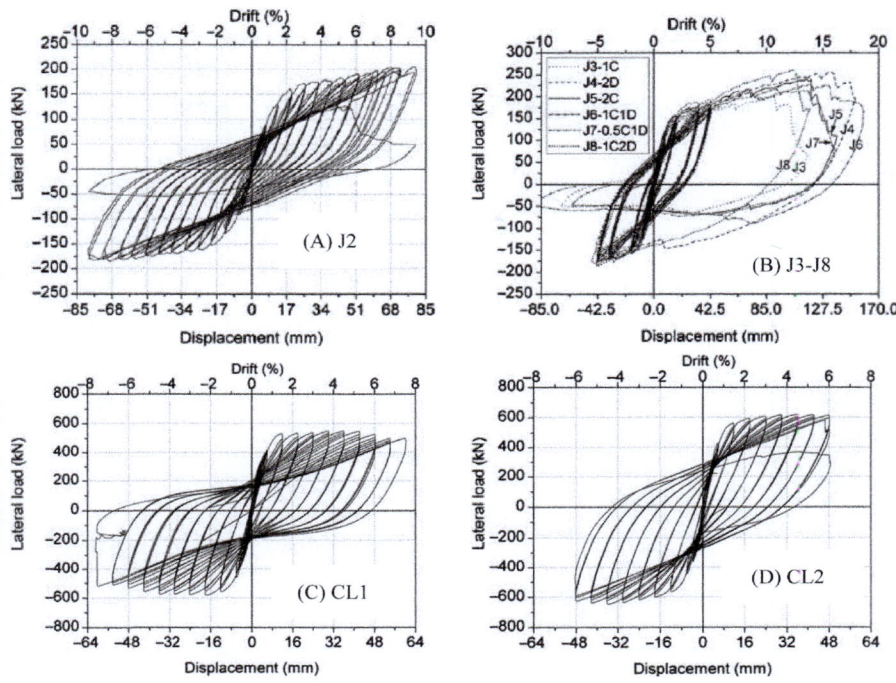

Figure 6.21 Column response curves (Gu et al., 2010).

The amount of confining FRP greatly affects the drift capacity of FRP-retrofitted columns. Confinement at low levels will increase the drift capacity of the columns; however, after a critical value, a further increase in confinement will reduce the deformation capacity of the columns.

6.6.2 Large-scale-reinforced concrete building confined with FRP

Wang et al. (2016) reported results from shake table tests of two ½-scale 4-story and 2-bay frames, as shown in Fig. 6.22. In the experimental program, one non-strengthened frame was tested first as the control specimen. The other one was ret-rofitted with CFRP according to the observations and results of the control frame tests and then tested using the shake table. The effectiveness of the seismic strengthening technique using FRP for substandard RC frames was examined by comparing the seismic performance of the two frames considering various terms. The retrofitting scheme was designed according to the failure mode of the control specimen and involved CFRP wrapping at and close to the beam—column joints at the lower two storys of the frame. The test results showed that the seismic performance of the substandard RC frame could be significantly improved by the CFRP strengthening technique. The maximum applied peak ground acceleration (PGA)

Table 6.6 Summary of test results (Gu et al., 2010).

Specimen	$\Delta_y'(+)$ (mm)	$\Delta_y'(-)$ (mm)	Δ_y (mm)	F_{max} (kN)	Δ_u (mm)	Δ_u/Δ_y	Drift ratio, θ_u (%)
J0	6.4	6.3	9.8	155.2	23.4	2.4	2.8
J1	6.9	3.7	9.8	179.2	72.3	7.4	8.5
J2	5.8	5	9.8	195.3	73.1	7.5	8.6
J3	5.7	5.1	9.8	199.4	91.5	9.3	10.8
J4	4.8	4.7	9.8	254.6	134.3	13.7	15.8
J5	5.3	4.8	9.8	238.2	119	12.1	14
J6	5.2	5.3	9.8	229	133	13.6	15.6
J7	5.2	4.9	9.8	213.6	115.7	11.8	13.6
J8	5.2	4.9	9.8	260.2	106.3	10.8	12.5
CH0	5.1	5	7.3	260	8.7	1.2	0.8
CH1	4.4	4.2	7.3	411.3	55	7.5	5
CH2	6.2	5.2	7.3	420.4	99	13.6	9
CH3	5.7	5.1	7.3	451.5	88	12.1	8
CL0	2.5	2.3	3.5	424.3	8.2	2.3	1
CL1	2.7	2.6	3.5	565.5	54	15.4	6.8
CL2	2.4	2.5	3.5	634.4	48	13.7	6
CL3	2.4	2.6	3.5	664.8	48	13.7	6

Note: Δ_y' = average of the measured displacements corresponding to F_y in the push and pull loading directions, which was obtained from the experimental force−displacement relationship. Δ_y' is determined the yield displacement $\Delta y = (F_i/F_y)(\Delta_y')$, where F_i = calculated ideal flexural capacity based on the extreme concrete compressive strain of 0.004 and F_y = first yield force obtained using standard sectional analysis procedures at the first point where the longitudinal reinforcement yields in tension or the extreme concrete reaches a maximum compressive strain of 0.002, whichever came first. The maximum lateral load, F_{max}, was defined as the average of the maximum lateral loads in the push and pull directions. The ultimate displacement, Δ_u, was defined as the maximum usable drift, beyond which the column failed. The ultimate drift ratio $\theta u = \Delta u/H$, where H is the column height.

that could be resisted by the control frame specimen was approximately 0.6 g, while the retrofitted frame specimen could endure seismic loading with a PGA value of 1.0 g without significant damage. The horizontal displacements, interstory drift ratios, and torsional displacements in the retrofitted frame specimens were also found to be much smaller than those of the control frame at the same earthquake levels.

6.7   Durability of FRP-confined concrete columns

Durability has been defined generally as the capability of the system to withstand detrimental strength, stiffness, and other unrequired performance alterations caused by several mechanisms, such as cracking, oxidation, chemical degradation, and delamination, in a specific period of time and under specific loading and environmental conditions (Karbhari et al., 2003; Karbhari, 2003). The long-term durability of FRP composites is one of the key factors for their successful application in

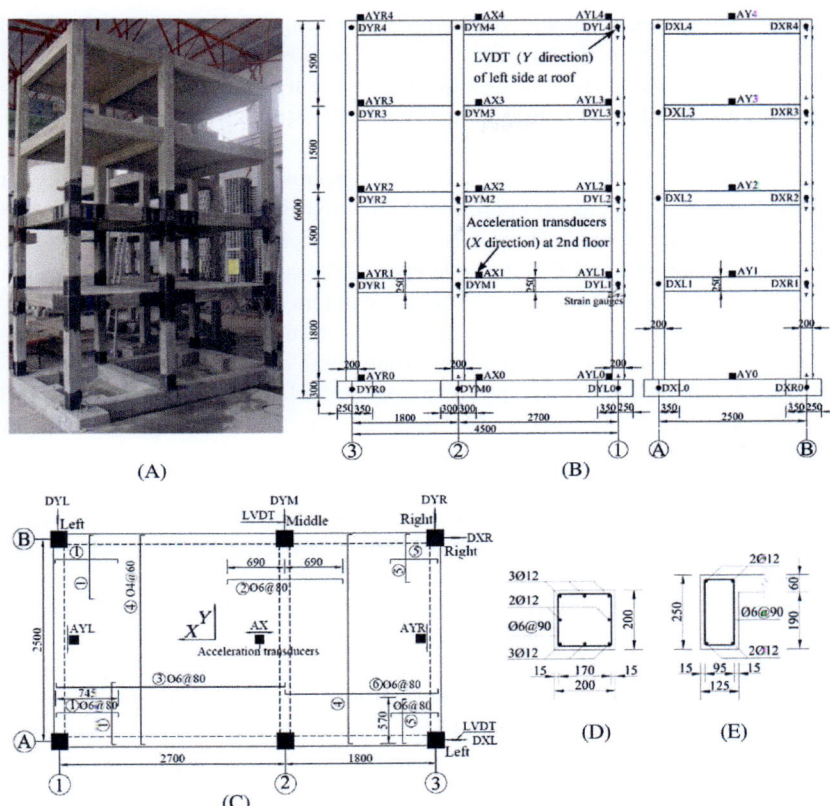

Figure 6.22 FRP retrofitted concrete building tests: (A) general view; (B) dimensions and locations of transducers in elevation; (C) reinforcement of slabs and locations of transducers in plan; (D) column reinforcement; (E) beam reinforcement (units: mm) (Wang et al., 2016).

concrete repairing or as reinforcing materials. Durability-related problems and doubts must be considered/solved before considering the field applications of FRP materials. For FRP-confined concrete, Siavashi et al. (2019) discussed that the definition of durability due to the degradation of bond strength at the concrete and FRP interface over time is important.

6.7.1 Laboratory evaluations for durability of FRP-confined columns

6.7.1.1 FRP-confined plain concrete

Karbhari and Eckel (1994) studied the effect of 60-day exposure at $-17.8°C$ on the concrete confined by composite materials (glass, carbon, and aramid). They concluded that exposure resulted in increased failure load and stiffness, and decreased the ultimate strain due to matrix hardening and change in damage modes of glass and carbon fibers. Soudki and Green (1997) exposed CFRP-confined concrete cylinders to 50 freeze—thaw cycles and thawed them in a water bath. They stated that cycling resulted in substantial reductions in the strength and stiffness, and failure was considered more catastrophic.

Toutanji and Balaguru (1998) studied three groups of normal strength concrete cylinders. Each group included eight specimens, that is, two unconfined specimens and six confined specimens. Three types of FRP were used in this study, that is, two types of carbon (Cl and C5) and one type of glass (GE), and one type of epoxy resin system was used. The FRP-confined cylinders were externally wrapped by unidirectional FRP sheets before applying the environmental cycles. In group 1, the specimens were not affected by any environmental conditions. Specimens of the second group were placed in a specially constructed environment chamber. They were exposed to 300 wet and dry cycles (mean temperature = 35°C, humidity = 90%). The third group included specimens that were subjected to approximately six freeze—thaw cycles per day for 50 days and they were subjected to 300 freeze—thaw cycles in a saltwater solution, similar to that of the wet and dry tests. The freeze—thaw tests included alternately lowering the temperature of the sample from 4.4 to −17.8°C and increasing it from −17.8 to 4.4°C within 4 h. Note that the seawater was simulated with 35 g of salt per liter of water. This is an approximate amount of salt found in the ocean.

Axial compressive strength of tested specimens: The FRP-confined specimens exposed to freeze—thaw cycles exhibited higher strength reduction than those exposed to wet—dry cycles. This should be expected because freezing and thawing exert more internal stresses. Owing to the influence of freezing and thawing conditions, the compressive strength of the samples wrapped with glass fibers decreased by 28%, while the samples of carbon Cl and C5 showed a decrease of 19%. As a result of wet and dry conditions, the strength of glass-confined concrete samples decreased by 10%, while that of carbon fiber-confined samples decreased minimally.

Axial stiffness of tested specimens: Freeze—thaw exposure had minimal effect on the stiffness of glass fiber and carbon fiber-wrapped columns. Dry and wet exposure had little effect on the stiffness of glass fiber-wrapped cylinders, while the stiffness of carbon fiber-wrapped concrete cylinders increased.

Ductility of tested specimens: The ductility of the samples wrapped with carbon fiber or glass fiber composites was significantly reduced when exposed to the freeze—thaw environment; however, the ductility of the samples wrapped with carbon fibers was not lost when exposed to the dry and wet environment; yet, the ductility of the samples wrapped with glass fiber decreased.

Toutanji and Balaguru (1999) argued that the failure of specimens tested by Toutanji and Balaguru (1998) under freeze—thaw cycles was due to the deterioration of concrete that was exposed instead of the failure of the wrapping system. Therefore the top and bottom parts (75 mm each) of the tested specimens were sawed off and tests were reconducted with a shorter length. The retested specimens reserved the strength of CFRP- and GFRP-confined specimens by approximately 95% and 88%, respectively. This indicates that if the uncovered concrete is protected from such conditions, the FRP-confined structure elements will suffer slight degradation. Toutanji and Balaguru (1999) stated that glass fibers are mainly calcium aluminum borosilicate with alkali contents. At low temperatures, the bonding between fibers and resins breaks down, resulting in penetration of the saltwater into the composite, which in turn causes a reduction in the axial strength. However, carbon fibers are not vulnerable to the same mechanism; thus the degradation degree is much smaller.

Karbhari et al. (2000) described the response of freeze—thaw (dry) exposure sequence to unidirectional carbon and glass FRP composite-confined unreinforced concrete cylinders (diameter = 152.4 and 304.8 mm, respectively), which had a compressive strength of 56 MPa for 28 days. Three different winding structures were adopted, namely, (1) three layers of carbon fiber fabric, where the fibers were in the circumferential direction; (2) two layers of carbon fiber fabric, where each layer of carbon fiber was in the circumferential and altitudinal directions, and the circumferential layer was the outermost layer (that is, 90/0 layup); and (3) seven layers of GFRP with fibers in the hoop direction. The cylinders were cured at 22.5°C for one month. Conditioned specimens were subjected to 201 freeze—thaw cycles between 22.5°C and −20°C, according to the freeze—thaw procedure of the United States of America. No solution was used during the freeze—thaw period, and the changes in the conditions were strictly controlled by the temperature of the conditioning room. A group of five unconfined concrete cylinders were placed in a 22.5°C test chamber and the freeze—thaw chamber as reference specimens. Test results showed that the average ultimate compressive strength of the tested control cylinders (unconfined specimens) exhibited a change of 5.7%, which was within the total dispersion range, but the initial stiffness decreased by 20.4% owing to the freeze—thaw cycles. For FRP-confined specimens, the changes in the ultimate strength level of the specimens exposed to freeze—thaw, in comparison to the wrapped specimens under the condition of constant 22.5°C can be neglected. Additionally, the changes in stiffness level were also small for the horizontally wrapped specimens. In the 90/0 CFRP jacket system, the stiffness was significantly reduced (11%). When unconfined concrete specimens were used as criteria to examine the conditional specimens, the stiffness of the two hoop-wrapped systems increased by 15% and 13%, respectively, when compared on this basis. Karbhari et al. (2000) stated that the stiffness of unconfined concrete specimens exposed to freeze—thaw conditions was reduced by 20.4% as compared to those exposed to 22.5°C, which emphasized the effect of exposure on the concrete itself. In fiber-reinforced composites, the ductility measured after the kinking point (defined as the point of change from initial linear response to nonlinear response) decreased due to the degradation of concrete stiffness and the increase in matrix hardening and brittleness. The results also showed that the freeze—thaw exposure affected the mechanical properties and the fracture failure mechanism. The effect of cycling on FRP composites can be assumed to be due to the effect of matrix hardening and stiffening, as well as the formation of microcracks parallel to the fibers and between the fibers and matrix (Karbhari et al., 2000).

6.7.1.2 FRP-confined reinforced concrete columns

Bae and Belarbi (2010): Most RC structures requiring strengthening, repairing, or upgrading are susceptible to changes in various environmental conditions throughout the seasonal change process, and the impact of each environmental condition is not independent of each other. Therefore it is essential to consider the durability of RC columns confined with FRP composites under combined environmental cycles, including all environmental conditions perceptible throughout the seasonal variations. A total of 36 small-scale RC columns and six midscale RC columns wrapped with FRP sheets were conditioned under environmental conditions, such as rapid temperature changes (freeze−thaw in winter and temperature cycles in summer), dry−wet cycles (rain or snow, humidity changes), ultraviolet (UV) radiation, and chemical erosion (deicing salts). Precisely, the environmental conditions were divided into six sets, that is, room temperature, freeze−thaw cycles, high temperature cycle, high humidity cycle, ultraviolet radiation, and saline solution effect. The test columns were wrapped with FRP sheets and treated under previously determined environmental cycles. Uniaxial compression tests were performed after the environmental adjustment. The freeze−thaw cycles did not adversely affect the performance of FRP-confined columns and the measured failure loads exhibited an increase of approximately 5%. However, the ductility index decreased by 3% for GFRP-confined columns. A combined environmental cycle, as shown in Fig. 6.23, did not cause a significant change in the failure load of CFRP-confined columns; however, the failure load of GFRP-confined columns decreased by 7%, which can be attributed to the plasticization of the matrix (softening of the polymer) under the effect of the high-temperature and high-humidity cycles. Note that the freeze−thaw cycles caused matrix hardening and in turn, an increase in the failure load. The high temperature

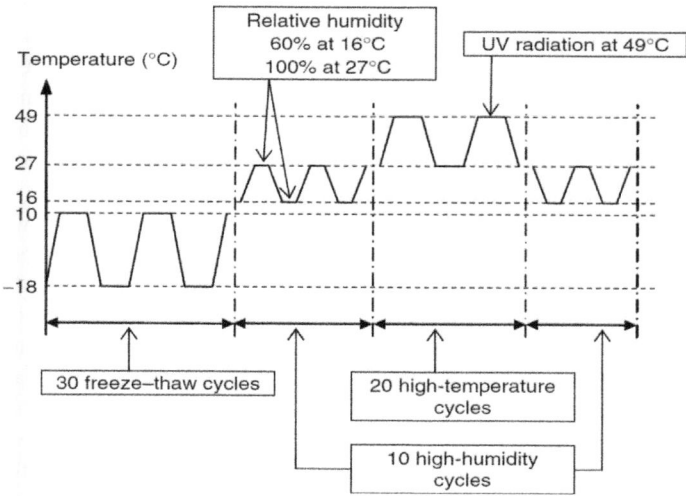

Figure 6.23 One cycle of the combined environmental cycles used (Bae and Belarbi, 2010).

and high humidity seemed to compromise this effect in the combined conditions. Bae and Belarbi (2010) stated that the decrease in failure load might be attributed to the microcracking at the matrix–fiber interface, which could be induced by the variations in temperature and relative humidity. Glass fibers are vulnerable to moisture effects because moisture extracts ions from the fibers, leading to degradation of the fibers, such as surface cracks. However, carbon fibers are not susceptible to this degradation mechanism. Because the thermal coefficients of the matrix and fibers were different, thermal variation caused shear stress between the matrix and fibers.

Among the environmental conditions tested by Bae and Belarbi (2010), saline solution seemed to have the most significant adverse effects, especially in GFRP-confined columns. Under freeze–thaw cycles and saline solution conditions, the average failure load of GFRP-confined columns was 14% lower than that of corresponding columns under freeze–thaw cycles. Under the same conditions, the average failure load of CFRP-confined columns was reduced by 5%. The average failure load of CFRP-confined columns exposed to the combined conditions and saline solution was 1% lower than that of the columns under the combined conditions. The ductility index of the GFRP-confined column was also adversely affected by the saline solution. In comparison to the columns under freeze–thaw cycles, the ductility index of the columns confined with a layer of GFRP decreased by 26%. In contrast, the saline solution did not affect the ductility index of CFRP-confined columns. Eventually, as compared to GFRP-confined columns under freeze–thaw cycles, the combined effects of high-temperature cycles with UV radiation and high-humidity cycles were the second most detrimental environmental conditions for GFRP-wrapped RC columns, which led to a reduction in the failure load.

6.7.2   Short-term field study on durability of RC bridges (3 years investigation)

Conducting field tests in sites to evaluate the structural behavior and durability of FRP composite systems in real service environments is vital to increase the assurance in designs with FRP composites. Hag-Elsafi et al. (2003) studied the bridge FRP strengthening systems. Results from both health monitoring and load testing showed that the degree of the bond between the FRP laminates and concrete and the efficiency of the retrofit system was not altered after 2 years of service. Reay and Pantelides (2006) conducted a nondestructive evaluation using electrical strain gauges, tiltmeters, thermocouples, and humidity sensors fixed on the bents of a bridge located in Utah (the State Street Bridge at Interstate 80 in Salt Lake City) for absolutely instantaneous health monitoring. They also measured the changes in the ultimate strength and strain capacity of the CFRP composite owing to the variations in temperature, moisture/humidity, ultraviolet light, and freezing/thawing cycles. Moreover, the long-term impacts of the environment on the bond strength between the CFRP composite and concrete and those on the confinement efficiency using CFRP composite were studied.

Real-time health monitoring produced a hypothesis that traffic loads and freeze–thaw cycling contributes to the interfacial moisture penetration of epoxy resin and concrete faces, which leads to concrete microcracking and deterioration of resin. In addition, several voids were observed in the CFRP composite system, whose size did not substantially increase over time, which gives a probability that they likely occurred during the application of the CFRP composite to the bridge.

Health monitoring has demonstrated that environmental conditions influence the CFRP composite-to-concrete bond strength and the efficiency of the CFRP composite in concrete confinement at ultimate loads. However, *after 3 years of typical application of FRP composite*, in general, the environmental impacts under service load conditions were moderate, and the general efficiency of the bridge seismic retrofit with CFRP composite is still in effect.

6.7.3   Long-term field study on durability of RC bridge columns (15 years investigation)

Short-term accelerated laboratory tests have proved that freeze−thaw exposures can cause matrix cracking, fiber-matrix debonding, and increase the brittleness, leading to different failure mechanisms under environmental conditions. Siavashi et al. (2019) studied the bond strength of an FRP system of a typical highway bridge's components in Michigan after 15 years of natural weathering. This is a comparatively harsh environment in the United States of America because the components of many civil infrastructures undergo annual freeze−thaw cycles. The study aimed to introduce a relationship relating the loss of bond strength between the FRP system and a typical highway bridge element over time. The data collected by Siavashi et al. (2019) are from two FRP-confined scale-model columns prepared by the Michigan Department of Transportation (MDOT) in July 1999 and tested in May 2015. These columns were located near the piers of an existing bridge in south-east Lansing, MI, which undergoes an average of 84 freeze−thaw cycles per year. Before testing these columns, different deterioration marks were observed on the column faces. Specifically, corrosion marks and other staining on the internal steel reinforcement were evident on Faces 1 and 2 of the columns (Fig. 6.24). This is not surprising because these faces are exposed to extreme environmental conditions.

Bond strength was defined through a pull-off adhesion test conducted using a portable automatic adhesion tester. Through testing, it was found that Face 1 of Column 2 showed a significantly inferior bond strength than the other column faces. This is not surprising because it is the utmost exposed face. Several failure modes were realized, involving failure in the concrete substrate, failure at the adhesive interface, and combined concrete/adhesive failures. In general, the failure modes were almost equally divided between the substrate and combined substrate/FRP interface failures.

Scale-model beams with concrete dimensions of $406 \times 51 \times 104$ mm were cast in March 2013 using an MDOT-certified ready-mix design, typical to that of field columns. One week after the FRP application (where the specimens were kept in the same temperature of 23°C), the specimens were tested for bond strength using the same method as the field columns. The mean bond strength is presented in Table 6.7 as the zero-time result. This value is considerably more than the bond strength defined in the weathered field columns at 186 months. To deeply recognize

Figure 6.24 Test columns under westbound Interstate 96 over Lansing Road (Siavashi et al., 2019).

Table 6.7 Bond strength test results (Siavashi et al., 2019).

Time (months)	Mean bond strength (MPa)	Sample size	COV
0	6.28	37	0.23
9	5.98	3	0.14
14	5.89	4	0.21
28	4.49	13	0.09
186 (Group 1)	3.41	7	0.4

how this strength weakened over time, more test specimens were arranged to simulate the in situ weathered results before the 186 months of exposure. These added specimens were left in the open air under exposure conditions comparable to Face 1 of Column 2 and tested at 9, 14, and 28 months of exposure. It was deduced that the bond strength decreased gradually from 6.27 MPa (time = 0; unweathered) to 3.41 MPa for tremendously exposed columns (at 186 months of weathering), demonstrating a reduction in the strength of approximately 42%.

For massively exposed elements, the best-fit regression curve predicting bond strength over time is given by ($b = -0.55 \ln(t) + 6.35$), where t denotes the time (months) and b denotes the bond strength (MPa). For broad application, normalizing this relationship such that it determines a dimensionless reduction factor (r) as a function of time in preference to direct bond strength results in ($r = 1.0 - 0.084 \ln(t)$).

6.7.4 Axial force assessment of FRP-confined columns based on durability

6.7.4.1 Concrete axial compressive strength after exposure to freeze−thaw cycles

Toutanji and Balaguru (1999) proposed two different damage factors based on the type of FRP composites (fiberglass and carbon fibers) to predict the strength of the wrapping system before and after *exposure to freeze−thaw cycles*.

$$f'_f = (1 - D)f_f \tag{6.2}$$

where f_f is the tensile strength of the fiber composite, f'_f is the residual strength, and D is the damage factor, which is given by

$$D = 1 - e^{\left(\frac{-T}{2C}\right)} \tag{6.3}$$

where T is the exposure test time in hours and C is a material constant mainly depending on FRP composites (matrix ductility and fiber type). For GFRP, the constant C is 110,027 and for CFRP, it is presumed that it is equal to 1800.

The ultimate strength of FRP-confined concrete after exposure to freeze/thaw environmental condition is calculated by

$$f'_{cc} = f'_{co}\left(1 + 3.5\left(\frac{2te^{\left(\frac{-T}{2C}\right)}f_f}{df'_{co}}\right)^{0.85}\right) \tag{6.4}$$

6.7.4.2 FRP-confined RC column compressive strength under various environmental conditions

Bae and Belarbi (2010): The axial compression capacity of FRP-confined RC columns under several environmental conditions can be defined using the following form.

$$\phi P_n = 0.85\phi\left[0.85\psi_f\phi_{env}f'_{cc}\left(A_g - A_{st}\right) + f_yA_s\right] \tag{6.5}$$

where ϕ is the strength reduction factor, ψ_f is the strength reduction factor proposed by the ACI Committee 440 to consider the uncertainty of modern technology and is equal to 0.95, ϕ_{env} is the strength reduction factor recommended by Bae and Belarbi (2010) to reflect the impacts of different combinations of environmental conditions, A_g is the gross-sectional area of the strengthened column, A_s is the cross-sectional area of the main steel reinforcement, and f_y is the yield strength of the main steel reinforcement. A model originally developed by Fam and Rizkalla (2001) was nominated and marginally revised by Bae and Belarbi (2010) to determine the compressive strength of FRP-confined concrete, f_{cc}.

The strength reduction factor ϕ_{env} can be defined by multiplying three subfactors (ϕ_{FT} reflects the impacts of the freeze−thaw cycles, ϕ_{Na} considers the influences of the saline solution, and ϕ_H accounts for the combined effects of high temperature with UV radiation and high-humidity cycles) as follows.

$$\phi_{env} = \phi_{FT}\phi_{Na}\phi_{HT} \tag{6.6}$$

Bae and Belarbi (2010) assumed that the changes in failure load of the tested specimens after exposure to different combinations of environmental conditions

Table 6.8 Strength reduction factor (Bae and Belarbi, 2010).

FRP sheet type/adhesive types	Strength reduction factor			Remark
	$\phi_{env} = \phi_{FT} \phi_H \phi_{Na}$			
	ϕ_{FT}	ϕ_H	ϕ_{Na}	
CFRP/epoxy	1	0.95	0.95	If one of the environmental conditions can be ignored, the strength reduction factor for the environmental condition can be taken as unity
GFRP/epoxy	1	0.85	0.85	

occurred owing to the variations in the mechanical properties of concrete and FRP sheets. Although the main steel reinforcements were intact, the strength reduction factor (ϕ_{env}) was suggested, as presented in Table 6.8.

Ultimately, to define a more statistically reliable strength reduction factor, further test results will be necessary.

6.7.5 Design specifications and recommendations

In design practice, the impacts of environmental conditions are considered by adopting prescribed environmental reduction factors on the mechanical characteristics of FRP material. For example, ACI 440.2R (ACI 2017) suggests an environmental reduction factor to reduce the strength and strain capacity of FRP composites according to the environment and fiber types. ACI recommends that if FRP systems are used in very harsh environments for a long time in high humidity, freeze—thaw cycles, saltwater, or alkalinity, lowering the reduction factors may be more suitable. ACI does confirm that FRP systems are further investigated for the impacts of environmental degradation, including freeze—thaw behavior. In comparison to ACI 440.2R, AASHTO guidelines (AASHTO 2012) do not explicitly state the environmental reduction factors. However, to account for possible bond degradation, AASHTO introduces an upper limit for the shear transfer strength of the FRP—concrete interface.

ACI 440.R2 (ACI 2017), as well as CNR DT-200 R1/2013 (CNR 2013), recommends an environmental reduction coefficient of 0.85 for CFRP in harsh environments. Other design guides, such as TR-55 (Concrete Society Working Party, 2013) and ISIS (Intelligent Sensing for Innovative Structures, 2008), determined that other variabilities may be related to different application methods. For instance, TR-55 proposed a reduction factor of 0.83 and 0.95 for wet lay-up strengthening systems and machine-controlled strengthening techniques, respectively. Similarly, ISIS Intelligent Sensing for Innovative Structures (2008) presented a total reduction factor of 0.75 for pultruded CFRP and 0.5625 for hand-applied, wet lay-up CFRP (consideration of uncertainties in material strength and concern of environmental degradation).

Table 6.9 Bond strength reduction factors (Siavashi et al., 2019).

Time		Reduction factor	
Years	**Months**	**Group 1**	**Group 1**
0	0	1	1
0.75	9	0.82	0.85
1.17	14	0.78	0.83
2.33	28	0.72	0.78
10	120	0.6	0.68
15.5	186	0.56	0.65
Extrapolated			
30	360	0.51	0.61
40	480	0.48	0.59
50	600	0.46	0.58
75	900	0.43	0.55

As presented in Table 6.9, by Siavashi et al. (2019), the values are smaller than the reduction coefficients of ACI, CNR, and TR55 when the medium duration (i.e., 10 years or more) is considered.

Siavashi et al. (2019) concluded that owing to the (mostly) nonexistence of long-term FRP deterioration data, a significantly large amount of further work is required to better understand bond deterioration, with explicit consideration of other environmental conditions, FRP reinforcement systems, and types of the substrate material.

6.8   Design methods of FRP jacket for deficient RC columns

The procedures for the design of FRP jackets for the plastic hinge confinement of RC columns are limited. The most commonly used design procedures are based on the strain energy (Seible et al., 1997), multivariate regression analysis index upgrading (Monti et al., 2001), and target displacement and confining pressure (Tastani and Pantazopoulou, 2006). Sheikh and Li (2007) investigated a method that allows a designer to calculate the amount of confining FRP required for a certain ductility performance, given the axial load on the column and the properties of the FRP for rectangular columns. In addition, a design approach was developed for circular columns based on the required lateral drift by Binici (2008). Another method was based on the enhancement in compressive strength and strain ductility provided by FRP jacket by Pantelides and Moran (2013).

6.8.1   Ductility design method by Seible et al. (1997)

Seible et al. (1997) introduced a design approach that can be applied to design FRP jacket thickness to provide confinement, clamping, and buckling restraints.

Table 6.10 summarizes all design equations required to design the thickness of the FRP jacket to prevent premature flexural failure, shear failure, and lap-splice failure.

6.8.1.1 Shear

Seible et al. (1997) assumed that hoop or horizontal reinforcement in the form of 90 degrees (from the column axis) could add shear strength to concrete columns. Hence, the loss of aggregate interlock in the shear cracks can be controlled by limiting the column dilation in the loading direction to the dilation strains of $\varepsilon_d < 0.004$ or 0.4% (Priestley, 1996).

6.8.1.2 Flexural hinge confinement

Seible et al. (1997) proposed a method that can be applied to design the amount of FRP suitable to increase the inelastic deformation capacity of the flexural plastic

Table 6.10 Summary of the design equations of FRP jacket based on column deficiency.

Deficiency	Rectangular		Circular
Shear	$t_f = \dfrac{\frac{V_o}{\phi_\phi} - (V_c + V_s + V_p)}{2 \times 0.004 E_f . D}$		$t_f = \dfrac{\frac{V_o}{\phi_\phi} - (V_c + V_s + V_p)}{\frac{\pi}{2} \times 0.004 E_f . D}$
	Proportional relationship: $t_f^\phi \sim \frac{1}{E_f.D} x C_\phi$		
Flexural	$t_j = 0.09 \dfrac{D(\varepsilon_{cu} - 0.004) f_{cc}'}{\phi_f f_{fu} \varepsilon_{fu}}$		
	$\varepsilon_{cu} = 0.004 + 2.8 \dfrac{\rho_f f_{fu} \varepsilon_{fu}}{f_{cc}}$		
	$\varepsilon_{cu} = \Phi_u . C_u$		
	$\mu_\Delta = 1 + 3\left(\frac{\Phi_u}{\Phi_y} - 1\right) \frac{L_p}{L}\left(1 - \frac{L_p}{L}\right)$		
	$L_p = 0.08L + 0.022 f_{sy} . d_b$		
	Proportional relationship : $t_j^c \sim \frac{D}{f_{fu} . \varepsilon_{cu}} x C_c$		
Lap-splice clamping	$t_f = 500 \dfrac{D(f_l - f_h)}{E_f}$		
	$f_l = \dfrac{A_s f_{sy}}{[\frac{p}{2n} + 2(d_b + cc)] L_s}$		
	Proportional relationship : $t_f^s \sim \frac{D}{E_f} x C_s$		

Note: V_o = column shear demand based on full flexural overstrength in the potential plastic hinges; and ϕ_v = shear capacity reduction factor (typically taken as 0.85). V_c, V_s and V_p = three shear capacity contributions from the concrete, horizontal steel reinforcement, and axial load based on the UCSD three-component shear model (Priestley, 1996) with reduction for the concrete component V_c in the flexural plastic hinge region, based on the ductility demand, and E_f and D = composite jacket modulus and the column dimension in the loading direction respectively. f_{cc}' = confined concrete compression strength that depends on the effective lateral continuing stress and the nominal concrete strength and can be conservatively taken as $1.5 f_{co}'$ for most retrofit designs (Priestley, 1996); f_{fu} and ε_{fu} = strength and deformation capacity of the composite jacket in the hoop direction; Φ_f = flexural capacity reduction factor (typically taken as 0.9); and ε_{cu} = ultimate concrete strain that depends on the level of confinement provided by the composite jacket. ρ_j representing the volumetric jacket reinforcement ratio, the ultimate section curvature Φ_u, neutral axis depth cu. L represents the shear span to the plastic hinge; Φ_y = section yield curvature; and f_{sy} and d_b = yield strength and bar diameter of the main column reinforcement (Priestley, 1996). The column dimension D in the loading direction, M and V the maximum column moment and shear. f_h represents the horizontal stress level provided by the existing hoop reinforcement in a circular column at a strain of 0.1%; and f_l the lateral clamping pressure over the lap splice L_s. p = perimeter line in the column cross-section along the lap-spliced bar locations; n = number of spliced bars along p; A_s = area of one main column reinforcing bar; and cc = concrete cover to the main column reinforcement with diameter db. The elastic jacket modulus E_f in the hoop direction; the ultimate unidirectional tensile strength f_{fu}; and the ultimate unidirectional tension failure strain ε_{fu}.

hinge regions in circular RC columns. For rectangular columns, Seible et al. (1997) recommended that an oval jacket should be provided via the added precast concrete segments with changing the radius of curvature in different loading directions. An equivalent circular column diameter (D_e) can be derived from the average of the oval jacket principal radii of curvature. Jacket thickness calculations can follow those for circular columns using the equivalent column diameter D_e. In cases where the column side aspect ratio of depth/width ≤ 1.5 and the sides of the columns have a depth of 0.75 m and width of 0.5 m, the rectangular composite jackets should have twice the theoretical thickness required for an equivalent circular column of diameter D.

6.8.1.3 Lap-splice clamping

Finally, lap-splice clamping requires sufficient lateral pressure f_l on the splice region to prevent the starter bars and column reinforcement to slip relative to each other. Furthermore, experimental test results showed that lap-splice debonding or relative slippage starts when the measured hoop or dilation strain levels are between 1000 and 2000 $\mu\varepsilon$. At strain levels of 2000 $\mu\varepsilon$, lap-splice debonding was in progress, as indicated by a loss in the lateral load-carrying capacity of the test columns (Seible et al., 1997). By limiting the dilation strain levels to 1000 $\mu\varepsilon$, the composite jacket thickness can be derived to ensure lap-splice clamping in accordance with Table 6.10. For rectangular columns with lap splices, changing the shape of the column cross-section to circular or oval shape is required to provide the necessary lateral clamping pressure.

6.8.2 Ductility design method by Monti et al. (2001)

Monti et al. (2001) proposed a design equation to calculate the suitable thickness of FRP jackets to enhance the ductility of existing RC bridge piers with circular cross-sections. The design aims to achieve an upgrading index, given as the ratio of the target to the available ductility at the pier base section, to be attained through FRP jacketing. The available ductility can be obtained through the usual assessment procedures for RC members, whereas the target ductility is evaluated based on the expected actions on the bridge. The upgrading index is initially defined in general terms and is subsequently extended to the case of piers built in seismic regions. It results in a simple expression in terms of easily computable quantities, such as the ultimate strain and the peak strength of concrete, before and after upgrading.

6.8.2.1 Ductility upgrading of piers in seismic regions

When designing upgrading interventions on structures in seismic areas, one deals with spectral ordinates that measure the seismic action imposed on structures (usually, in terms of forces). In ductility-based designs, it is well known that the spectral ordinates used for evaluating the forces on a structure are related to the amount of available ductility. From this point of view, a pier upgrading index I_{pier} can be defined in a similar way to what has already been defined in terms of ductility and linked to the spectral ordinates corresponding to a given ductility value.

Considering an elastic–plastic oscillator, the noncollapse requirement is expressed as

$$\delta_d^{tar} = \frac{mR(T)a_g}{F_y} \leq \delta_d^{ava} \tag{6.7}$$

The above form refers to that of a single degree of freedom (SDOF) system with mass m, elastic period T, and yield force F_y, under response acceleration $R(T)a_g$ [where $R(T)$ denotes the response spectrum and a_g is the peak ground acceleration] is required to attain a target displacement ductility δ_d^{tar}. It continues without collapse when δ_d^{tar} is lower than the available ductility δ_d^{ava}.

The pier upgrading index I_{pier} can be given by

$$I_{pier} = \frac{\delta_d^{tar}}{\delta_d^{ava}} = \frac{mR(T)a_g}{F_y \delta_d^{ava}} \tag{6.8}$$

For well-designed structures, Eq. (6.8) has a value <1 and thus upgrading is not required. For under-designed or damaged structures, the above equation yields $I_{pier} >1$.

6.8.2.2 Design procedure for designing an FRP jacket for circular columns

The design procedure can be summarized as follows.

1. Assess (through survey) the quantities f_{co}', n, μ_s, and μ_{st}.where $n =$ axial load ratio; μ_s, $\mu_{st} =$ mechanical ratios of longitudinal and transverse reinforcements, respectively, given by $\mu_s = (A_s f_{sy})/(A_{gr} f_{co}')$, $\mu_{st} = \rho_{st} f_{sy}/f_{co}'$, and
2. Compute $f_{cc}'^{ava} = f_{cc}'$, f_l, and ε_{cu}^{ava}(when only the steel hoops confinement is present) as follows.

$$\frac{f_{cc}'}{f_{co}'} = 2.25\sqrt{1 + 7.94\frac{f_l}{f_{co}'}} - 2\frac{f_l}{f_{co}'} - 1.254 \tag{6.9}$$

$$f_l = \frac{1}{2}k_e\rho_{st}f_{sy} \tag{6.10}$$

$$\varepsilon_{cu}^{ava} = 0.004 + 1.4\frac{\rho_{st}f_{sy}\varepsilon_{su}}{f_{cc}'} \tag{6.11}$$

where $\rho_{st} = 4A_{st} /(s_{st}d_{st}) =$ the volumetric ratio of steel hoops (spiral) with area A_{st}, spacing (pitch) s_{st}, and diameter d_{st}. f_{sy} and ε_{su} denote the steel yield strength and ultimate strain (usually 0.10), respectively. k_e is the tie-by-tie arching-effect coefficient (usually 0.8).

3. Assess the available curvature ductility δ_{χ}^{ava} at the pier base section through a section model using f_{co}', n, μ_s, and μ_{st}.
4. Evaluate the target ductility δ_{χ}^{tar} based on expected loads (for the case of seismic action).
5. Compute the upgrading index $I = I_{sec} = \delta_{\chi}^{tar} / \delta_{\chi}^{ava}$; verify that the available (AVA) quantities times the desired upgrading index I ($I.AVA$) > 0.01.

$$I_{sec} = \frac{\varepsilon_{cu}^{tar}}{\varepsilon_{cu}^{ava}} \cdot \frac{\sqrt{f_{cc}'^{tar}/f_{co}'}}{\sqrt{f_{cc}'^{ava}/f_{co}'}} \tag{6.12}$$

where $\varepsilon_{cu}^{tar} = (\varepsilon co(2 + 1.25(Ec/f_{co}')\varepsilon_{fu}\sqrt{f_l})$ is the target concrete ultimate strain to be attained through FRP confinement, E_c is the initial tangent modulus of concrete, and $f_l = (1/2)\ \rho_f\ f_{fu}$, and f_{fu} and ε_{fu} are the ultimate strength and the ultimate strain of the jacket, respectively. $P_f = 4t_j/d_f$ is the volumetric confinement reinforcement ratio of an FRP jacket with thickness t_f and diameter d_f. $f_{cc}'^{tar}$ is obtained through Eq. (6.9) and $f_l = (1/2)\ \rho_f f_{fu}$;
6. Select the material for FRP jacketing (f_{fu} and ε_{fu}).
7. Determine the FRP jacket thickness, t_f, from the following form.

$$\rho_f = 0.8 I^2 \cdot \frac{\left(\frac{f_{cc}'}{f_{co}'}\right)}{\left(\frac{f_{fu}}{f_{co}'}\right)} \cdot \frac{\varepsilon_{cu}^{ava\ 2}}{\sqrt{\varepsilon_{fu}^3}} \tag{6.13}$$

where $I = I_{sec}$ and $t_f = \rho_j d_j/4$.

6.8.3 Ductility design method by Tastani and Pantazopoulou (2006)

To design the upgrading scheme, the seismic demand needs to be determined in terms of displacement, whose prerequisites involve the idealization of the structure as an equivalent single degree of freedom system (ESDOF) through a selected empirical approximation of the predominant shape of lateral vibration and the calculation of the corresponding stiffness (secant to yield). For immediate results, the ESDOF properties may be used with the yield points spectra of the design earthquake to evaluate the anticipated displacement demand and corresponding displacement ductility (Aschheim and Black, 2000). Based on the equal displacement rule, the elastic spectral displacement is also the target displacement of the inelastic system, that is, $\Delta u = Sd$ (refer to Tastani and Pantazopoulou, 2006).

6.8.3.1 Strength assessment of FRP rehabilitated RC members

To redesign a substandard RC element considering seismic resistance, all other failure modes except flexural, which is the least undesirable, should be mitigated. The design forces must satisfy the following qualitative relationship.

$$Vu, lim = min[V_{iflex}, V_{shear}, V_{anch}, V_{buckl}] \tag{6.14}$$

where $V_{iflex} = M_u/L_s$ is the seismic shear force required to develop the ideal flexural resistance of the member, L_s is the shear span, V_{shear} is the nominal shear resistance, V_{anch} is the shear force when the anchorage/lap splice reaches their development capacity, and V_{buckl} is the shear force when the compression bars reach instantaneous buckling conditions at the critical section. The strength components in the above equation may be estimated from the variables of σ_{lat} calculated by the following forms.

1. *For shear*

    In any given direction of action y, the total transverse pressure, $\sigma_{lat,y}$, comprises the contributions of the FRP jacket and the occasional embedded stirrups.

$$\sigma_{lat,y} = \sigma_{lat,y}^f + \sigma_{lat,y}^{st} = 2k_{f,y}nt_f E_f \varepsilon_f^{eff}/b_y + k_{st,y}A_{st}f_{y,st}/(sb_y) \tag{6.15}$$

Parameters $k_{f,y}$ and $k_{st,y}$ are the effectiveness coefficients for the two transverse confining systems, ε_f^{eff} is the effective tensile strain that develops in the jacket near failure (which may occur either by debonding or by rupture, whichever prevails), E_f, n, and t_f are the elastic modulus, number, and thickness of one ply of FRP, respectively, b is the cross-section width at the splitting plane (orthogonal to the applied jacket force), A_{st} is the total cross-sectional area of stirrup legs crossing the splitting plane provided by a single stirrup layer, s is the longitudinal spacing of stirrups, and $f_{y,st}$ is their yield stress. $k_{f,y} = 1$ for fully wrapped columns and $k_{st,y} = 0.5$ and 1 for open shear links and well-anchored closed stirrups, respectively.

2. *For confinement*

The confining pressure is the average value σ_{lat}^{ave} obtained from the following form in the two principal directions of the cross-section as σ_{lat-x} and σ_{lat-y}.

$$\sigma_{lat}^{ave} = 0.5\left(\sigma_{lat,y} + \sigma_{lat,x}\right) = 0.5\left(k_f^c\rho_{fv}E_f\varepsilon_f^{eff} + k_{st}^c\rho_{sv}f_{y,st}\right) \tag{6.16}$$

where ρ_{fv} and ρ_{sv} are the volumetric ratios of FRP and stirrup reinforcement, respectively. The familiar expression for k_f^c approximates the volume fraction of core concrete that is effectively restrained (similar to the approach used to evaluate the confinement effectiveness of stirrups k_{st}^c (Priestley, 1996). Therefore $k_f^c = 1 - (b'^2 + d'^2)/[3A_g(1 - \rho_s)]$, where A_g is the gross cross-section of the element, ρ_s is the ratio of longitudinal reinforcement, b' and d' are the straight sides of the rectangular cross-section encased by the jacket after chamfering the corners (FIB Bulletin 14, 2001; ACI(440.2-02), 2002). For a cross-section with a side aspect ratio of 3, the confinement effectiveness coefficient becomes negligible ($k_f^c \approx 0$), whereas for square and circular sections, $k_f^c \approx 0.5$ and 1, respectively.

Shear strength calculations

The shear resistance of the RC members subjected to displacement reversals degrades with the number of cycles and the magnitude of imposed displacement ductility. Strength reduction is accounted for through a ductility dependent softening coefficient λ (Moehle et al., 2002). In redesigning substandard RC members for shear resistance, the residual $V_{n\text{-}res}$ before FRP-jacketing intervention and postupgrading resistance should be evaluated as follows:

$$V_n(q_{\text{old}}) = \lambda(q_{\text{old}})(V_s + V_c) \tag{6.17}$$

$$V_n(q_{\text{new}}) = min\{\lambda(q_{\text{old}}), \lambda(q_{\text{new}})\}(V_s + V_c) + V_f^w \tag{6.18}$$

$$V_s = \sigma_{lat,y}^{st} bd = (k_{st,y}A_{st}f_{y,st}/(sb_y))bd \tag{6.19}$$

$$V_c = 2\sqrt{f_{co}'}bd \tag{6.20}$$

$$V_w^f = \sigma_{lat,y}^f bh = 2k_{f,y}nt_f E_f \varepsilon_f^{eff}/b_y(bh) \tag{6.21}$$

$$\lambda = 1.15 - 0.075\mu_\Delta; \quad \lambda \leq (0.7, 1) \tag{6.22}$$

where μ_Δ is the imposed displacement ductility, q is the behavior index (or R, FEMA 273 1997), and $\sigma_{lat,y}^f$ is the transverse pressure in concrete owing to the jacket in the direction of lateral sway. The shear strength of the jacketed member is the sum of the jacket contribution, V_f^f, and the contribution of the existing mechanisms, namely concrete V_c and transverse steel V_s. In deriving the shear equation, it has been assumed that the target μ_Δ used in the redesign of the member is equal to the behavior index, q_{new} (or R_{new}). Shear equation recognizes that the existing mechanisms may have sustained damage during previous loadings. Thus the residual contributions of the core concrete and web reinforcement are considered, instead of the full contribution, by taking the minimum value of λ for these terms, based on the ductility demand either suffered during previous events or used as the target value for redesign. Based on the experiments, the softening coefficient is not applied to V_w^f as diagonal cracking is suppressed by the application of the jacket (Tastani and Pantazopoulou, 2003).

Ideal flexural capacity calculations

Flexural resistance is influenced by the increase in concrete strength due to confinement and containment of the cover that would otherwise spall-off at ultimate.

The confined concrete strength f'_{cc} and the corresponding strain can be calculated from the following equations.

$$f'_{cc} = f'_{co} + 1.5\left(k_f^c \rho_{fv} E_f \varepsilon_f^{eff} + k_{st}^c \rho_{sv} f_{y,st}\right) \tag{6.23}$$

$$\varepsilon_{cc} = 0.002 + 0.015\left(k_f^c \rho_{fv} E_f \varepsilon_f^{eff} + k_{st}^c \rho_{sv} f_{y,st}\right)/f'_{co} \tag{6.24}$$

The failure strain $\varepsilon_{cc,u}$ corresponding to a compression strength reduction above 15% is a lower bound expression (FIB Bulletin 24, 2003). For closed jackets the ε_f^{eff} is taken as $0.5\varepsilon_{fu,d}$.

$$\varepsilon_{cc,u} = \varepsilon_{c,u} + 0.075 \left(\frac{\left(k_f^c \rho_{fv} E_f \varepsilon_f^{eff} + k_{st}^c \rho_{sw} f_{y,st} \right)}{f_{co}'} - 0.1 \right) \geq \varepsilon_{c,u}; 0.003 \leq \varepsilon_{c,u} \leq 0.004$$

$$(6.25)$$

Anchorage/lap-splice strength calculations

A direct consequence of member upgrading through FRP jacketing is to increase the deformation demand in the lap-splice/anchorage regions. To remedy the anchorage problems, FRP jackets are wrapped orthogonal to the anticipated splitting cracks. The development capacity of a given anchorage length L_b is calculated from $F = \mu \, \sigma_{lat} \, \pi D_b \, L_b$, where μ is the coefficient of friction at the steel–concrete interface (the frictional coefficient is equal to 1.4 for concrete) and σ_{lat} is the pressure exerted upon the lateral surface of the bar by the cover, transverse stirrups, and FRP jacket. The average bond stress f_b is given by

$$f_{b,d} = \mu \left(C f_t' + \frac{k_{st}^{anch} A_{st} f_{y,st}}{(N_b s)} + \frac{2k_f^{anch} n t_f E_f \varepsilon_f^{eff}}{N_b} \right) / (\pi D_b) \qquad (6.26)$$

N_b is the number of bars (or pairs of spliced bars) laterally restrained by the transverse pressure. The value of ε_f^{eff} is the surface strain value associated with the attainment of bond strength along the bar, and it is in the order of $0.0015-0.002$ (Priestley, 1996). $C = p/\pi D_b$, where p is the perimeter line in the column cross-section along the lap-spliced bar locations and f_t' is the concrete tensile strength. $k_{st}^{anch} = k_f^{anch} = 1$ for fully wrapped columns.

Resistance to longitudinal bar buckling in FRP-wrapped RC elements

In plastic hinge regions with severe shear demand sideways, bar buckling is the likely failure pattern. The critical buckling stress $f_{s,crit}$ is related to s/D_L through $s/D_b = 0.785(E_r/f_{s,crit})^{1/2}$, where E_r is the double modulus of steel at the stress level considered (FIB Bulletin 24, 2003). From this relationship, given the full stress–strain diagram of the bar, the limiting strain–ductility curve ($\mu_{\varepsilon c} = \varepsilon_{s,crit}/\varepsilon_y$) may be plotted as a function of s/D_b. $\varepsilon_{s,crit}$ is the axial strain at the onset of instability for the given s/D_b.

By increasing the strain capacity of concrete through jacketing to levels higher than $\varepsilon_{s,crit}$, the effective s/D_b is reduced. The dependable strain ductility of compression reinforcement is

$$\mu_{\varepsilon c} = \max \left(\varepsilon_{s,crit}/\varepsilon_y, \varepsilon_{s,cu}/\varepsilon_y \right) \qquad (6.27)$$

In detailing the jacket, it is important to ensure that the target displacement ductility of the member after upgrading, $\mu_{\Delta req} = \Delta_u^{target}/ \Delta_y$, may be attained before buckling the primary reinforcement. To check this, the resulting curvature ductility demand $\mu_{\varphi,req}$ ($= \varphi_{u,req}/\varphi_y$) in the plastic hinge region of the member is obtained from $\mu_{\Delta,\rho\varepsilon\theta}$.

$$\mu_{\Delta,req} = 1 + 3\left(\mu_{\varphi,req} - 1\right)\left(1 - 0.5l_p/L_s\right)l_p/L_s; l_p = 0.08L_s + 0.022f_yD_b \quad (6.28)$$

From $\mu_{\varphi,req}$, the compression strain ductility demand, $\mu_{\varepsilon c,req}$, of the compression reinforcement may be estimated and compared to the dependable value resulting from Eq. (6.28). For example, for symmetric displacement reversals, $\mu_{\varepsilon c, req} = 1.1\mu_{\varphi req} - 1$ (FIB Bulletin 24, 2003). To preclude rebar buckling before the realization of the target displacement ductility, the value of $\mu_{\varepsilon c,req}$ should be less than the value of $\mu_{\varepsilon c}$ extracted from Eq. (6.27). It should be noted that $\mu_{\Delta, req} = \Delta_u^{target}/ \Delta_y$.

Deformation capacity assessment for FRP-encased members

An empirical lower bound expression for the available displacement ductility is given in the following form based on the results from over 70 published tests on the response of FRP jacketed RC prismatic members under reversed cyclic loading. μ_Δ is the function of transverse confining pressure σ_{lat}^{ave} ($\mu_\Delta \geq 1.3$ for poorly detailed members), given by

$$\mu_\Delta = 1.3 + 12.4\left(\frac{k_f^c\rho_{fv}E_f\varepsilon_f^{eff} + k_{st}^c\rho_{sv}f_{y,st}}{2f_{co}'} - 0.1\right) \quad (6.29)$$

where $\varepsilon_f^{eff} = \varepsilon_{fu,d}$. The yield displacement, as proposed by Priestley (1996), ($\Delta_{y=}\varphi_y(L_s + 0.022f_yD_b)^2/3)$ includes the flexural-slip component.

6.8.4 Ductility design method by Japan society of civil engineers (Japan Society of Civil Engineers JSCE, 2001)

According to Japan Society of Civil Engineers JSCE (2001), the ductility ratio of members upgraded with continuous fiber sheets, μ_{fd}, is given by

$$\mu_{fd} = \left[1.16 \bullet \frac{(0.5 \bullet V_c + V_s)}{V_{mu}} \bullet \left\{1 + \alpha_o \frac{\varepsilon_{fu} \bullet \rho_f}{V_{mu}/B \bullet Z}\right\} + 3.58\right]/\gamma_{bf} \leq 10 \quad (6.30)$$

where
V_c is the shear contribution due to concrete.

$$V_c = \beta_d \bullet \beta_p \bullet \beta_n \bullet f_{vcd} \bullet b_w \bullet d/\gamma_b \quad (6.31)$$

(both material factor and member factor γ_b are calculated as 1.0, that is, the "design shear capacity for bar members" in this code)

$$f_{vcd} = 0.20^3 \sqrt{\smash[b]{f'_{cd}}} \leq 0.72 (\text{N/mm}^2)$$

$$\beta_d = \sqrt[4]{1/d}(d: m), 1.5 \text{ when } \beta_d > 1.5$$

$$\beta_p = \sqrt[3]{100p_w}(d: m), 1.5 \text{ when } \beta_p > 1.5$$

$$\beta_n = 1 + M_o/M_d(N'_d \geq 0) \text{when } \beta_n > 2.0$$

$$\beta_n = 1 + 2M_o/M_d(N'_d \geq 0) \text{when } \beta_n > 0.0$$

N'_d: design axial compressive force
M_d: design bending moment
M_0: decompression moment
b_w: web width
d: effective depth
p_w: $A_s/(b_w \times d)$
A_s: cross-sectional area of reinforcing bars in tension side
f'_{cd}: design compressive strength of concrete (unit: N/mm^2)
γ_b: member factor (in general, may be set to 1.3)
V: shear contribution due to shear reinforcing bar members (both material factor and member factor are calculated as 1.0 for the "design shear capacity for bar members" in this code)

$$V_{sd} = \left[A_w \cdot f_{wyd}(\sin\alpha_s + \cos\alpha_s)/s_s \right] \cdot z/\gamma_b \tag{6.32}$$

A_w: total cross-sectional area of shear reinforcement in space s
f_{wyd}: design tension yield strength of shear reinforcement (400 N/mm^2 max.)
α_s: angle formed by shear reinforcement about the member axis
s: spacing of shear reinforcement

$$\varepsilon_{fu} = f_{fud}/E_f = (f_{fuk}/\gamma_{mf})/E_f \tag{6.33}$$

f_{fud}: design tensile strength of continuous fiber sheet (unit: N/mm^2)
f_{fuk}: characteristic value of tensile strength of continuous fiber sheet (unit: N/mm^2)
E_f: characteristic value of modulus of elasticity of continuous fiber sheet (unit: N/mm^2)
γ_{mf}: material factor of continuous fiber sheet (generally set to 1.2)
V_{mu}: maximum shear force when a member reaches the existing flexural load-carrying capacity M_u
1.0 is used as the material factor, material correction factor, and member factor for the reinforcement and concrete.)
γ_{bf}: member factor used for the calculation of μ_{fd} (generally set to 1.3)

ε_{fu}: ultimate strain of continuous fiber sheet (design tensile strength of continuous fiber sheet divided by the characteristic value of modulus of elasticity)

$$\rho_f = A_f/(S_f \bullet B) = 2n_f t_f' S_f/(S_f \bullet B) \qquad (6.34)$$

ρ_f: shear reinforcement ratio of continuous fiber sheet
S_f: spacing of continuous fiber sheets (unit: mm)
t_f: thickness of one ply of continuous fiber sheet (unit: mm)
n_f: number of plies of continuous fiber sheets
S_f': width of continuous fiber sheet (unit: mm)
α_0: coefficient used to calculate member ductility ratio (for columns shear-reinforced with lateral ties, α_0 may be used as the modulus of elasticity for the lateral ties)
B: member width (unit: mm)
z: lever arm length (generally set to $d/1.15$)

6.9 Numerical simulation of columns under axial and lateral loads

The simulation of RC columns retrofitted with FRP jackets using commercial software based on the finite element method (FEM) has been reported in the literature. One of these studies was reported by Wu et al. (2010) to evaluate the seismic performance of FRP-RC columns through numerical simulations of the load−deformation response using two-dimensional finite element analysis (2D-FEA). The mesh configuration is determined through convergence analysis and the crack models and the constitutive relationships of concrete are validated. The seismic performance of three RC columns strengthened with FRP sheets is assessed through a series of parametric studies, as presented by Wu et al. (2010). The investigation of the rational mesh configuration showed that a good mesh configuration has element dimensions approximating the spacing of cracks as they occurred in the test, and a highly refined mesh leads to localized early compressive failure using the strength theory.

As shown in Fig. 6.25A, the 2D-FEA results of column 1, using the standard specification (Japan Society of Civil Engineering JSCE, 2002), match fairly well with the experimental results, and while columns 2 and 3 have similar capacities with a good level of ductility, the analytically determined postpeak responses do not decrease in a manner similar to that seen in tests. The 2D-FEA results in Fig. 6.25B, derived using the Darwin−Pecknold equivalent uniaxial strain model, Saenz's equation, and failure criteria under biaxial stresses, considering the lateral confinement effects from both hoop steel and wrapped CFRP sheets, and the compressive strength reduction after cracking, exhibit good correlation with the experimental results in terms of capacities and postpeak behavior. In addition, the differences due to the number of FRP layers used are evident. In contrast, the use of Saenz's equation in the total strain model without

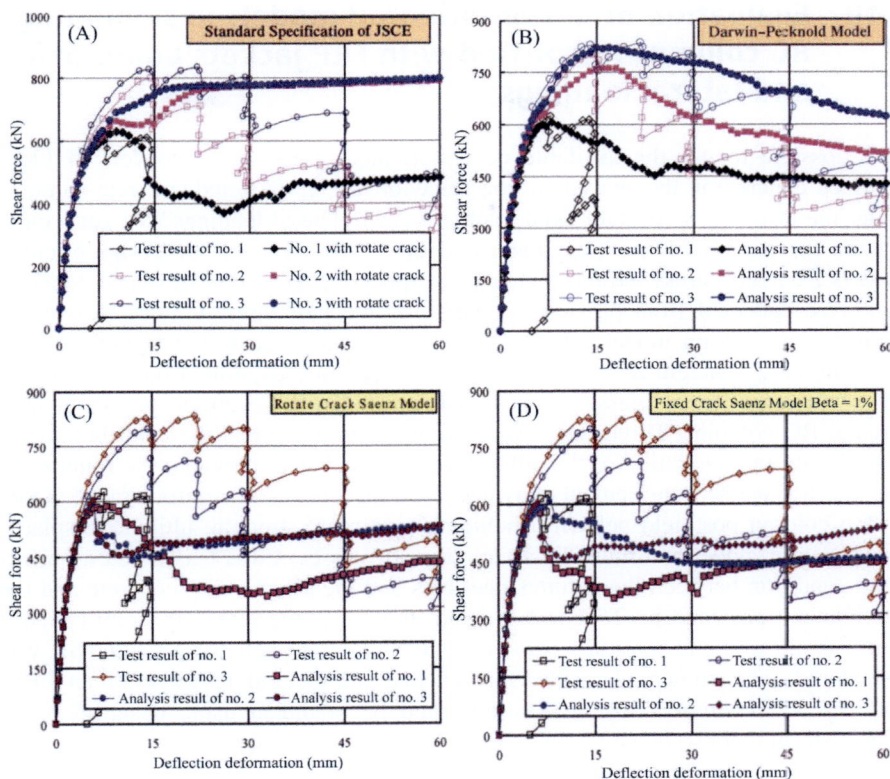

Figure 6.25 Analysis results using (A) standard specification of Japan Society of Civil Engineering JSCE (2002), (B) Darwin–Pecknold (1974) model, (C) total strain formulation and the Saenz (1964) equation (Wu et al., 2010).

considering the failure criteria of the concrete and the lateral confinement effect results in large differences between the analytical and experimental results, as shown in Fig. 6.25C.

The failure modes of the columns can be simulated, mimicking those observed in the experiments. These results confirm that the equivalent uniaxial strain model and failure criterion can be used to accurately simulate the nonlinear behavior and failure of concrete under a biaxial stress state, thereby validating the 2D-FEA model. Moreover, it is shown that a simple modified confinement model can be adopted to evaluate the confinement effects from hoop steel and FRP on concrete, which generally works in three-dimensional confinement. Through the numerical simulations and parametric study, the performance of the FRP retrofitting scheme is verified by analyzing the load–deformation responses and the stress and strain at inflection points and bottoms of the columns.

6.10   Evaluation of design-oriented models in simulating RC columns retrofitted with FRP jackets under axial and lateral loadings

Is it possible to use the available design-oriented stress—strain models of FRP-confined concrete in the analysis of the behavior of FRP-jacketed concrete columns under the effect of a constant axial load and cyclic lateral loadings? To get a clear answer for the raised question, Fahmy et al. (2016) selected 14 design-oriented models from 110 stress—strain models (see Appendix A) of FRP-confined concrete that were collected from previous studies over the past three decades. All of the studied stress—strain models of concrete were implemented in the OpenSees software, treating concrete as uniaxial material. This study examined the accuracy of the 14 design-oriented stress—strain models to predict the cyclic responses of nine circular RC columns. Three cases were simulated to determine the contributions of both the strain penetration of the column longitudinal steel bars into the foundation and the transverse confinement provided by transverse steel reinforcement (TSR) on the column postyield behavior including maximum strength, ultimate displacement, failure mode, and unloading/reloading stiffnesses. Case 1 assumed a perfect bond condition between the column steel bars and the surrounding concrete and the confinement provided by TSR was ignored. In case 2, the effect of strain penetration was allowed and in case 3, the effect of both strain penetration and TSR was considered. The numerical results showed that the general response of an FRP-confined RC column to cyclic loading could be predicted using design-oriented stress—strain models; however, the local stress—strain law obtained from concentric compression tests did not very well reflect the local behavior of the compression zone of members in flexure with axial force. The higher the column axial load ratio, the lower the ratio between the numerical and experimental ultimate column lateral displacements, and thus the predicted failure mode does not match well with the experimental results; as shown in Fig. 6.26.

Fahmy and Wu (2010) showed that several available design-oriented models failed to reflect the fact that when different FRP materials provide the same lateral stiffness, the predicted stress—strain relationship of the confined concrete should be identical, and the concrete axial strain capacity should be dependent on the strain capacity of the FRP used. Thus they proposed a stress—strain model based on the FRP lateral stiffness to overcome this problem. However, they did not explain whether this issue would have a significant effect on the flexural strength and ductility of FRP-RC columns under axial and lateral loads. Therefore Fahmy et al. (2016) selected an RC circular column confined with GFRP (providing a confinement ratio of 0.84) and named CF-R3, which was experimentally tested by Haroun and Elsanadedy (2005a,b). To present a comparative study between the three different FRP types providing the same lateral stiffness, additional two different FRP materials (CFRP and PET) were assumed for the confinement of the CF-R3 column, and they were designed to provide the same lateral stiffness as the GFRP jacket; for the concrete dimensions and reinforcement details of this column, refer

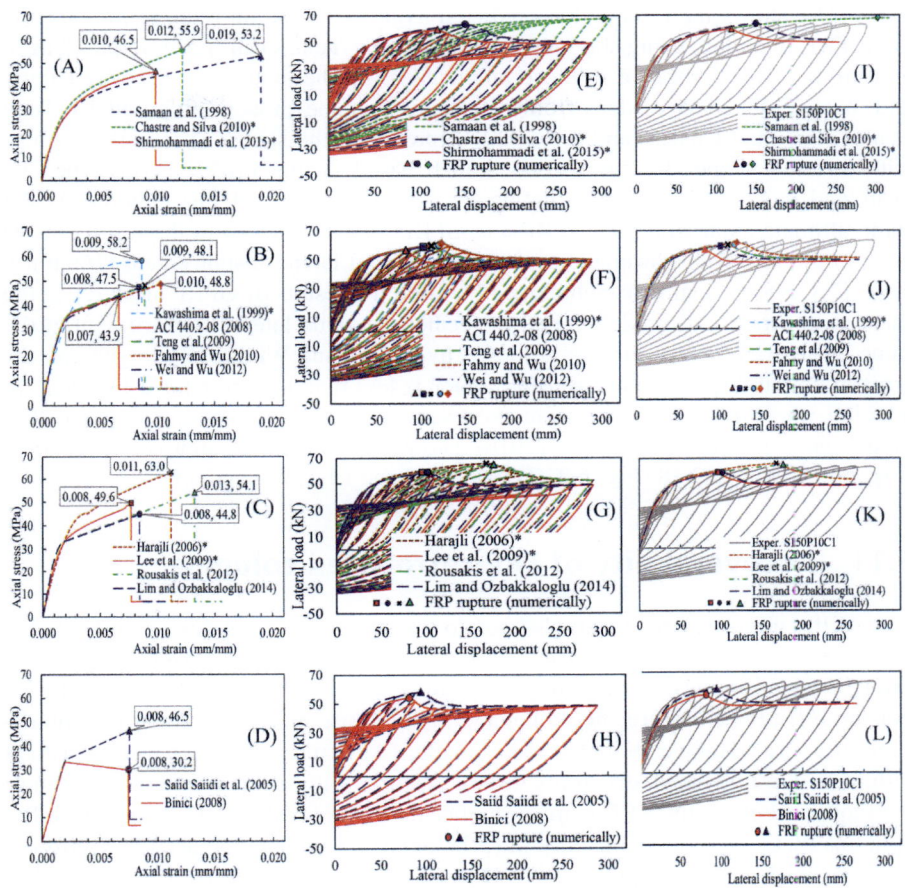

Figure 6.26 Predicted axial stress−strain curves of groups: (a) A, (b) B, (c) C, and (d) D (regarding these groups, refer to Section 6.8.1.1). Predicted cyclic load−displacement relationships for column S150P10C1 by all models of groups (e) A, (f) B, (g) C, and (h) D. The envelop curves of the predicted load−displacement relationships with the experimental results of column S150P10C1 by all models of groups (i) A, (j) B, (k) C, and (l) D. (* refers to stress−strain models with confinement provided by both FRP and TSR.) (Fahmy et al., 2017).

to Haroun and Elsanadedy (2005a,b). For the GFRP used, the elastic modulus was 36.5 GPa, and the jacket thickness was 11.4 mm. The tensile strength of the CFRP and PET were 4170 MPa and 923 MPa, and the elastic moduli were 232 GPa and 6.7 GPa, respectively, as given by Fahmy and Wu (2010). The design for the CFRP and PET jackets provided a thickness of 1.79 and 62.1 mm, respectively. Column CF-R3 was simulated using the OpenSees software under the effect of a constant axial load and a monotonic lateral load for the three types of FRP jackets.

Fig. 6.27A shows the error in the predicted axial stress—strain relationships for the three FRP materials using the models proposed by Teng et al. (2009) and Chastre and Silva (2010) compared to those predicted using the model proposed by Fahmy and Wu (2010). It is evident from the figure that the models proposed by Teng et al. (2009) and Chastre and Silva (2010) predicted the stress—strain relationships with different slopes for the second branch based on the FRP material type. All examined models were used to define the lateral response of this column. At the same ultimate axial strain for CFRP and GFRP, Fig. 6.27B shows the results of the error in the lateral displacement to the experimental lateral displacement (166 mm). Fig. 6.27C shows a similar comparison between the results of GFRP and PET at the same axial strain. All models failed to predict the same lateral displacement at the same axial strain except for that proposed by Fahmy and Wu (2010). In general, the error in lateral displacement for using CFRP is less than the error for using PET. In addition, there is a discrepancy in the evaluated error by the same model. The model proposed by Lee et al. (2010) showed an error of approximately 0.11 (Fig. 6.27B) and −3.18 (Fig. 6.27C) for CFRP and PET, respectively.

6.11   Recoverability of FRP-retrofitted columns

One of the important proposed changes to the current seismic design codes is the consideration of two earthquake designs under specific circumstances, that is, safety evaluation earthquake (SEE) and functional evaluation earthquake (FEE). For the FEE, the structure should behave in an elastic manner without any significant structural damage. For the SEE, standard structures should have no critical failure, while important structures should perform with limited damages. Because the requirement for the FEE is much higher than that for the SEE, the existing RC columns that satisfy the SEE only must be enlarged and/or increased in reinforcement to meet the new requirements. However, if a suitable retrofitting is used to assure postyield stiffness, the FEE design criteria may be met without dramatically increasing the column size or the amount of reinforcement. The main disadvantage of having a zero or negative postyield stiffness is that it results in a large residual displacement from the FEE. This large residual displacement significantly complicates the repair work after an earthquake. Earthquake-Resistant Design Codes in Japan specify that the residual displacement should not be greater than 1% of the pier height.

6.11.1   Residual deformations as a seismic performance index

Seismic Design Code for Railway Structures in Japan reflects the recent advances in earthquake engineering. Some new concepts of seismic design were adopted in this code by drawing on the lessons of the 1995 Kobe earthquake that caused devastating damages. During the 1995 Kobe earthquake, the Hanshin Expressway Kobe Route was subjected to near-field strong ground motions, resulting in serious damage to a large portion of the Kobe Route. Based on the postearthquake visual

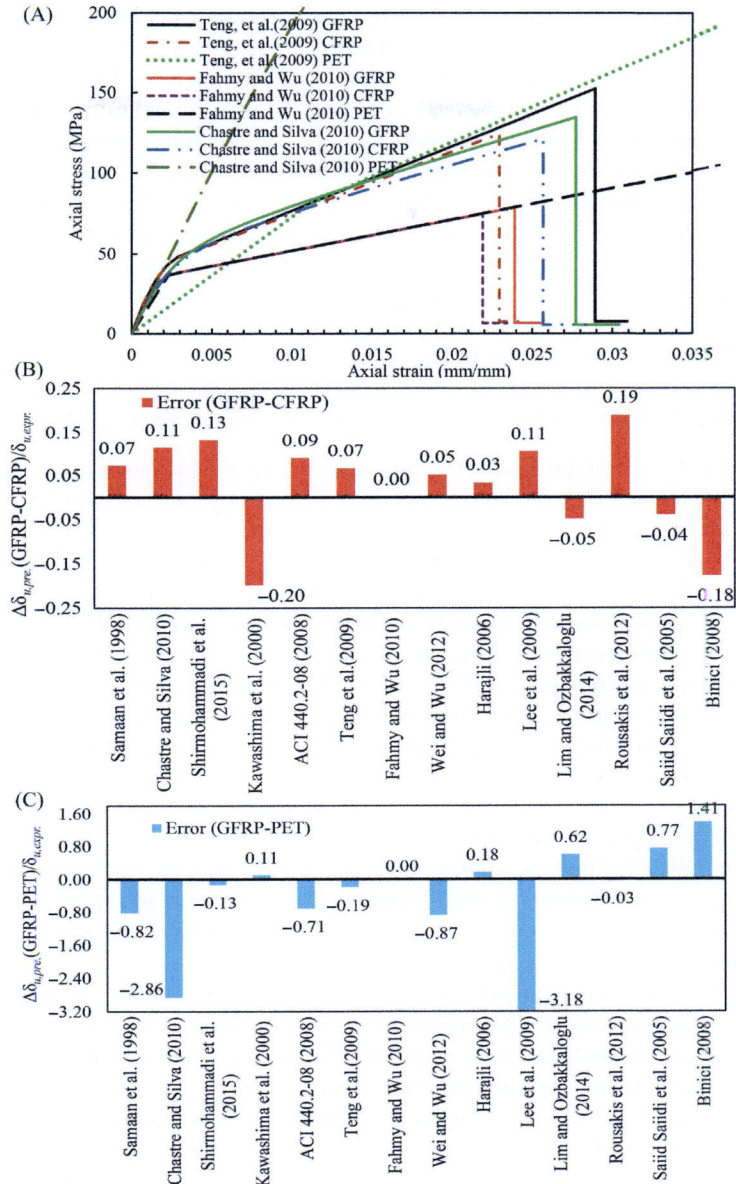

Figure 6.27 (A) Predicted stress−strain behavior for the CF-R3 column using different models when three different FRP materials can be applied for strengthening and providing the same lateral stiffness, (B) results of the error in the lateral displacement to the experimental lateral displacement at the same ultimate axial strain for CFRP and GFRP, and (C) results of the error in the lateral displacement to the experimental lateral displacement at the same ultimate axial strain for PET and GFRP (Fahmy et al., 2017).

inspection of the aboveground damage, the damage of an RC pier was classified as follows: As = collapsed or almost collapsed; A = very severe; B = severe; C = mild; and D = no damage (JSCE 2000). For the Hanshin Expressway Kobe Route, one of the measurements checked during the investigation of the reconstruction and repair was the amount of inclination of the columns (residual drift ratio), where large residual inclinations made it difficult to readjust the bridge girders and caused visual uneasiness, even if repairs were possible. The piers classified as As and A were demolished because it was impossible to repair them for reuse. Yozo et al. (2005) reported that the residual inclination tends to be large in severely damaged piers, and it also exists in many lightly damaged or even nondamaged piers. Hence, piers with an inclination larger than a certain limit were demolished even if the visually judged damage was mild.

6.11.2   Postyield stiffness as a seismic performance index

Because the strength requirement for very rare earthquakes is much higher than rare earthquakes, the existing RC bridge columns that satisfy rare earthquakes must be enlarged and/or increased in reinforcement to meet the new requirements. However, if a suitable FRP retrofitting is used effectively to assure postyield stiffness, the very-rare earthquake design criteria may be met without dramatically increasing the column size or the amount of reinforcement. Moreover, minimum irrecoverable deformation would be achieved in case the structure has positive postyield stiffness. The study by Kawashima et al. (1998) showed that the residual displacement ratio spectrum depends significantly on a bilinear factor r (ratio between the structure second postyield stiffness and the initial elastic stiffness). The residual displacement ratio response spectrum is small when $r > 0.0$, and sharply increases as r approaches zero. If the residual deformations are found to be excessive, it was suggested by Christopoulos and Pampanin (2004) that the following design considerations are applicable, that is, appropriately changing the material properties (e.g., steel with high strain hardening) and changing the reinforcement properties and section design to increase the postyield stiffness at a section level because the residual deformations are primarily a function of postyield stiffness.

Fig. 6.28 shows three potential responses of a structure under the action of an earthquake. The key difference between these responses is the inelastic performance, that is, negative, zero, or positive postyield stiffness. At the same lateral drift, unloading stiffnesses are parallel in accordance with the Takeda model (Takeda et al., 1970), where the unloading stiffness K is a function of column first stiffness K_1 and ductility μ (Eq. 6.35).

It is evident from Fig. 6.28 that negative postyield stiffness results in a large residual displacement, which in turn is a disadvantage that should be avoided to quickly recover the structure. This large residual displacement significantly complicates the repair work after an earthquake (Yozo et al., 2005).

$$K = \frac{K_1}{\sqrt{\mu}} \qquad\qquad\qquad (6.35)$$

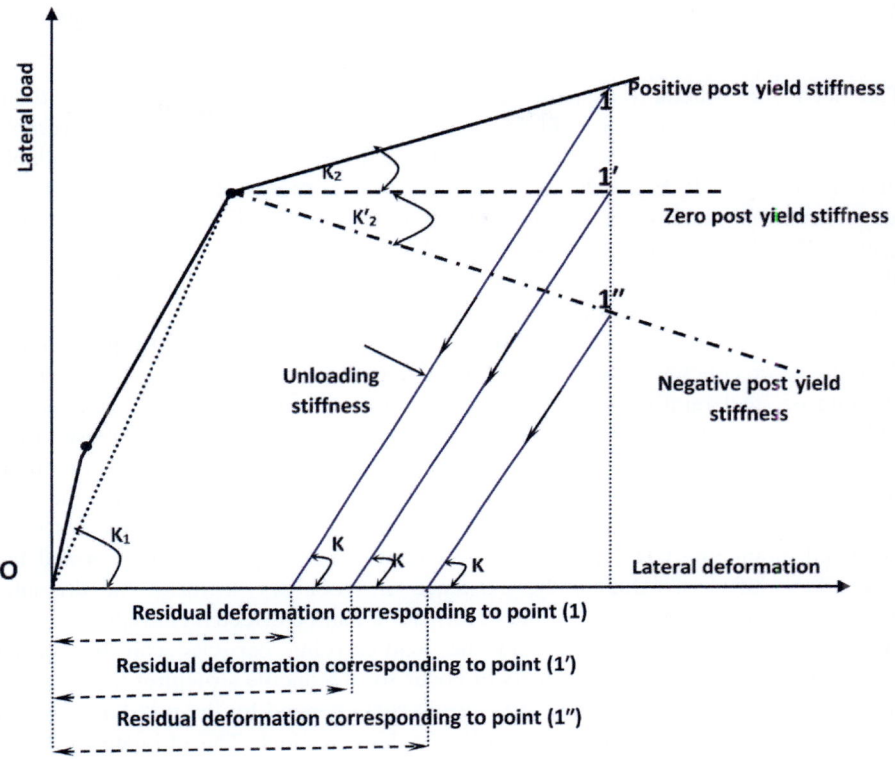

Figure 6.28 Effect of postyield stiffness on the residual deformation (Fahmy et al., 2010).

The 1996 Seismic Design Specifications of Highway Bridges in Japan specifies that the residual displacement should not be greater than 1% of the pier's height (JSCE, 2000), and it provides the following equation for evaluation.

$$\delta_{res} = C_R(\mu_R - 1)(1 - r)\delta_y \tag{6.36}$$

where δ_{res} is the residual displacement of a pier after the earthquake, μ_R is the response ductility factor of piers, $r = (K_2/K_1)$ is the bilinear factor defined as a ratio between K_2 (postyield stiffness) and K_1, C_R = factor depending on the bilinear factor r, and δ_y is the yield displacement.

The equation explicitly verifies that as the ratio r increases, the residual displacement will decrease accordingly, and it can be concluded that the piers with high r values have a higher seismic performance.

Based on that, the following sections address the recoverability of existing and new structures using both postyield stiffness and residual deformations as seismic performance indices, measuring the required recoverability of important bridges and define their limit states under the action of a strong earthquake.

6.11.3   Idealized load—deformation model of FRP-RC damage-controllable structures

The need for structural systems to withstand large earthquake forces without compromising safety has brought the challenge of designing a quickly recoverable structure. Here, the authors propose a mechanical model for a damage-controllable structure using FRP. Fig. 6.29 shows the mechanical model of the proposed structure, where the lateral response proceeds along O-A-B-C-D-E-F. The behavior of a general RC flexural structure whose lateral response is along O-A'-B'-C'-D'-F' is also given for comparison. Before the cracking of the concrete, lines OA and OA' corresponding to both types of structures share similar stiffnesses. The stiffness of the proposed structure, K_1, is slightly greater than that of K_1' of the general RC structure after concrete cracking. The most remarkable difference occurs after the yielding of steel reinforcement, that is, after point C and C'. For the general RC structure, the deformation increases dramatically almost without any increase in the load-carrying capability, that is, along line C'D', no postyield stiffness is demonstrated. However, with the proposed approach, the structure can still carry the load even after the steel reinforcement is yielded and the hardening behavior has been exhibited along line CD. The stiffness K_2 between points C and D is termed the "secondary stiffness." Due to the existence of secondary stiffness, the dramatic increase in deformation and residual deformation can be effectively controlled after the reinforcement is obtained, and the load-carrying capacity can be further improved. Based on the requirements of codes for the ductile structures to withstand strong earthquakes, the proposed structure is characterized by the part DE after the hardening zone, where favorable ductility is demonstrated. The ultimate drift (δ_u) corresponding to point F or F' of the proposed structure and the general RC structure, respectively, is defined for both structures to be at 20% strength decay. The use of 20% strength decay as the failure criterion is consistent with that employed by previous researchers because it is reasonable to accept some strength decay during the seismic response of a structure before it can be considered to have failed (Park and Paulay, 1975).

According to the mechanical behavior shown in Fig. 6.29, the load—deformation of the proposed structure can be divided into four main zones; Zone 1: from point O to B; Zone 2: from point B to D; Zone 3: from point D to E; and Zone 4: after point E. Zone 1 corresponds to a stage of no damage or concrete cracking. Under a small earthquake, the mechanical behavior should be controlled in this zone and the original function of the structure can be maintained without any repair and replacement of elements. Zone 2 corresponds to the hardening behavior after the yielding of steel reinforcements, where a distinct secondary stiffness is demonstrated and the dramatic deformation can be effectively controlled. Under medium or strong earthquakes, the mechanical behavior of the proposed structure should be within Zone 2. Thus damage can be effectively controlled by the secondary stiffness. The original function of the structures can be quickly recovered through repairs after medium or large earthquakes. Zone 3 corresponds to the ductile behavior after hardening, where favorable ductility is demonstrated under a

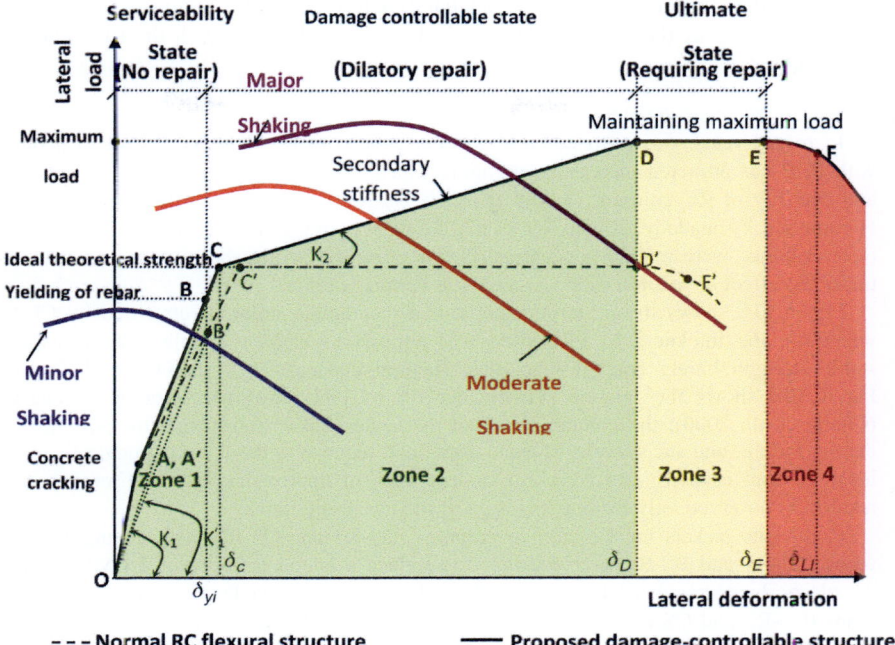

Figure 6.29 Idealized load—deformation behavior of proposed damage-controlled structures (Fahmy, 2010).

large earthquake. The proposed structure can be kept in place for a relatively long time without collapse during large earthquakes, though severe damage may occur. The original function of the structures may be recovered through the replacement of some elements. During a severe earthquake, the mechanical behavior may enter Zone 4 and collapse.

The proposed mechanical model can satisfy the seismic design philosophy that the structure suffers no damage under small earthquakes, exhibits prompt recoverability under medium earthquakes, and does not collapse under large earthquakes.

6.11.4 Enhancing recoverability and controllability of deficient RC columns using of FRP confinement as a retrofitting technique

6.11.4.1 Postyield performance of FRP-confined rectangular and circular columns

This part aims to present experimental results from the literature of several columns after retrofitting with FRP composites. The authors focus on retrofitting schemes that successfully enhanced the performance of the postyielding stage of tested columns through the achievement of a clear positive postyield stiffness and a considerable increase in the column deformability to finally capture a ductile flexural failure mode.

a. *Columns with shear deficiency*

In recent years, owing to certain excellent properties of FRP, retrofitting structures with FRP has attracted increasing attention, especially for improving the shear strength and ductility of RC columns (Seible et al., 1997; Xiao et al., 1999; Chang et al., 2001; Haroun and Elsanadedy, 2005a; Wu et al., 2006). Existing studies on FRP-confined concrete columns were mainly concentrated on altering the brittle shear failure of columns under the effect of seismic action to a ductile flexural failure.

Fig. 6.30A shows the hysteretic loops of a rectangular column retrofitted by CFRP wrapping; the thickness of FRP used was enough to have a ductile column with a stable strength barely equal to the ideal flexural capacity (V_{yi}) (Seible et al., 1997). Fig. 6.30B shows the seismic performance of a CFRP-retrofitted rectangular column (Chang et al., 2000); the column achieved the same goal of retrofitting by exhibiting a ductile failure, and successfully showed a gradual increase in the lateral strength over the ideal flexural capacity until reaching a drift ratio of approximately 3.0; however, this result was observed only for one direction of the hysteretic loops.

Composite jackets for the circular columns (CS-R1 and CS-R2) and rectangular columns (RS-R1 and RS-R5) were designed to induce a lateral pressure of 2.1 MPa within the plastic hinge regions (Haroun and Elsanadedy, 2005a,b). Jackets for the circular columns (CS-R3 and CS-P1) and the rectangular columns (RS-R2, RS-R3, RS-R4, and RS-R6) provided a confinement pressure of 2.1 MPa, and the jacket for CS-R4 provided a lateral pressure of 1.0 MPa (Haroun and Elsanadedy, 2005a,b). In the cases of circular and rectangular columns, entirely wrapping the columns with FRP to give a minimum confinement pressure of 2.1 MPa is the best retrofitting to avoid shear failure and to ensure the appearance of the increasing straight envelop line for the hysteretic response after achieving the ideal flexural capacity.

The retrofit methods investigated in the study of Xiao et al. (1999) utilized multilayer prefabricated composite shells. The shells were fabricated either continuously to form a multilayer roll or cut into individual single-layer cylindrical shells with a longitudinal slit for each layer. Fig. 6.31 shows the envelop curves for the model columns retrofitted with individual and continuous prefabricated composites shell jackets. Excellent performance with significantly increased ductility and stable hysteresis loops are observed. An overstrength of approximately 10% compared with the ideal flexural strength was developed at the peaks of subsequent loading cycles; that is, the existence of postyielding stiffness was observed. Little difference can be seen between the responses of the column with an individual shell jacket and that with a continuous shell jacket, up to a displacement ductility factor of 10 or a drift ratio of approximately 5.30%.

b. *Columns with lap-splice deficiency*

Considerable research efforts are being made to develop and apply retrofitting strategies to upgrade the seismic performance of deficient lap-spliced columns (Seible et al., 1997; Xiao et al., 1999; Chang et al., 2001; Haroun and Elsanadedy, 2005a,b; Harries et al., 2006). Two retrofitted circular columns were tested by Xiao et al. (1999). The circular C2-RT4 column was retrofitted using an FRP layer, and successfully reached the ideal flexural capacity, showing a stable hysteretic response up to a ductility factor of approximately 6. The increase in FRP layers by one layer, for the C3-RT5 column, limited the FRP hoop strain to 0.001 (mm/mm), such that the column gained a gradual

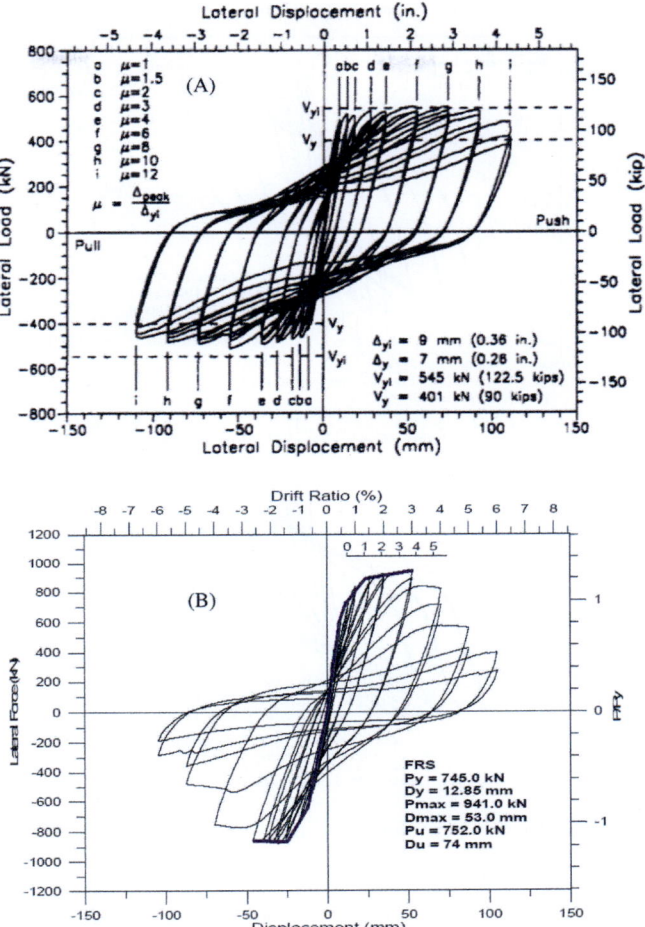

Figure 6.30 Load—displacement for (A) rectangular column (Specimen (a)) (Seible et al., 1997) and (B) rectangular column (Specimen FRS) (Chang et al., 2000).

increase in the lateral resistance over the ideal flexural capacity, which appears to be an increasing straight line up to a ductility factor of approximately 4 (Fig. 6.32).

Seible et al. (1997) tested two circular lap-spliced columns (Fig. 6.33A,B). They reported that the primary confinement jacket thickness of the first column was 20% less than the required design thickness. This design showed a stable hysteretic response over the ideal lateral strength, and the lap-splice debonding occurred at a displacement ductility greater than 5. The second column was wrapped with an FRP jacket of thickness equal to the required design details. This column showed a gradual increase in the lateral strength over the ideal flexural capacity, and the displacement ductility factor before the starter bar rupture was approximately 8.

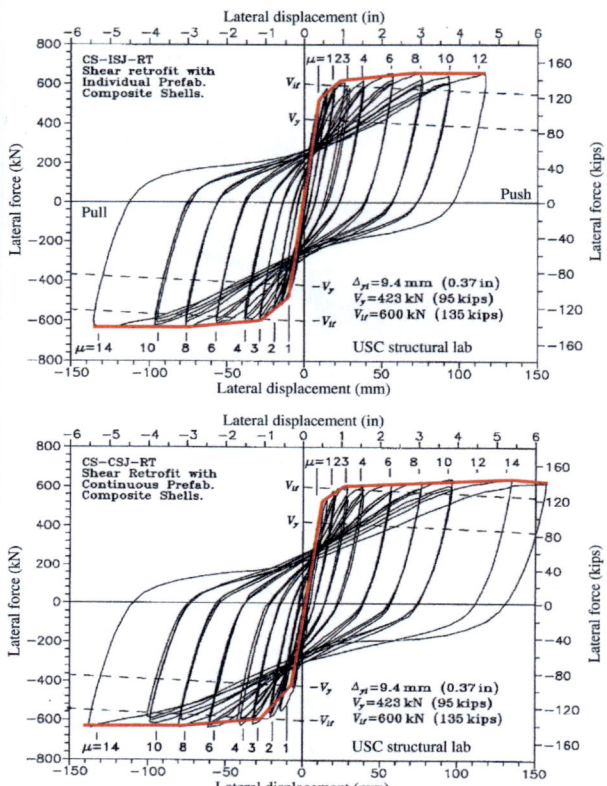

Figure 6.31 Load−displacement envelope for circular columns (CS-ISJ-RT and CS-CSJ-RT) (Xiao et al., 1999).

Haroun and Elsanadedy (2005b) tested columns of circular and square cross-sections. Of the FRP-retrofitted circular columns, the two CF-R1 and CF-R2 columns successfully achieved the theoretical strength, but the applied FRP retrofitting was insufficient to activate postyield stiffness. Meanwhile, the design considered for the remainder of the retrofitted circular columns CF-R3, CF-R4, and CF-R5 (Fig. 6.34A) behaved similarly and demonstrated a significant improvement in their cyclic performance with a gradual increase in the lateral capacity over the theoretical flexural strength until a ductility of approximately 5, which demonstrated the appearance of postyield stiffness. The hysteric response of all square-jacketed columns (one of these columns was a quasicircular section with continuous confinement) had very limited improvement in clamping on the lap-splice region (Fig. 6.34B).

c. *Columns with plastic hinge (flexural) deficiency*

Researchers (Seible et al., 1997; Fahmy, 2010) have applied retrofitting strategies to upgrade the seismic performance of deficient plastic hinge regions of columns. The design scenario considered by Seible et al. (1997) is to calculate the appropriate composite jacket thickness for upgrading the limited inelastic

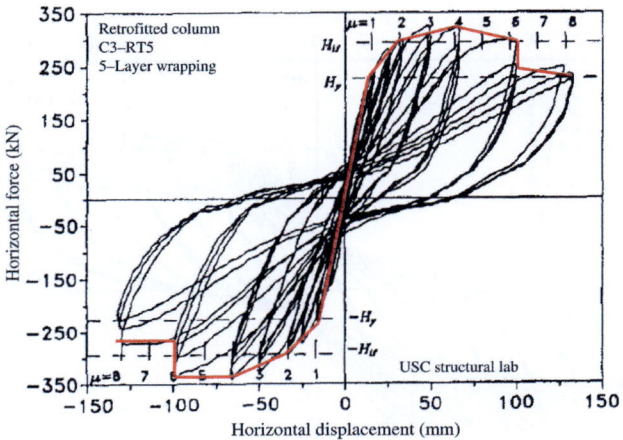

Figure 6.32 Load–displacement envelope for circular lap-spliced column (Specimen C3-RT5) (Xiao et al., 1997).

deformation at the plastic hinge region under the effect of seismic action; this approach was also appropriate for inducing a second stiffness after the elastic one until a displacement ductility greater than 7; (Fig. 6.35). Furthermore, from the viewpoint of the existence of postyield stiffness and the enhancement of the column deformability in the inelastic stage, load–displacement envelopes of a database available in the literature for retrofitted rectangular and circular columns with weak confinement in the plastic zones were evaluated by Fahmy (2010).

6.1.4.2 Residual inclination and limit states of FRP-confined RC columns

The aim of retrofitting columns with FRP is to achieve a more ductile member than the as-built one. However, from the standpoint of recoverability, ductile columns are not necessarily recoverable columns. The global performance of the structure is governed by the performance of individual members: it is necessary to control the behavior of structural elements according to their required performance to achieve a recoverable structure. Therefore scrutinizing the inelastic performance of tested retrofitted columns is necessary to define the recoverable state of FRP-retrofitted columns. Measuring the seismic performance of FRP-retrofitted columns based on postyield stiffness and residual deformation as metrics for the required recoverability shows that many columns successfully achieved the secondary stiffness; however, the residual deformation of these columns corresponding to the endpoint of the secondary stiffness can be more than 1.0%, as shown in Fig. 6.33. Based on these results, the limit state for recoverable columns should not be the endpoint of the postyield stiffness, and a redefinition of the endpoint of the recoverable state using the nonlinear pushover test results becomes necessary. The relationship

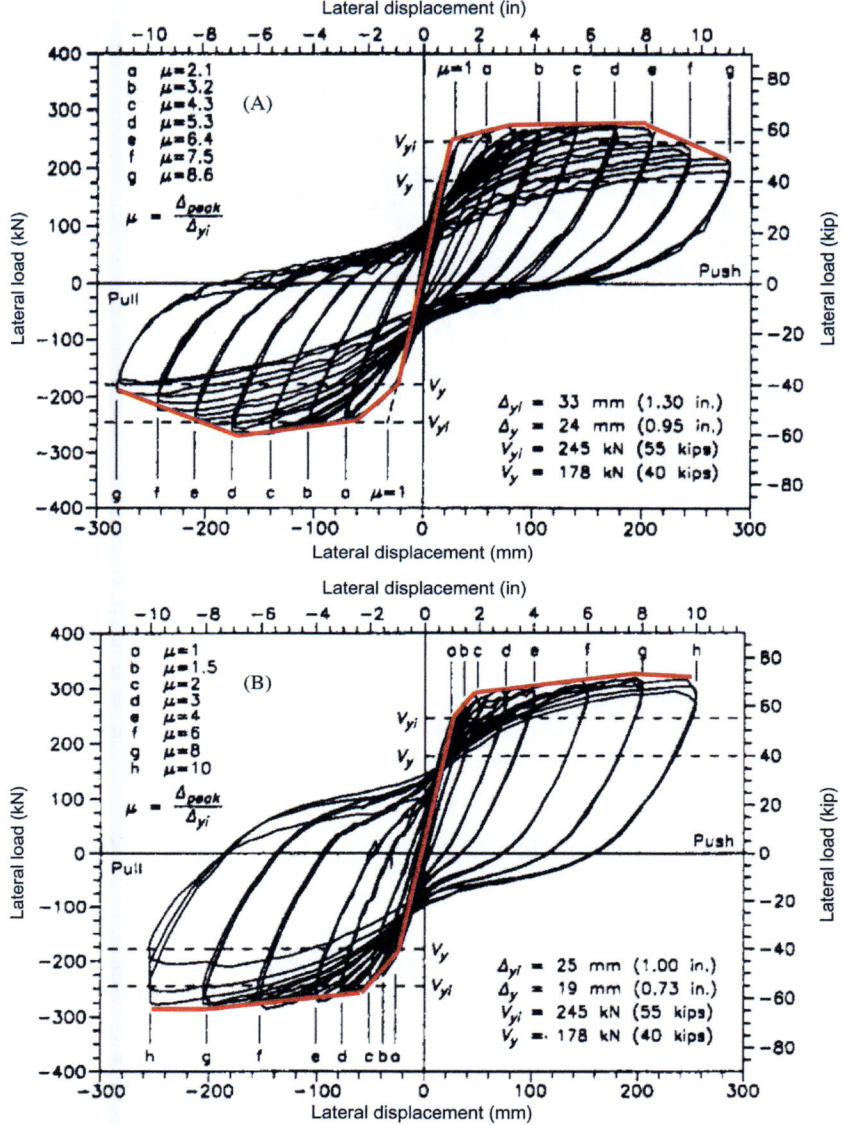

Figure 6.33 Load–displacement envelope for (A) circular column (test 1) (Seible et al., 1997) and (B) for circular column (test 2) (Seible et al., 1997).

between the column drift ratios and the corresponding column residual drift ratios at the column theoretical strength, recoverable limit, and maximum strength of these columns are plotted in Fig. 6.36. It is clear that a residual drift ratio of 1% does not correspond to a definite column drift ratio because many other factors

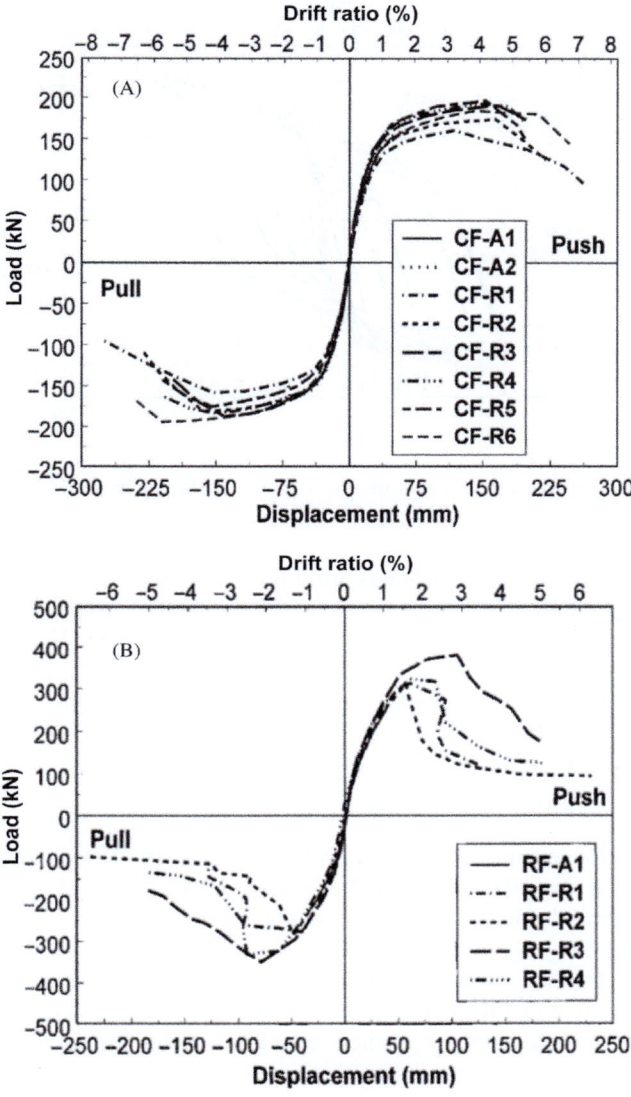

Figure 6.34 Load—displacement envelopes for (A) circular lap-spliced columns (Haroun and Elsanadedy, 2005b) and (B) square lap-spliced columns (Haroun and Elsanadedy, 2005b).

affect the performance of these columns. However, it is interesting to note that there is a range of the drift ratio of 2%−3.5%, within which the recoverability limit state should be checked by evaluating the residual inclination value within this range to define the endpoint of the recoverable state. Furthermore, an equation is proposed to define the column residual deformations within this range. In conclusion, the

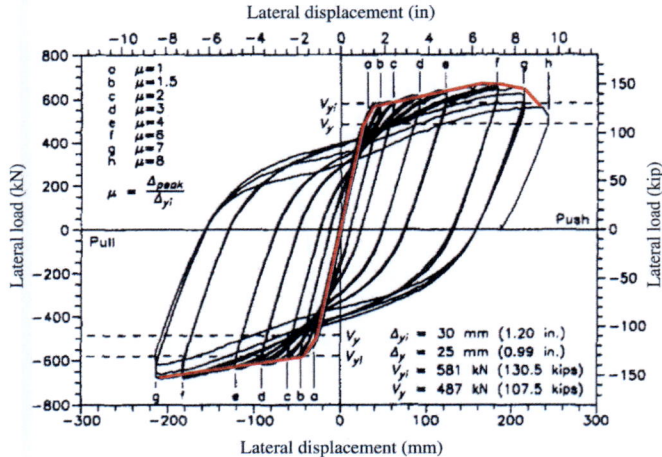

Figure 6.35 Load−displacement envelope for rectangular (Seible et al., 1997).

authors recommend three limit states for FRP-RC bridge columns, as shown in Fig. 6.36. The first state is the state of pure recoverability, whose endpoint corresponds to a column drift ratio of 2%, as shown in Fig. 6.36. Here, the residual deformation of all the represented columns is below the recoverability limit. The second state is the state of checking the recoverability limit, which corresponds to column drift ratios between 2% and 3.5%. The third state is the irrecoverable state, where the residual deformations exceed the recoverability limit.

a. *Recoverability of FRP-retrofitted RC bridge bent*

The design of a seismic retrofit of a bridge bent using CFRP composite was presented in detail by Pantelides and Gergely (2002). Based on the lateral load−displacement hysteretic response of the as-built and FRP-retrofitted bents (Pantelides and Gergely, 2002), a comparison between the experimental performances of the structure in the two configurations is given in Fig. 6.37. It is clear that externally applied CFRP composites did not significantly affect the initial stiffness; this is an advantage of the applied retrofitting technique because the number of fibers used will not attract additional force under seismic action (Saiidi et al., 2009). In addition, it is evident that the applied retrofitting technique was able to increase the displacement ductility significantly. That is, the bent-cap column joint shear capacity was enhanced, and the overall damage was controlled. Fig. 6.38 shows the relationship between the bent lateral displacement and residual deformation for both tested bents. Superimposing the recoverability limit on this figure confirms the findings of Fahmy et al. (2009); that is, the end of the recoverable zone of the FRP-retrofitted system falls between drift ratios of 2% and 3.5%.

The ends of the recoverable zones of the CFRP-retrofitted bent are 2.3% and 2.8% in the push and pull directions of loading, respectively (see Figs. 6.37 and 6.38). In

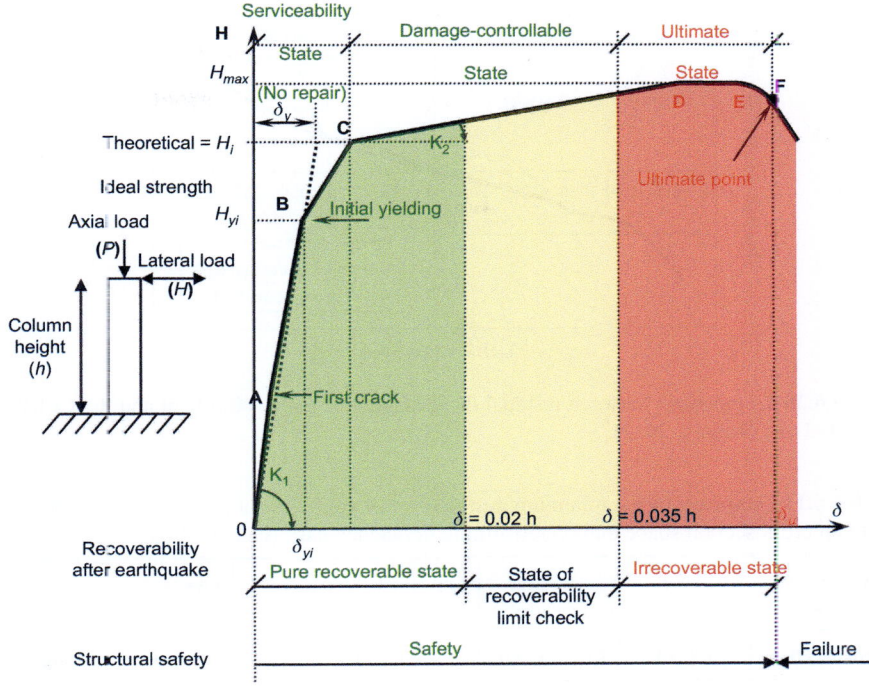

Figure 6.36 Recoverable and irrecoverable states of damage-controlled FRP-RC columns (Fahmy, 2010).

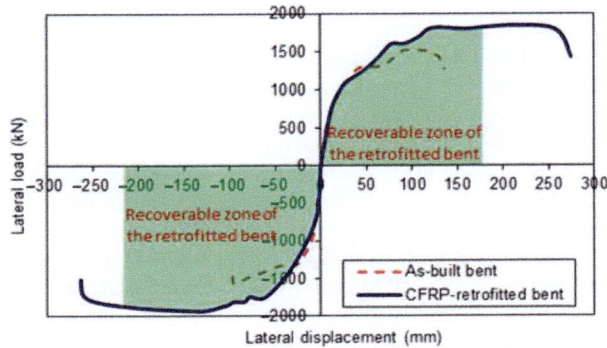

Figure 6.37 Envelope curves of the lateral load–displacement response of both tested bents (Fahmy, 2013).

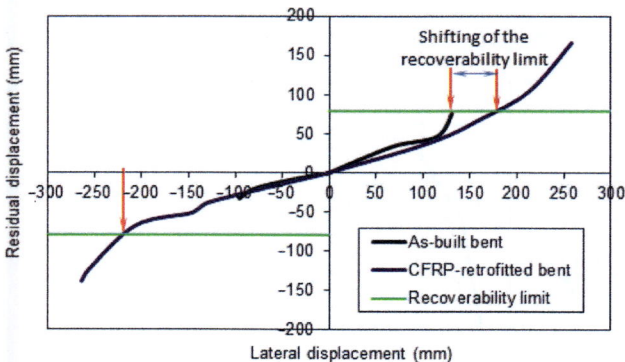

Figure 6.38 Comparison between residual displacement of the as-built bent and the CFRP retrofitted one (Fahmy, 2013).

Table 6.11 Summary of the minimum design level of confinement ratio based on both column cross-section shape and preretrofitting deficiency case (Fahmy, 2010).

Column deficiency	Shear		Lap splice		Flexural	
Cross-section	**Circular**	**Rectangular**	**Circular**	**Rectangular**	**Circular**	**Rectangular**
f_l/f'_{co}	0.105	0.066	0.216	0.247	0.112	0.066

addition, it is interesting to note that the applied technique successfully shifted the recoverability limit to a 2.3% drift, which was 1.68% in the as-built bent (Fig. 6.38).

a. *Design provisions satisfying required recoverability*
  i. *FRP design requirements for plastic hinge confinement*
    It is widely accepted that confining concrete columns with FRP composites acts to provide support to longitudinal reinforcement and restrains the lateral expansion of the core concrete, which in turn enables higher strains sustained by the compression concrete before failure. This means that there is a high probability of steel hardening, and consequently an increase in the lateral load capacity envelope of the columns. Coupled with this, confining the core concrete enables increasing the concrete compression strength, which is a further reason for the existence of the postyield stiffness. The region of the column over which enhanced confinement should extend is designated as the plastic end region. Fahmy et al. (2010) used a large database for FRP-retrofitted RC columns that successfully realized the enhancement in both the flexural deformability and the lateral strength, and they suggested the confinement design requirements that are necessary to improve the flexural deficiency of this region; see Table 6.11.
    The confinement pressure in Table 6.10 is defined as

$$f_l = 0.5\rho_f f_f \tag{6.37}$$

$$\rho_f = 4t_f/D \text{ for circular column} \tag{6.38}$$

$$\rho_f = 2t_f(b + h)/(bh) \text{ for rectangular column} \tag{6.39}$$

where f_l is the confinement strength of FRP, f_f is the ultimate tensile strength of FRP, t_f is the total thickness of the FRP jacket, and ρ_f is the volumetric ratio of FRP to concrete.

ii. *FRP design for column with shear deficiency*

One of the key shear force transfer mechanisms is the aggregate interlock in inclined cracks, which can be controlled by horizontal FRP wrapping to limit the column dilation in the loading direction to the experimentally determined dilation strain of $\varepsilon_d < 0.004$ (mm/mm) (Seible et al., 1997; Xiao et al., 1999). For circular and rectangular columns, composite jackets should be designed to provide a minimum confinement pressure of 2.1 MPa for the entire height without exceeding a jacket strain of 0.004 mm/mm. The required jacket thickness of the rectangular columns should be increased by a factor of 1.5, as in the design considered by Haroun and Elsanadedy (2005a,b). These two assumptions for the dilation strain and the confinement pressure value are proposed to be suitable for retrofitted columns to ensure the existence of positive postyield stiffness.

iii. *FRP design for columns with lap-spliced reinforcement*

Another objective of external FRP jacket retrofitting is to improve the behavior of the splice, by improving the distribution of bond stress along the splice and effectively increasing the average bond stress. Additionally, for a recoverable structure, it is necessary to postpone the onset of splitting and to reduce the severity of the subsequent deterioration. By using a database of columns with lap-spliced reinforcement that successfully achieved the aim of ductile-recoverable structures, appropriate design conditions are found.

For circular columns, a hoop strain of 1,000 $\mu\varepsilon$ is appropriate for the design of the composite jacket to ensure lap-splice clamping (Seible et al., 1997; Haroun and Elsanadedy, 2005a,b). A lateral clamping pressure $f_{l(lapsplice)}$ over the lap splice L_s was taken to be 2.0 MPa for the samples tested by Haroun and Elsanadedy (2005a, b) and by Seible et al., 1997) based on the following equation:

$$f_{l(lapsplice)} = \frac{A_s f_y}{\left[\frac{p}{2n} + 2(d_b + cc)\right]L_s} \tag{6.40}$$

where p is the perimeter line in the column cross-section along the lap-spliced bar location, n is the number of spliced bars along p, A_s is the area of the main column reinforcing bar, f_y is the yield strength of the longitudinal steel, and cc is the concrete cover to the main column reinforcement with diameter d_b. However, the author recommends the supposed criterion by Harajli (2008) for seismic FRP bond strengthening, where a minimum development stress $f_s = 1.25f_y$ is used instead of f_y as in Eq. (6.41), to obtain the final following form:

Table 6.12 Ratio between lateral drift corresponding to recoverability limit and column drift by the end of postyield stiffness (Fahmy, 2010).

Column deficiency	Shear		Lap splice		Flexural	
Cross-section	Circular	Rectangular	Circular	Rectangular	Circular	Rectangular
$\varphi = (\delta_{\delta res=0.01L}/ \delta_D)$	0.53	0.89	0.591	0.672	0.565	0.686

$$f_{l(lapsplice)} = \frac{1.25 A_s f_y}{\left[\frac{P}{2n} + 2(d_b + cc)\right]L_s} \geq 2.0 \text{ MPa} \tag{6.41}$$

In order to prevent sliding in the lap splices length for rectangular columns, the ultimate strain of FRP should be limited to 0.0015 (Chang et al., 2004; Haroun and Elsanadedy, 2005a,b). The lateral clamping pressure over the lap splice could also be determined from Eq. (6.41). Steel plates should also be used in the wrapper region of columns to improve the lateral strength and displacement ductility. The dimensions of the steel plates are given by Chang et al. (2005) as follows:

A. Thickness of the steel plates (t_1 and t_2): $t_1 \geq 0.01\ b$, $t_2 \geq 0.01\ h$,where b and h are width and depth of column, respectively, and t_1 and t_2 are thicknesses of the steel plates in the short and long directions of the rectangular column, respectively.
B. Width of the steel plates (S_{w1} and S_{w2}): $S_{w1} \geq (b - 10)$, $S_{w2} \geq (h - 10)$ (mm),where S_{w1} and S_{w2} are the widths of the steel plates in the short and long directions of the rectangular column, respectively.
C. Height of the steel plate (S_h): $S_h \geq (L_s + 10)$ (cm)
   iv. *Effect of column deficiency and cross-section shape on recoverability limit*

Fahmy et al. (2009) and Fahmy et al. (2010) addressed that the recoverability limit of the examined columns, which were confined with external FRP jackets and were able to achieve postyield stiffness, lies between drift ratios of 2% and 3.5%. In this section, the impact of both column deficiency and cross-section shape on the end of the recoverable state of FRP-retrofitted columns is considered. The deformation corresponding to a residual deformation equal to 1% of the column height is experimentally determined from the hysteretic curves of 39 scaled-model tests, which are listed in Table 6.12.

A ratio (φ) between the lateral column deformation corresponding to the recoverability limit ($\delta_{\delta res=0.01L}$) and the deformation by the end of the postyield stiffness (δ_D) is determined based on the test results of 39 columns because this ratio has considerable importance in the FRP-jacketing design. With respect to both column deficiency and cross-section shape, the averages of these values are listed in Table 6.12

b. *Conclusions*

This chapter provides design engineers and the research community with an overview of different topics addressing the confinement of existing deficient RC columns with external FRP sheets. Several key conclusions could be drawn as follows:

i. External FRP confinement of existing deficient RC columns is a superior retrofitting tool for several reasons: (1) external confinement could ensure a maximum lifespan compared with the other traditional retrofitting techniques, (2) external FRP jackets improve the concrete behavior through controlled increases in both the strength and the strain capacity based on the amount and type of FRP used, and (3) enhanced concrete behavior could be guaranteed for low- to high-strength concrete of different cross-section shapes.

ii. External FRP confinement is an indirect source for enhancing the behavior of strengthened columns, and retrofitted columns achieve reasonable improvements in their axial and lateral behaviors.

iii. External FRP confinement can be adopted as a seismic proof tool providing required ductility and an enhancement in the column lateral strength. It can provide an enhancement in the recoverability of the retrofitted columns under lateral loading up to drift ratios of 2%−3%.

c. *Future research*

During the last three decades, the research community has successfully conducted comprehensive studies on the confinement of concrete with FRP composites. We presently have a comprehensive database of FRP-confined concrete covering a wide range of different influential design parameters, such as the grade of concrete, cross-section shape, concrete dimensions, loading conditions (static/dynamic, monotonic/cyclic, axial/eccentric, etc.), and type and amount of FRP materials. Many design-oriented models have been proposed to predict the stress−strain behavior of FRP-confined concrete. Several studies have attempted to provide design engineers with a rational design guideline for the external confinement of deficient RC columns (buildings and bridges) under the effect of seismic action, and certain serious investigations examined the real enhancement in the performance of a large scale or even a real scale of existing deficient structures before and after confinement of the weak zones with FRP sheets. However, several gaps should be theoretically and experimentally addressed:

i. Rather few experimental studies are available for rectangular columns confined with FRP under the effect of eccentric axial loads.

ii. Eccentricity-based stress−strain models should be a focus of future research, because the available stress−strain laws developed for axially loaded concrete may not be applicable for evaluating the performance of concrete locating in compressed zones from the action of axial and bending loading (Wu and Jiang, 2013).

iii. Analysis-oriented models should be comprehensively enhanced as many studies proved that the accuracy of the available design-oriented models is higher than that of the analysis-oriented models.

iv. A unified advanced design guideline considering a certain seismic performance behavior should be developed and evaluated in light of experimental studies of large-scale samples focusing on the behavior of RC bridge columns, which, to the best knowledge of the author, has not yet been examined.

References

ACI(440.2-02), 2002. Building Code Requirement for Structural Concrete (ACI 440.2-02), in ACI Committee 318. American Concrete Institute, Farmington Hills, MI.

ACI(440.2-08), 2008. Building Code Requirement for Structural Concrete (ACI 440.2R-08) and Commentary (440.2R-08), in ACI Committee 318. American Concrete Institute, Farmington Hills, MI.

ACI440R-96, 1996. State-of-the-Art Report on Fiber Reinforced Plastic (FRP) Reinforcement for Concrete Structures. American Concrete Institute, Farmington Hills, MI.

Aire, C., Gettu, R., Casas, J.R., Marques, S., Marques, D., 2010. Concrete laterally confined with fiber-reinforced polymers (FRP): experimental study and theoretical model. Mater. Constr. 60 (297), 19−31.

Al-Salloum, Y., Siddiqui, N., 2009. Compressive strength prediction model for FRPconfined concrete. In: Proc. 9th Int Symp on Fiber Reinforced Polymer Reinforcement for Concrete Structures.

Al-Tersawy, S.H., Hodhod, O.A., Hefnawy, A.A., 2007. Reliability and code calibration of RC short columns confined with CFRP wraps. In: Proc. 8th Int. Symp. on Fiber Reinforced Polymer Reinforcement for Concrete Structures. Univ. of Patras, Patras, Greece.

Albanesi, T., Nuti, C., Vanzi, I., 2007. Closed form constitutive relationship for concrete filled FRP tubes under compression. Constr. Build. Mater. 21 (2), 409−427.

Almusallam, T.H., 2007. Behavior of normal and high-strength concrete cylinders confined with E-glass/epoxy composite laminates. Compos. Part B: Eng. 38 (5−6), 629−639.

Aschheim, M., Black, E.F., 2000. Yield point spectra for seismic design and rehabilitation. Earthq. Spectra 16 (2), 317−336.

Bae, S.W., Belarbi, A., 2010. Effects of various environmental conditions on RC columns wrapped with FRP sheets. J. Reinf. Plast. Compos. 29 (2), 290−309.

Ballinger, C., Maeda, T., Hoshijima, T., 1993. Strengthening of reinforced concrete chimneys, columns and beams with carbon fiber reinforced plastics. Spec. Publ. 138, 233−248.

Becque, J., Patnaik, A.K., Rizkalla, S.H., 2003. Analytical models for concrete confined with FRP tubes. J. Compos. Constr. 7 (1), 31−38.

Beddiar, A., Zitoune, R., Collombet, F., Grunevald, Y.H., Abadlia, M.T., Bourahla, N., 2015. Compressive behaviour of concrete elements confined with GFRP-prefabricated bonded shells. Eur. J. Environ. Civ. Eng. 19 (1), 65−80.

Benzaid, R., Chikh, N.E., Mesbah, H., 2008. Behaviour of square concrete column confined with GFRP composite warp. J. Civ. Eng. Manag. 14 (2), 115−120.

Benzaid, R., Mesbah, H., Chikh, N.E., 2010. FRP-confined concrete cylinders: axial compression experiments and strength model. J. Reinf. Plast. Compos. 29 (16), 2469−2488.

Berthet, J.F., Ferrier, E., Hamelin, P., 2005. Compressive behavior of concrete externally confined by composite jackets. Part A: Exp. Study. Constr. Build. Mater. 19 (3), 223−232.

Berthet, J.F., Ferrier, E., Hamelin, P., 2006. Compressive behavior of concrete externally confined by composite jackets. Constr. Build. Mater. 20 (5), 338−347.

Binici, B., 2005. An analytical model for stress−strain behavior of confined concrete. Eng. Struct. 27 (7), 1040−1051.

Binici, B., 2008. Design of FRPs in circular bridge column retrofits for ductility enhancement. Eng. Struct. 30 (3), 766—776.

Bisby, L.A., Dent, A.J.S., Green, M.F., 2005. Comparison of confinement models for fiber-reinforced polymer-wrapped concrete. ACI Struct. J. 102 (1).

Breña, S.F., Schlick, B.M., 2007. Hysteretic behavior of bridge columns with FRP-jacketed lap splices designed for moderate ductility enhancement. J. Compos. Constr. 11 (6), 565—574.

Campione, G., Miraglia, N., 2003. Strength and strain capacities of concrete compression members reinforced with FRP. Cem. Concr. Compos. 25 (1), 31—41.

Cevik, A., 2011. Modeling strength enhancement of FRP confined concrete cylinders using soft computing. Expert. Syst. Appl. 38 (5), 5662—5673.

Chaallal, O., Shahawy, M., Hassan, M., 2003. Performance of axially loaded short rectangular columns strengthened with carbon fiber-reinforced polymer wrapping. J. Compos. Constr. 7 (3), 200—208.

Chai, Y.H., Priestley, M.N., Seible, F., 1991. Seismic retrofit of circular bridge columns for enhanced flexural performance. ACI Struct. J. 88 (5), 572—584.

Chang, K.C., Chang, S.B., Liu, K.Y., Wang, P.H., 2005. Analysis and design of rectangular RC columns with lap-spliced longitudinal reinforcement. NCREE, Res. Prog. Accomplish 29—32. Available from: http://www.ncree.org/.

Chang, K.C., Chung, L.L., Lee, B.J., Tsai, K.C., Hwang, J.S., Hwang, S.J., 2000. Seismic Retrofit Study of RC Bridge Columns. International Training Program for Seismic Design of Building Structures.

Chang, K.C., Liu, K.Y., Chang, S.B., 2001. Seismic retrofit study of RC rectangular bridge columns lap-spliced at the plastic hinge zone. In: Proc. of the International Conference on FRP composites in Civil Engineering.

Chang, K.C., Chang, S.P., Liu, K.Y., 2004. Seismic retrofit study of rectangular RC columns lap spliced at plastic hinge zone. In: Proc. of 16th KKCNN Symposium on Civil Engineering, pp. 221—227.

Chastre, C., Silva, M.A.G., 2010. Monotonic axial behavior and modeling of RC circular columns confined with CFRP. Eng. Struct. 32 (8), 2268—2277.

Chines, C.G., 2010. Technical Code for Infrastructure Application of FRP Composites. GB-50608China.

Christopoulos, C., Pampanin, S., 2004. Towards Performance-Based Design of MDOF Structures with Explicit Consideration of Residual Deformations.

Ciupala, M., Pilakoutas, K., Mortazavi, A., 2007. Effectiveness of FRP composites in confined concrete. In: Proc. 8th Int. Symp. on Fiber Reinforced Polymer Reinforcement for Concrete Structures.

CNR-DT(200/2004), 2004. Guide for the Design and Construction of Externally Bonded FRP Systems for Strengthening Concrete Structures. Reported by ACI Committee. 440(2002).

Cole, C., Belarbi, A., 2001. Confinement characteristics of rectangular FRP-jacketed RC columns. In: Proc. of the Fifth International Symposium on Fiber Reinforced Polymer for Reinforced Concrete Structures (FRPRCS-5), pp. 823—832.

Concrete Society Working Party, 2013. Design guidance for strengthening concrete structures using fibre composite materials. TR 55 (Concrete Society Technical Rep. 55). Concrete Society, Crowthorne.

Darwin, D., Pecknold, D.A.W., 1974. Inelastic Model for Cyclic Biaxial Loading of Reinforced Concrete. Report on Research Project Sponsored by the NSF, University of Illinios, UILU-ENG-74-2018, 169.

De Lorenzis, L., Tepfers, R., 2003. Comparative study of models on confinement of concrete cylinders with fiber-reinforced polymer composites. J. Compos. Constr. 7 (3), 219−237.

De Luca, A., Nardone, F., Matta, F., Nanni, A., Lignola, G.P., Prota, A., 2010. Structural evaluation of full-scale FRP-confined reinforced concrete columns. J. Compos. Constr. 15 (1), 112−123.

Demers, M., Neale, K.W., 1999. Confinement of reinforced concrete columns with fiber-reinforced composite sheets − an experimental study. Can. J. Civ. Eng. 26 (2), 226−241.

Doran, B., Koksal, H.O., Turgay, T., 2009. Nonlinear finite element modeling of rectangular/square concrete columns confined with FRP. Mater. Des. 30 (8), 3066−3075.

Eid, R., Paultre, P., 2007. Plasticity-based model for circular concrete columns confined with fiber-composite sheets. Eng. Struct. 29 (12), 3301−3311.

Eid, R., Dancygier, A.N., Paultre, P., 2007. Elastoplastic confinement model for circular concrete columns. J. Struct. Eng. 133 (12), 1821−1831.

Elsanadedy, H.M., 2002. Seismic Performance and Analysis of Ductile Composite-Jacketed Reinforced Concrete Bridge Columns. University of California, Irvine.

Fahmy, M.F., 2010. Enhancing Recoverability and Controllability of Reinforced Concrete Bridge Frame Columns using FRP Composites. Ph.D. Thesis, Ibaraki Univ., Hitachi, Japan.

Fahmy, M.F.M., 2013. Preventing failure in advanced composites, Chapter 21, advanced fiber-reinforced polymer (FRP) composites to strengthen structures vulnerable to seismic damage. In: Bai, Jiping (Ed.), Advanced Composites in Civil Engineering. Woodhead Publishing Limited, United Kingdom, pp. 511−551.

Fahmy, M.F., Farghal, O.A., 2016. Eccentricity-based design-oriented model of fiber-reinforced polymer-confined concrete for evaluation of load-carrying capacity of reinforced concrete rectangular columns. J. Reinf. Plast. Compos. 35 (23), 1734−1758.

Fahmy, M.F.M., Wu, Z., 2010. Evaluating and proposing models of circular concrete columns confined with different FRP composites. Compos. Part B: Eng. 41 (3), 199−213.

Fahmy, M.F.M., Zhishen, W., Gang, W., 2009. Seismic Performance Assessment of Damage-Controlled FRP-Retrofitted RC Bridge Columns Using Residual Deformations.

Fahmy, M.F.M., Wu, Z., Wu, G., 2010. Post-earthquake recoverability of existing RC bridge piers retrofitted with FRP composites. Constr. Build. Mater. 24 (6), 980−998.

Fahmy, M.F., Ismail, A.M., Wu, Z., 2017. Numerical study on the applicability of design-oriented models of FRP-confined concrete for predicting the cyclic response of circular FRP-jacketed RC columns. J. Compos. Constr. 21 (5), 04017017.

Fakharifar, M., Chen, G., Sneed, L., Dalvand, A., 2015. Seismic performance of post-mainshock FRP/steel repaired RC bridge columns subjected to aftershocks. Compos. Part B: Eng. 72, 183−198.

Fam, A.Z., Rizkalla, S.H., 2001. Confinement model for axially loaded concrete confined by circular fiber-reinforced polymer tubes. ACI Struct. J. 98 (4), 451−461.

Fardis, M.N., Khalili, H.H., 1982. FRP-encased concrete as a structural material. Mag. Concr. Res. 34 (121), 191−202.

Faustino, P., and Chastre, C. (2015). Analysis of load-strain models for RC square columns confined with CFRP. Compos. Part B: Engineering.

Faustino, P., Chastre, C., Paula, R., 2014. Design model for square RC columns under compression confined with CFRP. Compos. Part B: Eng. 57 (0), 187−198.

Feng, P., Cheng, S., Bai, Y., Ye, L., 2015. Mechanical behavior of concrete-filled square steel tube with FRP-confined concrete core subjected to axial compression. Compos. Struct. 123, 312−324.

Ferracuti, B., Savoia, M., 2005. Cyclic behaviour of FRP-wrapped columns under axial and flexural loadings. In: Proc. of the International Conference on Fracture, IOS Press, 5696.

FIB Bulletin 14, 2001. Externally bonded FRP reinforcement for RC structures. Bulletin 14, 138.

FIB Bulletin 24, 2003. Seismic assessment of reinforced concrete buildings. Bulletin 14, 138.

Gallardo-Zafra, R., Kawashima, K., 2009. Analysis of carbon fiber sheet-retrofitted RC bridge columns under lateral cyclic loading. J. Earthq. Eng. 13 (2), 129−154.

Gambarelli, S., Nisticò, N., Ožbolt, J., 2014. Numerical analysis of compressed concrete columns confined with CFRP: microplane-based approach. Compos. Part B: Eng. 67 (0), 303−312.

Gerstle, K.H., 1981. Simple formulation of triaxial concrete behavior. J. Proc. 78 (5), 382−387.

Ghorbi, E., Soltani, M., Maekawa, K., 2013. Development of a compressive constitutive model for FRP-confined concrete elements. Compos. Part B: Eng. 45 (1), 504−517.

Ghosh, K.K., Sheikh, S.A., 2007. Seismic upgrade with carbon fiber-reinforced polymer of columns containing lap-spliced reinforcing bars. ACI Struct. J. 104 (2), 227.

Girgin, Z.C., 2009. Modified failure criterion to predict ultimate strength of circular columns confined by different materials. ACI Struct. J. 106 (06).

Gu, D.-S., Wu, G., Wu, Z.-S., Wu, Y.-F., 2010. Confinement effectiveness of FRP in retrofitting circular concrete columns under simulated seismic load. J. Compos. Constr. 14 (5), 531−540.

Guralnick, S.A., Gunawan, L., 2006. Strengthening of reinforced concrete bridge columns with FRP wrap. Pract. Period. Struct. Des. Constr. 11 (4), 218−228.

Hadi, M.N.S., 2006. Comparative study of eccentrically loaded FRP wrapped columns. Compos. Struct. 74 (2), 127−135.

Hadi, M.N.S., Widiarsa, I.B.R., 2012. Axial and flexural performance of square RC columns wrapped with CFRP under eccentric loading. J. Compos. Constr. 16 (6), 640−649.

Hag-Elsafi, O., Kunin, J., Alampalli, S., 2003. In-service Evaluation of a Concrete Bridge FRP Strengthening System. Special Rep. No. FHWA/NY/SR-03/139, New York State Department of Transportation, Albany, NY.

Harajli, M.H., 2005. Behavior of gravity load-designed rectangular concrete columns confined with fiber reinforced polymer sheets. J. Compos. Constr. 9 (1), 4−14.

Harajli, M.H., 2006. Axial stress−strain relationship for FRP confined circular and rectangular concrete columns. Cem. Concr. Compos. 28 (10), 938−948.

Harajli, M.H., 2008. Seismic behavior of RC columns with bond-critical regions: criteria for bond strengthening using external FRP Jackets. J. Compos. Constr. 12 (1), 69−79.

Harajli, M.H., Hantouche, E., Soudki, K., 2006. Stress−strain model for fiber-reinforced polymer jacketed concrete columns. ACI Struct. J. 103 (5).

Haroun, M.A., Elsanadedy, H.M., 2005a. Behavior of cyclically loaded squat reinforced concrete bridge columns upgraded with advanced composite-material jackets. J. Bridge Eng. 10 (6), 741−748.

Haroun, M.A., Elsanadedy, H.M., 2005b. Fiber-reinforced plastic jackets for ductility enhancement of reinforced concrete bridge columns with poor lap-splice detailing. J. Bridge Eng. 10 (6), 749−757.

Harries, K.A., Kharel, G., 2003. Experimental investigation of the behavior of variably confined concrete. Cem. Concr. Res. 33 (6), 873−880.

Harries, K.A., Ricles, J.R., Pessiki, S., Sause, R., 2006. Seismic retrofit of lap splices in non-ductile square columns using carbon fiber-reinforced jackets. ACI Struct. J. 103 (6), 874.

Hollaway, L.C., Head, P.R., 2001. Advanced Polymer Composites and Polymers in the Civil Infrastructure. Elsevier Science Ltd, Oxford.

Hoshikuma, J., Kawashima, K., Nagaya, K., Taylor, A.W., 1997. Stress—strain model for confined reinforced concrete in bridge piers. J. Struct. Eng. 123 (5), 624—633.

Hu, B., 2013. An improved criterion for sufficiently/insufficiently FRP-confined concrete derived from ultimate axial stress. Eng. Struct. 46 (0), 431—446.

Hu, H., Seracino, R., 2014. Analytical model for FRP-and-steel-confined circular concrete columns in compression. J. Compos. Constr. 18 (3), A4013012.

Ilki, A., Kumbasar, N., Koc, V., 2002. Strength and deformability of low strength concrete confined by carbon fiber composite sheets. In: Proc. of the Pavement Mechanics Symposium at the 15th ASCE Engineering Mechanics Conference.

Ilki, A., Kumbasar, N., Koc, V., 2004. Low strength concrete members externally confined with FRP sheets. Struct. Eng. Mech. 18, 167—194.

Ilki, A., Peker, O., Karamuk, E., Demir, C., Kumbasar, N., 2006. Axial behavior of RC columns retrofitted with FRP composites. Advances in Earthquake Engineering for Urban Risk Reduction. Springer, pp. 301—316.

Ilki, A., Peker, O., Karamuk, E., Demir, C., Kumbasar, N., 2008. FRP retrofit of low and medium strength circular and rectangular reinforced concrete columns. J. Mater. Civ. Eng. 20 (2), 169—188.

ISIS (Intelligent Sensing for Innovative Structures), 2008. FRP rehabilitation of reinforced concrete structures. Design Manual 4 Version. ISIS Canada Corporation, Winnipeg, MB, p. 2.

Ismail, A.M., Fahmy, M.F., Wu, Z., 2019. Prediction accuracy of seismic behavior of RC noncircular columns retrofitted with FRP sheets and the impact of local concrete compression behavior. J. Struct. Eng. 145 (10), 04019098.

Issa, M.A., Alrousan, R.Z., Issa, M.A., 2009. Experimental and parametric study of circular short columns confined with CFRP composites. J. Compos. Constr. 13 (2), 135—147.

Japan Society of Civil Engineering (JSCE), 2002. Standard Specification for Concrete Structure—2002, 24.

Japan Society of Civil Engineers (JSCE), 2001. Recommendations for Upgrading of Concrete Structures with Use of Continuous Fiber Sheets, Research Committee on Upgrading of Concrete Structures with Use of Continuous Fiber Sheets, Japan Society of Civil Engineers, Japan.

Jiang, J.-F., Wu, Y.-F., 2012. Identification of material parameters for Drucker—Prager plasticity model for FRP confined circular concrete columns. Int. J. Solids Struct. 49 (3-4), 445—456.

Jiang, J.-F., Wu, Y.-F., 2014. Characterization of yield surfaces for FRP-confined concrete. J. Eng. Mech. 140 (12), 04014096.

Jiang, J.F., Wu, Y.F., 2016. Plasticity-based criterion for confinement design of FRP jacketed concrete columns. Mater. Struct. 49 (6), 2035—2051.

Jiang, T., Teng, J., 2006. Strengthening of short circular RC columns with FRP jackets: a design proposal. In: Proceedings of 3rd International Conference on FRP Composites in Civil Engineering (CICE 2006), Miami, FL, 13—15 December 2006, pp. 187—192.

Jiang, T., Teng, J.G., 2007. Analysis-oriented stress—strain models for FRP—confined concrete. Eng. Struct. 29 (11), 2968—2986.

Jiang, J., Wu, Y., Zhao, X., 2011. Application of Drucker—Prager plasticity model for stress—strain modeling of FRP confined concrete columns. Procedia Eng. 14 (0), 687—694.

Jolly, C., Lilistone, D., 1998. The stress—strain behavior of concrete confined by advanced fiber composites. In: Proc. 8th BCA Conference Higher Education and the Concrete Industry.

Jolly, C.K., Lillistone, D., 2000. An innovative form of reinforcement for concrete columns using advanced composites. Struct. Eng. 78 (23/24).

Kabir, M.Z., Shafei, E., 2012. Plasticity modeling of FRP-confined circular reinforced concrete columns subjected to eccentric axial loading. Compos. Part B: Eng. 43 (8), 3497—3506.

Karabinis, A.I., Rousakis, T.C., 2002. Concrete confined by FRP material: a plasticity approach. Eng. Struct. 24 (7), 923—932.

Karabinis, A.I., Rousakis, T.C., Manolitsi, G.E., 2008. 3D finite-element analysis of substandard RC columns strengthened by fiber-reinforced polymer sheets. J. Compos. Constr. 12 (5), 531—540.

Karbhari, V.M., 2003. Durability of FRP composites for civil infrastructure—myth, mystery or reality. Adv. Struct. Eng. 6 (3), 243—255. Available from: https://doi.org/10.1260/136943303322419250.

Karbhari, V.M., Eckel, D.A., 1994. Effect of cold regions climate on composite jacketed concrete columns. J. Cold Reg. Eng. ASCE 8 (3), 73—86.

Karbhari, V.M., Gao, Y., 1997. Composite jacketed concrete under uniaxial compression—verification of simple design equations. J. Mater. Civ. Eng. 9 (4), 185—193.

Karbhari, V.M., Eckel, D.A., Tunis, G.C., 1993. Strengthening of concrete column stubs through resin infused composite wraps. J. Thermoplast. Compos. Mater. 6 (2), 92—107.

Karbhari, V.M., Rivera, J., Dutta, P.K., 2000. Effect of short-term freeze—thaw cycling on composite confined concrete. J. Compos. Constr. 4 (4), 191—197.

Karbhari, V.M., Chin, J., Hunston, D., Benmokrane, B., Juska, T., Morgan, R., et al., 2003. Durability gap analysis for fiber-reinforced polymer composites in civil infrastructure. J. Compos. Constr. 7 (3), 238—247. Available from: https://doi.org/10.1061/(ASCE)1090-0268(2003). 7:3(238).

Kawashima, K., Hosotani, M., Yoneda, K., 2000. Carbon fiber sheet retrofit of reinforced concrete bridge piers. Proc. of the International Workshop on Annual Commemoration of Chi-Chi Earthquake, Vol. II-Technical Aspect. National Center for Research on Earthquake Engineering, Taipei, Taiwan, ROC, pp. 124—135.

Kawashima, K., MacRae, G.A., Hoshikuma, J.-i., Nagaya, K., 1998. Residual Displacement Response Spectrum.

Kono, S., Inazumi, M., Kaku, T., 1998. Evaluation of confining effects of CFRP sheets on reinforced concrete members. In: Proc. 2nd Int. Conf. on Composites in Infrastructures, pp. 345—355.

Kumutha, R., Vaidyanathan, R., Palanichamy, M.S., 2007. Behaviour of reinforced concrete rectangular columns strengthened using GFRP. Cem. Concr. Compos. 29 (8), 609—615.

Lam, L., Teng, J.G., 2002. Strength models for fiber-reinforced plastic-confined concrete. J. Struct. Eng. 128 (5), 612—623.

Lam, L., Teng, J.G., 2003a. Design-oriented stress—strain model for FRP-confined concrete in rectangular columns. J. Reinforced Plast. Compos. 22 (13), 1149—1186.

Lam, L., Teng, J.G., 2003b. Design-oriented stress—strain model for FRP-confined concrete. Constr. Build. Mater. 17 (6), 471—489.

Lee, J.-Y., Yi, C.-K., Jeong, H.-S., Kim, S.-W., Kim, J.-K., 2010. Compressive response of concrete confined with steel spirals and FRP composites. J. Compos. Mater. 44 (4), 481−504.

Li, G., 2006. Experimental study of FRP confined concrete cylinders. Eng. Struct. 28 (7), 1001−1008.

Li, J.H., Wang, X.T., 2006. Behavior of steel reinforced high-strength concrete columns under low cyclic reversed loading. In: 8th International Conference on Steel-Concrete Composite and Hybrid Structures, Proceedings, pp. 741−747.

Li, Y.F., Sung, Y.Y., 2004. A study on the shear-failure of circular sectioned bridge column retrofitted by using CFRP jacketing. J. Reinforced Plast. Compos. 23 (8), 811−830.

Li, Y.-F., Lin, C.-T., Sung, Y.-Y., 2003. A constitutive model for concrete confined with carbon fiber reinforced plastics. Mech. Mater. 35 (3-6), 603−619.

Lim, J.C., Ozbakkaloglu, T., 2014a. Confinement model for FRP-confined high-strength concrete. J. Compos. Constr. 18 (4), 04013058.

Lim, J.C., Ozbakkaloglu, T., 2014b. Design model for FRP-confined normal- and high-strength concrete square and rectangular columns. Mag. Concr. Res. 1020−1035.

Lin, H.-J., Chen, C.-T., 2001. Strength of concrete cylinder confined by composite materials. J. Reinforced Plast. Compos. 20 (18), 1577−1600.

Maalej, M., Tanwongsval, S., Paramasivam, P., 2003. Modeling of rectangular RC columns strengthened with FRP. Cem. Concr. Compos. 25 (2), 263−276.

Mandal, S., Hoskin, A., Fam, A., 2005. Influence of concrete strength on confinement effectiveness of fiber-reinforced polymer circular jackets. ACI Struct. J. 102 (3), 383−392.

Mander, J.B., Priestley, M.J.N., Park, R., 1988. Theoretical stress−strain model for confined concrete. J. Struct. Eng. 114 (8), 1804−1826.

Manfredi G., 2001. R. R. Models of concrete confined by fiber composites. In: FRPRCS-5, Thomas Telford, P. 865e874.

Marques, S.P.C., Marques, D.Cd.S.C., Lins da Silva, J., Cavalcante, M.A.A., 2004. Model for analysis of short columns of concrete confined by fiber-reinforced polymer. J. Compos. Constr. 8 (4), 332−340.

Matthys, S., Toutanji, H., Audenaert, K., Taerwe, L., 2005. Axial load behavior of large-scale columns confined with fiber-reinforced polymer composites. ACI Struct. J. 102 (2), 258.

Matthys, S., Toutanji, H., Taerwe, L., 2006. Stress−strain behavior of large-scale circular columns confined with FRP composites. J. Struct. Eng. 132 (1), 123−133.

Megalooikonomou, K.G., Monti, G., Santini, S., 2012. Constitutive model for fiber-reinforced polymer-and tie-confined concrete. ACI Struct. J. 109 (4), 569−578.

Mirmiran, A., 1996. Analytical and Experimental Investigation of Reinforced Concrete Columns Encased in Fiberglass Tubular Jackets and Use of Fiber Jacket for Pile Splicing. Final Report, Contact number B9135, Florida Dept. of Tallahassee, FL.

Mirmiran, A., Shahawy, M., 1997a. Behavior of concrete columns confined by fiber composites. J. Struct. Eng. 123 (5), 583−590.

Mirmiran, A., Shahawy, M., 1997b. Dilation characteristics of confined concrete. Mech. Cohesive Frict. Mater. 2 (3), 237−249.

Mirmiran, A., Shahawy, M., Samaan, M., Echary, H.E., Mastrapa, J.C., Pico, O., 1998. Effect of column parameters on FRP-confined concrete. J. Compos. Constr. 2 (4), 175−185.

Mirmiran, A., Zagers, K., Yuan, W., 2000. Nonlinear finite element modeling of concrete confined by fiber composites. Finite Elem. Anal. Des. 35 (1), 79−96.

Miyauchi, K., 1997. Estimation of strengthening effects with carbon fiber sheet for concrete column. In: 3rd Int. Symp. of Non-Metallic Reinforcement for Concrete Structures, pp. 217—224.

Miyauchi, K., Inoue, S., Kuroda, T., Kobayashi, A., 1999. Strengthening effects with carbon fiber sheet for concrete column. Japan concr. inst. 21, 143—150.

Moehle, J., Elwood, K., Sezen, H., 2002. Gravity load collapse of building frames during earthquakes. In: S.M. Uzumeri Symp. Behavior and Design of Conc. Str. for Seismic Performance, ACI-SP 197.

Mohamed, H.M., Masmoudi, R., 2010. Axial load capacity of concrete-filled FRP tube columns: experimental versus theoretical predictions. J. Compos. Constr. 14 (2), 231—243.

Monti, G., Nisticò, N., Santini, S., 2001. Design of FRP jackets for upgrade of circular bridge piers. J. Compos. Constr. 5 (2), 94—101.

Moran, D.A., Pantelides, C.P., 2002. Variable strain ductility ratio for fiber-reinforced polymer-confined concrete. J. Compos. Constr. 6 (4), 224—232.

Moran, D.A., Pantelides, C.P., 2012. Elliptical and circular FRP-confined concrete sections: a Mohr—Coulomb analytical model. Int. J. Solids Struct. 49 (6), 881—898.

Mostofinejad, D., Moshiri, N., Mortazavi, N., 2015. Effect of corner radius and aspect ratio on compressive behavior of rectangular concrete columns confined with CFRP. Mater. Struct. 48 (1—2), 107—122.

Nanni, A., Bradford, N.M., 1995. FRP jacketed concrete under uniaxial compression. Constr. Build. Mater. 9 (2), 115—124.

Nanni, A., Norris, M.S., 1995. FRP jacketed concrete under flexure and combined flexure-compression. Constr. Build. Mater. 9 (5), 273—281.

Neale, K.W., Labossiere, P., 1997. State-of-the-art report on retrofitting and strengthening by continuous fiber in Canada. In: Proc. 3rd Int. Symposium, in Non-Metallic (FRP) Reinforcement for Concrete Structures, pp. 25—39.

Nisticò, N., Monti, G., 2013. RC square sections confined by FRP: analytical prediction of peak strength. Compos. Part B: Eng. 45 (1), 127—137.

Nisticò, N., Pallini, F., Rousakis, T., Wu, Y.-F., Karabinis, A., 2014. Peak strength and ultimate strain prediction for FRP confined square and circular concrete sections. Compos. Part B: Eng. 67 (0), 543—554.

Ozbakkaloglu, T., 2013a. Behavior of square and rectangular ultra high-strength concrete-filled FRP tubes under axial compression. Compos. Part B: Eng. 54, 97—111.

Ozbakkaloglu, T., 2013b. Compressive behavior of concrete-filled FRP tube columns: assessment of critical column parameters. Eng. Struct. 51, 188—199.

Ozbakkaloglu, T., Lim, J.C., 2013. Axial compressive behavior of FRP-confined concrete: experimental test database and a new design-oriented model. Compos. Part B: Eng. 55 (0), 607—634.

Ozbakkaloglu, T., Saatcioglu, M., 2006. Seismic behavior of high-strength concrete columns confined by fiber-reinforced polymer tubes. J. Compos. Constr. 10 (6), 538—549.

Ozbakkaloglu, T., Lim, J.C., Vincent, T., 2013. FRP-confined concrete in circular sections: review and assessment of stress—strain models. Eng. Struct. 49 (0), 1068—1088.

Pantelides, C.P., Gergely, J., 2002. Carbon-fiber-reinforced polymer seismic retrofit of RC bridge bent: design and in situ validation. J. Compos. Constr. 6 (1), 52—60.

Pantelides, C.P., Moran, D.A., 2013. Design of FRP jackets for plastic hinge confinement of RC columns. J. Compos. Constr. 17 (4), 433—442.

Pantelides, C.P., Yan, Z., 2007. Confinement model of concrete with externally bonded FRP jackets or posttensioned FRP shells. J. Struct. Eng. 133 (9), 1288—1296.

Park, R., Paulay, T., 1975. Reinforced Concrete Structures, 1975. John Wiley & Sons, Inc., New York, p. 769.

Park, T.W., Na, U.J., Chung, L., Feng, M.Q., 2008. Compressive behavior of concrete cylinders confined by narrow strips of CFRP with spacing. Compos. Part B: Eng. 39 (7), 1093–1103.

Park, J.-H., Jo, B.-W., Yoon, S.-J., Park, S.-K., 2011. Experimental investigation on the structural behavior of concrete filled FRP tubes with/without steel re-bar. KSCE J. Civ. Eng. 15 (2), 337–345.

Parvin, A., Brighton, D., 2014. FRP composites strengthening of concrete columns under various loading conditions. Polymers 6 (4), 1040–1056.

Pellegrino, C., Modena, C., 2010. Analytical model for FRP confinement of concrete columns with and without internal steel reinforcement. J. Compos. Constr. 14 (6), 693–705.

Pessiki, S., Harries, K.A., Kestner, J.T., Sause, R., Ricles, J.M., 2001. Axial behavior of reinforced concrete columns confined with FRP jackets. J. Compos. Constr. 5 (4), 237–245.

Pham, T.M., Hadi, M.N.S., 2013. Strain estimation of CFRP-confined concrete columns using energy approach. J. Compos. Constr. 17 (6), 04013001.

Pham, T.M., Hadi, M.N.S., 2014a. Confinement model for FRP confined normal- and high-strength concrete circular columns. Constr. Build. Mater. 69 (0), 83–90.

Pham, T.M., Hadi, M.N.S., 2014b. Predicting stress and strain of FRP-confined square/rectangular columns using artificial neural networks. J. Compos. Constr. 18 (6).

Pham, T.M., Hadi, M.N.S., 2014c. Stress prediction model for FRP confined rectangular concrete columns with rounded corners. J. Compos. Constr. 18 (1), 04013019.

Picher, F., Rochette, P., Labossiere, P., 1996. Confinement of concrete cylinders with creep. In: First International Conference on Composites in Infrastructure.

Priestley, M.J.N., 1996. Seismic Design and Retrofit of Bridges. John Wiley & Sons.

Ramkrishna Dandapat, A.D., Bhattacharyya, S.K., 2012. Localized failure in fiber-reinforced polymer-wrapped cylindrical concrete columns. ACI Struct. J. 109 (4).

Realfonzo, R., Napoli, A., 2011. Concrete confined by FRP systems: confinement efficiency and design strength models. Compos. Part B: Eng. 42 (4), 736–755.

Reay, J.T., Pantelides, C.P., 2006. Long-term durability of state street bridge on interstate 80. J. Bridge Eng. 11 (2), 205–216.

Restrepol, J., DeVino, B., 1996. Enhancement of the axial load carrying capacity of reinforced concrete columns by means of fiberglass-epoxy jackets. In: Proc. of the 2nd International Conference on Advanced Composite Materials in Bridges and Structures, pp. 547–553.

Richart, F.E., Brandtzæg, A., Brown, R.L., 1929. Failure of plain and spirally reinforced concrete in compression. University of Illinois. Engineering Experiment Station. Bulletin; no. 190.

Rochette, P., Labossiere, P., 1996. A plasticity approach for concrete columns confined with composite materials. In Second International Conference on Advanced Composite Materials in Bridges and Structures, Montreal.

Rochette, P., Labossière, P., 2000. Axial testing of rectangular column models confined with composites. J. Compos. Constr. 4 (3), 129–136.

Rousakis, T.C., 2005. Mechanical Behaviour of Concrete Confined by Composite Materials. Ph.D. Thesis, Democritus Univ. of Thrace, Xanthi, Greece (in Greek).

Rousakis, T.C., 2014. Elastic fiber ropes of ultrahigh-extension Capacity in strengthening of concrete through confinement. J. Mater. Civ. Eng. 26 (1), 34–44.

Rousakis, T.C., Karabinis, A.I., 2012. Adequately FRP confined reinforced concrete columns under axial compressive monotonic or cyclic loading. Mater. Struct. 45 (7), 957–975.

Rousakis, T.C., Tourtouras, I.S., 2015. Modeling of passive and active external confinement of RC columns with elastic material. ZAMM-J. Appl. Math. Mech./Z. Angew. Math. Mech. 95 (10), 1046–1057.

Rousakis, T.C., Karabinis, A.I., Kiousis, P.D., 2007. FRP-confined concrete members: axial compression experiments and plasticity modeling. Eng. Struct. 29 (7), 1343–1353.

Rousakis, T.C., Rakitzis, T.D., Karabinis, A.I., 2012a. Design-oriented strength model for FRP-confined concrete members. J. Compos. Constr. 16 (6), 615–625.

Rousakis, T., Rakitzis, T., Karabinis, A., 2012b. Empirical modeling of failure strains of uniformly FRP confined concrete columns. In: Proc. of 6th International Conference on FRP Composites in Civil Engineering – CICE, 13–15.

Saadatmanesh, H., Ehsani, M.R., Jin, L., 1996. Seismic strengthening of circular bridge pier models with fiber composites. ACI Struct.J. 93 (6), 639–738.

Saafi, M., Toutanji, H., Li, Z., 1999. Behavior of concrete columns confined with fiber reinforced polymer tubes. ACI Mater. J. 96 (4), 500–509.

Sadeghian, P., Fam, A., 2015. Improved design-oriented confinement models for FRP-wrapped concrete cylinders based on statistical analyses. Eng. Struct. 87 (0), 162–182.

Sadeghian, P., Rahai, A.R., Ehsani, M.R., 2008. Numerical modeling of concrete cylinders confined with CFRP composites. J. Reinf. Plast. Compos. .

Saenz, L.P., 1964. Discussion of equation for the stress–strain curve of concrete by Desayi and Krishnan. J. Am. Concr. Inst. 61 (9), 1229–1235.

Saiid Saiidi, M., Sureshkumar, K., Pulido, C., 2005. Simple carbon-fiber-reinforced-plastic-confined concrete model for moment-curvature analysis. J. Compos. Constr. 9 (1), 101–104.

Saiidi, M.S., O'Brien, M., Sadrossadat-Zadeh, M., 2009. Cyclic response of concrete bridge columns using superelastic nitinol and bendable concrete. ACI Struct. J. 106 (1), 69.

Samaan, M., Mirmiran, A., Shahawy, M., 1998. Model of concrete confined by fiber composites. J. Struct. Eng. 124 (9), 1025–1031.

Seible, F., Priestley, M.J.N., Hegemier, G.A., Innamorato, D., 1997. Seismic retrofit of RC columns with continuous carbon fiber jackets. J. Compos. Constr. 1 (2), 52–62.

Shahawy, M., Mirmiran, A., Beitelman, T., 2000. Tests and modeling of carbon-wrapped concrete columns. Compos. Part B: Eng. 31 (6-7), 471–480.

Sharma, S.S., Dave, U.V., Solanki, H., 2013. FRP wrapping for RC columns with varying corner radii. Procedia Eng. 51 (0), 220–229.

Shehata, I.A.E.M., Carneiro, L.A.V., Shehata, L.C.D., 2002. Strength of short concrete columns confined with CFRP sheets. Mater. Struct. 35 (1), 50–58.

Shehata, I., Carneiro, L., Shehata, L., 2007. Strength of confined short concrete columns. In: Proc. 8th Int. Symp. on Fiber Reinforced Polymer Reinforcement for Concrete Structures.

Sheikh, S.A., Li, Y., 2007. Design of FRP confinement for square concrete columns. Eng. Struct. 29 (6), 1074–1083.

Sheikh, S.A., Yau, G., 2002. Seismic behavior of concrete columns confined with steel and fiber-reinforced polymers. ACI Struct. J. 99 (1), 72–80.

Shirmohammadi, F., Esmaeily, A., Kiaeipour, Z., 2015. Stress–strain model for circular concrete columns confined by FRP and conventional lateral steel. Eng. Struct. 84 (0), 395–405.

Siavashi, S., Eamon, C.D., Makkawy, A.A., Wu, H.C., 2019. Long-term durability of FRP bond in the midwest United States for externally strengthened bridge components. J. Compos. Constr. 23 (2), 05019001.

Soudki, K.A., Green, M.F., 1997. Freeze—thaw response of CFRP wrapped concrete. Concr. Int. 19 (8), 64—67.

Spoelstra, M.R., Monti, G., 1999. FRP-confined concrete model. J. Compos. Constr. 3 (3), 143—150.

Sundarraja, M.C., Prabhu, G.G., 2012. Experimental study on CFST members strengthened by CFRP composites under compression. J. Constr. Steel Res. 72, 75—83.

Tabbara, M., Karam, G., 2007. Modeling the strength of concrete cylinders with FRP wraps using the Hoek-Brown strength criterion. In: Proc. 8th Int. Symp. on Fiber Reinforced Polymer Reinforcement for Concrete Structures.

Takeda, T., Sozen, M.A., Nielsen, N.N., 1970. Reinforced concrete response to simulated earthquakes. J. Struct. Div. 96 (12), 2557—2573.

Tamuzs, V., Tepfers, R., Sparnins, E., 2006a. Behavior of concrete cylinders confined by carbon composite 2. Prediction of strength. Mech. Compos. Mater. 42 (2), 109—118.

Tamuzs, V., Tepfers, R., Zile, E., Ladnova, O., 2006b. Behavior of concrete cylinders confined by a carbon composite 3. deformability and the ultimate axial strain. Mech. Compos. Mater. 42 (4), 303—314.

Tan, K.H., 1997. Behaviour of hybrid FRP-steel reinforced concrete beams. In: Proc. 3rd Int. Symposium, FRPRCS. vol. 3, pp. 487—494.

Tan, K.H., 2002. Strength enhancement of rectangular reinforced concrete columns using fiber-reinforced polymer. J. Compos. Constr. 6 (3), 175—183.

Tan, K.H., Bhowmik, T., Balendra, T., 2013. Confinement model for FRP-bonded capsule-shaped concrete columns. Eng. Struct. 51 (0), 51—59.

Tao, Z., Yu, Q., Zhong, Y.Z., 2008. Compressive behaviour of CFRP-confined rectangular concrete columns. Mag. Concr. Res. 60 (10), 735—745.

Tastani, S., Pantazopoulou, S.J., 2003. Strength and deformation capacity of Brittle RC members jacketed with FRP Wraps. In: CD-ROM Proceedings of the FIB-Symposium on Concrete Structures in Seismic Regions.

Tastani, S.P., Pantazopoulou, S.J., 2006. Fiber reinforced polymers in seismic upgrading of existing reinforced concrete structures. In: Proc. 8th US National Conference on Earthquake Engineering (8NCEE), Citeseer, pp. 18—22.

Teng, J.G., Huang, Y.L., Lam, L., Ye, L.P., 2007. Theoretical model for fiber-reinforced polymer-confined concrete. J. Compos. Constr. 11 (2), 201—210.

Teng, J.G., Jiang, T., Lam, L., Luo, Y.Z., 2009. Refinement of a design-oriented stress—strain model for FRP-confined concrete. J. Compos. Constr. 13 (4), 269—278.

Teng, J.G., Hu, Y.M., Yu, T., 2013. Stress—strain model for concrete in FRP-confined steel tubular columns. Eng. Struct. 49 (0), 156—167.

Teng, J.G., Lin, G., Yu, T., 2014. Analysis-oriented stress—strain model for concrete under combined FRP-steel confinement. J. Compos. Constr. 19 (5), 04014084.

Thériault, M., Neale, K.W., 2000. Design equations for axially loaded reinforced concrete columns strengthened with fiber reinforced polymer wraps. Can. J. Civ. Eng. 27 (5), 1011—1020.

Thériault, M., Neale, K.W., Claude, S., 2004. Fiber-reinforced polymer-confined circular concrete columns: investigation of size and slenderness effects. J. Compos. Constr. 8 (4), 323—331.

Toutanji, H., 1999. Stress—strain characteristics of concrete columns externally confined with advanced fiber composite sheets. ACI Mater. J. 96 (3).

Toutanji, H., Balaguru, P., 1998. Durability characteristics of concrete columns wrapped with FRP tow sheets. J. Mater. Civ. Eng. 10 (1), 52—57.

Toutanji, H.A., Balaguru, P., 1999. Effects of freeze—thaw exposure on performance of concrete columns strengthened with advanced composites. Mater. J. ACI 96 (5), 505—611.

Toutanji, H., Han, M., Gilbert, J., Matthys, S., 2010. Behavior of large-scale rectangular columns confined with FRP composites. J. Compos. Constr. 14 (1), 62—71.

Triantafyllou, G.G., Rousakis, T.C., Karabinis, A.I., 2015. Axially loaded reinforced concrete columns with a square section partially confined by light GFRP straps. J. Compos. Constr. 19 (1), 04014035.

Vandoros, K.G., Dritsos, S.E., 2008. Concrete jacket construction detail effectiveness when strengthening RC columns. Constr. Build. Mater. 22 (3), 264—276.

Vasumathi, A.M., Rajkumar, K., Ganesh Prabhu, G., 2014. Compressive behaviour of RC column with fiber reinforced concrete confined by CFRP strips. Adv. Mater. Sci. Eng. 2014.

Vintzileou, E., Panagiotidou, E., 2007. An empirical model for predicting the mechanical properties of FRP-confined concrete. In: Proc. 8th Int. Symp. on Fiber Reinforced Polymer Reinforcement for Concrete Structures.

Vintzileou, E., Panagiotidou, E., 2008. An empirical model for predicting the mechanical properties of FRP-confined concrete. Constr. Build. Mater. 22 (5), 841—854.

Wang, L.-M., Wu, Y.-F., 2008. Effect of corner radius on the performance of CFRP-confined square concrete columns: test. Eng. Struct. 30 (2), 493—505.

Wang, Y.-f, Wu, H.-l, 2010. Size effect of concrete short columns confined with aramid FRP jackets. J. Compos. Constr. 15 (4), 535—544.

Wang, D.Y., Wang, Z.Y., Yu, T., Li, H., 2016. Shake table tests of large-scale substandard RC frames retrofitted with CFRP wraps before earthquakes. J. Compos. Constr. 04016062.

Wei, Y., Wu, Y., 2011. Stress—strain modeling of rectangular concrete columns confined by FRP jacket. Advances in FRP Composites in Civil Engineering. Springer, pp. 618—621.

Wei, Y.-Y., Wu, Y.-F., 2012. Unified stress—strain model of concrete for FRP-confined columns. Constr. Build. Mater. 26 (1), 381—392.

Wu, G., Lü, Z.T., Wu, Z.S., 2006. Strength and ductility of concrete cylinders confined with FRP composites. Constr. Build. Mater. 20 (3), 134—148.

Wu, H.-L., Wang, Y.-F., 2010. Experimental study on reinforced high-strength concrete short columns confined with AFRP sheets. Steel Compos. Struct. 10 (6), 501—516.

Wu, Y.-F., Jiang, C., 2013. Effect of load eccentricity on the stress—strain relationship of FRP-confined concrete columns. Compos. Struct. 98, 228—241.

Wu, Y.-F., Wang, L.-M., 2009. Unified strength model for square and circular concrete columns confined by external jacket. J. Struct. Eng. 135 (3), 253—261.

Wu, Y.F., Wei, Y., 2014. General stress—strain model for steel- and FRP-confined concrete. J. Compos. Constr. 19 (4), 04014069.

Wu, Y.-F., Wei, Y.-Y., 2010. Effect of cross-sectional aspect ratio on the strength of CFRP-confined rectangular concrete columns. Eng. Struct. 32 (1), 32—45.

Wu, Y.-F., Zhou, Y.-W., 2010. Unified strength model based on Hoek-Brown failure criterion for circular and square concrete columns confined by FRP. J. Compos. Constr. 14 (2), 175—184.

Wu, G., Wu, Z.S., Lü, Z.T., 2007. Design-oriented stress—strain model for concrete prisms confined with FRP composites. Constr. Build. Mater. 21 (5), 1107—1121.

Wu, H.-L., Wang, Y.-F., Yu, L., Li, X.-R., 2009. Experimental and computational studies on high-strength concrete circular columns confined by aramid fiber-reinforced polymer sheets. J. Compos. Constr. 13 (2), 125—134.

Wu, Z., Zhang, D., Karbhari, V.M., 2010. Numerical simulation on seismic retrofitting performance of reinforced concrete columns strengthened with fiber reinforced polymer sheets. Struct. Infrastruct. Eng. 6 (4), 481–496.

Wu, Y.-F., Yun, Y., Wei, Y., Zhou, Y., 2014. Effect of predamage on the stress–strain relationship of confined concrete under monotonic loading. J. Struct. Eng. 140 (12), 04014093.

Xiao, Y., Ma, R., 1997. Seismic retrofit of RC circular columns using prefabricated composite jacketing. J. Struct. Eng. 123 (10), 1357–1364.

Xiao, Y., Wu, H., 2000. Compressive behavior of concrete confined by carbon fiber composite jackets. J. Mater. Civ. Eng. 12 (2), 139–146.

Xiao, Y., Wu, H., 2003. Compressive behavior of concrete confined by various types of FRP composite jackets. J. Reinf. Plast. Compos. 22 (13), 1187–1201.

Xiao, Y., Wu, H., Martin, G.R., 1999. Prefabricated composite jacketing of RC columns for enhanced shear strength. J. Struct. Eng. 125 (3), 255–264.

Xiao, Q.G., Teng, J.G., Yu, T., 2010. Behavior and modeling of confined high-strength concrete. J. Compos. Constr. 14 (3), 249–259.

Youssef, M.N., Feng, M.Q., Mosallam, A.S., 2007. Stress–strain model for concrete confined by FRP composites. Compos. Part B: Eng. 38 (5–6), 614–628.

Youssf, O., ElGawady, M.A., Mills, J.E., Ma, X., 2014. Finite element modeling and dilation of FRP-confined concrete columns. Eng. Struct. 79 (0), 70–85.

Yozo, F., Satoko, H., Masato, A., 2005. Damage Analysis of Hanshin Expressway Viaducts during 1995 Kobe Earthquake. I: Residual Inclination of Reinforced Concrete Piers.

Yu, T., Teng, J.G., 2011. Design of concrete-filled FRP tubular columns: provisions in the chinese technical code for infrastructure application of FRP composites. J. Compos. Constr. 15 (3), 451–461.

Reinforcing spalling resistance of concrete structures with bonded fiber−reinforced polymer composites

7

7.1 Introduction

Many tunnels, as well as road and railway bridges, require taking measures to prevent the peeling and spalling of their concrete pieces caused by the action of earth pressure, increased weight of vehicles and traffic density, corrosion of steel bar, deterioration of concrete, and poor original construction. Fiber-reinforced polymer (FRP) sheets have been increasingly used to reinforce and repair tunnel linings (Fig. 7.1A) and the slabs of viaducts (Fig. 7.1B) because of their advantages. In these FRP-strengthened structures, both in-plane shear stress and out-of-plane normal stress are imposed on the interface between FRP sheets and concrete.

Concrete spalls from the concrete surface owing to different failure mechanisms. To study the bonding and debonding mechanism behind spalling failure, a punching−peeling test against the spalling of the concrete piece is an important tool. A useful analytical technique for this problem is introduced in this chapter.

The punching−peeling test has been used as a convenient and relatively simple means of characterizing spalling resistance. The test consists of measuring the force required to peel a sheet from a substrate, as shown in Fig. 7.1C. This force is termed the peel force. Gent and Hamed (1975) discussed the dependence of the peel force upon the peel angle by limiting the stress analysis to small bending deformations of the bonded sheet. The small bending models for the peeling of an elastic sheet from a rigid substrate were extended to large bending by Nicholson (1977). However, the interfacial shear stress was neglected in their study (Gent and Hamed, 1975; Nicholson, 1977). Recently, researchers used an energy balance approach to deduce the relationship between the toughness of the interface and the peel force (Thouless and Jensen, 1992; Karbhari and Engineer, 1996; Kimpara et al., 1998). Karbhari and Engineer (1996) developed a peel test for investigating the bond between composites and concrete. A methodology for understanding the different mechanisms and modes of interfacial fracture was also developed. In the as discussed earlier peel test, the peel angle can be adjusted by an instrument. There are a few studies on strengthening against punching shear failure (Ozbolt and Eligehausen, 2001; Wang and Tan, 2001; Binici and Bayrak, 2002). However, in

Structures Strengthened with Bonded Composites. DOI: https://doi.org/10.1016/B978-0-12-821088-8.00007-2

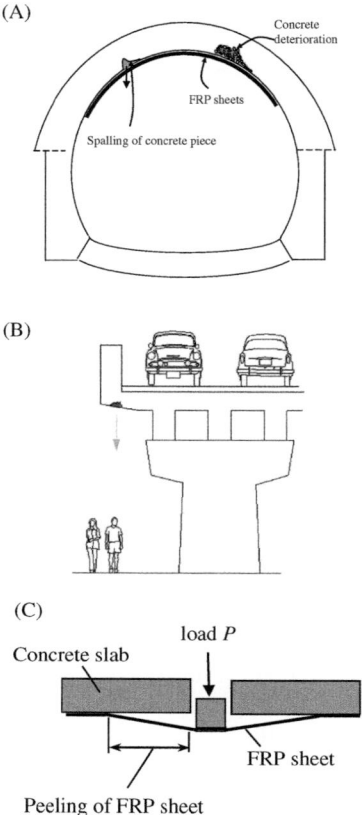

Figure 7.1 Schematic diagram of spalling of concrete piece and punching–peeling test
(Wu et al., 2005b). (A) Spalling of concrete piece from concrete lining in tunnel.
(B) Spalling of concrete piece from slabs of viaducts. (C) Punching–peeling test.

their study, the load was directly applied to the slab and spalling resistance was not
considered. Kimpara et al. (1998) proposed a peeling test method of FRP sheets
bonded on mortar and concrete to characterize the peeling strength and examine the
effects of different surface treatments and primers. In such a test, namely a spalling
resistance test or punching–peeling test, the peel angle depends only on the interfacial behavior. The relation between sheet deflection, debonding length, and load
was obtained from the geometric consideration and equilibrium condition of a thin
elastic membrane, from which the energy release rate due to peeling was expressed
as a function of the deflection-to-debonding-length ratio by a simple formula. In
this chapter, the peeling behavior and the spalling resistance of FRP sheets externally bonded to concrete plates and beams are introduced based on available experimental and analytical investigations, as shown in Fig. 7.1C.

7.2 Spalling resistance of beams with fiber-reinforced polymer sheets

7.2.1 Experimental study on peeling and spalling resistance of unidirectional fiber—reinforced polymer sheets

Wu et al. (2005a) studied the spalling resistance of unidirectional FRP sheets externally bonded to concrete beams. The punching—peeling test setup and specimen were designed as shown in Fig. 7.2. In the test, an FRP sheet is forced to peel under the action of a pin. The pin is intimately contacted with the sheet near the contacting point. The FRP sheet is subjected to (1) tensile stress of the sheet, (2) compressive stress from the pin, and (3) out-of-plane shear stress. The FRP sheet ruptures as a result of (2) and (3).

A total of 42 beam specimens (650 mm × 150 mm × 120 mm) were fabricated in 14 series, with each series having three specimens named No. 1, No. 2, and No. 3. The specimens in series 2 were the control specimens composed of a concrete beam externally bonded with a one-layer bidirectional carbon fiber—reinforced polymer (CFRP) sheet by an epoxy adhesive. All the specimens were tested in a displacement-controlled mode with an average displacement increment of 0.05 mm. Electrical strain gauges and a linear variable displacement transducer (LVDT) were used to measure FRP strains and pin deflections, respectively. The width of the FRP sheet is 120 mm. The span of the specimen is 600 mm. The length of the FRP sheet is 570 mm. A notch (30 mm × 30 mm) was made in the center at the bottom of the beam, where loading pin is situated.

A series of experimental parameters, such as different FRP materials, adhesives, surface treatment, and concrete strength, were investigated to understand the effect of spalling resistance. The material properties of the FRP sheets used in the test are presented in Table 7.1. Various fibers were used in the test. In the test, the average strengths of the 28-day concrete beams for which the design strengths are 10 and 22.5 MPa are 11.3 and 21.4 MPa, respectively. Concrete beams with strength of 21.4 MPa are fabricated for specimens in the first 13 series, but the concrete strength of the specimen in series 14 was 11.3 MPa (Table 7.2). Fast-hardening methyl methacrylate (MMA), modified fast-hardening epoxy, and acrylate epoxy are utilized for comparison with the usual epoxy in series 10, 11, and 12, respectively. The specimens in series 13 and 14 are those without surface treatment.

7.2.1.1 Failure modes

Three debonding failure modes were identified from this study: debonding within surface concrete, debonding in concrete—adhesive interface, and debonding in FRP composite—adhesive interface. Moreover, a typical failure due to FRP rupture was also observed. The debonding of the specimens begins at the midspan within the vicinity of the loading point and propagates further along the length of the specimens with increased loading.

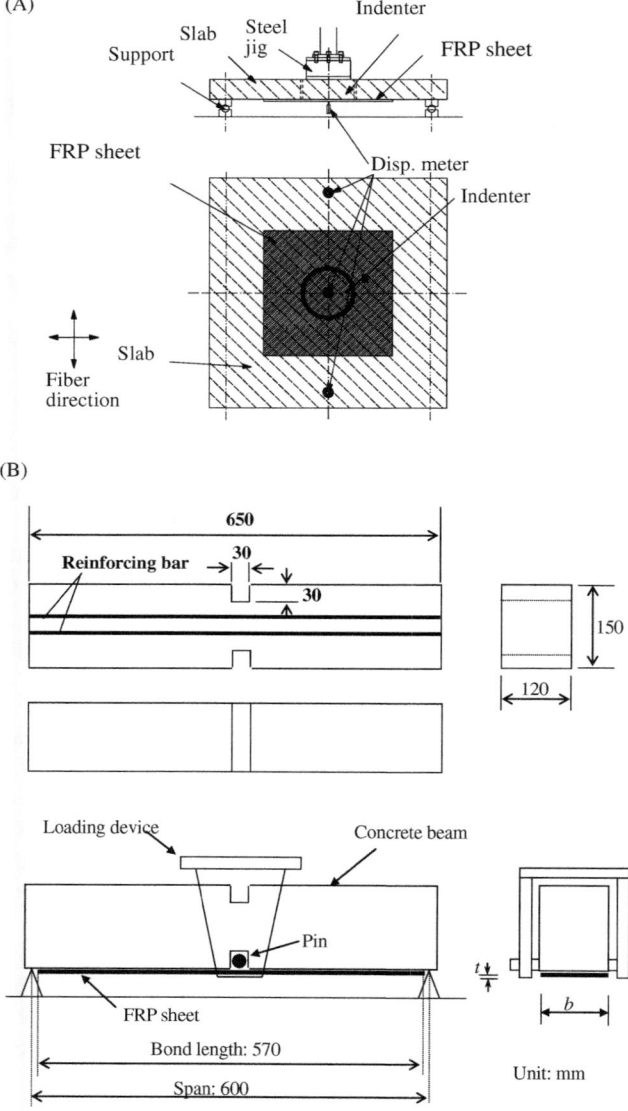

Figure 7.2 Test set-up and specimen shape (Wu et al., 2005a). (A) Bidirectional FRP sheets. (B) Unidirectional FRP sheets.

7.2.1.2 Load–deflection relationships

The experimental load–deflection relation is shown in Fig. 7.3. Fig. 7.3A shows the results of specimens in series 1 for the unidirectional CFRP sheet. The load increases with the increased deflection up to a turning point, at which debonding

Table 7.1 Material properties of FRP sheets (provided by manufacturer) (Wu et al., 2005a).

FRP materials	Young's modulus E (GPa)	Tensile strength (MPa)	Design thickness t (mm)	Fiber arial weight (g/m^2)
Unidirectional carbon fiber	230	3500	0.111	200
Bidirectional carbon fiber	230	3500	0.0556	200
Aramid fiber 1	110	2000	0.121	175
Aramid fiber 2	80	2300	0.117	162
Glass fiber	130	2200	0.079	200
Vinyl fiber mesh	–	–	–	–
Polyester fiber	14.6	1100	0.116	160
Nylon fiber	2.14	600	0.0965	110
PBO fiber	235	3500	0.128	200
Prepreg with short glass fiber	9.10	100	1.4	–

Table 7.2 Test parameters (Wu et al., 2005a).

Series	FRP materials	Adhesives	Test variation
1	Unidirectional carbon fiber	Epoxy	Old data (available from the literature)
2	Bidirectional carbon fiber	Epoxy	Control specimen
3	Bidirectional aramid fiber 1	Epoxy	Fiber type
4	Bidirectional aramid fiber 2	Epoxy	Fiber type
5	Glass fiber	Epoxy	Fiber type
6	Vinyl fiber mesh	Epoxy	Fiber type
7	Polyester fiber	Epoxy	Fiber type
8	Nylon fiber	Epoxy	Fiber type
9	PBO fiber	Epoxy	Fiber type
10	Bidirectional carbon fiber	Fast hardening (MMA)	Adhesive type
11	Bidirectional carbon fiber	Fast hardening (modified epoxy)	Adhesive type
12	Prepreg with short glass fiber	Acrylate epoxy	Material
13	Bidirectional carbon fiber	Epoxy	Surface treatment
14	Bidirectional carbon fiber	Epoxy	Low concrete strength (11.3 MPa)

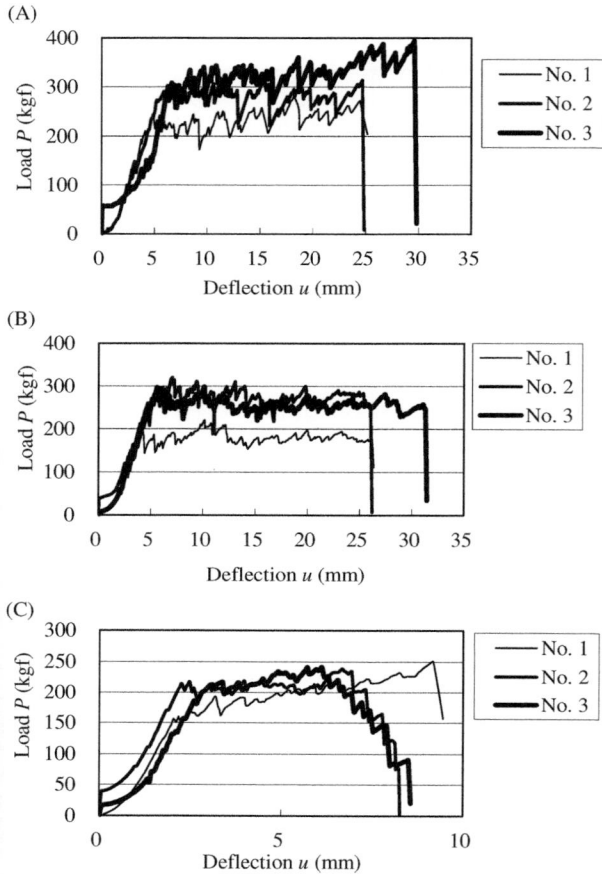

Figure 7.3 Experimental load—deflection curves (Wu et al., 2005a). (A) Load—deflection curves of specimens in series 1. (B) Load—deflection curves of specimens in series 2. (C) Load—deflection curves of specimens in series 3.

occurs. Because of inhomogeneities and asperities of adhesive and concrete surface behavior, the load—deflection curves are not straight lines. The load fluctuates with the propagation of macrodebonding between the turning point and failure point. From the perspective of microdebonding, the load decreases when microdebonding occurs, after which the load increases until new microdebonding occurs in front of the previous instance. Debonding propagated and stopped repeatedly, which could be observed in the tests. A debonding propagation length of approximately 10 mm was seen each time. Because of interface unevenness and differences in adhesive thickness, the load—deflection curves are not horizontal lines. For the specimens in series 1, debonding occurs in concrete near the interface.

Table 7.3 Peeling load and failure pattern (Wu et al., 2005a).

Series	Peel load (kgf)					Failure pattern
	No. 1	No. 2	No. 3	Average	Max.	
1	244	294	338	292	394	Debonding in concrete
2	186	290	269	248	320	Debonding in concrete
3	212	205	214	210	251	Debonding in concrete
4	201	257	226	228	282	Debonding in FA
5	132	172	188	164	189	FRP sheet rupture
6	44	36	30	37	59	FRP sheet rupture
7	121	153	152	142	179	Debonding in FA
8	108	131	119	119	148	Debonding in FA
9	230	283	265	259	313	Debonding in concrete
10	123	149	149	140	187	Debonding in CA
11	130	121	145	132	164	Debonding in CA
12	134	161	169	155	227	Debonding in concrete
13	177	203	201	193	257	Debonding in CA
14	155	196	180	177	228	Debonding in concrete

CA, Interface between concrete and adhesive; FA, interface between FRP sheet and adhesive.

The load—deflection curves for the control specimens in series 2 are shown in Fig. 7.3B. The debonding load is lower than that of the specimens of series 1 although the CFRP sheet has the same thickness. In this case, the fibers are bidirectional and therefore the actual thickness in one direction is only half of that in the specimens of series 1. Debonding also occurs in concrete near the interface in these specimens.

The load—deflection curves for the specimens in series 3 are shown in Fig. 7.3C. A crack appears in the epoxy at the contacting point between the loading pin and adhesive after debonding propagates for 50 mm. Finally, the experiment ends with the rupture of the FRP sheet after the debonding extension stops. The failure patterns and debonding loads are shown in Table 7.3 for all groups of specimens.

7.2.1.3 Parameters affecting spalling resistance

Effect of different continuous fibers

The experimental load—deflection curves for series 1, 2, 3, 4, 5 and series 2, 6, 7, 8, 9 are given in Fig. 7.4A and B, respectively. It can be seen that the larger the FRP sheet stiffness is, the higher is the maximum peeling load. The peeling load will be large if debonding occurs in concrete because of the good interfacial bonding. It is considered that there are other factors influencing the peeling load. The peeling load for specimens in series 1 is much higher than that for specimens in series 9 although both groups have almost the same stiffnesses and failure patterns (failure in concrete). It is likely that the local aggregate interlock caused this difference.

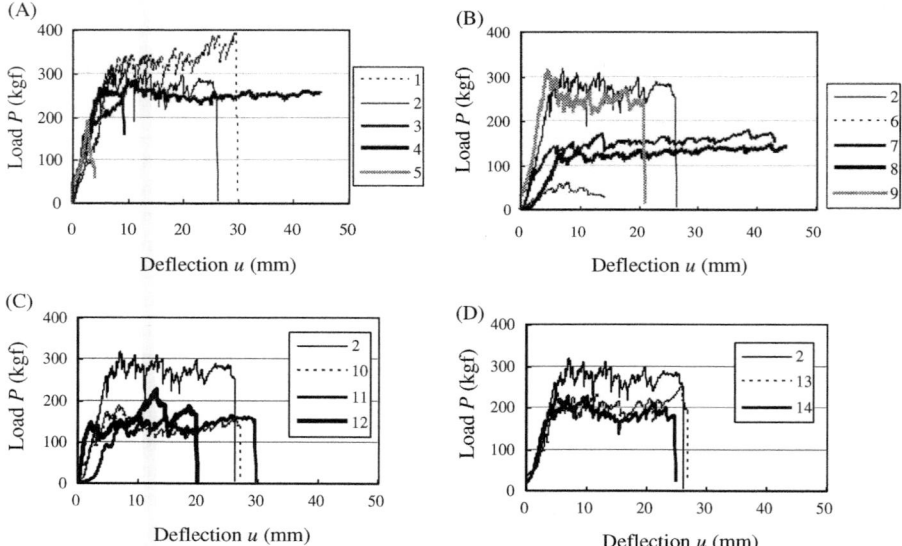

Figure 7.4 Comparison of experimental load—deflection curves (Wu et al., 2005a). (A) Effect of different fiber series (1—5). (B) Effect of different fiber series (2, 6—9). (C) Effect of different adhesives. (D) Effect of surface treatment and concrete strengths.

Effect of different adhesives

The effect of various adhesives is shown in Fig. 7.4C. When fast-hardening epoxy is used, the maximum value of the peeling load decreased markedly compared to their counterpart specimens. It is found that the failure patterns changed from debonding in concrete into debonding at the interface between FRP sheet and adhesive. The interfacial fracture energy is decreased because of the poor bonding resulting from fast hardening. The stiffness of ultraviolet ray hardened prepreg sheet is significantly lower. Therefore the peeling load is consequently lower. Compared with FRP sheets with high stiffness values, the peeling load in specimens with low stiffness is less influenced by adhesives.

Effect of surface treatment and concrete strength

The effect of surface treatment and concrete strength is shown in Fig. 7.4D. The load—deflection curves are almost the same before debonding initiates. The peeling load is lower if the concrete surface is not treated. It is found that the failure patterns change from debonding in concrete into debonding at the concrete—adhesive interface. The interfacial fracture energy is decreased because of the poor bonding resulting from no treatment. The failure patterns return from debonding at the concrete—adhesive interface to debonding in concrete if the strength of the concrete beam is too low, as in the specimen in series 14.

It can be concluded that the peeling load and failure pattern are closely related to the FRP sheet stiffness, adhesive strength, surface treatment, and concrete

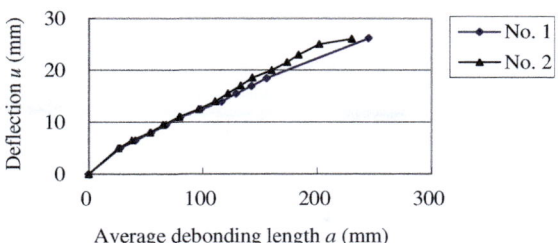

Figure 7.5 Deflection–debonding length curves of specimens in series 2 (Wu et al., 2005a).

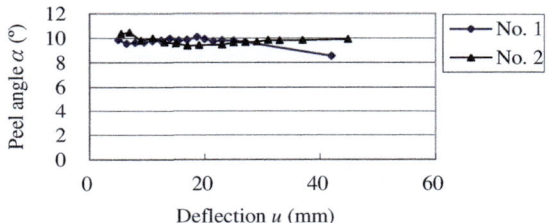

Figure 7.6 Peel angle–deflection curves of specimens in series 4 (Wu et al., 2005a).

strength. The higher FRP sheet stiffness, higher concrete and adhesive strength, and better surface treatment will give a larger peeling load. On the other hand, the higher stiffness of FRP sheet, higher adhesive strength, better surface treatment, and lower concrete strength will result in a higher probability of failure in concrete.

Deflection–average debonding length curves are recorded in the experiment. The typical curves for specimens in series 2 are shown in Fig. 7.5; the curves are approximately linear. The average debonding length is obtained by taking the mean value of the left and right debonding lengths.

The peel angle is measured for various values of deflection. The peel angle–deflection curves for specimens in series 4 are shown in Fig. 7.6. The peel angle is approximately 10 degrees for the two specimens. The peel angles vary from 4 to 12 degrees, except for the specimens in series 6, 7, and 8, for which peel angle is larger than 12 degrees because of the significantly lower stiffness of the FRP sheets.

7.2.2 Theoretical evaluation for spalling resistance of beams with fiber-reinforced polymer sheets

The schematic debonding extension is shown in Fig. 7.7 (Wu et al., 2005a). In Figs. 7.2B and 7.7, b_f is the width of the FRP sheet. α_{max} is the peel angle when debonding appears and propagates. u and Δu are the deflection and deflection increment of the pin, respectively. u_m is the maximum deflection of the pin before debonding. L and ΔL are debonding length and debonding length increment,

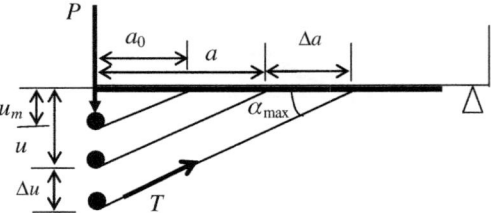

Figure 7.7 Schematical debonding extension (Wu et al., 2005a).

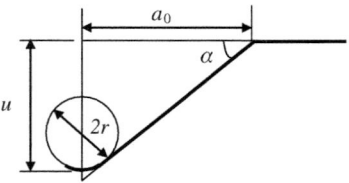

Figure 7.8 Initial loading before debonding (Wu et al., 2005a).

respectively. r is the radius of the pin. T is the peel force per unit width of the FRP sheet. P is the load applied to the pin through the loading device.

1. Before debonding (Fig. 7.8)

When a load is applied to the pin, the FRP sheet of length $2a_0$ is first subjected to deformation because of the initial debonding length, as shown in Fig. 7.8. The peel angle increases from zero to the maximum value with increasing load. Then the interfacial debonding between the FRP sheet and concrete beam appears. The load and peel angle remain constant until complete debonding occurs. It is assumed that the diameter of the pin is small compared with the debonding length. The equilibrium of the pin gives

$$P = 2Tb_f \sin\alpha \tag{7.1}$$

where α is the angle between the beam's bottom and the FRP sheet. According to the geometric relationship, the tensile strain in the FRP sheet ε_f is obtained as

$$\varepsilon_f = \frac{1}{\cos\alpha} - 1 \tag{7.2}$$

Assuming that the FRP sheet is a linear elastic material obeying Hook's law, E_f is Young's modulus of the FRP sheet, and t_f is the thickness of the FRP sheet, the peel force is given by

$$T = E_f t_f \varepsilon_f \tag{7.3}$$

Substituting Eq. (7.3) into Eq. (7.1) yields

$$P = 2E_f t_f b_f \varepsilon_f \sin\alpha \tag{7.4}$$

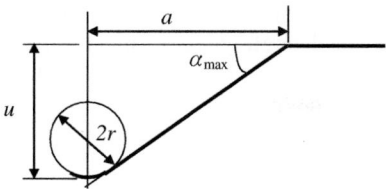

Figure 7.9 Debonding propagation (Wu et al., 2005a).

Substituting Eq. (7.2) into Eq. (7.4) yields

$$P = 2E_f t_f b_f x(1 - \frac{1}{\sqrt{1+x^2}}), \quad x = \tan\alpha = \frac{u}{a_0} \quad \text{for } 0 \le u \le u_m \text{ or } 0 \le \alpha \le \alpha_{max} \tag{7.5}$$

2. Debonding propagation (Fig. 7.9)

By using an energy balance with no other external work being done apart from peeling the specimen from the concrete, the input energy must equal the total work done:

$$P \cdot \Delta u = 2\Delta a \, b_f t_f U + 2\Delta a \, b_f G_f \tag{7.6}$$

where U is the strain energy density and G_f is the interfacial fracture energy. The tensile strain in the FRP sheet is obtained as

$$\varepsilon_f = \frac{1}{\cos\alpha_{max}} - 1 \tag{7.7}$$

The strain energy density can be calculated as

$$U = \frac{1}{2} E_f \varepsilon_f^2 \tag{7.8}$$

A similar derivation as is Eq. (7.4) gives

$$P = 2E_f t_f b_f \varepsilon_f \sin\alpha_{max} \tag{7.9}$$

and the geometric relationship yields

$$\Delta u = \Delta a \tan\alpha_{max} \tag{7.10}$$

Substituting Eqs. (7.7–7.10) into Eq. (7.7) yields

$$G_f = E_f t_f \left(\frac{1}{2}x^2 + \frac{1}{\sqrt{1+x^2}} - 1\right) \quad x = \tan\alpha_{max} \tag{7.11}$$

Eq. (7.11) can be deduced by a different approach (Kimpara et al., 1998). Consider the fact that the peel angle α_{max} is much smaller than 1. By using the Taylor power series expansion, Eq. (7.11) is simplified to the following approximate but simple analytical formula:

$$G_f = \frac{3}{8}E_f t_f x^4 \qquad\qquad (7.12)$$

Eq. (7.9) can be simplified to

$$P = E_f t_f b_f x^3 \qquad\qquad (7.13)$$

Solving Eq. (7.12) for x and then substituting into Eq. (7.13) gives a simple expression:

$$P = E_t b_f \left(\frac{8G_f}{3E_f t_f}\right)^{0.75} = 2.087 b_f G_f^{0.75}(E_t)^{0.25} \qquad\qquad (7.14)$$

It can be concluded that the peel load P is proportional to the FRP sheet stiffness E_t raised to the power of 1/4 and is proportional to the interfacial fracture energy G_f raised to the power of 3/4. Therefore the peel load increases with increasing interfacial fracture energy and FRP sheet stiffness. In the case of a lower interfacial fracture energy G_f, choosing FRP sheets with high Young's modulus or more layers can increase the peel load.

Eq. (7.14) can be used to identify the interfacial fracture energy if the peeling load and stiffness of the FRP sheet are known. The peeling load can be determined according to the load−deflection curve. In fact, the peeling load changes constantly as the result of unevenness of the concrete surface, variable adhesive thickness, loading disturbance, and defects of the FRP sheet. Here, the maximum peeling load is considered as the average value of the maximum peak values at points P_1, P_2, P_3, and P_4 as shown in Fig. 7.10. The interfacial fracture energies calculated from the maximum peeling load are shown in Table 7.4 and Fig. 7.11A. These values of the interfacial fracture energy fall within the range of 0.344−1.543 N/mm. The mean value of the maximum and minimum of the extreme points P_1, P_2, P_3, P_4, P_5, P_6,

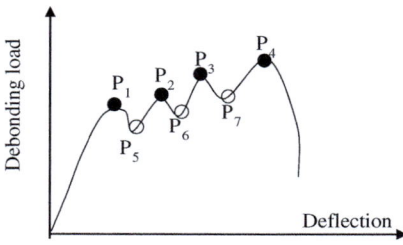

Figure 7.10 Schema of determining maximum and mean peeling loads (Wu et al., 2005a).

Table 7.4 Maximum values of interfacial fracture energy (mean values) (Wu et al., 2005a).

Series	Interfacial fracture energy			
	No. 1	No. 2	No. 3	Average
1	0.694(0.652)	0.890(0.829)	1.072(1.034)	0.88(0.83)
2	0.610(0.592)	1.105(1.029)	0.998(0.935)	0.9(0.85)
3	0.715(0.685)	0.685(0.663)	0.727(0.632)	0.71(0.66)
4	0.754(0.717)	1.048(1.012)	0.878(0.844)	0.89(0.86)
5	0.414(0.414)	0.592(0.591)	0.666(0.666)	0.55(0.55)
6	–	–	–	–
7	0.683(0.652)	0.936(0.861)	0.928(0.901)	0.85(0.80)
8	1.199(1.095)	1.543(1.503)	1.360(1.333)	1.37(1.31)
9	0.605(0.571)	0.798(0.758)	0.732(0.673)	0.71(0.67)
10	0.349(0.339)	0.453(0.393)	0.451(0.437)	0.42(0.39)
11	0.379(0.385)	0.344(0.305)	0.436(0.417)	0.39(0.36)
12	0.393(0.378)	0.502(0.491)	0.536(0.499)	0.48(0.45)
13	0.569(0.532)	0.684(0.663)	0.674(0.596)	0.64(0.60)
14	0.477(0.445)	0.654(0.618)	0.583(0.532)	0.57(0.53)

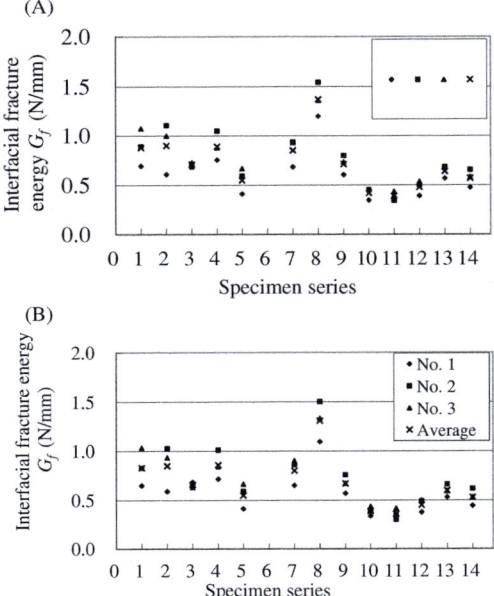

Figure 7.11 Identified interfacial fracture energy (Wu et al., 2005a). (A) Using the maximum peeling load. (B) Using the mean peeling load.

and P_7 through the whole debonding process is considered as an acceptable value of the peeling load. Using this method, the identified values of the interfacial fracture energy are also shown in Table 7.4 and Fig. 7.11B. The values of the interfacial fracture energy fall within the range of $0.305-1.503$ N/mm.

Theoretical results can be compared with the experimental values by using the identified values of the interfacial fracture energy. The theoretical and experimental load–deflection curves for specimens 1 and 3 in series 2 are shown in Fig. 7.12A and B, respectively. The theoretical load–deflection curves based on the approach proposed by Wu et al. (2005a) are obtained using the identified interfacial fracture energy. A close agreement between the peeling load predicted from the proposed analytical method and the experimental debonding load is observed. There is little difference between the peeling load–deflection curves obtained by using the identified interfacial fracture energy from the maximum and mean peeling loads. It can also be observed that the theoretical and experimental loads are different before debonding occurs. It is believed that the experimental debonding length is larger than the predicted initial debonding length of 15 mm. Therefore the slope of the theoretical load–deflection curves is larger than the experimental value.

The deflection–debonding length curves of specimen 1 in series 2 are shown in Fig. 7.12C. The theoretical curves are linear and pass the origin of the coordinate. The experimental curve is approximately linear. The angle–deflection curves of specimen 1 in series 2 are given in Fig. 7.12D. In the early stage of debonding, the experimental peel angle is much larger than the theoretical angle. The theoretical and experimental peel angles become closer with the further propagation of debonding. The theoretical curves are horizontal and linear lines.

7.3 Spalling resistance of structures with bidirectional fiber–reinforced polymer sheets

7.3.1 Experimental studies

Wu et al. (2005b) carried out experiments using a static load tester on a concrete slab externally bonded with FRP sheets, as shown in Fig. 7.2A. Seven small-scale specimens (500 mm in length $\times$ 500 mm in width $\times$ 80 mm in thickness) and seven large-scale specimens (2300 mm $\times$ 2300 mm $\times$ 200 mm) were tested. Each slab was made with a hole in the center, in which an indenter made of concrete was fitted. The FRP sheets were bonded to the lower surface of the concrete slab and the indenter. The load was applied to press the indenter moving downward, and then the indenter forced the FRP sheet to peel off from the slab. In this way, the FRP sheets bonded to concrete slabs were subjected to a punching–peeling test. The compression strengths of the concrete used for the small-scale specimens were 22.1 and 11.8 N/mm^2. The compression strength of the concrete used for the large-scale specimens was 28 N/mm^2. The displacement was measured at three points, as

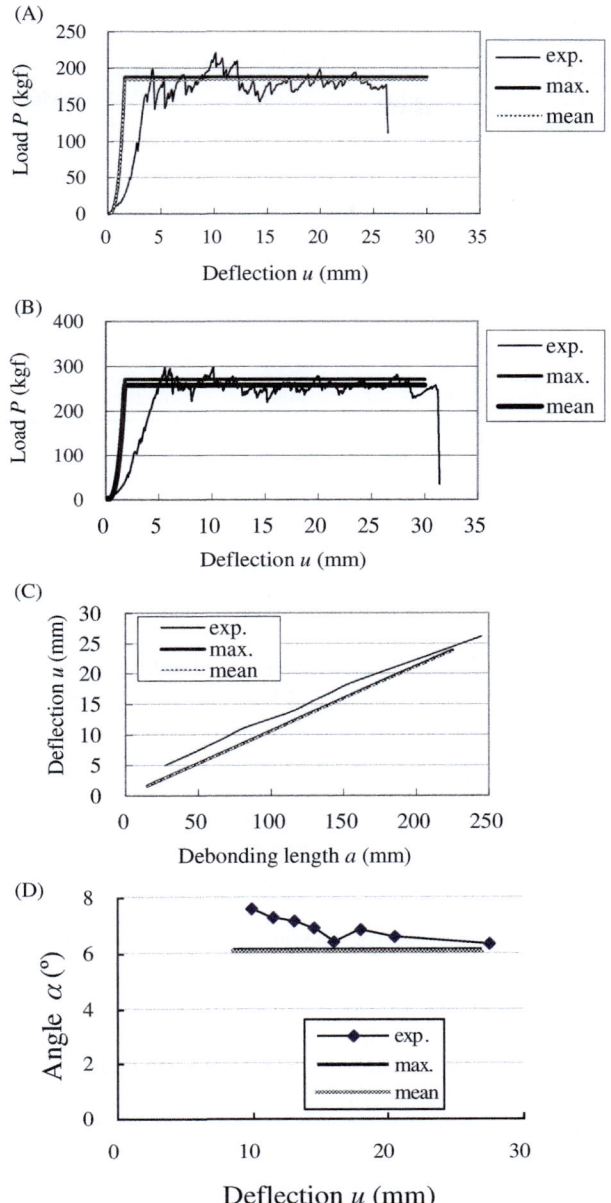

Figure 7.12 Comparison of theoretical with experimental results (Wu et al., 2005a). (A) Load—deflection curves of specimen 1 in series 2. (B) Load—deflection curves of specimen 3 in series 2. (C) Deflection—debonding length curves of specimen 1 in series 2. (D) Angle—deflection curves of specimen 1 in series 2.

Table 7.5 Properties of FRP sheets (Wu et al., 2005b).

FRP materials	Fiber area weight (g/m²)	Young's modulus E_f (GPa)	Tensile strength (MPa)	Design thickness t_f (mm)	Stiffness of FRP sheet $Et \times 10^3$ (N/mm)
High-strength carbon fiber	200	235	4150	0.111	26.1
Aramid fiber	260	118	2450	0.193	22.8
Glass fiber	300	74	1650	0.118	8.7
PBO fiber	200	265	4450	0.128	33.9

shown in Fig. 7.2. To avoid the stress relaxation immediately after the loading, the displacement and the load were measured 2 min after an increment of loading.

Four types of FRP sheets made of different fibers were used in the tests. They are high-strength carbon fiber, aramid fiber, glass fiber, and poly-p-phenylene-benzobisoxazole (PBO) fiber. The mechanical properties of the FRP sheets used in the tests are shown in Table 7.5.

Each specimen was subjected to surface preparation, primer coating, and smoothing with epoxy putty before the FRP sheet was bonded to it with epoxy resin. The seven small-scale specimens were numbered as S-1, S-2, S-3, S-4, S-5, S-6, and S-7. The seven large-scale specimens were numbered as L-1, L-2, L-3, L-4, L-5, L-6, and L-7. The FRP sheet was bonded in two orthogonal layers (0/90 degrees) to the lower surface of the slab, except for the specimen S-3, where only one layer of carbon FRP sheet was bonded in the support direction. The diameter of the hole at the center of the slab is 50 mm for the small-scale specimens. The bond length for all specimens is 190 mm. S-1 is the control specimen. The compressive strength of concrete for specimen S-2 is 11.8 N/mm². Carbon FRP sheet is bonded to the concrete slab using fast-hardening epoxy and cured for 3 h for specimen S-4. Specimens S-5, S-6, S-7 were fabricated with glass, aramid, and PBO FRP sheets, respectively. The specimen details are shown in Table 7.6.

The diameter of the hole at the center of the slab varied between 200 and 1000 mm for the large-scale specimens. The compressive strength of concrete for all the large-scale specimens is 28 N/mm². L-1 is the control specimen. The bond length of specimen L-2 is 900 mm. Specimen L-3 is constrained by four steel plates, the dimensions of which are $50 \times 600 \times 3.2$ mm³, which were arranged in a square at a distance of 300 mm from the center of the circular hole. Each of the four plates was fixed with three M10 anchor bolts at a 250-mm pitch. The hole diameters for specimens L-4 and L-5 are 500 and 1000 mm, respectively. Specimens L-6 and L-7 were fabricated with glass and aramid sheets, respectively. The specimen details are shown in Table 7.7 and Fig. 7.13.

Table 7.6 Specimen parameters for small-scale specimens (Wu et al., 2005b).

Specimen	FRP type	Sheet direction	Concrete strength (MPa)	Adhesives	Fiber unit weight (g/m²)	Test variation
S-1	Carbon fiber	0/90° two layers	22.1	Epoxy	200	Control specimen
S-2	Carbon fiber	0/90° two layers	11.8	Epoxy	200	Lower concrete compressive strength
S-3	Carbon fiber	One layer	22.1	Epoxy	200	Effect of number of FRP layers
S-4	Carbon fiber	0/90° two layers	22.1	MMA	200	Type of epoxy (fast hardening)
S-5	Aramid fiber	0/90° two layers	22.1	Epoxy	260	Fiber type
S-6	Glass fiber	0/90° two layers	22.1	Epoxy	300	Fiber type
S-7	PBO fiber	0/90° two layers	22.1	Epoxy	200	Fiber type

Table 7.7 Specimen parameters for large-scale specimens (Wu et al., 2005b).

Specimen	FRP type	Hole diameter (mm)	Bond length (mm)	Fiber area weight (g/m²)	Test variation
L-1	Carbon fiber	Φ200	500	200	Control specimen
L-2	Carbon fiber	Φ200	900	200	Increase in the bond length
L-3	Carbon fiber	Φ200	500	200	Using additional plate constraint
L-4	Carbon fiber	Φ500	500	200	Size effect of the hole (diameter)
L-5	Carbon fiber	Φ1000	400	200	Size effect of the hole (diameter)
L-6	Aramid fiber	Φ200	500	260	Fiber type
L-7	Glass fiber	Φ200	500	300	Fiber type

Fig. 7.14 shows the loading pattern in the punching—peeling test. All the specimens were tested in a displacement-controlled mode with an average displacement increment of 0.5 mm and with an average loading speed of 1 mm/min. Loads were recorded 2 min after each loading increment. Electrical strain gauges and a LVDT were used to measure the FRP strains and indenter deflections, respectively.

Figs. 7.15 and 7.16 give the debonding patterns of several specimens. The white lines indicate the shape of the peeling zones of FRP sheets for every 2 mm of

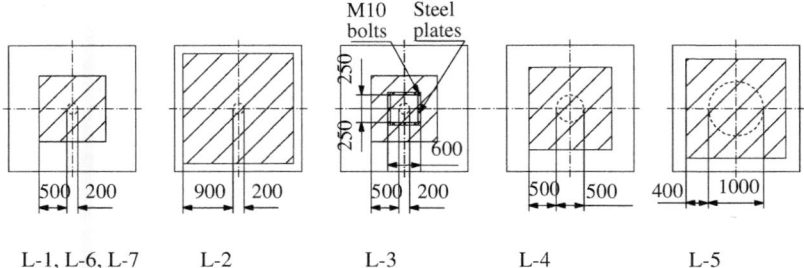

Figure 7.13 Dimensions of large-scale specimens (Wu et al., 2005b).

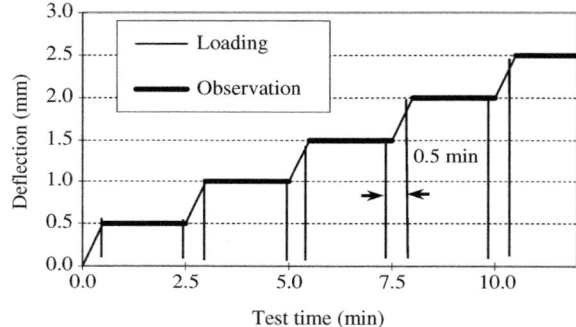

Figure 7.14 Loading pattern (Wu et al., 2005b).

downward movement. It can be seen that the shape of the debonding zone is approximately elliptical for specimen S-3 bonded with one layer of carbon FRP sheet to the lower surface of the slab (Fig. 7.15B). In the early stages of peeling, the shape of the debonding zone is approximately circular for all specimens bonded with two orthogonal layers (0/90 degrees). As the FRP sheets continued peeling, the debonding developed more quickly in the fiber direction than in the 45-degree direction, and the peeling zone gradually became square-shaped because of the orthogonal alignment. Test results for the small-scale and large-scale specimens are summarized in Tables 7.8 and 7.9, respectively.

Except for the specimens S-5 and L-7, maximum load occurred when the FRP sheets were being peeled off completely. The small-scale and large-scale specimens will be discussed in the following sections.

7.3.1.1   Small-scale specimens

The experimental load−deflection curves of the small-scale specimens are shown in Fig. 7.17.

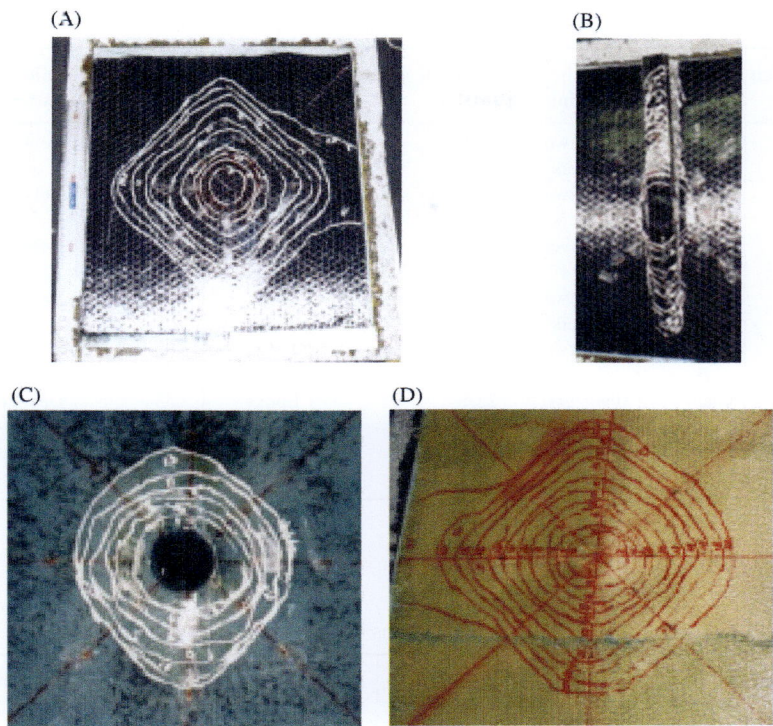

Figure 7.15 Debonding pattern of small-scale specimens (Wu et al., 2005b). (A) S-1 CFRP sheet (0/90°) (B) S-3 CFRP sheet (0°). (C) S-5 GFRP sheet (0/90°). (D) S-6 AFRP sheet (0/90°).

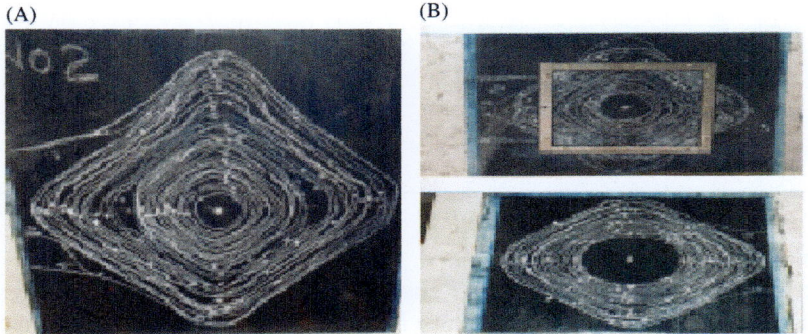

Figure 7.16 Debonding pattern of large-scale specimens (Wu et al., 2005b). (A) L-2 $L = 900$ mm. (B) L-3 plate constraint.

Table 7.8 Test results for small-scale specimens (Wu et al., 2005b).

Specimen	FRP type	Sheet direction	Concrete strength (MPa)	Adhesives	Max. load (kN)	Failure pattern
S-1	Carbon fiber	0/90° two layers	22.1	Epoxy	7.95	Debonding
S-2	Carbon fiber	0/90° two layers	11.8	Epoxy	8.18	Debonding
S-3	Carbon fiber	One layer	22.1	Epoxy	1.78	Debonding
S-4	Carbon fiber	0/90° two layers	22.1	MMA	5.65	Debonding
S-5	Aramid fiber	0/90° two layers	22.1	Epoxy	3.08	Fiber rupture
S-6	Glass fiber	0/90° two layers	22.1	Epoxy	9.33	Debonding
S-7	PBO fiber	0/90° two layers	22.1	Epoxy	7.96	Debonding

Table 7.9 Test results for large-scale specimens (Wu et al., 2005b).

Specimen	FRP type	Hole diameter (mm)	Bond length (mm)	Max. deflection (mm)	Max. load (kN)	Failure pattern
L-1	Carbon fiber	Φ200	500	43.0	17.0	Debonding
L-2	Carbon fiber	Φ200	900	71.0	23.9	Debonding
L-3	Carbon fiber	Φ200	500	46.0	33.7	Bolt pulled out debonding
L-4	Carbon fiber	Φ500	500	42.5	26.1	Debonding
L-5	Carbon fiber	Φ1000	400	32.0	29.1	Debonding
L-6	Aramid fiber	Φ200	500	48.0	24.6	Debonding
L-7	Glass fiber	Φ200	500	48.0	16.4	Fiber rupture

Parameters affecting spalling resistance
Effect of fiber-reinforced polymer sheet layers and direction In the case of two-layer orthogonal alignment as shown in Figs. 7.15A and 7.17A for specimen S-1, the load increases with increasing deflection. The debonding extends in all directions. The peeling load attains its maximum value of 7.95 kN when the tip of the CFRP sheet propagates to the position where the bond area is too small and bond

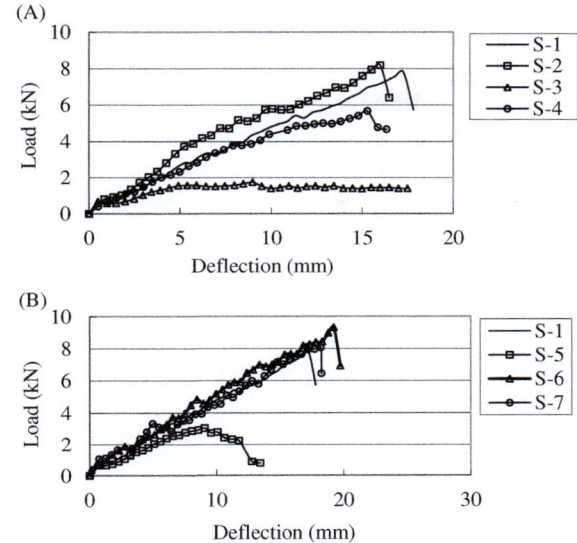

Figure 7.17 Experimental load–deflection curves for small-scale specimens (Wu et al., 2005b). (A) Effect of concrete strength, FRP sheet plated, adhesives. (B) Effect cf fiber types.

stress between the FRP sheets and concrete substrate cannot balance the stress in the FRP sheets. Then, debonding develops rapidly in the 0-degree fiber direction and the load decreases. In the case of one-layer unidirectional alignment as shown in Figs. 7.15B and 7.17A for specimen S-3, the support is in the fiber direction. The debonding extends in the 0-degree fiber direction with the debonding width equal to the hole diameter. The load maintains a nearly constant value of 1.78 kN with increasing deflection. The maximum load in S-1 is more than four times that in S-3 although only twice as many FRP sheets are used.

Effect of concrete strength By comparing specimens S-1 and S-2 in Fig. 7.17A, the debonding loads and maximum load for specimen S-2 with a lower concrete strength of 11.8 MPa are slightly higher than that for specimen S-1 with a higher concrete strength of 22.1 MPa. The difference between the maximum loads of both specimens is considered to be within the experimental error. The effect of the concrete strength on the spalling resistance cannot be concluded from the experiment for the following two reasons: (1) the failure patterns of both specimens S-1 and S-2 are interface debonding between the FRP sheet and concrete and are not debonding in concrete. The maximum load depends on the concrete strength only if a failure occurs in concrete; (2) the concrete compressive strength inside the two specimens was different at different places. It seems that there is no significant difference between the surface concrete strengths of the two specimens.

Effect of adhesives The epoxy in the control specimen S-1 was cured for at least 7 days at a constant room temperature of 23°C before the test. The MMA adhesive in specimen S-4 was cured only for 3 h at the same room temperature of 23°C before the test. Compared with ordinary epoxy, MMA adhesive can harden in a much shorter period. MMA adhesive can harden at the temperature of −10°C, but ordinary epoxy is difficult to harden at temperatures below 5°C. It can be seen in Fig. 7.17A that the debonding loads for both specimens are very similar if the deflection is smaller than 10 mm although the maximum load for specimen S-1 is larger than that for specimen S-4. Therefore the MMA adhesive is considered to be an adequate adhesive for the rapid repair of tunnels and bridges against concrete spalling.

Effect of fiber types Fig. 7.17B shows the experimental load−deflection curves for specimens S-1, S-5, S-6, and S-7 bonded with CFRP, glass fiber−reinforced polymer (GFRP), aramid fiber−reinforced polymer (AFRP), and PBO fiber−reinforced polymer (PFRP) sheets, respectively. Glass fiber rupture in S-5 occurred at the indenter bottom when the deflection of the concrete indenter reached 8 mm, as shown in Fig. 7.17B. The maximum load in S-5 was much lower than that in the control specimen S-1. The load increased with increasing deflection if the deflection is smaller than 7 mm. After that, the load decreased with increasing deflection. A theoretical analysis of the reason that rupture is more likely to appear in glass fiber will be given in the following section. The maximum loads in the AFRP sheet specimen S-6 and the PFRP sheet specimen S-7 have the same order of magnitude as that in the CFRP sheet specimen S-1.

7.3.1.2 Large-scale specimens

The experimental load−deflection curves are shown in Fig. 7.18.

Parameters affecting spalling resistance

Effect of bond length As shown in Fig. 7.18A, specimens L-1 and L-2 have the same carbon fiber and hole diameter. The slopes of the load−displacement curves are almost the same. However, the maximum load in L-2 with a bond length of 900 mm is much higher than that in L-1 with a bond length of 500 mm, increasing from 17.0 to 23.9 kN. Therefore the load-carrying capacity increases with increasing bond length.

Effect of plate constraint As shown in Fig. 7.18B, specimens L-1 and L-3 have the same carbon fiber, hole diameter, and bond length. The maximum load of specimen L-3 is much higher than that of specimen L-1, which shows the efficiency of plate constraint. However, debonding is observed outside the plate constraint area because of the large bolt pitch of 250 mm. A smaller bolt pitch can achieve higher specimen stiffness and smaller deflection.

Effect of hole diameter As shown in Fig. 7.18C, the larger the hole diameter is, the higher is the initial debonding load (approximately 1 kN for L-1, 9 kN for L-4, and 17 kN for L-5). The slope of the load−deflection curves remains almost the

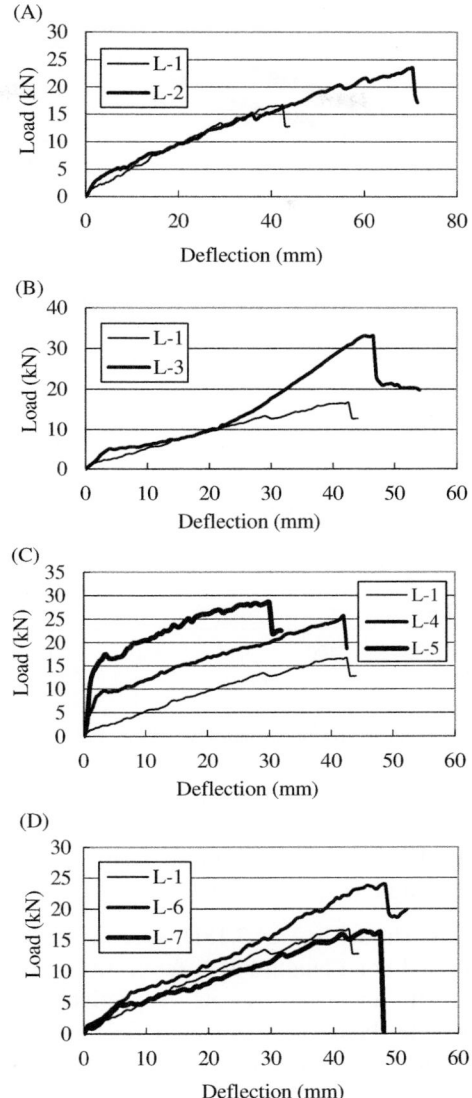

Figure 7.18 Experimental load−deflection curves for small-scale specimens (Wu et al., 2005b). (A) Effect of bond length. (B) Effect of plate constraint. (C) Effect of the hole diameters. (D) Effect of fiber types.

same despite different hole diameters. The larger the sum of the debonding length and the hole radius $a + r_0$ (SDLHR) is, the higher is the maximum load. The relationship between the indenter deflection and SDLHR, as shown in Fig. 7.19, is almost linear. Under the same deflection, the SDLHR in specimen L-4 with a hole

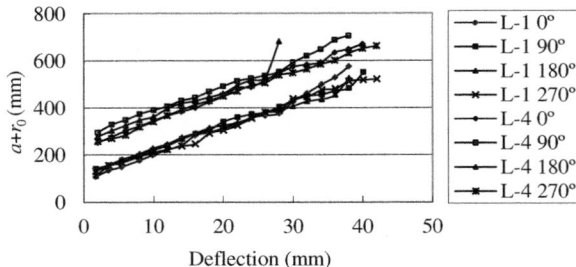

Figure 7.19 SDLHR—deflection curves for specimens L-1 and L-4 (Wu et al., 2005b).

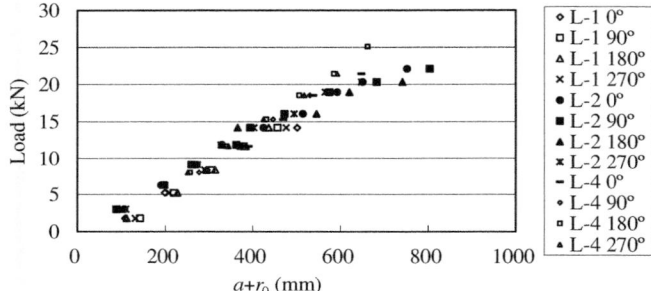

Figure 7.20 Load—SDLHR curves for specimens L-1, L-2, and L-4 (Wu et al., 2005b).

diameter of 500 mm is larger than that in specimen L-1 with a hole diameter of 200 mm. The load—SDLHR curves for specimens L-1, L-2, and L-4 are shown in Fig. 7.20. The load—SDLHR relation is almost independent of the hole diameter (L-1 and L-4) and bond length (L-1 and L-2).

Effect of fiber type As shown in Fig. 7.18D, specimens L-1, L-6, and L-7 have the same hole diameter and bond length but different fibers. L-1, L-6, and L-7 were bonded with CFRP, AFRP, and GFRP sheets, respectively. Glass fiber rupture in L-7 occurred at the indenter bottom. The load—deflection curves are almost linear for all fiber sheets. The maximum load in specimen L-6 is the highest of the three specimens.

7.3.2 Analytical studies

In the early stages of peeling, the shape of the debonding zone is approximately circular for specimens bonded in two orthogonal layers (0/90 degrees). As the FRP sheets continued peeling, the debonding developed more quickly in the fiber direction than in the 45-degree direction, and the peeling zone gradually became square-shaped because of the orthogonal alignment (Wu et al., 2005b).

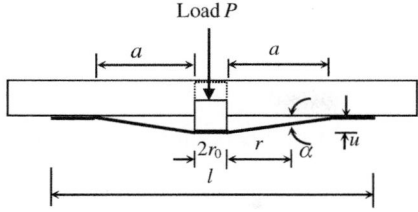

Figure 7.21 Equilibrium of forces between FRP sheet and cylinder (Wu et al., 2005b).

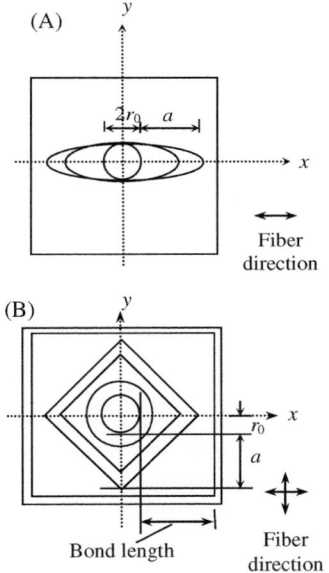

Figure 7.22 Failure pattern of specimens (Wu et al., 2005b). (A) One layer of FRP sheet. (B) Two layers of FRP sheet.

The peeling load is applied to the cylinder indenter of radius r_0. An equilibrium state due to extension of the debonding is schematically shown in Fig. 7.21. The debonding patterns of both one- and two-layer FRP sheets are shown in Fig. 7.22.

7.3.2.1 One-layer fiber-reinforced polymer sheet

According to the test, the debonding area for one layer of the FRP sheet is elliptical:

$$\frac{x^2}{(a+r_0)^2} + \frac{y^2}{r_0^2} = 1 \tag{7.15}$$

The circular cylindrical indenter on which the load is applied can be expressed by

$$x^2 + y^2 = r_0{}^2 \tag{7.16}$$

The peeling angle and strain of fibers along the x direction are obtained by a geometrical analysis

$$\tan\alpha' = \frac{u}{a\sqrt{1 - \frac{y^2}{r_0{}^2}}} \tag{7.17}$$

α' is the peel angle between the plate bottom and the FRP sheet in the fiber direction at y for the unidirectional case or the angle the linking line with the origin makes with a horizontal plane for the two-layer bidirectional case. u is the deflection of the indenter, a is the debonding length in the fiber direction, and r_0 is the hole radius.

$$\varepsilon_f = \frac{1}{\cos\alpha'} - 1 \tag{7.18}$$

The equilibrium of forces between the sheet and indenter gives

$$P = 2\int_{-r_0}^{r_0} E_f\varepsilon_f\sin\alpha' t_f \, dy = 4Etr_0\beta\arctan\beta \ , \qquad \beta = \frac{u}{a} \tag{7.19}$$

Because the as discussed earlier load—deflection curve is not linear, the conventional compliance method is not appropriate to evaluate the energy release rate. A modification of the compliance method for evaluating the energy release rate was proposed by Kimpara et al. (1998). The elastic strain energy can be obtained by integrating Eq. (7.19) with respect to u, under the condition that the debonding length a is constant.

$$U = 2Etar_0\left[(1 + \beta^2)\arctan\beta - \beta\right] \tag{7.20}$$

Finally, the energy release rate is derived by differentiating the strain energy with respect to a, under the condition that the u value of the FRP sheet is constant.

$$G_e = -\frac{\partial U}{\pi r_0 \partial a} = \frac{2}{\pi}E_t\left[\beta - (1 - \beta^2)\arctan\beta\right] \tag{7.21}$$

where G_e is completely determined by β. The derivation method as discussed earlier is called the membrane peeling method. Setting $G_e = G_f$ in Eq. (7.21) yields a relation between the interfacial fracture energy G_f and β.

$$G_f = \frac{2}{\pi} E_t \left[\beta - (1 - \beta^2) \arctan\beta \right] \tag{7.22}$$

Consider the fact that β is much smaller than 1. By using the Taylor power series expansion, Eq. (7.22) is simplified to the following approximate but simple analytical formula:

$$G_f = \frac{8}{3\pi} E_t \beta^3 \tag{7.23}$$

Eq. (7.19) can be simplified to

$$P = 4E_t r_0 \beta^2 \tag{7.24}$$

Solving Eq. (7.23) for β and then substituting into (7.24) gives a simple expression:

$$P = (3\pi)^{2/3} r_0 G_f^{2/3} (E_t)^{1/3} = 4.462 r_0 G_f^{2/3} (E_t)^{1/3} \tag{7.25}$$

It can be concluded that P is proportional to the FRP sheet stiffness E_t raised to the power of 1/3 and is proportional to the interfacial fracture energy G_f raised to the power of 2/3. Therefore the peeling load increases with increasing interfacial fracture energy or FRP sheet stiffness. In the case of a lower interfacial fracture energy G_f, choosing FRP sheets with high Young's modulus or more layers of FRP sheets can increase the peeling load.

7.3.2.2 Two layers of fiber-reinforced polymer sheet

The equilibrium of forces between the FRP sheet and cylindrical indenter is shown in Fig. 7.21. According to the debonding pattern of two layers of FRP sheet as shown in Fig. 7.23 and the assumption that the peel angle α is constant along the fiber direction for a given debonding length, the resultant force in the FRP sheets in the x-direction can be written as

$$P_x = 4 \int_0^{r_0 + a} \sigma t \cos\theta \sin\alpha'(y) dy \tag{7.26}$$

The approximation that r_0 and α' are small is introduced in Eq. (7.26). Differentiating the geometric relation

$$y = (r_0 + a) \frac{\tan\theta}{\tan\theta + 1} \tag{7.27}$$

yields

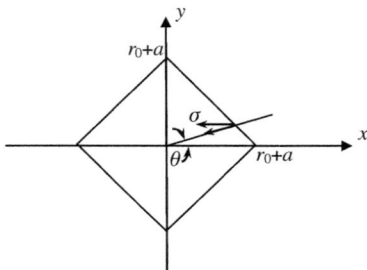

Figure 7.23 Debonding pattern of two layers of FRP sheets (Wu et al., 2005b).

$$dy = \frac{r_0 + a}{2\sin\theta\cos\theta + 1} d\theta \tag{7.28}$$

Additionally,

$$\sin\alpha' = \sqrt{2}\sin\alpha\cos\left(\frac{\pi}{4} - \theta\right), 0 \le \theta \le \frac{\pi}{2} \tag{7.29}$$

Substituting (7.28) and (7.29) into (7.26) and integrating with respect to θ from 0 to $\pi/2$ gives

$$P_x = \pi\sigma \ t_f(r_0 + a)\sin\alpha \tag{7.30}$$

The resultant force in the y-direction can be obtained similarly as $P_y = P_x$, such that

$$P = P_x + P_y = 2\pi\sigma \ t_f(r_0 + a)\sin\alpha \tag{7.31}$$

Eq. (7.31) can be rewritten as

$$P = 2\pi E_f \ t_f(r_0 + a)\beta\left(1 - \frac{1}{\sqrt{1 + \beta^2}}\right), \ \beta = \frac{u}{a} = \tan\alpha \tag{7.32}$$

Integration of Eq. (7.32), with respect to u under the condition that the debonding length a is constant, gives

$$U = 2\pi E t a (r_0 + a)\left(\frac{1}{2}\beta^2 - \sqrt{1 + \beta^2} + 1\right) \tag{7.33}$$

Differentiating U with respect to a, under the condition that the sheet deflection u is constant, gives the energy release rate

$$G_e = -\frac{\partial U}{4(r_0 + a)\partial a} = \frac{\pi}{2}E_t\left(\frac{1}{2}\beta^2 + \frac{1}{\sqrt{1+\beta^2}} - 1\right)$$
$$-\frac{\pi}{2}E_t\frac{a}{a+r_0}\left(\frac{1}{2}\beta^2 - \sqrt{1+\beta^2} + 1\right)$$

(7.34)

When a is relatively large, $a/(a + r_0) \approx 1$. Eq. (7.34) simplifies to

$$G = \frac{\pi}{2}E_t\left(\frac{1}{\sqrt{1+\beta^2}} + \sqrt{1+\beta^2} - 2\right)$$

(7.35)

Setting $G = G_f$ in Eq. (7.35) yields a relation between the interfacial fracture energy G_f and β.

$$G_f = \frac{\pi}{2}E_t\left(\frac{1}{\sqrt{1+\beta^2}} + \sqrt{1+\beta^2} - 2\right)$$

(7.36)

The value of β can be found from Eq. (7.20) for a given a and r_0. For a large value of a, the peeling angle α is completely determined by the interfacial fracture energy and stiffness, see Eq. (7.36). By considering the fact that β is much smaller than 1 and using the Taylor power series expansion, Eq. (7.36) is simplified to the following approximate but simple analytical formula:

$$G_f = \frac{\pi}{8}E_t\beta^4$$

(7.37)

Eq. (7.32) can be simplified to

$$P = \pi E t(r_0 + a)\beta^3$$

(7.38)

Solving Eq. (7.37) for β and then substituting into (7.38) gives a simple expression:

$$P = \pi\left(\frac{8}{\pi}\right)^{3/4}(r_0 + a)G_f^{3/4}(E_t)^{1/4} = 6.333(r_0 + a)G_f^{3/4}(E_t)^{1/4}$$

(7.39)

It can be concluded that P is proportional to the FRP sheet stiffness E_t raised to the power of 1/4 and is proportional to the interfacial fracture energy G_f raised to the power of 3/4. By defining the unit debonding strength P_0 as P divided by the debonding perimeter, the following relation can be obtained:

$$P_0 = \frac{P}{4\sqrt{2}(r_0 + a)} = 1.12G_f^{3/4}(Et)^{1/4}$$

(7.40)

Table 7.10 Comparison between theoretical and experimental results ($G_f = 0.25$ N/mm) (Wu et al., 2005b).

Specimen no.	Theory		Experiment		Experiment/theory	
	Max. load (kN)	Max. deflection (mm)	Max. load (kN)	Max. deflection (mm)	Max. load	Max. deflection
L-1	17.1	35.1	17.0	43	0.99	1.22
L-2	28.5	63.2	23.9	71	0.84	1.12
L-4	21.4	35.1	26.1	43	1.22	1.23
L-5	25.7	28.1	29.1	32	1.13	1.14
L-6	16.5	36.3	24.6	48	1.49	1.32
L-7	13.0	46.2	16.4	48	1.23	1.04

The normal stress in the FRP sheet simplifies to

$$\sigma_f = \sqrt{\frac{2E_f G_f}{\pi\ t_f}} \tag{7.41}$$

The deflection of the indenter can be written as

$$u = 1.263 a G_f^{1/4}(Et)^{-1/4} \tag{7.42}$$

7.3.3   Theoretical results

Table 7.10 shows a comparison of the experimental results obtained from the large-scale specimens and the theoretical results with $G_f = 0.25$ kN. In the analysis, it was assumed that the maximum load and maximum displacement were reached when the sheet peeled off completely. In terms of both the maximum load and maximum displacement, the theoretical values agree well with the experimental values, except for specimen L-6. It is believed that the interfacial fracture energy of specimen L-6 bonded with the FRP sheet is greater than 0.25 kN.

The experimental and theoretical load−SDLHR curves for specimen L-1 bonded with the CFRP sheet are shown in Fig. 7.24A. The theoretical values were obtained by taking $G_f = 0.25$ N/mm. The theoretical values agree well with the experimental results. The experimental and theoretical load−SDLHR curves for specimen L-6 bonded with the FRP sheet are shown in Fig. 7.24B. Again, the interfacial fracture energy was taken as 0.25 N/mm. An underestimated interfacial fracture energy caused the difference between the experimental and theoretical values.

The strain−deflection plot for specimen L-1 is shown in Fig. 7.25. Theoretically, the strain in the CFRP sheet is a constant value of 2469 after the appearance of the debonding. In the experiment, the strain still increases after the

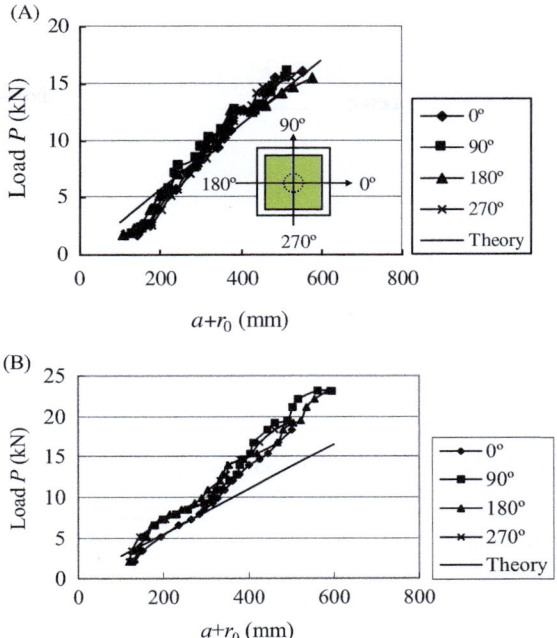

Figure 7.24 Comparison of load−SDLHR curves (G_f = 0.25 N/mm) (Wu et al., 2005b). (A) Specimen L-1. (B) Specimen L-6.

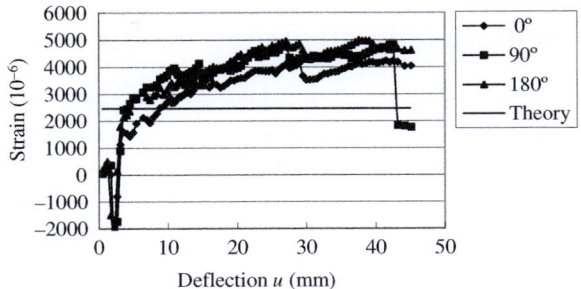

Figure 7.25 Comparison of strain−deflection curves for L-1 (G_f = 0.25 N/mm) (Wu et al., 2005b).

debonding. The peeling angle−deflection plot is given in Fig. 7.26. Most of the experimental values fall within the range of 4−6 degrees. The theoretical value is 4.03 degrees. Stresses in the FRP sheets can be calculated from Eq. (7.41). However, if the stress is set as equal to the tensile strength in the FRP sheets, the minimum interfacial fracture energy for the occurrence of fiber rupture can be obtained as shown in Table 7.11.

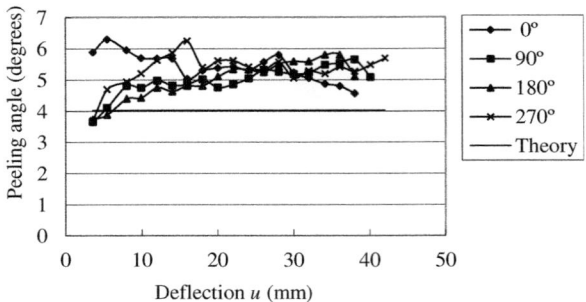

Figure 7.26 Comparison of peeling angle−deflection curves $G_f = 0.25$ N/mm (Wu et al., 2005b).

Table 7.11 Minimum interfacial fracture energy for the appearance of fiber rupture (Wu et al., 2005b).

FRP type	G_f (N/mm)
Carbon	12.78
Aramid	15.42
Glass	6.82
PBO	15.02

7.4 Experimental study on arched beams

It is necessary to recognize the extent to which applying fiber sheet as a counter-measure against peeling off can cope with peeling cases of various scales. Kojima et al. (2004) conducted a thorough experimental study on scaled model slabs. The test was important to explain the mechanisms of small and large separation of concrete pieces. In the large-scale specimens, the curvature of the arch of a tunnel lining was considered as a parameter that would affect the resistance of the FRP to the fall-down of concrete parts. The large-scale test specimen was a square slab as shown in Fig. 7.27, which had a width of 2300 mm, a thickness of 150 mm, and a central hole of diameter approximately 200 mm.

As shown in Table 7.12, experiments were conducted on three cases for the Shinkansen (a network of high-speed railway lines in Japan) cross-section: radius of curvature $(R) = 4750$ mm, $R = 2220$ mm, and without curvature $(R = \infty)$. Continuous CFS were bonded to the bottom surface, and the sheets were applied in two directions (weight per unit area: 200 g/m^2). For the specimens with curvature, the two directions of the continuous fibers were parallel to the tunnel axis direction and the circumferential direction. The characteristic concrete compressive strength $f_{ck} = 20$ N/mm^2 at 28 days.

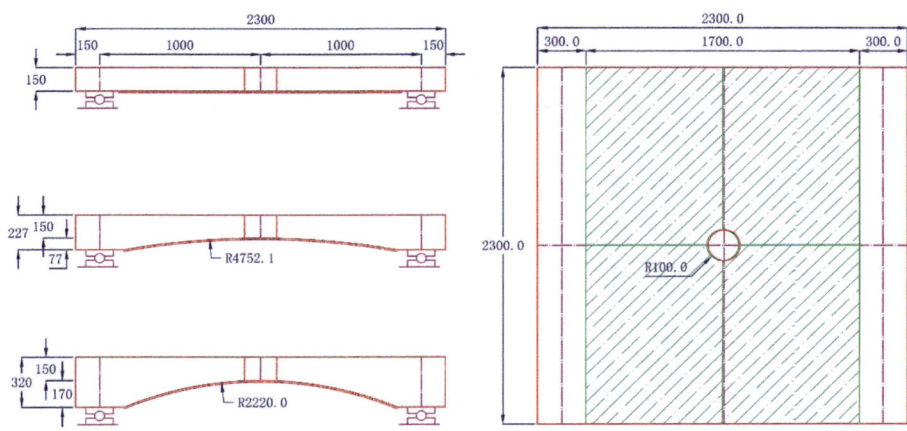

Figure 7.27 Dimensions of large-scale specimens (Kojima et al., 2004).

Table 7.12 Large-scale test (Kojima et al., 2004).

No.	Name	Radius of curvature (m)	Remarks
1	Flat slab	∞	Without curvature
2	Curved slab $R = 4750$	4.75	The Shinkansen line standard section
3	Curved slab $R = 2220$	2.22	The Shinkansen line standard section

The loading (punching) on the specimen by the indenter was under displacement control, at a rate of 0.2 mm/min until the start of peeling, and 1.0 mm/min after the start of peeling. The measured quantities are the punching load, loading point displacement, edge perimeter (peeling perimeter: L) of the peeling of the fiber sheet, peel length of the fiber sheet, and surface strain of the fiber sheet. The measurements were carried out at intervals of 0.05 mm before the start of peeling and at intervals of 0.5 mm after the start of peeling, and the condition of the fiber sheet adhesion was observed during the subsequent 2 min. A hammering test was carried out every 2 mm of indentation, followed by marking the peeled area on the surface of the fiber sheet.

Fig. 7.28A shows the relationship between the loading point displacement and the punching force. A linear slope was observed in the first stage, irrespective of the magnitude of the radius of curvature. The start of peeling corresponded to a displacement of 8.2 mm for the specimen without curvature (flat slab) and was 8.0 and 4.3 mm for the R4750 and R2220 specimens, respectively. It can be seen that the slope of the second linear stage after peeling decreases as the radius of curvature decreases.

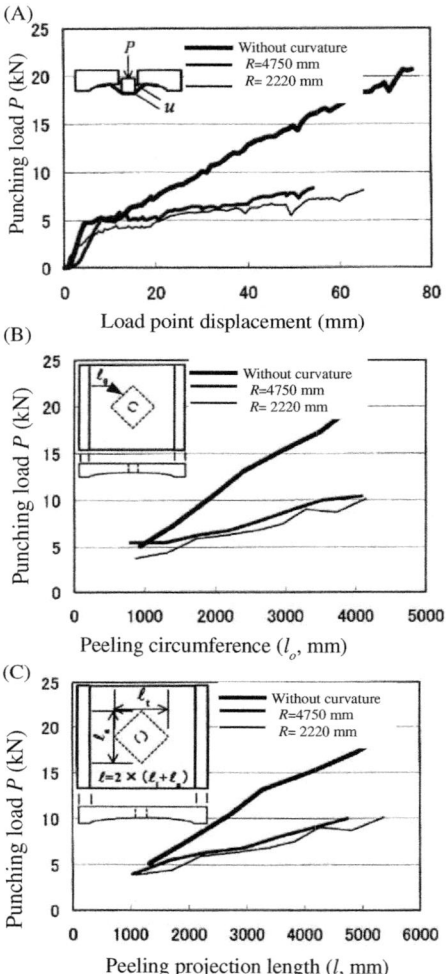

Figure 7.28 Punching test results (Kojima et al., 2004). (A) Punching Load versus load point displacement. (B) Punching Load versus peeling circumference. (C) Punching Load versus peeling projection length.

Fig. 7.28B and C shows the relationship between the punching load and peel circumference, and the punching load and peeled projected length, respectively. Based on the test results of small specimens tested by the authors in the same study, the punching load at the second stage (the stage where the fiber sheet peels off) was extrapolated as shown in Fig. 7.28B and C. Note that the test specimens were tested for a clear span of 2000 mm. It is concluded that the influence of the scale-model size on the peel strength is small. Incidentally, in the case where there is no curvature on the surface, the peeling proof strength is increased because the peeling circumference of the fiber sheet becomes longer.

7.5 Spalling prevention design method

7.5.1 Basic assumption of the design

The following assumptions are made to simplify the design.

1. The spalling strength is proportional to the spalling perimeter (l_o) or spalling projection length ($l = 2(l_1 + l_2)$), and thus the spalling strength of the unit length is treated as unit spalling strength (s_{po}), to eliminate the effect of different shapes of the spalling chunk.
2. The unit spalling strength (s_{po}) is determined by the tensile stiffness of the fiber sheet, and the spalling angle is determined as a certain value for a certain type of fiber sheet.
3. The effect of the arch curvature on the spalling strength is considered to be a reduction factor to the strength of the case without any curvature.
4. Fiber sheet debonding takes place in concrete, that is, the bonding strength between the fiber sheet and concrete is larger than the tensile strength of concrete and it is assumed that the debonding occurs owing to the tensile failure of concrete. Thus the unit debonding resistance is determined by the tensile property of concrete. The tensile property of concrete is expressed by its compressive strength (f'_{ck}) and the diameter of the maximum aggregate (d_{max}), as shown in Eq. (7.43).

$$G_F = 10 \cdot d_{\max}{}^{1/3} \cdot f'_{ck} 1/3 \tag{7.43}$$

where f'_{ck} should be larger than 15 N/mm^2.

7.5.2 Design method

Based on the assumption (1) as discussed earlier, the design method for the fiber sheet is proposed. In the design, when the fiber sheet is bonded on the surface of concrete, two points must be checked. First, whether the fiber sheet will debond owing to the self-weight of the spalling chunk. Second, whether the fiber sheet will rupture when debonding takes place.

1. Check of fiber sheet debonding
   This check is conducted as follows:

$$\gamma_i \cdot \frac{W_d}{P_{ud}} \leq 1.0 \tag{7.44}$$

$$P_{ud} = 2 \cdot \frac{S_{po}}{\gamma_m} \cdot \frac{(l_t + \gamma_a \cdot l_a)}{\gamma_b} \tag{7.45}$$

Here, W_d is the design spalling load (refer to Fig. 7.29), P_{ud} is the design spalling resistance, γ_i is a structural factor, and s_{po} is the unit spalling strength. γ_m is a safety factor, γ_b is a material factor, γ_a is a reduction factor considering the effect of curvature, l_a is the project length in the circumferential direction, and l_t is the project length in the axial direction (refer to Fig. 7.29).

2. Check of fiber sheet rupture

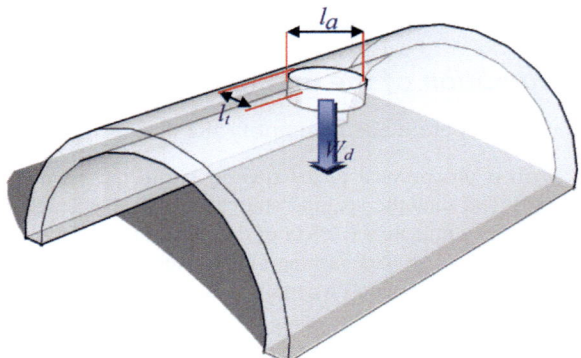

Figure 7.29 Definition of l_a, l_t, and W_d of spalling chunk (Kojima et al., 2004).

If the stress level in the fiber sheet at debonding is underestimated, fiber sheet rupture may occur before the debonding proceeds, which is an unsafe scenario. The check of fiber sheet rupture uses a modification factor ρ_m and is conducted as follows:

$$\rho_m \cdot \frac{\sigma_{p0}}{f_{ud}} \leq 1.0 \tag{7.46}$$

$$\sigma_{p0} = E \cdot \left(\frac{1}{\cos\theta} - 1\right) \tag{7.47}$$

where σ_{p0} is the stress level in the fiber sheet upon debonding, f_{ud} is the design strength of the fiber sheet (equal to the tensile strength divided by the material safety factor γ_m), E_f is the tensile elastic modulus of the fiber sheet, θ is the spalling angle (rad), and ρ_m is a modification factor.

There are several parameters in the as discussed earlier equations such as the unit strength (s_{po}) and reduction factor (γ_a) in Eq. (7.45), and the spalling angle in Eq. (7.47). It would be complicated to analyze and test these parameters for every case, and thus a monogram method is needed.

7.5.3   Unit spalling strength (s_po) and spalling angle (θ)

7.5.3.1   Spalling model and spalling resistance

According to the results of the punching test, the unit spalling strength (s_{po}) and spalling angle (θ) are significantly affected by the properties of the fiber sheet. The design model is shown in Fig. 7.30, which is a 2D beam model. This model shows an ideal balanced mechanical state in which the chunks spall from the lining and the fiber sheet starts to debond. The bending stiffness of the fiber sheet is ignored.

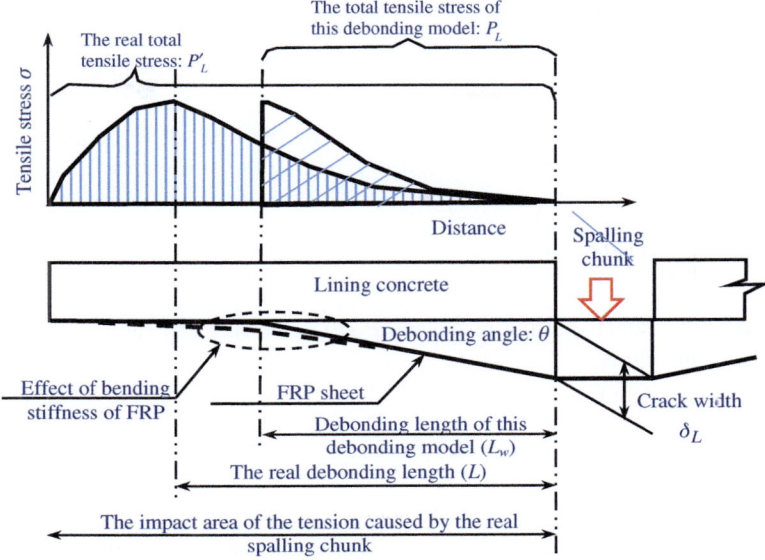

Figure 7.30 Debonding model of concrete chunk (Kojima et al., 2004).

According to the conditions of the concrete (f'_{ck} and d_{max}), the failure energy G_F should be calculated with Eq. (7.43) and f_{tk}, as given by Eq. (7.48):

$$f_{tk} = 0.23f'_{ck}2/3(N/mm^2)$$ (7.48)

Kojima et al. (2004) proposed a bilinear relationship describing the concrete tensile softening behavior, whereby the crack width δ_L can be predicted from the following form:

$$\delta_L \cdot \frac{f_{tk}}{G_f} = 5$$ (7.49)

Because δ_L is infinitesimal,

$$L_w = \frac{\delta_L}{\theta}$$ (7.50)

The tensile stress distribution (curve) on the surface of the fiber sheet and concrete is approximated as Eq. (7.51) (Kojima et al., 2004):

$$\frac{\sigma(\delta)}{f_{tk}} = e^{-19\delta}$$ (7.51)

Thus if the crack width at x is δ, and the spalling angle is θ, then $\delta = \theta x$.

$$\sigma(x) = f_{tk}e^{-19\theta x} \tag{7.52}$$

Second, the total tensile stress P_l undertaken by the fiber sheet on the left side is equal to the integral of Eq. (7.52) from 0 to L_w along the x-direction.

$$P_l(\theta) = \int_0^{L_w} \sigma(x)dx = \int_0^{L_w} f_{tk} \cdot e^{-19\theta x}dx \tag{7.53}$$

Thus the fiber sheet spalling resistance $P_u(\theta)$ determined by the tensile softening property of concrete is

$$P_u(\theta) = 2P_l(\theta) = 2\int_0^{\delta_L/\theta} f_{tk} \cdot e^{-19\theta x}dx \tag{7.54}$$

7.5.3.2 Relationship between the weight of the spalling chunk and the elongation of the fiber sheet

If the bending stiffness of the fiber sheet is ignored, the weight of the spalling chunk W and tensile load T, as functions of θ, are as shown in Eqs. (7.55) and (7.56).

$$W(\theta) = 2T(\theta) \cdot \sin\theta \tag{7.55}$$

$$T(\theta) = E_f \cdot t_f \cdot \frac{(L_w/\cos\theta - L_w)}{L_w} \tag{7.56}$$

Thus

$$W(\theta) = 2\left(\frac{1}{\cos\theta} - 1\right) \cdot E_f \cdot t_f \cdot \sin\theta \tag{7.57}$$

Here, E_f is the tensile elastic modulus of the fiber sheet and t_f represents the thickness or cross-sectional area per unit width of the fiber sheet (mm^2/mm).

7.5.3.3 Unit spalling strength

To calculate the debonding resistance of different types of fiber sheets, according to Eqs. (7.54) and (7.57), the weight of the spalling chunk and spalling resistance P_u with respect to the spalling angle θ are shown in Fig. 7.31 (Kojima et al., 2004). For example, for the CFS (400 g/m^2, $E.t = 27,195$ N/mm), when the angle is 2.89 degrees, the spalling resistance is calculated to be 3.48 N/mm. Furthermore,

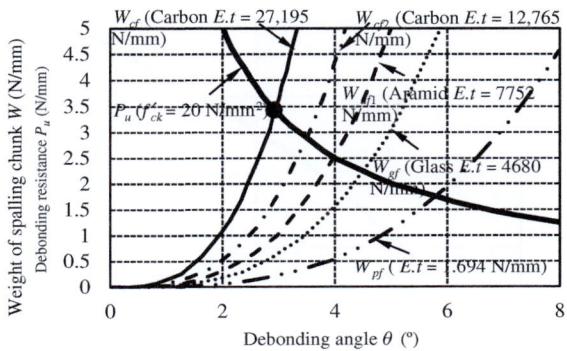

Figure 7.31 Relationship between debonding angle (θ) and weight of spalling chunk (W) together with debonding resistance (P_u) ($f'_{ck} = 20$ N/mm^2) (Kojima et al., 2004).

because a 2D beam model is adopted, the spalling resistance P_u is the unit spalling strength (s_{po}) of different types of fiber sheet, which can be calculated if the intersection for each type of fiber sheet is obtained.

7.5.3.4 Plotting monogram

For a lining with a spalling problem, as the strength of concrete is commonly low, the compressive strength of 15 N/mm^2 is a basic condition. Then,

$$\delta_L = 0.239 \text{mm} \tag{7.58}$$

$$\sigma(x) = 1.40e^{-20\theta x} \tag{7.59}$$

$$P_u(\theta) = \frac{0.139}{\theta} \tag{7.60}$$

Kojima et al. (2004) reported that if the intersection of Eqs. (7.58) and (7.55) is solved using a graphical solution, and the monograms of unit spalling strength (s_{po}) and spalling angle (θ) are obtained, the results are as shown in Figs. 7.32 and 7.33.

7.5.4 Reduction factor of curvature

Because the tunnel wall exhibits curvature, the effect of the curvature on the debonding resistance should be considered. The debonding angle is divided into the angle between the tangent of the curve and the horizontal line (θ_R) and the angle between the horizontal line and the FRP sheet (θ_F), as shown in Fig. 7.34.

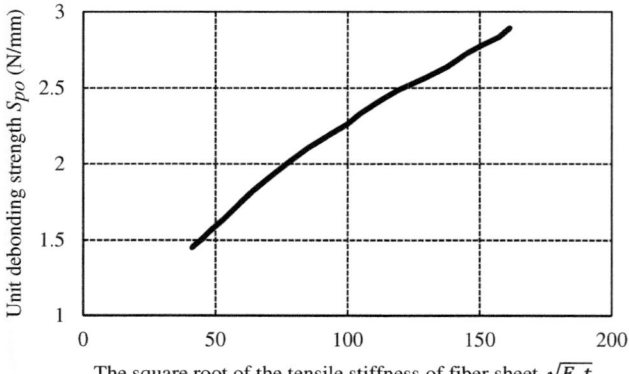

Figure 7.32 Relationship between the square root of the tensile stiffness of fiber sheet and unit debonding strength (Kojima et al., 2004).

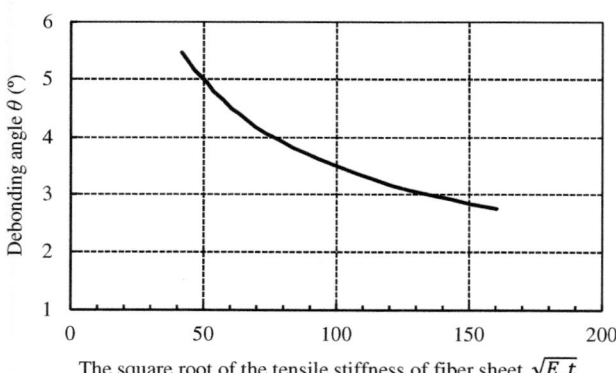

Figure 7.33 Relationship between the square root of the tensile stiffness of fiber sheet and debonding angle (Kojima et al., 2004).

The crack width $\delta_L(x)$ at location x is given by

$$\delta_L(x) = x \cdot \tan(\theta_F) + R \cdot \left(\cos\left(\sin^{-1}\left(\frac{L_R - x}{R} \right) - \cos\left(\sin^{-1}\left(\frac{L_R}{R} \right) \right) \right) \right) \quad (7.61)$$

When $f'_{ck} = 15$ N/mm^2, the maximum crack width that can transfer the tensile stress is 0.239 mm in Eq. (7.61) with reference to Eq. (7.58). Setting x equal to L_R, L_R can be calculated:

$$\delta_L(L_R) = L_R \cdot \tan(\theta_F) + R \cdot \left(\cos\left(\sin^{-1}\left(\frac{L_R}{R} \right) \right) \right) = 0.239 \quad (7.62)$$

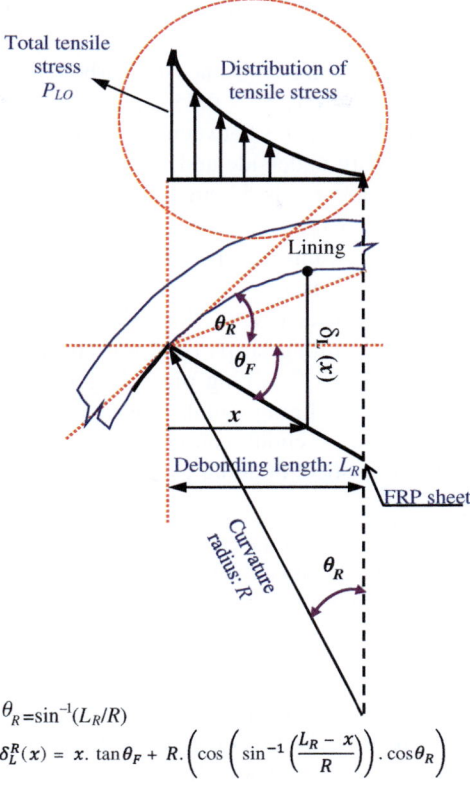

$$\theta_R = \sin^{-1}(L_R/R)$$
$$\delta_L^R(x) = x \cdot \tan\theta_F + R \cdot \left(\cos\left(\sin^{-1}\left(\frac{L_R - x}{R}\right)\right) \cdot \cos\theta_R\right)$$

Figure 7.34 Debonding model of lining surface with curvature (Kojima et al., 2004).

Here, substituting the calculated L_R for different radii of curvature (R) and fiber sheet spalling angles (θ_F) from Eq. (7.62) into Eq. (7.61), the spalling resistance P_u can be calculated using Eq. (7.63):

$$P_u = 2P_{Lo} = 2\int_0^{L_r} \sigma\left(\delta_L(x)\right)dx \tag{7.63}$$

where $\sigma(\delta)$ equals $1.40e^{-20\delta}$ according to Eq. (7.59) and δ is the crack width.

However, if the bending stiffness of the fiber sheet is ignored, assuming the weight of the spalling chunk at the spalling angle θ_F is W, then W is expressed as in Eq. (7.64).

$$W(\theta) = 2\left(\frac{1}{\cos\theta_F} - 1\right) \cdot E_f \cdot t_f \cdot \sin\theta_F \tag{7.64}$$

7.6 Selection of method to prevent spalling

Fig. 7.35 shows the design spalling resistance of carbon fiber sheet (400 g/mm^2, $s_{po} = 2.94$ N/mm, $\gamma_m = 1.1$, $\gamma_b = 1.1$, $\gamma_a = 0.95$, $l_a:l_t = 1:1$), existing spalling accidents (dot points), and the relationship between the weight of a hypothetical spalling chunk (a cuboid with square underside and 50 cm thickness) and the debonding perimeter. When the lining thickness is 50 cm and the perimeter is less than 3.4 m, the design spalling resistance of the fiber sheet is sufficient up to the weight of a spalling chunk of 8.3 kN. If the chunk size increases, the weight will become significantly larger and debonding will occur.

Based on the as discussed earlier results, the failure mode and scale of spalling are divided into four types: ① lining strips (spalling chunk size up to 0.825 m at one side and weight up to 7.8 kN, ② lining piece larger than 0.825 m at one side and weight lower than 7.8 kN, ③ lining chunk larger than 0.825 m at one side and weight higher than 7.8 kN, and ④ chunk caused by a landslide. Based on the four types, the selection of the fiber sheet bonding is considered as follows:

1. Based on Fig. 7.35, using only fiber sheet bonding (400 g/m^2 carbon fiber sheet) can solve the problem in ①, ②, and a portion of the conditions in ③. However, the spalling caused by surrounding rock stress should be considered if it is significant. Additionally, for ③, FRP bonding combined with other measures (such as strengthening by cement or lock bolt) would be better.
2. Furthermore, because carbon fiber is electrically conductive, care should be taken with respect to its use in railway tunnels. Additionally, the fiber sheet should be bonded to the concrete surface with a sufficient amount of resin. Thus to apply the proposed method, the following steps must be performed: (1) removal of leakage, (2) removal of weak or dirty concrete, and (3) avoiding unevenness. If the as discussed earlier problems exist, other appropriate measures should be considered and selected.

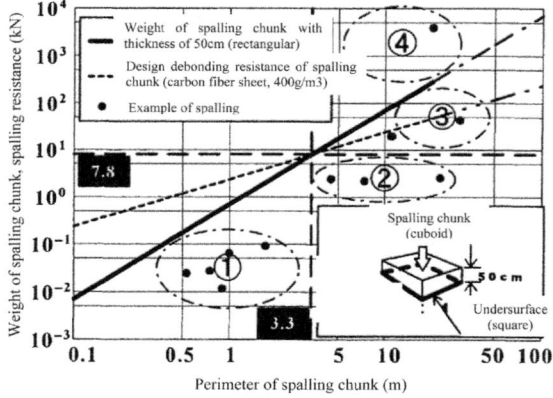

Figure 7.35 Case studies (Kojima et al., 2004).

7.7 Conclusions

To study the effect of FRP sheets bonded to the surface of the concrete to prevent concrete spalling from a concrete structure, a series of punching—peeling tests were conducted, and a new analytical model that uses the interfacial fracture energy G_f to calculate the spalling load was proposed. By comparing the analysis results with the experimental results, the following conclusions can be drawn.

1. When FRP sheets are bonded in two mutually orthogonal layers, the load of spalling increases significantly. The whole FRP sheets can carry the load, which is different from the case of unidirectional fiber sheets.
2. By applying the new analytical technique that uses the interfacial fracture energy G_f between the FRP sheets and concrete, it is possible to accurately calculate the maximum load and maximum displacement by determining the value of G_f experimentally.
3. The maximum load increases with increasing interfacial fracture energy and stiffness of the FRP sheets.
4. It is concluded that only two material parameters, the interfacial fracture energy of the FRP—concrete interface and stiffness of the FRP sheets, are necessary to model the interfacial spalling behavior.
5. Based on the information presented as discussed earlier, a design method is proposed. To simplify the design method, a monogram can be used for the unit spalling strength of the fiber sheet (s_{po}), debonding angle of the fiber sheet (θ), and reduction factor considering the radius of curvature (γ_a). Furthermore, the scope of application of the design method is divided into four parts: ① lining strip, ② lining piece, ③ lining chunk, and ④ landslide chunk. The fiber sheet bonding method can be used alone for ①, ②, and a portion of ③. However, the spalling caused by rock stress should be considered in combination with certain other measures for rock stress. Additionally, for the condition of ①, there is no strict requirement for the selection of the fiber sheet material.

References

Binici, B., Bayrak, O., 2002. Strengthening reinforced concrete flat plates using fiber reinforced polymer. In: Proceedings of the Tenth U.S.-Japan Conference on Composite Materials, California, pp. 377—386.

Gent, A.N., Hamed, G.R., 1975. Peel mechanics. J. Adhes. 7, 91—95.

Karbhari, V.M., Engineer, M., 1996. Investigation of bond between concrete and composites: use of a peel test. J. Reinf. Plast. Compos. 15, 208—227.

Kimpara, I., et al., 1998. Characterization of peeling strength of FRP sheets bonded on mortar and concrete. In: Proceedings of the Eighth JAPAN-U.S. Conference on Composite Materials, pp. 1010—1019.

Kojima, Y., Yoshikawa, K., Muguruma, T., Kobayyashi, A., Wakana, K., Asakura, T., et al., 2004. A design method of fiber reinforced plastic methods as a countermeasure for concrete spalling from tunnel lining. 土木学会論文集 No. 756/VI-62, pp. 101—116.

Nicholson, D.W., 1977. Peel mechanics with large bending. Int. J. Fract. 13 (3). 279—287.

Ozbolt, J., Eligehausen, R., 2001. Punching failure-influence of material properties and size effect. In: Proceedings of the Fourth International Conference on Fracture Mechanics of Concrete and Concrete Structures, Tokyo, pp. 719–725.

Thouless, M.D., Jensen, M.D., 1992. Elastic fracture mechanics of the peel-test geometry. J. Adhes. 39, 185–197.

Wang, J.W., Tan, K.H., 2001. Punching shear behaviour of RC flat slabs externally strengthened with CFRP system. In: Proceeding FRPRCS-5, Cambridge, pp. 997–1005.

Wu, Z., Yuan, H., Kojima, Y., Ahmed, E., 2005a. Experimental and analytical studies on peeling and spalling resistance of unidirectional FRP sheets bonded to concrete. Compos. Sci. Technol. 65 (7), 1088–1097.

Wu, Z., Yuan, H., Asakura, T., Yoshizawa, H., Kobayashi, A., Kojima, Y., et al., 2005b. Peeling behavior and spalling resistance of bonded bidirectional fiber reinforced polymer sheets. J. Compos. Constr. 9 (3), 214–226.

Further reading

Brosens, K., Gemert, D.V., 1998. Plate end shear design for external CFRP laminates, Fracture Mechanics of Concrete Structures. Proceedings FRAMCOS-3. AEDIFICATIO Publishers, Freiburg, pp. 1793–1804, D-79104.

Täljsten, B., 1996. Strengthening of concrete prisms using the plate-bonding technique. Int. J. Fract. 82, 253–266.

Triantafillou, T.C., 1998. Fracture mechanics approaches to concrete strengthening using FRP materials. Proceedings FRAMCOS-3. AEDIFICATIO Publishers, Freiburg, pp. 1761–1770, D-79104.

Wu, Z.S., Yoshizawa, H., 1999. Analytical/Experimental study on composite behavior in strengthening structures with bonded carbon fiber sheets. J. Reinf. Plast. Compos. 18 (2), 1131–1155.

Wu, Z.S., Niu, H.D., 2000a. Study on debonding failure load of RC beams strengthened with FRP sheets. J. Struct. Eng. JSCE 46A, 1431–1441.

Wu, Z.S., Niu, H.D., 2000b. Shear transfer along FRP-concrete interface in flexural members. J. Mater. Concr. Struct. Pavements JSCE 49 (662), 1431–1441.

Wu, Z.S., Yuan, H., Yoshizawa, H., Kanakubo, T., 2001. Experimental/analytical study on interfacial fracture energy and fracture propagation along FRP-concrete interface. Fracture Mechanics for Concrete Materials: Testing and Applications. ACI International, Farmington Hills, MI, pp. 133–152, SP-201.

Wu, Z.S., Yuan, H., Niu, H.D., 2002. Stress transfer and fracture propagation in different kinds of adhesive joints. J. Eng. Mech. ASCE 128 (5), 562–573.

Yuan, H., Wu, Z.S., 2000. Energy release rates for interfacial crack in laminated structures. J. Struct. Mech. Earthq. Eng. JSCE 17 (1), 19–31.

Yuan, H., Wu, Z.S., Yoshizawa, H., 2001. Theoretical solutions on interfacial stress transfer of externally bonded steel/composite laminates. J. Struct. Mech. Earthq. Eng. JSCE 18 (1), 27–39.

Yuan, H., Teng, J.G., Seracino, R., Wu, Z.S., Yao, J., 2004. Full-range behavior of FRP-to-concrete bonded joints. Eng. Struct. 26 (5).

Index